Horse Pasture Management

Horse Pasture Management

Paul Sharpe
University of Guelph, Guelph, ON, Canada (retired)

ACADEMIC PRESS
An imprint of Elsevier

Academic Press is an imprint of Elsevier
125 London Wall, London EC2Y 5AS, United Kingdom
525 B Street, Suite 1650, San Diego, CA 92101-4495, United States
50 Hampshire Street, 5th Floor, Cambridge, MA 02139, United States
The Boulevard, Langford Lane, Kidlington, Oxford OX5 1GB, United Kingdom

Library of Congress Cataloging-in-Publication Data
A catalog record for this book is available from the Library of Congress

British Library Cataloguing in Publication Data
A catalogue record for this book is available from the British Library

ISBN: 978-0-12-812919-7

For Information on all Academic Press publications visit our website at
https://www.elsevier.com/books-and-journals

Working together
to grow libraries in
developing countries

www.elsevier.com • www.bookaid.org

Publisher: Andre Gerhard Wolff
Senior Acquisition Editor: Anna Valutkevich
Editorial Project Manager: Pat Gonzalez
Production Project Manager: Poulouse Joseph
Designer: Mark Rogers

Typeset by TNQ Technologies

This book is dedicated to my wife, Helen MacGregor, for her unlimited support of me and my career. We shared a joy in teaching biological and agricultural sciences to students whose first love was horses. Helen organized most of the lab sessions and all of the educational tours for a degree program in equine management. She found many excuses to bake pans of brownies for the students. When our campus was closed, Helen was soon working three part-time jobs and encouraging me to pursue my dream of writing and editing this book, which would fill a need for a single source of information about proper management of the most natural environment for horses. I am most thankful to her, and I think many students and horses are too.

Contents

Contributors   xiii
Biography   xv
Preface   xxi
Acknowledgements   xxiii

1. Forage Plant Structure, Function, Nutrition, and Growth

Daniel J. Undersander

Structure and Physiology of Pasture Grasses and Legumes   1
  Plant Organic Compounds   1
  The Cell: The Most Basic Structural Unit of Plant Organization   3
  Organs   4
  How Grasses Develop   6
  How Legumes Develop   8
Review Questions   10
Reference   10

2. Identification of Temperate Pasture Grasses and Legumes

Michael D. Casler and Daniel J. Undersander

Introduction   11
Grasses   11
  Seedlings   11
  Sod Versus Bunch Grasses   12
  Glossary   12
  Is It a Grass?   13
  The Grass Whorl   13
  Ryegrasses   13
  Tall and Meadow Fescue   14
  Orchardgrass   15
  Timothy   17
  Kentucky Bluegrass   18
  Reed Canarygrass   19
  Smooth Bromegrass   20
  Quackgrass   20
  Legumes   22
  Glossary   24
  Alfalfa   24
  Alsike Clover   25
  Birdsfoot Trefoil   25
  Crown Vetch   28
  Hairy Vetch   28
  Kura Clover   30
  Red Clover   32
  White Clover   33
Review Questions   35
Reference   35

3. Nutritional Value of Pasture Plants for Horses

Paul Sharpe

Introduction   37
Carbohydrates   38
  Diversity in Carbohydrates   38
  Structural Versus Nonstructural Carbohydrates   42
  What About Fructans?   43
Lipids   43
  Fatty Acids of Special Interest   45
  Vitamin A   46
  Vitamin D   46
  Vitamin E   46
  Vitamin K   46
Proteins   47
Water-Soluble Vitamins   48
Important Minerals   49
Nutrient Analysis of Feeds   50
Digestibility of Feeds   52
  Link Between Digestibility and Energy Evaluation of Feeds   53
Useful Energy Terminology   53
Chemical Analysis of Plants   54
  Uses and Relative Advantages of NDF and ADF Measurements   56
  Further Uses of ADF and NDF   56
  Alternative Methods of Analysis of Forage Components   57
  Factors Influencing Forage Dry Matter Intake   57
Concentrations of Nutrients in Typical Pasture Plants   57

Digestion of Plant Nutrients by Horses   60
    Carbohydrate Digestion in the Stomach and
        Small Intestine   60
    Carbohydrate Digestion in the Cecum and
        Colon   60
    Fat Digestion   61
    Nitrogen Digestion   61
Nutrient Requirements of Horses   61
Feed Intake   62
Review Questions   62
References   62

4. Soils for Horse Pasture Management

Paul Voroney

Introduction   65
What Is Soil?   65
Composition of Mineral Soils   66
Organic Matter and Soil Humus   68
Aggregation   70
Nature of Soil Pore Space   71
Soil Water   71
Plant Available Water   73
Water Drainage   73
Soil Aeration   74
Soil Chemistry   74
Soil Acidity and Alkalinity   75
Soil Fertility   75
Soil Biology   76
Secrets to Soil Management for Sustainable
    Pasture Production   77
Review Questions   78

5. Introduction to Pasture Ecology

Edward B. Rayburn and Paul Sharpe

Optimal Environment Versus Limiting Factors   81
Plants   82
Respiration   82
Light Interception: Canopy Height and Time
    of Year   83
Energy Reserves Cycle With Growth   83
Growth Under Rotational Grazing   84
Root Growth   85
Growth Has Two Phases   85
Growing Points   85
Cell Wall Content Changes With Season and
    Plant Type   86
Forage Quality, Antiquality, and Palatability   87
Competition Between Plants   87
Plant Diversity, Morphology, and Tolerance to
    Grazing   88
The Grazing Animals   88

Forage Mass and Dry Matter Intake   90
Selective Grazing   90
Review Questions   91
References   91
Further Reading   91

6. Pasture Plant Establishment and Management

S. Ray Smith and Krista L. Lea

Characteristics of a Healthy Pasture   93
    Establishment   93
    Management Following Establishment   97
    Pasture Renovation   99
Resting Pastures   100
Review Questions   102
References   103

7. Forage Yield and Its Determination

Paul Sharpe and Edward B. Rayburn

Introduction and Questions About Forage
    Yield   105
Answering the Horse Owner's First Three
    Questions   106
Preparing to Answer the Fourth Question   106
Methods and Tools for Determining
    FDM Yield   106
    Direct Measurement of Forage Height and
        Yield   107
    Indirect Measurement to Estimate Yield   107
The Relationship Between Forage Height and
    Forage Yield in Pasture   113
Using Numbers From Sample Measurements
    to Estimate Forage Yield for an Acre and a
    Whole Field   116
Converting Forage Yields Between Pounds
    per Acre and Kilograms per Hectare   116
Answering the Horse Owner's
    Fourth Question   117
Review Questions   118
References   118
Further Reading   119

8. Grazing Behavior, Feed Intake, and Feed Choices

Paul Sharpe and Laura B. Kenny

Introduction   121
Grazing by Horses Compared to Other
    Herbivores   121
Factors Regulating Feed Intake   123
The Diets of Feral Horses   123

Timing of Grazing Patterns in Feral and
  Captive Horses   123
  Feral Horses   124
  Captive Horses   124
Measuring the Amount and Rate of Feed
  Intake   124
  Factors Influencing Bite Rate   124
  Horses Prefer Fresh Pasture to Dried Hay   125
  Forage Quality Affects Intake Rates   125
How Young Animals Learn to Make Feed
  Choices   125
Animals Make Feeding Decisions Based on
  Feedback From Past Choices   126
Making Choices While Grazing   127
  Factors Affecting the Bite and Plant Level 129
  Factors Affecting the Patch and Feeding Site
    Level   130
  Factors Affecting the Camp and Home Range
    Level   132
Effects of Grazing Behavior on the
  Environment   132
  Animal Waste   132
  Trampling   132
  Overgrazing   133
  Positive Effects of Grazing on Plants   133
Managing Grazing Behavior   133
  Rotational Grazing   133
  Multispecies Grazing   134
  Manure and Parasites   134
  Maximizing Pasture Intake and Nutrition   134
  Limiting Dry Matter and NSC Intake   135
Conclusions   135
Review Questions   135
References   135

9. Managing Equine Grazing for Pasture
   Productivity

   Laura B. Kenny, Amy Burk and Carey A. Williams

Introduction   141
Grazing Behavior   141
Stocking Rate and Density   142
Over- and Understocking   143
Grazing Systems   144
  Continuous Grazing   144
  Rotational Grazing   146
  Number of Grazing Units   147
  Size of Grazing Units   147
  Total Acreage Required   148
  When to Graze and Remove Horses/Length
    of Grazing Periods   148
Grazing Season   151
Winter Pasture Management   151

Summer Pasture Management   151
More Intensive Management Strategies   152
Additional Management   152
Continuous Versus Rotational Grazing   153
Grazing Plans for Linda and Emily   153
Conclusion   154
Review Questions   154
References   155
Further Reading   155

10. Mixed Species Grazing

    Paul Sharpe

What Is Mixed Species Grazing?   157
What Benefits Can Mixed Species Grazing
  Provide?   157
  Improved Forage Utilization Efficiency   158
  Mixed Species Grazing Helps to Control
    Parasites   160
  Reducing Ecologic Risks of Avermectins   161
  Improved Weed Control, Especially with
    Training   162
Why/How Does Multispecies Grazing Work?   163
  Utilization of Forage Affected by Feces of
    Other Animal Species   165
  Other Influences on Workability of Multi-
    species Grazing   166
What Are Potential Disadvantages to Mixed
  Species Grazing on a Horse Farm?   166
How Is Information About Nonequine
  Species Useful in Managing Horses?   167
Economic Considerations of Mixed Species
  Grazing   167
  Potential Positive Economic Responses   167
  Potential Negative Economic Consequences   168
Managing Mixed Species Grazing   168
Mixed Species Grazing in Action   170
What Research Still Needs to Be Done?   172
Review Questions   172
References   172

11. Production and Management of Hay
    and Haylage

    Jimmy Henning and Laurie Lawrence

Introduction   177
What Is Quality Hay?   177
Factors Affecting the Nutrient Value of Hay   177
  Plant Species   177
  Stage of Maturity   178
  Visual and Physical Characteristics of
    High-Quality Hay   179
  Chemical Composition and Forage Quality   182

Producing Quality Hay 185
Harvesting 186
Swath and Windrow Manipulation 187
Baling 190
Hay Preservatives 190
Bale Handling 190
Bale Conditioning and Storage 192
Bale Packages for Horses 192
Storing Hay 195
Choosing Hay for Horses 196
Cleanliness First 196
Matching Hay Type to Horse Type 196
Is Hay a Nutritionally Balanced Diet? 197
Feeding Hay to Individually Housed Horses 198
Horses Fed in Groups 199
Baling Forage Crops for Silage 201
Advantages and Disadvantages of Silage Baling 202
Forage Requirements 203
Machinery Requirements for Baleage 203
Bale-Wrapping and Bagging Equipment 204
Review Questions 207
References 207
Further Reading 208

12. Climate, Weather, and Plant Hardiness

Paul Sharpe and Edward B. Rayburn

Introduction 209
Solar Radiation 209
Air Temperature 210
Growing Degree Days 212
Alfalfa 212
Grasses 213
Diversity Among Varieties 214
Elevation and Topographic Position 214
Precipitation 214
Seasonal Changes in Forage Quality 215
Regional Climatic Effects 215
Cold Temperature Effects 216
North to South and East to West Effects in North America 216
The Transition Zone 217
Restrictions to Growth 217
Drought 218
Flood 221
Winter Hardiness 222
High Temperature Stress 223
Establishment 223
Climate Change 223
Pests 225
Breeding Cultivars for the New Climate 225
Effects of Climate Change on Horses 225

Effect of Animals on Climate Change 226
Grazing and Cycling of Greenhouse Gases 227
Professional Help 227
Review Questions 228
References 228
Further Reading 231

13. Matching Plant Species to Your Environment, Weather, and Climate

Edward B. Rayburn

Introduction 233
Adaptation to the Site and Management 233
Forage Yield 239
Forage Quality 241
Review Questions 243

14. Managing Manure, Erosion, and Water Quality in and Around Horse Pastures

Laura B. Kenny, Michael Westendorf and Carey A. Williams

Introduction 245
Erosion 245
Contaminated Water Leaching/Runoff 247
Parasite Concerns 248
Other Environmental Concerns 249
Benefits of Manure on Pasture 250
Spreading Manure on Pasture 250
Managing Pastures to Alleviate Environmental Concerns and Protect Water Quality 252
Grazing Near Streams: Riparian Buffers 257
Maintaining and Managing Riparian Forest Buffers 257
Conclusion 258
Review Questions 258
References 259

15. Fencing and Watering Systems

Paul Sharpe

Purposes and Desired Features of Fences 261
Planning Fences 262
Locations of Fences 263
Materials to Consider 265
Stones and Trees 265
Wood 265
Metals 267
Plastics 269
Fiberglass 271
Concrete 271
Plastic Coatings 271

Fence Planning, Types, Designs, and
 Descriptions   271
  Post and Rail Fences   275
  Wood Posts and Steel Strand Wire   277
  Wood Posts and Steel Woven Wire   280
  Wood Posts and Polymer-Coated Wire
   "Boards" or "Rails"   280
  Wood Post and All Polymer Strand   282
  All Plastic Polymer Posts and Rails   282
  Rubber Belt   282
  Permanent Electric   283
  Electric Fence Connections and Current   284
  Safety in Electric Fence Construction and
   Maintenance   285
  Training Horses to Electric Fencing   285
  Maintaining the Current   285
  Semipermanent and Temporary Electric
   Fences   287
 Gates   292
 Maintenance of Fences   295
 Drinking Water Systems   297
  Drinking Water Quality   298
  Water Sources   299
  Moving Water from Source to Pasture   299
  Watering Devices   303
 Information Sources   307
  Manufacturers of Electric Fencing   307
  Manufacturers of Waterers   307
 Review Questions   307
 References   308

**16. Pasture-Related Diseases and
 Disorders**

*Bridgett McIntosh, Tania Cubitt and
Sherrene Kevan*

 Introduction   311
 Colic   311
 Pasture-Associated Laminitis   313
 Pasture-Associated Obstructive Pulmonary
  Disease   315
 Chronic Obstructive Pulmonary Disease   315
 Summer Pasture-Associated Obstructive
  Pulmonary Disease   315
 Equine Grass Sickness   316
 Seasonal Pasture Myopathy   317
 Nitrate Poisoning in Horses   318
 Pasture-Associated Liver Disease in the
  Horse   319
 Pasture-Associated Stringhalt   319
 Other Plants That Are Toxic to Horses   320
  Trees   320
  Forbs   321

  Legumes   321
  Grasses   324
 Other Conditions That Are Toxic to Horses   324
  Selenium Deficiency   324
 Selenium Poisoning   326
 Getting Help   326
 Review Questions   327
 References   327

17. Coexisting With Wildlife

Paul Sharpe and Daniel J. Undersander

 Introduction   329
 The Relationship Between the Abundance
  of Species of Wild Plants and Animals
  and the Health of an Ecosystem   329
  Habitat   330
  Biodiversity and Ecosystem Services   331
  Wildlife Population Imbalance   332
 Benefits of a Diverse Ecosystem   333
  Accommodating Hunters or Other
   Consumers of a Diverse Ecosystem   333
 Effects of Adding or Subtracting Wild
  Species on an Ecosystem   334
  Subtracting and Adding Animal Species   334
  Subtracting and Adding Plant Species   335
  Habitat Fragmentation   335
  Horse Parasite Control and Unintended
   Effects   336
 Techniques of Encouraging More Wildlife
  to Visit or Inhabit Your Farm   336
  Diversity of Forage Species   336
  Forage Management and Habitat   336
  Haying Practices   337
  Grazing Management Techniques   338
  Providing Drinking Water, Feed, Cover,
   and Breeding Habitat   339
  Subtracting and Adding Prolific Omnivores   339
  Tools and Resources to Help You Coexist
   With Wildlife   340
 Explaining Situations in Which Wild
  Animals Can Be Detrimental to the Goals
  of Your Farm, Including Predation and
  Crop Damage   340
  Predation, Injuries, and Scaring   342
  Crop Damage   342
  Arthropod Pests   343
 Determining Which Detrimental Species May
  Be Present or Responsible for Damage   345
  Predation and Signs of Predators   345
 Techniques to Discourage Certain Animal
  Species From Visiting Your Farm   345
  Scaring Nuisance Animals   346

Integrated Methods   346
Fences   346
Trapping and Hunting   347
Guardian Animals   349
Eliminating Attractive Plants, Water, Feed,
  or Cover   350
Dead Animal Disposal   350
A Model to Watch   350
Review Questions   351
References   351
Further Reading   354

18. University of Kentucky Horse Pasture Evaluation Program

Krista L. Lea, S. Ray Smith and Thomas (Tom) Keene

Introduction   355
**History of the University of Kentucky Horse
  Pasture Evaluation Program**   355
  Mare Reproductive Loss Syndrome   356
  University of Kentucky Equine Initiative   356
  University of Kentucky Horse Pasture
    Evaluation Program Today   356
  Tall Fescue Toxicosis   357
Pasture Sampling   358
Data Reporting and Recommendations   359
  Soil Maps   359
  Pasture Maps, Data, and Photographs   359
  Recommendations   359
  Farm Data Summary Sheet   359
  Publications   359
Case Studies   361
  Farm #1: Large-Scale Commercial Breeding
    Farm   361
  Farm #2: Medium-Scale Commercial
    Breeding Farm   362
  Farm #3: Small Private Farm   363

**Challenges Facing University of Kentucky
  Horse Pasture Evaluation Program**   363
  Soil Sampling   363
  Competition With Commercial Businesses   364
  Extension Versus Fee for Service   364
  Securing Funding   364
  Labor and Quality Control   365
**Other Impacts of University-Based Farm
  Services**   365
  Training Students   365
  Relationships With Farms   365
  Developing Resources   365
  Future Grants   366
Review Questions   366
References   366

Appendix 1 Units of Measurement and
Conversion Factors   369
Appendix 2 Measuring Forage Dry Matter Yield
Using Clipped Forage Samples   371
Appendix 3 Graphic Representation of Changes in
Sward Density and Forage Yield With Increasing
Forage Height   373
Appendix 4 Independent Evaluation of Falling Plate
Meter and Rising Plate Meter   375
Appendix 5 Soil Maps from University of Kentucky
Pasture Evaluation Program   377
Appendix 6 2016 Field Recommendations: Central
Kentucky Horse Farm   381
Appendix 7 Answers to Review Questions   383
Appendix 8 Metric Equivalents for Hay Bale Sizes,
as Described in Chapter 11   403
Appendix 9 Environmental Risk Assessment Survey
for Farms   405
Index   411

Contributors

Amy Burk, University of Maryland, Department of Animal and Avian Sciences, College Park, MD, United States

Michael D. Casler, USDA-ARS, U.S. Dairy Forage Research Center, Madison, WI, United States

Tania Cubitt, Performance Horse Nutrition, Jeffersonton, VA, United States

Jimmy Henning, Department of Plant and Soil Sciences, College of Agriculture, Food and Environment, University of Kentucky, Science Center North, Lexington, KY, United States

Thomas (Tom) Keene, Department of Plant and Soil Sciences, College of Agriculture, Food and Environment, University of Kentucky, Science Center North, Lexington, KY, United States

Laura B. Kenny, Extension Division of the College of Agricultural Sciences, Penn State University, University Park, PA, United States

Sherrene Kevan, Enviroquest Ltd., Cambridge, ON, Canada

Laurie Lawrence, Department of Animal and Food Sciences, College of Agriculture, Food and Environment, University of Kentucky, Garrigus Building, Lexington, KY, United States

Krista L. Lea, Department of Plant and Soil Sciences, College of Agriculture, Food and Environment, University of Kentucky, Science Center North, Lexington, KY, United States

Bridgett McIntosh, Virginia Tech, Department of Animal and Poultry Sciences, Middleburg, VA, United States

Edward B. Rayburn, West Virginia University Extension Service, Morgantown, WV, United States

Paul Sharpe, University of Guelph, Guelph, ON, Canada (retired)

S. Ray Smith, Department of Plant and Soil Sciences, College of Agriculture, Food and Environment, University of Kentucky, Science Center North, Lexington, KY, United States

Daniel J. Undersander, College of Agriculture and Life Sciences, University of Wisconsin, Madison, WI, United States

Paul Voroney, School of Environmental Sciences, University of Guelph, Guelph, Ontario, Canada

Michael Westendorf, Rutgers, the State University of New Jersey, Rutgers, the State University of New Jersey, Department of Animal Science, School of Environmental and Biological Sciences, New Brunswick, NJ, United States

Carey A. Williams, Rutgers, the State University of New Jersey, Department of Animal Science, School of Environmental and Biological Sciences, New Brunswick, NJ, United States

Biography

Dr. Amy O. Burk

Amy Burk is an Associate Professor in the Animal and Avian Sciences Department at the University of Maryland in College Park, Maryland. Dr. Burk earned the following degrees: BS Biology from James Madison University, plus MS and PhD in Equine Nutrition from Virginia Polytechnic Institute and State University. She is active on many professional and state boards and university committees, including the National Association of Equine Affiliated Academics. She performs research on pasture and manure best management practices, with consideration given to soil erosion and grazing management. Topics of courses taught by Dr. Burk include pasture management, hay production, horse management, equine science, and experiential learning in equine breeding. Her extension duties include the title of Extension Horse Specialist for the state of Maryland. She has demonstrated and communicated that maintaining productive, dense pasture is effective at reducing sediment and nutrient runoff from horse farms, which earned her an Excellence in Extension award in 2013. In the area of equine nutrition, Dr. Burk communicates awareness of how the digestive tract functions, how forages can be used to improve horse health, and how obesity negatively affects horse health. She also provides leadership in the Maryland 4-H horse program.

Dr. Michael Casler

Michael Casler is Research Geneticist with the USDA Agricultural Research Service of the United States. He has also been Professor of Agronomy at the University of Wisconsin since 1980. He conducts research on agronomy, breeding, genetics, and genomics of perennial grasses for pastures and as potential biomass crops for conversion to bioenergy. He has worked with graziers on the development of new varieties and new species for grazing since 1988. The greatest impacts of his pasture research are the development of "Spring Green" festulolium and the identification of meadow fescue as a high-value grass for grazing in the northern United States and Canada. His pasture research has focused on cattle and sheep pastures, where the focus is on high-quality forage to enhance weight gains or milk production.

Dr. Tania A. Cubitt

Dr. Tania A. Cubitt is a native of Queensland Australia. She received her Bachelor of Science from the University of Queensland in Animal Science. Dr. Cubitt received her Master of Science from Virginia Tech in Equine Nutrition and Growth; this work focused on environmental influences on hormonal and growth characteristics in Thoroughbred fillies. She received her Doctor of Philosophy in Equine Nutrition and Reproduction also from Virginia Tech; this work focused on nutritional effects on ovarian function. Dr. Cubitt currently holds a position as a nutrition consultant with Performance Horse Nutrition (PHN). Her interests are focused on developing feeding strategies for horses with special needs horses including metabolic syndrome, developmental orthopedic disease, gastric ulcers, and senior horses, as well as feeding the broodmare. PHN is an international equine nutrition consulting company. PHN works with horse owners, veterinarians, and feed manufacturers worldwide in designing feeding programs, solving feed-related issues, and formulating feeds that complement local forages. PHN has two PhD equine nutritionists that are well respected as scientists, equine nutrition consultants, and horsemen. In 2010, PHN was the official nutrition partner of Alltech for the Alltech FEI World Equestrian Games.

Dr. Jimmy C. Henning

Jimmy Henning is Extension Professor and Extension Forage Specialist in the Department of Plant and Soil Science at the University of Kentucky. His extension program focuses on hay and haylage production and nutritional quality, as well as pasture establishment and management. He is a cofounder of the Kentucky Grazing Schools and the UK Forage Variety Testing program. He led in the implementation of forage variety trials for grazing tolerance to cattle and for preference by horses. He is part of a forage team that is actively serving the Kentucky horse industry through the Equine Pasture

Evaluation program. Dr. Henning is a graduate of the University of Georgia and the University of Kentucky, College of Agriculture.

He began his career at the University of Missouri as a Forage Extension Specialist and has worked at the University of Kentucky since 1990. While at the University of Missouri, he led the educational program on hay quality using a mobile forage testing lab. Dr. Henning is a Fellow of the American Society of Agronomy and has received the Merit and Medallion Awards of the American Forage and Grassland Council. In addition, he was awarded the Whiteker Award for Excellence in Extension by the UK College of Agriculture, which is the highest honor given by the College for extension programming.

Dr. Henning served as Extension Agriculture and Natural Resource Program leader (2003–07) and then Associate Dean for Cooperative Extension (2007–17). In 2017, he returned to the faculty and resumed his work in forage extension.

He is a native of Guymon, Oklahoma, and was reared in several towns and communities in Georgia. His outside interests include music and photography. He is married to the former Faye Fleming of Tifton, Georgia, and they have one daughter and one grandson.

Thomas (Tom) Keene

Tom Keene was born and raised in Springfield, Kentucky, along with his seven other siblings. He graduated from the University of Kentucky in 1979 with a BS in Production Agriculture and spent 10 years managing two large Thoroughbred farms. From there, he worked in the commercial hay business moving hay nationwide for 16 years. He joined the staff at the University of Kentucky as a Forage Agronomist in 2005 and earned his Master's Degree in Plant and Soil Sciences in 2014. He and Dr. Ray Smith initiated the University of Kentucky Pasture Evaluation Program in 2005. Tom lives in Lexington with his wife, Margaret (Muggs), and has three grown children and one granddaughter.

Laura Kenny

Laura Kenny is an Equine Natural Resources Extension Educator with Penn State Extension. A lifelong horse enthusiast, she completed her BS in Animal Science with an Equine focus at Rutgers University. Her MS was in Plant Biology, also at Rutgers, studying the effects of rotational grazing compared to continuous grazing of horses. Her extension programming helps horse owners to understand how and why to improve pastures and overall environmental stewardship on farms. Professional interests include forage biology and management, grazing behavior and management, nutrient/manure management and composting, and business planning for horse farms.

Sherrene D. Kevan, MSc, MRSB

Sherrene D. Kevan, MSc, MRSB, is primary consultant and president of her company Enviroquest Ltd. She has over 42 years of experience in the fields of biology, ecology, toxicology, apiculture, ornithology, and botany. She lives in Ontario, Canada, with her husband, two dogs, and two horses. As an equestrian, she continues to compete in dressage and will begin endurance riding next year.

Dr. Laurie Lawrence

Laurie Lawrence is a Provost's Distinguished Service Professor in the Department of Animal and Food Sciences at the University of Kentucky, where she is active in equine nutrition research and teaching. She has advised more than two dozen graduate students, including many who are currently employed as equine nutritionists in academia and industry. She is the author of more than 80 refereed research publications, several book chapters, and over 100 abstracts, proceedings papers, and popular press articles. She is a past president of the Equine Nutrition and Physiology Society and a past director of the American Society of Animal Science. In 1998, she received the AFIA Award for research in nonruminant nutrition, and in 2015, she received the AFIA Award for equine nutrition research. She has also received the Distinguished Service Award from the Equine Nutrition and Physiology Society, the Great Teacher Award from the University of Kentucky, the Thomas Poe Cooper Award for research in the University of Kentucky's College of Agriculture, and the public service award from the Kentucky Forage and Grassland Council. In 2008, she received the Equine Science Award from the American Society of Animal Science, and in 2011, she was named a Fellow of the American Society of Animal Science. She was the chair of the National Research Council committee to revise the current National Research Council publication the "Nutrient Requirements of Horses." Dr. Lawrence teaches equine nutrition, equine science, and equine evaluation to students at the University of Kentucky and serves as faculty advisor to the Horse Racing Club. Dr. Lawrence serves on the board of the Kentucky Equine Management Internship (KEMI) program, which brings college students to Central Kentucky to complete 22-week internships on Thoroughbred farms. Dr. Lawrence presents numerous invited lectures throughout the United States each year, and she has been invited to present lectures and/or consult with equine nutritionists in Japan, Australia, Argentina, Canada, Brazil, and Dubai.

Krista L. Lea

Krista Lea was raised in Amarillo Texas and moved to Kentucky to complete the Kentucky Equine Management Internship in 2007. From there, she completed an Animal Science BS in 2009 and an Integrated Plant and Soil Sciences MS in 2014, both from the University of Kentucky. She currently works as a Research Analyst for the Department of Plant and Soil Sciences at the University of Kentucky and has coordinated the UK Horse Pasture Evaluation Program since 2010. Krista and her husband own and show American Quarter Horses and have a daughter.

Dr. Bridgett McIntosh

Dr. Bridgett McIntosh is the Equine Extension Specialist for Virginia Tech and is headquartered at the Middleburg Agricultural Research and Extension Center (M.A.R.E. Center) in Middleburg, Virginia. While she works in many areas of equine research and education, her main focus is on pasture and grazing management to improve horse health and environmental stewardship. Bridgett received her PhD (2006) in Animal Science with a concentration in equine nutrition from Virginia Tech. She was awarded the John Lee Pratt Fellowship in Animal Nutrition to study carbohydrate profiles in feeds and forages, and the avoidance of equine laminitis. She graduated with her MS (2003) in Animal Science with a concentration in equine nutrition, also from Virginia Tech, where she studied feed intake and digestibility in horses. Her BA in Biology is from Hollins College (1997), where she was a member of the I.H.S.A. team and competed in the hunters and equitation. Bridgett has been involved with the horse industry throughout her life. She grew up on a horse farm in upstate New York where she began showing at 7 years old and her family ran a small boarding stable. Bridgett currently competes in the hunters, and she and her husband are avid foxhunters.

Dr. Edward B. Rayburn

Ed Rayburn is an Extension Specialist at West Virginia University. He was educated at Cornell University (BS, Wildlife Management) and Virginia Tech (MS, Wildlife Management, PhD Pasture Agronomy). He works with other WVU specialists, county agents, farmers, Conservation District, and NRCS staff in developing and implementing on-farm research and teaching programs to support pasture-based livestock production and to help landowners develop improved pasture production systems on their farms. He previously worked for the USDA Soil Conservation Service in western New York as an RC&D Grassland Specialist serving dairy, beef, sheep, and equine producers in the western counties of New York. Ed's second job for close to 40 years was running a cow-calf operation on the Appalachian Plateau region in southwestern New York and northcentral West Virginia.

Dr. Paul Sharpe

Paul Sharpe studied zoology and botany before doing a second bachelor's degree in Agriculture. Following a working holiday in Western Australia, where he worked on large sheep farms, he observed different forages and weeds than in his home province of Manitoba. His MS at University of Guelph was on fertility of dairy cows in Jamaica, and his PhD at University of Saskatchewan was on reproduction in beef cattle and sheep. Postdoctoral work on ruminant reproduction at the University of Adelaide in South Australia also revealed "strange" plants and ways of managing them. Each time he came back to Canada from Australia, he stopped in New Zealand, where he noticed grazing techniques that were different than in Australia. Rotational grazing of sheep and beef cattle were two of his first research topics as a Senior Lecturer at the New Liskeard College of Agricultural Technology in "northern" Ontario. He expanded the sheep research facility and modernized the pastures for sheep grazing research. Research topics involved intensity of rotation, annual brassica crops for autumn grazing, forage system pros and cons, and species mixtures for rotational grazing of sheep. Half of his students were in equine management programs and half were in agriculture. Closure of that college led him to Kemptville College of Agricultural Technology, which later became University of Guelph, Kemptville Campus, near Ottawa, Ontario. Research topics included improvements to fertility of dairy cattle, alternative forages and grains (pearl millet and sorghum) for livestock, and use of a water trough in a pasture to alter locations of drinking and defecating beside a stream. Distractions from research included a term as Secretary/Treasurer of the Canadian Society of Animal Science and a 4-year term as Associate Director Academic of Kemptville Campus. The main program for teaching was the Agriculture Diploma, supplemented by the Equine Certificate and Diploma. In 2007, when the University of Guelph started a bachelor's program in Equine Management at Kemptville, Paul began teaching first-year biology, which taught him how much had changed since he took such a course in 1971. In his 28-year teaching career, he taught 23 different courses. The next challenge was to develop and teach pasture management in the Equine Management program, and this continued until the closure of the campus in 2015. The Equine Management program coordinator, Dr. Katrina Merkies, introduced Paul to the National Association of Equine Affiliated Academics (NAEAA) and the Equine Science Society (ESS), which have biannual symposia together. Having searched unsuccessfully for a relevant textbook in pasture management for horses,

Paul consulted members of ESS and NAEAA about producing such a book. Now that he was retired, Paul recruited coauthors from NAEAA and ESS and became the editor of this book.

Dr. Ray Smith

Dr. Ray Smith completed his Biology BS from Asbury College in 1983 and Agronomy MS and PhD from the University of Georgia in 1987 and 1991, respectively. Dr. Smith has been the Forage Extension Specialist at the University of Kentucky since 2004 and was promoted to Extension Professor in 2010.

Dr. Ray Smith is a native of Georgia and received his undergraduate degree from Asbury University in Kentucky in 1983. After teaching high school biology for 2 years, he entered a graduate degree program in agronomy and plant breeding at the University of Georgia. From 1991 to 2001, Ray held a research, teaching, and extension position at the University of Manitoba, Canada, with a focus on alfalfa and native grass breeding, seed production, and forage management. He was the Forage Extension Specialist at Virginia Tech from 2001 to 2004 and is now Professor and Extension Specialist at the University of Kentucky. Ray is the current chair of the Continuing Committee for the International Grassland Congress and past President of the American Forage and Grassland Council. He has published 40 articles in refereed journals, presented 155 papers at professional conferences, written over 100 extension publications, and given over 600 extension presentations. Ray has been the advisor for 16 master's, 3 PhD's, 5 Post-docs, and 24 senior research students. His current extension activities include working closely with county agents and producers; conducting applied forage research for Kentucky and the transition zone; helping organize state, regional, national, and international forage conferences; and writing applied agricultural publications. His current research projects include evaluating forage varieties for grazing tolerance and yield, developing forage production systems, pasture evaluation methods, and developing computer and time-lapse photography teaching tools.

Dr. Daniel J. Undersander

Dr. Undersander received his bachelor's degree from the University of Minnesota and MS and PhD from Purdue University. He has worked at Texas A&M and Clemson University, where he did grazing research. For the last 30 years, he has coordinated the multidepartment extension forages program at the University of Wisconsin. He has authored or coauthored over 1500 publications. He was senior editor of the CD entitled "Pastures for Horses." He is a member of several interagency committees within Wisconsin developing environmental regulations, on the Board of Directors of two national organizations, and on the advisory board of four nationally distributed industry magazines.

Dr. Paul Voroney

Dr. Paul Voroney is Professor of Soil Biology and Biochemistry in the School of Environmental Sciences, Ontario Agricultural College, at the University of Guelph, Guelph, Ontario. His area of specialization is soil science, with an emphasis on plant and soil management effects on the nature and dynamics of the constituents of soil organic matter. His research areas include soil biology and biochemistry, with a focus on the biogeochemical cycling of soil carbon, nitrogen, and phosphorous. His research program uses tracer techniques (13C, 14C, and 15N) to assess rates of decomposition of plant residues, formation of stable soil organic matter (soil humus), and on soil fertility. He has a special interest in the utilization of nonhazardous organic wastes (e.g., composts) as amendments to enhance soil fertility and restore effects of soil degradation. He has published 153 articles in scientific journals and 21 book chapters, and he has supervised/cosupervised 20 PhD and 29 MSc students.

Dr. Michael L. Westendorf

Michael L. Westendorf is an Associate Professor and the Extension Specialist in Livestock and Dairy in the Department of Animal Sciences at Rutgers, The State University of New Jersey, in New Brunswick, New Jersey. He earned his BS in Animal Science from the University of Idaho, plus MS and PhD degrees in Animal Science and Ruminant Nutrition from the University of Kentucky. Dr. Westendorf conducts research related to animal agriculture and creates outreach and extension programs that serve the needs of New Jersey horse and livestock farmers. He specializes in waste management, by-product utilization, on-farm composting, and optimization of animal growth. Dr. Westendorf teaches and coordinates courses in production animal management, farm productivity analysis, artificial insemination, and animal evaluation and selection.

Dr. Carey A. Williams

Carey A. Williams, PhD, joined Rutgers University in July 2003 as its Equine Extension Specialist and Associate Director of Outreach for the Equine Science Center, taking an active role in teaching, conducting research, and working with the equine and academic communities to ensure the viability of the horse industry in New Jersey. Wisconsin native, Dr. Williams earned her doctorate degree in animal and poultry sciences (with an emphasis on equine nutrition and exercise physiology) in June 2003 from Virginia Polytechnic Institute and State University. She holds a master's degree in equine nutrition, also from Virginia Tech, and a bachelor's degree from Colorado State University. At Rutgers, Dr. Williams maintains a herd of Standardbred mares for nutrition, pasture, and exercise physiology research. Her main focus of research was antioxidant supplementation and decreasing the stress of exercise and competition in performance horses. However, currently she is performing research on best management practices with horse farms and finding ways for farmers to keep their horse farms and pastures environmentally friendly while maintaining optimal horse health and economic viability. Along with this, she is currently investigating gastrointestinal health of horses on pasture and in competition. As a hobby, she trains and competes with her off-the-track Thoroughbred mare at various local and regional dressage shows.

Preface

This book is about providing a better life for horses. Horses evolved from small forest-dwelling creatures to much larger creatures of the grasslands, with systems for locomotion, digestion, and environmental awareness that allowed them to thrive. The grasslands probably experienced seasonal growth patterns, so digestive systems needed to provide enough nutrients to support athletic, resilient prey animals through pregnancies and lactations while the nutrient supply rose and fell with the seasons. Feral horses still have high levels of fertility and excellent body condition in unfenced grassland environments where management by humans is minimal. Perhaps this is an indication that domesticated horses should be provided with environmental conditions that are as close as possible to their natural environment to promote their welfare.

The management of a horse pasture can be simplistic and minimal, or it can be oriented toward provision of an environment as similar as possible to a range in which feral horses thrive and in which their quality of life is optimal. The goal of providing a better life for horses is in line with the 1990s idea of the "five freedoms" and the more recent "five domains" for animal welfare. Several chapters in this book describe practices that are likely to help horse pasture managers ensure that horses in their care are free of thirst, hunger, malnutrition, discomfort, exposure, pain, injury, disease, fear, and distress, plus able to express normal behavior, in line with the five freedoms. Achievement of the first two of these five domains is, at least indirectly, addressed in this book by recommending opportunities for ideal intakes of water and food, plus provision of an environment that allows horses to choose shelter from extreme temperature, sun, wind, or precipitation. Under the health domain, there are recommendations that help to reduce parasitism. The maintenance of a healthy turf provides both a comfortable resting place and a running surface that cushions and absorbs impact, thus protecting limbs from injuries. Diversity of pasture plants is promoted, and normal grazing behavior of horses is discussed, so managers can provide opportunities for horses to explore, move freely, and make choices in what they eat, satisfying the behavior domain. The fifth domain, mental state, is ideally promoted in a well-managed pasture, where thirst, hunger, poor air quality, barn noises, physical exhaustion, predation, frustration, and boredom are minimized, and pleasant tastes and smells, satiety, freedom, physical comfort, thermal comfort, the ability to socialize, and opportunities for play are promoted.

Horses are trained to provide great service to humankind by carrying us, pulling our loads, and entertaining us with their athletic abilities. Horse owners and managers have learned a great deal about maintaining these animals, so they can continue to serve us. Yet, some of the basic origins and needs of horses are often forgotten or given lower priorities than competitions, tack, lessons, grooming, convenience for clients, and the financial considerations of running a business. It is not always practical to run a business managing horses for clients if the horses are part of a semiwild herd living in an unfenced rangeland. Clients, trainers, sales agents, farriers, and horse health practitioners need the horses to be readily accessible. Conversely, the horses may experience better welfare living freely on the unfenced grassland.

Am I suggesting that there is something wrong with the way some horses are maintained? Without giving away the answer, let me describe what I think is a fairly typical environment in many parts of North America and Australia. There is a fence around an area that is mostly green, but the green plants are mostly very short (like a golf green), except for some tall ones that are flowering and producing seeds because no horse wants to eat them. The horses are in that same fenced pasture for the whole growing and grazing season and maybe through part of the nongrowing season also. A casual observer would likely conclude that horses graze plants closer to the ground than cattle do. The chapters on grazing behavior and management explore this idea. These same horses appear to avoid grazing what looks like nutritious abundant forage in at least one portion of the pasture, and the horses can be observed to do much of their defecating among that tall grass too. Not far from that longer grass is an area of bare soil. It happens to be near a gate and a water trough big enough to provide all the water those horses need for several days. By the time a reader is halfway through this book, she or he should be able to describe several improvements to make that pasture a better place for horses to live.

A course in comparative anatomy teaches how much we can learn about one species that is valuable in our understanding of other species. Some equine students have indicated a desire to learn only about horses, and some agriculture students have wanted to only learn about sheep, beef, or dairy cattle. Each of them can benefit, either now or in the future, from learning something about more than just her or his target species. A colleague specializes in dairy cattle nutrition and feeding, yet he makes a point of attending poultry nutrition conferences because many nutrition innovations are tried on poultry first.

There are many areas of interest in horse management and pasture management for horses that have not been intensively researched yet. However, much has been learned about grazing animal behavior, the nutritional value of pasture plants, pasture ecology, and grazing management with sheep or cattle as the experimental animals. Most of this knowledge is relevant to horse pasture managers. A necessary skill set is learning how to adapt knowledge about other animals to use on a horse farm. Unfortunately, most research is expensive. Much research involving livestock has been financed by funds for agriculture, food, and environmental protection. When more funds become available for horse-specific research, there can be more information on grazing management that is specific to horses.

Both print and online publications help me to increase my knowledge of and maintain my enthusiasm for pastures and grazing when I am not teaching or researching them, so I recommend supplementing your textbook studies by regular reading of relevant periodicals. One article discussed the practical dilemma of confining horses to a small enough paddock that they eat the forage down to about half its starting height within 3 days while also having grazing paddocks large enough that horses get a reasonable amount of exercise. This textbook will help readers gather the appropriate information and use it to tackle this and related problems in grazing management.

My teaching career involved diploma and degree programs in agriculture and equine studies. When part of a 4-year degree in equine management came to our campus in 2007, I had the opportunity to teach biology and to develop and teach a course in pasture management, based partly on the biology material.

In my search for a pasture management textbook, I discovered that Gillian McCarthy had written *Pasture Management for Horses and Ponies* in England in 1987, and it was out of print by 2007. Later, I discovered another book, *The Pony Club Guide to Pasture Management*, by Elizabeth O'Beirne-Ranelagh, published in 2010. It has many beautiful photographs and is an excellent book for pony club members in England, but my students agreed with me that it is not suitable for a degree-level program in Canada. There are many books on pasture management with cattle and sheep as their focus. There are also many extension publications from Canadian provincial ministries or departments of agriculture and from American university extension branches on aspects of pasture management for horses. For several years, I searched for a textbook that would bring all the resource material I wanted into one publication. Eventually, I realized that I needed to start this process myself.

One component of a pasture management course that I value highly is the laboratory and field trip portion, which includes tours to farms where horses graze. If you have any desire to learn how to manage pastures for horses, I encourage you to visit as many pastures as you can. Horse farms in your area that have a reputation for excellent pasture management are probably the best places to go. If those are not available, then tour any well-managed pastures, especially where geographic conditions are similar to yours, and tour horse farms anywhere, preferably, some that have advanced grazing practices. Make notes and take pictures for future reference.

My colleague and coordinator of the equine management program, Dr. Katrina Merkies, introduced me to the National Association of Equine Affiliated Academics (NAEAA) and the Equine Science Society (ESS), and through their conferences, I met a number of the authors of chapters in this book, and they referred me to others. This book is the result of 3 years of planning, recruiting, writing, editing, and encouraging. I hope it encourages you to keep learning.

Paul Sharpe
Email address: psharpe@uoguelph.ca
Telephone number: 613 258 3177

Acknowledgements

This book was made possible by people of diverse skills in many different places. My parents encouraged my interest in and love of animals and our natural environment. Dr. David Stewart, at Brandon University, introduced me to agricultural researchers in a course in reproductive physiology. Dr. Phillips, at University of Manitoba, provided contacts that helped me to have a working holiday in Australia and observe the diversity of forages there. Dr. Gordon King, at University of Guelph, sent me to Jamaica to study fertility in dairy cows, which expanded my appreciation of the importance of the feed supply for cattle and horses. Dr. Katrina Merkies, from the University of Guelph, agreed that I should develop and teach a course in pasture management to students in a degree program of equine management. She introduced me to the National Association of Equine Affiliated Academics (NAEAA) and their meetings. In particular, alternate year meetings of NAEAA are held with the Equine Science Society (ESS), which exposed me to research on pasture management for horses. I am indebted to many members of NAEAA and ESS. Some of them became coauthors of this book or referred me to others who joined us. Some others declined to join the book project because they knew that they would not have time to write a chapter for it, and I respect them for realizing this and telling me so up front.

I am grateful to all those who encouraged my coauthors and me in this project and to those who helped with writing, reviewing, providing figures, and making suggestions. Ed Rayburn of West Virginia University Extension confirmed my interpretation of his reports on forage yields and then became a coauthor, with involvement in three chapters.

I thank reviewers of chapters, including Jerry Holechek, Professor of Range Science at New Mexico State University, Heather Smith Thomas, a rancher near Salmon, Idaho, and author of books and magazine articles on horses and cattle, Greg Wall, a retired English teacher who proofread my chapters, Martin K. Nielsen (U. Kentucky), Ann Swinker (Penn State U.), Masoud Heshemi (U. Massachusetts), Tanja Hess and Anthony Knight (Colorado State U.), and Bryan Stegelmeier (ARS, USDA).

Advice, answers to questions, and referrals to other sources of info were gratefully received from Bob Coleman (U. Kentucky), Donald Ball (Auburn U.), Joel Salatin (Editor of *The Stockman Grass Farmer* magazine), Christine O'Reilly (Pasture and Forage Specialist, Ontario Min of Ag., Food and Rural Affairs), Laurie Lawrence (U. Kentucky), Deb Bennet (Equine Studies Institute, Livingston, CA), Krishona Martinson (U. Minnesota), Ray Smith and Krista Lea (U. Kentucky), Kathy Voth and Rachel Gilker (Editors of OnPasture.com).

Figures, information, and/or permission to use them came from John Worley (U. of Georgia), Karen Launchbaugh (U. of Idaho), Andy MacDonald (Highland Fence), Ellen Brisendine (*The Cattleman* magazine), Randy Lenz (Stay-Tuff Fence), Wendy Etheridge (Bowser Supply Ltd, England), Carrie Byrum (Gallagher, North America), Stephanie Church (TheHorse.com), Vern Baron (Agriculture and Agri-Food Canada), Savannah Petrachenko and Dwayne Job, System Fence, Rockwood, ON), Steve Kenyon (rancher, *The Stockman Grass Farmer* contributor), Becky Koch and Agnes Vernon (NDSU Extension Service), Sandra Mark (Ontario Min. Gov. and Consumer Services), Pam Devore (U. Wisconsin Extension), Tammy Parish (Tru-Test Group, Speedrite), Thomas Griggs (W. Virginia U.), David Ellis (Am. Horticultural Soc.)

Help with the UK Extension Pasture Evaluation Program was provided by Julia Becker, Thane Anderson, and Savannah Taylor. Robin Sakowski, Manager of Access Services at the University of Guelph Library, allowed me to extend my loan of several books throughout my writing and editing period. Kristi Gomez, Pat Gonzalez, Swapna Praveen, and Poulouse Joseph at Elsevier corresponded politely and patiently with me through all of their duties related to this book.

Paul Sharpe, Editor

Chapter 1

Forage Plant Structure, Function, Nutrition, and Growth

Daniel J. Undersander
College of Agriculture and Life Sciences, University of Wisconsin, Madison, WI, United States

STRUCTURE AND PHYSIOLOGY OF PASTURE GRASSES AND LEGUMES

First, we will describe the plant organic compounds found in pasture grasses and legumes, then the cell structure, then plant structures, and lastly, how plants use these to grow.

Plant Organic Compounds

Carbohydrates (Carbon-Containing Compounds: Either Structural or Nonstructural)

Carbohydrates (CHOs) contain the elements carbon (C), hydrogen (H), and oxygen (O). Their structures are the simplest of the organic compounds and mostly include sugars and polymers or chains of sugars. In plants, there are nonstructural and structural CHOs.

1. Nonstructural carbohydrates (NSCs) sometimes are called total nonstructural carbohydrates. Three classes of NSCs are distinguished.
   a. **Monosaccharides** (single sugars) contain either six carbon atoms (hexose sugars) or five carbon atoms (pentose sugars). Fig. 1.1 shows structures of a hexose called **glucose** and a pentose called **fructose**.
   b. **Disaccharides** are two sugar molecules linked together, for example, glucose linked to fructose produces the disaccharide **sucrose** (Fig. 1.2). Two glucoses bonded together can produce either maltose or cellobiose, depending on the bonds between them.
   c. **Polysaccharides** are composed of three or more linked sugar molecules. Large polysaccharides include **starch**, a nonstructural carbohydrate made of several α-D-glucose sugars joined by α-1,4 linkages (Fig. 1.3). Starch is a storage compound in legumes and C4 grasses. **Fructan** is a polysaccharide made of fructose units joined by various linkages. It is a storage compound in cool season grasses.

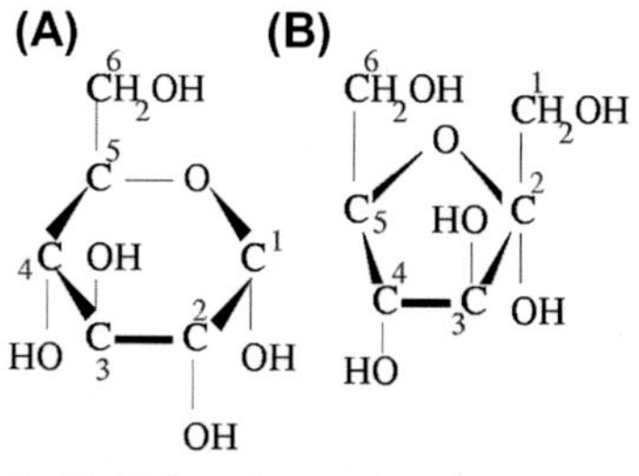

FIGURE 1.1 (A) Alpha-D-glucose, a hexose, on the left. (B) Alpha-D-fructose, a pentose, on the right. *(A) Source: https://commons.wikimedia.org/wiki/File:Alpha-D-glucose_Haworth.png. GNU Free Documentation License, Version 1.2, Free Software Foundation. Creative Commons Attribution-Share Alike 3.0 Unported. (B) https://commons.wikimedia.org/wike/File:Alpha-d-fructose.png. GNU Free Documentation License, Version 1.2, Free Software Foundation. Creative Commons Attribution-Share.*

Horse Pasture Management. https://doi.org/10.1016/B978-0-12-812919-7.00001-9

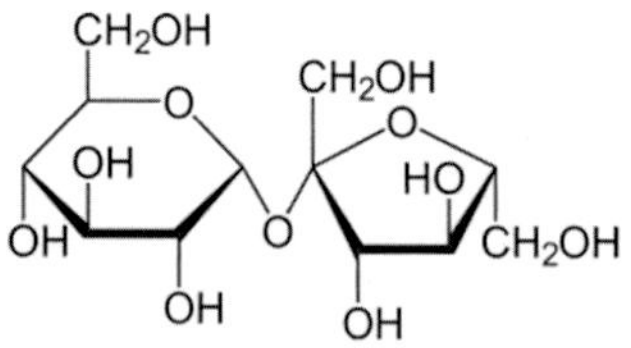

FIGURE 1.2 A disaccharide called sucrose, also known as table sugar, is composed of a glucose bonded to a fructose. *Source: https://commons. wikimedia.org/wiki/File:Sucrose_structure-formula.png. Author: Bas. GNU Free Documentation License, Version 1.2, Free Software Foundation. Creative Commons Attribution-Share Alike 3.0 Unported.*

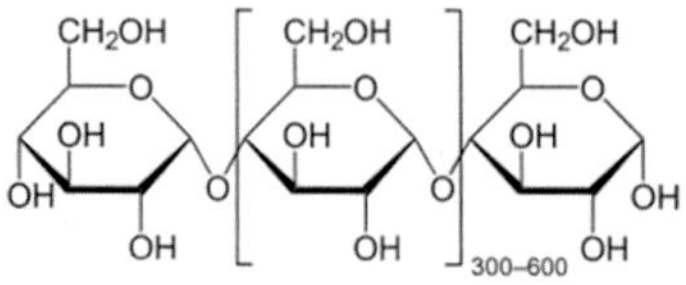

FIGURE 1.3 A polysaccharide called starch of the amylose type, having a linear arrangement of glucose molecules with α-1,4 linkages. *Source: https:// commons.wikimedia.org/wiki/File:Amylose.png. GNU Free Documentation License, Version 1.2, Free Software Foundation. Creative Commons Attribution-Share Alike 3.0 Unported.*

2. Structural carbohydrates provide strength to cell walls of plants, being concentrated in specialized cells for structural support. Three types of structural carbohydrates are pectins, hemicellulose, and cellulose.
   a. **Pectins** are polymers of α-D-galacturonic acid, and they act like glue between cells. They gelatinize in water.
   b. **Hemicelluloses** are complexes of the sugars xylose (4C and 5C forms; abundant in cobs, hulls, shells, wood, and straw) and mannose (5C and 6C forms; involved in protein metabolism) and other polysaccharides.
   c. **Cellulose** is composed of β-D-glucose units, joined by β-1,4 linkages (Fig. 1.4). Cellulose is a very important component of primary cell walls of plants and possibly the most abundant organic polymer on Earth.

Lignin

Lignin is a polymer of phenolic acids. Lignin binds to cellulose in cell walls, adding rigidity. It increases the ability of plant tissues to transport water internally without loss. Since neither mammals nor the microbes in their gastrointestinal tracts make enzymes that digest lignin, it is indigestible by horses and ruminants. Thus, lignin continues to bind other cell wall components, even in the digestive tracts of herbivores, limiting their digestibility.

Nitrogenous Compounds

Plants use nitrogen that they obtain from the soil and metabolites of carbohydrates to synthesize the following:

1. inorganic molecules, including nitrate (NO_3^-) and ammonia (NH_3);
2. amino acids (simple nitrogen-containing organic compounds). These are subunits of proteins, responsible for transporting nitrogen within plants. For example, the amino acids glutamine and aspartate can each react with ammonium, (NH_4^+) to gain an extra amino group that can be released elsewhere;
3. proteins (linked chains of amino acid molecules) have many functions, and in plants they primarily function as enzymes;
4. DNA and RNA.

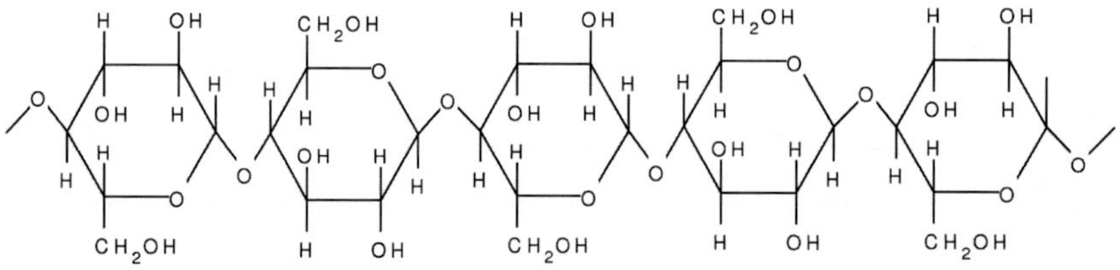

FIGURE 1.4 A segment of a cellulose polymer, showing β-1,4 linkages between glucose units. *Source: Cellulose strand-es.jpg. Author: I. Laghi. Licensed under the Creative Commons Attribution-Share Alike 3.0 Unported license.*

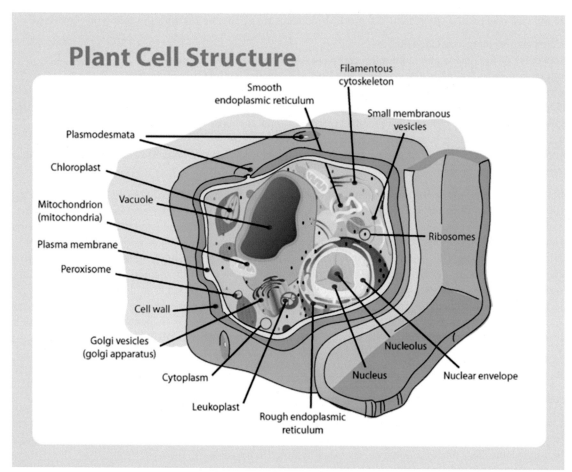

Plant Cell Structure

FIGURE 1.5 Plant Cell Structure. Note the plasmodesmata penetrating the cell membrane and cell wall for exchanging nutrients and signalling chemicals with other cells, ribosomes for synthesizing proteins, endoplasmic reticulum for synthesizing lipids and proteins, vacuoles and vesicles for storing things made in the cell, one chloroplast here to represent many that exist in leaf mesophyll cells, and a thick cell wall outside the cell membrane. *Source: https://commons.wikimedia.org/wiki/File:Plant_cell_structure.png. Public Domain. Author: LadyofHats. Creative Commons attribution-ShareAlike License.*

The Cell: The Most Basic Structural Unit of Plant Organization

Plant cells differ from animal cells in two ways:

1. Plant cells have a cell wall that provides "rigidity" and "shape" to the cell.
2. Plant cells have chloroplasts to convert solar energy into chemical energy.

The cell wall consists of the primary cell wall and the secondary cell wall. The primary cell wall is formed first and composed primarily of cellulose and hemicellulose. It provides some rigidity while allowing the cell wall to increase in size as the cell grows. Young plants, with only the primary cell wall, are lush, tender, and highly digestible (Fig. 1.5).

The secondary cell wall forms inside the primary cell wall when the cell is full size. It is composed of cellulose, hemicellulose, and lignin. This wall gives the cell its final rigidity. It makes the cell impermeable to water and also reduces digestibility of the cell.

The area between the cell walls of adjacent cells is called the middle lamella. It is high in pectin (a complex polymer of sugar acids) which is gelatinous and binds the cells together.

Mono and disaccharides and amino acids can diffuse through the cell wall for transport to locations of growth or storage. Forage plants also have microscopic channels between cells and through cell walls called plasmodesmata for transport of large molecules such as proteins, RNA, viroids, and viral genomes from cell to cell. Nutrient molecules to be used for growth and signalling molecules to stimulate responses to attack are examples of chemicals that pass through plasmodesmata.

Inside the cell walls is the cytoplasm of the cell. It contains many organelles with the following functions:

1. mitochondria for energy metabolism
2. chromosomes in the nucleus for telling the cell what to do and for reproduction
3. smooth endoplasmic reticulum for synthesis of lipids and metabolism of carbohydrates
4. ribosomes and rough endoplasmic reticulum for synthesis of proteins
5. Golgi apparatus for modifying products of endoplasmic reticulum and synthesizing some carbohydrates such as pectins
6. vacuoles for storing ions, proteins, by-products, pigments, and chemical weapons against herbivory
7. peroxisomes for converting fatty acids to sugar in germinating seeds
8. chloroplasts, which are specialized compartments containing chlorophyll that convert energy from sunlight into chemical energy (photosynthesis)

In photosynthesis, sunlight is absorbed to produce energy-rich molecules (ATP and NADPH), which are then used to synthesize carbohydrates from carbon dioxide (CO_2) and water. Plants adapted to cool, wet environments fix the carbon dioxide into a three-carbon compound and are called C3 (cool season) plants. Plants adapted to hot, sunny environments fix the CO_2 into a four-carbon compound and are called C4 (warm season) plants. The sugar becomes dissolved in phloem sap and circulates in the plant. The amount of sugar produced is greater than the amount used up in plant growth and development through the day, so sugar accumulates during the sunny hours. During the dark hours, sugar will continue to be used up in plant metabolism, without being replaced, so the concentration of sugar in the plant declines.

C3 pasture grasses store energy as fructan, while legumes and C4 plants store energy as starch. This is significant as C3 grasses are sweeter and more palatable than legumes and C4 grasses, but there is evidence that too much fructan may adversely affect horse health.

Organs

The Leaf Is the Principle Photosynthetic Organ

The leaf has an upper and lower epidermis. This outer layer of cells is coated with waxy **cutin,** forming a layer called the **cuticle** to prevent undesired water loss from the leaf. The waxy layer also protects against physical damage and damage from other organisms. The wax is a polymer of omega hydroxyl fatty acids.

Below the upper epidermis is a tightly packed layer of **palisade parenchyma** (palisade mesophyll) cells. These cells have a high chlorophyll content and are the site where light energy is converted into high energy molecules (photosynthesis). Below the palisade parenchyma cells are the loosely packed **spongy mesophyll** cells that use the high-energy molecules to incorporate CO_2 into C3 or C4 compounds.

Stomates (stoma) are located primarily on the bottom of the leaves. They open during the day to allow CO_2 to enter, oxygen to leave, and water to evaporate from the leaf and cool the plant. Stomates close in dark or drought to reduce water loss.

Stem: Supports and Connects Other Organs With the Root

The stem is the conduit for water and nutrient transport. Minerals are carried with water from the root to leaves in the **xylem tissue**, which is primarily composed of dead cells. Movement of water through the xylem is driven by negative pressures, including tension from water evaporation from the leaf. The xylem contains cells called **tracheids** and **vessel elements** that conduct and disperse water. Carbohydrates are carried to growing points, to fruits or seeds, and to storage in the root via the **phloem tissue**, which is composed of living cells. The phloem includes **sieve cells**, which are tubes with perforated ends, and **companion cells** next to them that serve the sieve cells. Water in a plant with the dissolved organic compounds is called sap. Movement of sap is called translocation and is controlled by positive hydrostatic pressure (by active phloem loading and unloading).

The stem is green (has chlorophyll) but has little photosynthetic activity. Its epidermis may have a waxy coat. The stem consists of a mature and fully differentiated region (usually near the base of the plant) and an actively growing region (usually near the top of the plant). Dicots have a definite **cortex** region containing vascular bundles and **sclerenchyma** support cells. Softer ground tissue toward the middle contains mostly **parenchyma** cells referred to as **pith**, along with some **collenchyma** cells for flexible support. Monocots do not have separate cortex and pith regions. Fig. 1.6 shows cross-sections of a dicot (which includes legumes) stem and a monocot (which includes grasses) stem, indicating differences in the arrangement of vascular bundles of xylem, phloem, and sclerenchyma. Note that many cool season grasses require exposure to cold (vernalization) to produce a stem, so they do not produce a stem in the seeding year and only produce stems in the first spring growth of succeeding years.

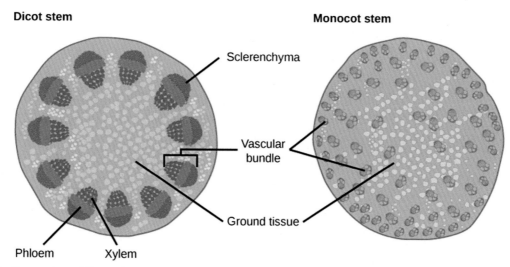

Dicot stem

Monocot stem

Sclerenchyma

Vascular bundle

Ground tissue

Phloem   Xylem

FIGURE 1.6 Cross-sections of dicot and monocot stems. The outer cortex region of dicots contains large **vascular bundles** of xylem, phloem, and supporting **sclerenchyma** cells. The soft inner pith region is called ground tissue, which consists mostly of **parenchyma** cells and some **collenchyma** cells. Monocot stems have smaller vascular bundles which are dispersed through the ground tissue and are slightly concentrated near the outer surface. *Source: https://commons.wikimedia.org/wiki/File:Figure_30_02_06.jpgCreative Commons Attribution 4.0 International license.*

Buds: Embryonic Stem, Leaf, Root, or Flower Tissue

A **bud** contains embryonic tissue (meristem) than can develop into various plant parts (leaves, stems, flowers) depending on hormonal conditions.

An **apical** (or terminal) **bud** is the primary growing point located at the apex (tip) of the stem and is the dominant bud. The **terminal bud** can cause all the **axillary buds** below it to remain dormant (Fig. 1.7).

An axillary bud is an embryonic shoot that lies dormant at the junction of the stem and petiole (stalk that attaches leaves to the stem). Axillary buds will grow in some plants and produce branches off the main stem. Some plants have strong **apical dominance**, which occurs when the apical meristem produces the hormone **auxin** to prevent **axillary (lateral)** buds from growing. In these plants, axillary buds will begin developing when they are exposed to less auxin, for example, if apical dominance is broken by removing the terminal bud, i.e., **grazing**.

A **crown** is the top part of a root system from which a stem(s) arises. A **crown bud** is an embryonic shoot that develops on the crown. A **tiller** is a stem produced by grass plants; tiller refers to all shoots that grow after the initial parent shoot grows from a seed. Tillers most commonly form at the crown but may also form in leaf axils (e.g., nodes on a main stem). Branching at axillary meristems is controlled by a combination of hormonal, environmental, developmental, and genetic control. The hormone cytokinin promotes axillary bud outgrowth. Another hormone from the roots called strigolactone inhibits bud outgrowth (McSteen, 2009).

Roots: Below Ground Portion of Plant

Roots are the underground portion of the plant that anchor the plant and absorb water and nutrients. Several root types exist:

1. A tap root is a large, central, dominant root growing downward from the crown. It is generally large in diameter at the crown and tapers with depth in the soil. Taproots frequently store significant carbohydrates and proteins for winter survival. Examples of plants with tap roots are alfalfa and turnips (Fig. 1.8).
2. Fibrous roots are thin, branched roots growing from the crown. This is the alternative to a tap root and occurs in all grasses and many legumes (Fig. 1.9).
3. Secondary roots are side branches off the main root (either a taproot or in fibrous root systems).
4. Adventitious roots are roots that form on any part of plant other than the roots. They may form at nodes on the stem or on leaves.
5. Root hairs (extensions of epidermal cells) increase root surface area for nutrient absorption and are sites of initial nodulation activity in legumes.

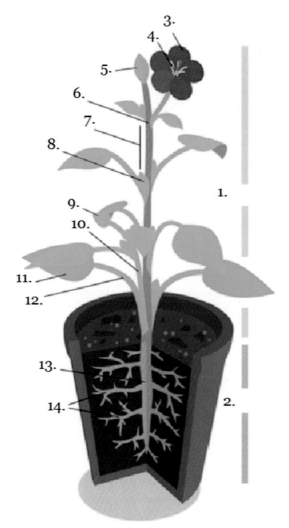

FIGURE 1.7 The active apical bud on the tip of a plant shoot produces auxin, which inhibits cell division at lateral (axillary) buds below it. Removal of the apical bud by grazing, pruning, or mowing removes the source of auxin, allowing cell division and growth from lateral buds, producing branches. Plant parts: 1. Shoot system. 1. Root system. 3. Flower and petals. 4. Stamens and carpels. 5. Apical bud. 6. Node. 7. Internode. 8. Axillary bud. 9. Second apical bud. 10. Vegetative shoot. 11. Leaf. 12. Petiole. 13. Taproot. *Source: https://commons.wikimedia.org/wiki/File: Plant morphology eudicot numbers.png Creative Commons Attribution-Share Alike 4.0 International license. Author: Sten.*

How Grasses Develop

Establishment

Grasses have hypogeal emergence, meaning that the growing point is below soil surface for some time after emergence. Hypogeal emergence protects the growing point from damage by grazing.

Recommended seeding rate for pastures is generally 50 to 75 seeds per square foot. The seeds per pound are shown in Table 1.1 for different pasture grasses and legumes. Note that, while many more seedlings may emerge, the stand will generally thin to 15–20 plants per square foot (0.0929 sq m) in 3–4 months.

Seedlings are considered perennial when a crown forms and, for grasses, when tillering begins from the crown. Grasses continue to form tillers throughout first year.

Many pasture cool season grasses do not head out in the seeding year (timothy, bluegrass, and many ryegrasses would be exceptions that head out). Cool season grasses generally form flower buds in the fall that require exposure to cold (vernalization) for development into stems the next spring.

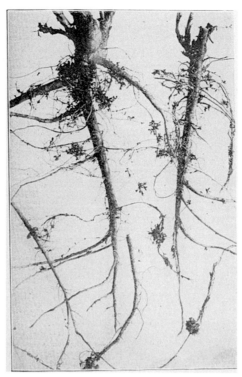

FIGURE 1.8 Tap roots of alfalfa plants typically can grow 6—16 feet (2—5 m) deep, with extremes reported longer than 50 feet (15 m). Creeping-rooted alfalfa varieties have been developed to better suit a pasture environment and management as opposed to production strictly for hay. These varieties have both types of roots. https://commons.wikimedia.org/wiki/File: Annual_report_of_the_Agricultural_Experiment_Station_of the_University_of_Minnesota_1902_(14755551846).jpgFlickr's The Commons (https://flickr.com/commons). No known copyright restrictions exist.

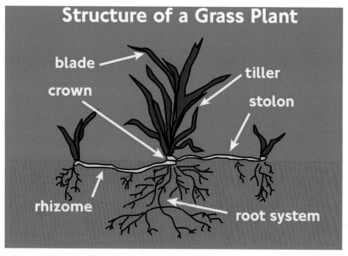

FIGURE 1.9 Structure of a grass plant, showing its fibrous root system, the slight difference between rhizomes (horizontal stems below ground) and stolons (horizontal stems on the soil surface), and a tiller shooting from the crown. *Source: https://commons.wikimedia.org/wiki/File:Grass-plant-structure.png. GNU Free Documentation License Version 1.2, Free Software Foundation. Author: Wackymacs at English Wikipedia.*

Growth Following Establishment

Early spring grass growth in years after seeding is all leaves. The growing point starts underground and becomes the seed head. Stem elongation begins when the plant is about 4 inches (10.16 cm) tall. Elongation of the stem is due to cell division at the nodes. Leaves and stem are pushed up from below.

New tillers begin to emerge from crown buds formed after stem elongation begins. In orchardgrass and ryegrass, this occurs early, and it occurs much later in timothy and bromegrass. Thus, the latter two species must have longer recovery periods between grazings. Removal of above ground growth favors tillering.

TABLE 1.1 Seed Weight and Seeds per Square Foot

Species	Seed Weight (Seeds/lb)	Seeds/ft^2/lb of Seed
Grasses		
Bluegrass, Kentucky	2,200,000	51
Meadow fescue	226,000	5
Festulolium	227,000	5
Orchardgrass	600,000	14
Reed canarygrass	526,000	12
Ryegrass, Italian	270,000	6
Ryegrass, perennial	230,000	5
Smooth bromegrass	136,000	3
Tall fescue	190,000	4
Timothy	1,234,000	28
Legumes		
Alfalfa	220,000	5
Alsike clover	680,000	16
Birdsfoot trefoil	372,000	9
Ladino clover	784,000	18
Red clover	252,000	6
White clover, intermediate	784,000	18
Kura clover	251,000	6

Regrowth after flowering or removal of seed head is mostly leaves. New tillers begin when the primary tiller is about 6–10 inches (15.24 − 25.4 cm) tall.

Buds formed in late summer become next year's flowers.

How Legumes Develop

Legumes have epigeal emergence, which means that the growing point moves above ground with plant emergence. If the growing point of a legume seedling is damaged or removed (i.e., grazing), that stem will cease growing and the seedling will die. About 3–4 weeks after emergence, the stem pulls back toward the root and forms an enlarged diameter that becomes the crown. New tillers arise both from the crown and axillary buds. Plants flower in the seeding year and all subsequent years. Defoliation encourages tillering.

New spring growth occurs from buds formed in the fall. Thus, if winter injury kills some of the buds, the plant has to start over in the spring growing new buds, which delays plant growth and yield.

Pasture Plant Growth

Plant growth depends on energy derived from photosynthesis, which occurs in the green chlorophyll, primarily in the leaves. The amount of photosynthesis depends on the amount of sunlight and temperature and whether the plant has adequate water and nutrients for growth. All plants are both photosynthesizing energy compounds and using energy compounds to respire and grow. At low temperatures, the rate of photosynthesis is slow, so growth is slow. At the optimum temperature for a species, photosynthesis maximally exceeds respiration and the plant is growing at the fastest rate. Above the optimum temperature, respiration increases faster than photosynthesis, so net growth is reduced. The optimum temperature for cool season grasses may be as low as 72°F (22°C) and for warm season grasses may be close to 85°F (30°C).

Generally, in a healthy plant, photosynthesis occurs faster in the leaves than can be transported to growing points of the plant or to plant storage areas. Daily sugar production above a plant's capacity to transport it out of the leaves is converted into long chain carbohydrates and stored temporarily in the leaves. This is starch in warm season grasses and legumes and fructans in cool season grasses. These long chain carbohydrates are broken down overnight (when photosynthesis is not occurring), and sugars are transported out of the leaves. Thus, forage leaves are low in sugar, starch, and/or fructan in the morning and higher in these highly digestible compounds in the afternoon.

The particular plant part for the storage of reserve carbohydrates (sugars, starches, and fructans) varies widely in the different species and may include the root, stem, leaves, and seed. Alfalfa, red clover, sweet clover, and birdsfoot trefoil store the largest proportion of their reserve carbohydrates in the roots, ladino clover in the stolons, bromegrass and reed canarygrass in the rhizomes, and orchardgrass, tall fescue, meadow fescue, ryegrasses, and timothy in the stem bases.

The reserve carbohydrates in the storage organ are used to start growth in the spring and after each cutting or grazing. They also are used to develop heat and cold resistance, to support life during periods of dormancy, to promote flower and seed formation, and for other processes that go on within the plant. Starch is the primary storage polysaccharide accumulated in the legumes and grasses of tropical and subtropical origin. Grasses of temperate origin accumulate fructans in their vegetative tissues. Species in both the legume and grass families accumulate starch in their seeds. Plants go through periods when carbohydrates are used and when they are stored in a cyclic pattern, between early growth and maturity.

Terminology

Apical meristem: a region of rapid cell division (meristem) at the top of a shoot or tip of a root that eventually becomes shoots, leaves, and buds.

Axil: upper angle between a petiole of a leaf and the stem from which it grows.

Axillary bud: bud formed in axil of a leaf.

Crown: area at base of stem with tightly packed nodes and internodes that function to generate vegetative growth.

Crown bud: bud formed on the crown.

Cuticle: a protecting layer covering the epidermis of leaves, young shoots, and other aerial plant organs consisting of lipid and hydrocarbon polymers impregnated with wax.

Epidermis: a single layer of cells that covers the leaves, flowers, roots, and stems of plants forming a boundary between the plant and the external environment. The epidermis of above ground plant parts has cutin in the cell walls and cells are covered with a cuticle.

Internode: length of stem between nodes.

Intercalary meristem: meristematic tissue derived from the apical meristem that becomes separated from the apex in the course of development of the plant by regions of more or less mature tissues.

Leaf: plant organ borne by the stem, responsible for photosynthesis and gas exchange; comprised of the blade and petiole.

Meristem: area of actively dividing cells; capable of differentiating into specialized tissue.

Mesophyll: the leaf tissue, located in between the layers of epidermis, that carries on photosynthesis, consisting of the palisade layer and the spongy parenchyma.

Node: solidified place on the stem that bears a leaf.

Pericarp: wall of the ovary (pod) that encloses seeds.

Petiole: the stalk that attaches the leaf to the stem.

Phloem: the vascular tissue that conducts sugars and other metabolic products downward from the leaves.

Plasmodesmata: microscopic channels through plant cell walls that enable transport and communication between cells.

Regrowth: vegetative bud and shoot elongation, either after shoot is cut or after shoot has attained sufficient maturity.

Rhizome: horizontal underground stem.

Solute potential (osmotic potential): a pressure that needs to be applied to a solution to prevent the inward flow of water across a semipermeable membrane.

Stem: aerial portion of plant with nodes and internodes.

Stoma (also **stomate**, plural **stomata**): an opening, mostly on the undersurface of plant leaves, used for gas exchange.

Transpiration: the loss of water by evaporation in terrestrial plants, especially through the stomata; accompanied by a corresponding uptake from the roots.

Xylem: the vascular tissue that conducts water and dissolved nutrients upward from the root and adds strength to the stem.

REVIEW QUESTIONS

1. Name two polysaccharides in plants and indicate whether they are structural or nonstructural.
2. Name an indigestible polymer that binds other cell wall components.
3. Define "plasmodesmata" and explain how they benefit a plant.
4. Explain two similarities and two differences between xylem and phloem cells.
5. Describe two places where buds are found in legume forage plants such as alfalfa. Explain their relationship and the role of auxin in it.
6. Explain the role of the cuticle, mesophyll cells, and stomata in leaves.
7. How does the concentration of carbohydrates change within a plant through a sunny day and the following night?
8. Explain two differences between C3 and C4 plants.
9. Name three mineral nutrients required by plants and briefly describe their role in plant metabolism.

REFERENCE

McSteen P. Hormonal regulation of branching in grasses. Plant Physiol. 2009;149(1):46−55. https://www.ncbi.nlm.nih.gov/pmc/articles/PMC2613715/.

Chapter 2

Identification of Temperate Pasture Grasses and Legumes

Michael D. Casler[1] and Daniel J. Undersander[2]

[1]USDA-ARS, U.S. Dairy Forage Research Center, Madison, WI, United States; [2]College of Agriculture and Life Sciences, University of Wisconsin, Madison, WI, United States

INTRODUCTION

Pasture plant identification is critical to maintaining healthy and high-quality pastures that provide healthy and nutritious forage for horses. Excessive stress applied to pastures, such as heat, drought, and overgrazing, can lead to changes in species composition. As perennials die out, annuals can invade pastures, reducing quality and resiliency of the pasture. Changes can also occur among perennials in pastures that may contain several species. This can be very critical if the changes in composition are in favor of low-quality plants that may contain antiherbivory compounds that may be toxic to horses. This can be especially serious, because it is common that these toxins are often linked to improved persistence of some pasture plants, especially under stress.

Owners should be aware of the fundamental characteristics of the most basic pasture plants, which plants are the most or least desirable, and how to identify them in a pasture setting. Most grasses and legumes are easy to identify in a hay production system or when they are allowed to go to seed ripening and the floral structures are present. The trick and challenge comes in trying to identify grasses and legumes in pastures that are heavily or frequently grazed, where there are no floral structures. This chapter is focused on identification of perennial grasses and legumes in temperate pastures. For identification of warm-season forages common to the United States, refer to *Southern Forages, Fifth Edition* (Ball et al., 2015).

GRASSES

Grass classification and identification is officially based on floral reproductive structures. These are either panicles, such as that found on oats, or spikes, such as can be seen on wheat. These are easily identified by comparing to photos available on the internet for just about any grass species.

These structures are almost universally absent in horse pastures, which tend to be heavily and/or frequently grazed. Floral structures originate within each stem, from the floral primordium, a small cluster of cells that contains all the information required to produce pollen, eggs, and seed in response to the correct day length and temperature stimuli. In heavily or frequently grazed pastures, livestock can be encouraged to consume enough forage that they will consume these stems before the floral primordia complete their transformation into floral reproductive structures. In these situations, we must rely on vegetative plant parts—leaves, stem bases, crowns, and roots—to identify pasture plants. Because these vegetative plant parts were never intended to be the main classification traits used to classify or identify individual species, our job can often be extremely difficult. To the uninitiated, differences between pasture plants for these characteristics are often subtle and indistinct.

Seedlings

When grasses are very young, usually at the two-leaf to four-leaf stage, it is fairly easy to determine if they are a perennial or annual. Most pastures are planted to perennials, but annuals tend to invade new seedings during establishment and older pastures that have reduced stands. Use the "pull test" to determine whether the grass seedling is a perennial or an annual. An annual grass will pull easily and will have many short roots. A perennial grass will be harder to pull and will have at least one long root that will likely break off when you pull the seedling.

Sod Versus Bunch Grasses

Usually, the first step to identify pasture grasses in the vegetative stage is to determine if they are bunch grasses or sod formers. Sod formers will have rhizomes, which are underground stems that grow horizontally, eventually emerging above the soil to form a new tiller or stem. Rhizomes can be as long as 4–5 feet (1.5 m), in the case of quackgrass, or less than 0.5 inches (1 cm) for grasses such as tall fescue. Sod-forming grasses form dense mats or irregular clumps. Bunch grasses are nearly always found in neat, round, and symmetrical clumps.

Glossary

The next steps involve knowing more about the structure and anatomy of a grass plant, so the glossary and Figs. 2.1 and 2.2 are very important.

- **Auricle:** a short extension of the leaf blade that wraps partway around the stem
- **Awn:** a stiff, hair-like extension on some glumes
- **Collar:** a light-colored band of tissue opposite the ligule, on the outer side of the grass leaf
- **Culm:** the jointed stem of grasses
- **Glume:** a tiny leaf-like structure enveloping the seed
- **Internode:** the area of the stem between the nodes
- **Leaf blade:** the flat, expanded portion of the grass leaf
- **Ligule:** a membrane or series of hairs on the inner side of the grass leaf where the blade joins the sheath
- **Node:** the place on the stem where a leaf attaches
- **Rachilla:** the point of attachment of a seed to the seed head or to another seed
- **Sheath:** the part of the leaf that wraps around the main stem

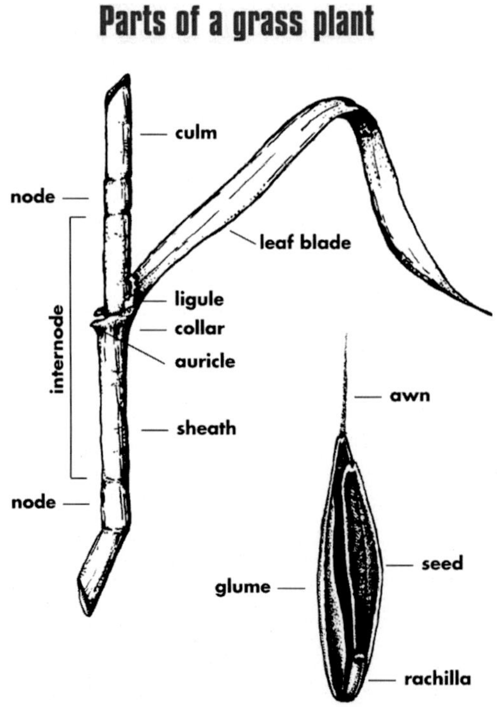

Parts of a grass plant

FIGURE 2.1 Structure and anatomy of grasses: stem and leaf on the left, seed on the right.

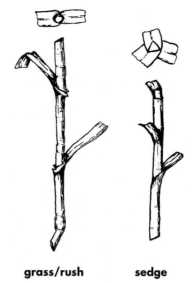

grass/rush   **sedge**

FIGURE 2.2   Grasses and rushes (left) have round stems, while sedges (right) have triangular stems.

Is It a Grass?

Grasses can sometimes be confused with sedges and rushes. To distinguish them, look for the following characteristics:

- Sedges have triangular stems that are filled with pith. The nodes are inconspicuous, and leaves grow from the stem in three directions when viewed from the top (Fig. 2.2).
- Rushes have round or flat stems. Stems are commonly leafy only at the base. Leaves grow from two directions when viewed from the top.
- Grasses have round or flat stems. Stems are leafy along the entire length. Leaves grow from two directions when viewed from the top (Fig. 2.2).

The Grass Whorl

Grass tillers or shoots emerge from the soil as a whorl. Corn (maize) is a grass, so just picture a young corn plant that is knee or waist high, only on a much smaller scale. Leaves are wrapped around each other in a sequential pattern, with the outside leaves emerging first, and the inside leaves emerging later to be formed higher up on the stem. For perennial pasture grasses, leaves are either folded or rolled. This trait can be observed by cutting a cross-section of the whorl and looking at the cross-section with a magnifying glass or scope. The other way to determine this is to feel the stem: folded leaves tend to have flattened stems and rolled leaves tend to have round stems (Fig. 2.3).

Ryegrasses

The most common ryegrasses are perennial (Fig. 2.4A) and annual, which is a variant of Italian ryegrass (Fig. 2.4B). Annual ryegrasses are not true annuals, but they were developed in the southern United States by breeding varieties that would establish quickly in autumn, produce high yields of quality forage during winter, and die out in summer when the temperatures become high. A few varieties of annual ryegrass can survive hot summer climates and cold winter climates, but all are very short lived. Annual ryegrass is distinguished from perennial ryegrass by having wider leaves, a sparser growth habit that is more stemmy and less leafy, and with short awns on the seeds.

Perennial ryegrass is a bunch grass with extremely rapid establishment. This grass has limited value in North America due to relatively low cold tolerance and winter hardiness. Perennial ryegrass survives only for one to two winters in very extreme climates with cold winters, especially if there is relatively low snow cover to act as insulation.

Ryegrass produces tillers from crown buds at the base of the plant. Carbohydrate storage is in stem bases. It establishes rapidly and yields well under cool, wet conditions. Ryegrass has low drought and heat tolerance. Perennial ryegrass grows less over summer than annual ryegrass. Ryegrass establishes rapidly, providing quick ground cover and a ready supply of forage. Establish by sod seeding, conventional tillage, interseeding, or frost seeding.

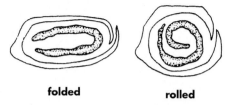

folded   **rolled**

FIGURE 2.3 Cross-section of grass stems, showing how unemerged leaves are either rolled or folded before they emerge.

FIGURE 2.4 Infloresences or seed heads of (A) perennial ryegrass and (B) Italian ryegrass.

Ryegrass may be grazed closely early in the season. A rest period should follow this first grazing to allow tiller development and carbohydrate production. Graze again when plants are 8 inches tall. Leave a stubble height of 3–4 inches. Ryegrass produces high-quality forage in cool, wet weather. Annual and perennial ryegrasses have relatively shallow root systems, so hot, dry weather will reduce yields. Perennial ryegrass should be included in most pasture seeding mixtures at a low seeding rare. Do not rely on this short-lived species as the only grass in a pasture mix.

Crown rust can severely defoliate plants, reducing forage yield, quality, and persistence. Plant resistant varieties whenever possible. Also, be sure to obtain seed that is certified to be endophyte-free. This will eliminate animal health problems associated with toxins produced by a fungus that often in association with ryegrass plants. Annual (Italian) ryegrass and perennial are available in two forms: diploid and tetraploid. Diploid varieties are more densely tillering. Tetraploid varieties have greater resistance to crown rust, giving them summer productivity and quality.

When purchasing annual ryegrass, select late maturing varieties to reduce head formation and maintain high quality. For perennial ryegrass, use forage type varieties rather than turf varieties, which are extremely low growing and low yielding.

Ryegrasses have an endophytic fungus that lives inside the plant, producing alkaloids that can be highly toxic to horses. Varieties should be either certified endophyte-free or contain a "friendly" endophyte that has been tested and certified to produce no toxic alkaloids (Table 2.1).

Tall and Meadow Fescue

Tall fescue is a bunch-type grass that spreads from short rhizomes. Tall fescue has poor palatability and should not be mixed with other pasture species. It is a good choice in areas where animal traffic is high. Tall fescue is also widely used for ditch embankments and grass waterways.

TABLE 2.1 Characteristics of Perennial and Annual (Italian) Ryegrasses

Seed and Seedling Characteristics	Adult Vegetative Characteristics
Seed about 6–8 mm; rachilla attached	Bunch-type growth habit
Leaves are slightly folded in shoot	Leaves strongly folded in shoot
Narrow leaf blade; underside glossy	Smooth sheath; sides overlap at top
Rapid emergence	Membranous ligule
	Long auricles
	Spiked seed head

Tall fescue also contains an endophytic fungus that produces toxic alkaloids. Varieties should be either certified endophyte-free or contain a "friendly" endophyte that has been tested and certified to produce no toxic alkaloids. Meadow fescue also contains an endophyte, but it does not produce alkaloids that are toxic to livestock.

Carbohydrate storage in tall fescue takes place in stem bases and short rhizomes. Tall and meadow fescues are very tolerant of drought and flooding. They are also tolerant of low fertility conditions, although they respond well to optimum soil fertility levels. Both fescues are more shade tolerant than other cool-season grasses.

Tall and meadow fescues are as easily established as orchardgrass, timothy, and perennial ryegrass. They may be interseeded, established by conventional methods, or no-till seeded in a killed sod.

Tall fescue can be grazed early in the spring, but avoid grazing once stem elongation begins. Plants may be grazed or cut for hay after growth is 10 inches tall and plant carbohydrate reserves have been replenished. Leave at least 4 inches of stubble to protect stem bases where carbohydrates are stored. Tall fescue will continue to grow more through the summer than most cool-season grasses (Fig. 2.5; Table 2.2).

FIGURE 2.5 (A) Tall fescue seed heads and (B) auricles wrapping around the stem at the leaf base.

Orchardgrass

Orchardgrass is a bunch-type grass that produces an open sod (Fig. 2.6A and B). Orchardgrass is best adapted to a wide range of soils with good moisture where management is intense and grazing/haying will be frequent.

TABLE 2.2 Characteristics of Tall and Meadow Fescues

Seed and Seedling Characteristics	Adult Vegetative Characteristics
Seed about 6–8 mm; club-shaped rachilla	Bunch-type growth habit
Shoot is rolled in the whorl	Prominent leaf veins; sharp edges
Leaf base appears early	Lower surface of leaf blade is glossy
Leaves rolled in sheath; sides overlap at top	Short ligule
	Blunt auricles with few hairs
	Seed head is a panicle (like oats)

FIGURE 2.6 Orchardgrass (A) seed heads and (B) tall membranous ligule.

Orchardgrass does not produce rhizomes. Instead, it forms bunches by profuse tiller formation at the base of the plant. Carbohydrate storage for regrowth is mainly in the stem bases. Orchardgrass has only fair drought and heat tolerance, winterhardiness, and persistence. Stands tend to thin, leaving large clumps of orchardgrass plants scattered throughout the paddock.

Orchardgrass is easier to establish than most of the other cool-season grasses. It is a good choice for interseeding into existing pasture with a no-till drill or by frost seeding, as well as for seeding into a killed sod or a conventionally tilled seedbed.

Orchardgrass is one of the earliest maturing grasses. Unlike smooth bromegrass and timothy, the main stem has less influence on tillering in orchardgrass. Tiller formation begins early and continues throughout the season. Regrowth following grazing comes from the production and elongation of new leaves and the elongation of cut leaves on the stubble. As tiller formation and regrowth is rapid, orchardgrass should be grazed frequently to maintain adequate quality. Grazing timing is important as orchardgrass matures more rapidly than other species and forage quality drops quickly. Following seedhead development or removal, the subsequent forage is nearly all leaves. Orchardgrass is a very aggressive species and is not compatible with low-growing legumes. Frequent grazing will help avoid loss of other desirable species. Even though orchardgrass has rapid regrowth following grazing, it is still important to allow a rest period to reestablish carbohydrate levels. For high yields and good quality, allow orchardgrass to regrow to 10 inches before grazing. Despite its vigor, orchardgrass is susceptible to close grazing; leave a stubble height of 3–4 inches (Table 2.3).

TABLE 2.3 Characteristics of Orchardgrass

Seed and Seedling Characteristics	Adult Vegetative Characteristics
Seed about 6–8 mm; glumes attached	Bunch-type growth habit
Seed is often curved	Leaves strongly folded
Leaves strongly folded in whorl	Sheath strongly flattened; sides overlap
Broad, V-shaped leaf	Stem is prominently flattened
Bluish-green color	Long ligule with cuts or splits; no auricles
	Panicle seed head; seeds strongly clumped

Timothy

Timothy (*Phleum pratense* L.) is a bunch-type grass that produces an open sod. It is best adapted to cool, moist soils. In the seeding year, timothy forms a shoot, which may or may not produce tillers, depending on environmental conditions. In spring of the second year, internodes elongate, and the seedhead forms. During this time, lower nodes form an enlarged food storage organ called a corm (Fig. 2.7). Corms supply energy for subsequent tiller formation. As secondary shoots develop, the corms on the primary shoot deteriorate, and a secondary corm is formed. Very few corms overwinter, and new spring growth develops from buds at the base of the plant. Unlike other cool-season grasses, timothy produces flowers and seedheads throughout the summer. Timothy has excellent winterhardiness but poor drought and heat tolerance and is not persistent under grazing.

Timothy seedlings are more vigorous than smooth bromegrass seedlings but less vigorous than most forage grasses. Timothy can be interseeded, sod seeded, or seeded using conventional methods.

Do not graze timothy during stem elongation. Grazing during this period, when food reserves are low, will slow regrowth and accelerate stand loss. It is more sensitive than most other species to grazing while stems are elongating. Wait until plants are 10 inches tall or until new basal tillers are visible before grazing. Leave at least 4 inches of stubble to keep from removing young tillers and developing corms. Timothy pastures can be stemmy due to constant seedhead production. This reduces palatability to animals, so graze routinely to maintain quality. Timothy grows best under cool, wet conditions. Stand production and persistence will decline severely under heat or drought stress. It is shallow rooted and not a good choice for sandy soils. Timothy responds to nitrogen fertilization, but performs better than most cool-season grasses with low nitrogen (Table 2.4).

FIGURE 2.7 Timothy (A) seed heads, (B) tall membranous ligule, and (C) corms. The corms are swollen stem bases that are a distinct and unique trait of timothy.

TABLE 2.4 Characteristics of Timothy

Seed and Seedling Characteristics	Adult Vegetative Characteristics
Seeds less than 3 mm; nearly round	Bunch-type grass
Rounded shoot; leaves rolled	Crown has corms, enlarged stem base
Oldest leaf twisted viewed from above	Flat leaf blade, rolled within whorl
	Smooth sheath; sides overlap at top
	White ligule; very tall
	Auricles absent or very tiny
	Seed heads are dense cylinders, very long and narrow
	Seed heads often present on regrowth

Kentucky Bluegrass

Kentucky bluegrass *(Poa pratensis* L.) is a sod-forming grass. It is widely grown as a pasture in many cool-season regions of North America. Kentucky bluegrass is well adapted to less-managed pastures with low animal stocking rates (Fig. 2.8) (Table 2.5).

New shoots of Kentucky bluegrass develop from rhizomes or from axillary buds in old shoots. Most of these shoots develop during short days in early spring or autumn. Rhizomes develop from buds on aboveground shoots. Carbohydrate storage is in roots in early spring and in roots and rhizomes later in the year. Bluegrass is winter hardy and persistent but has only fair drought and heat tolerance.

Kentucky bluegrass is slower to establish than orchardgrass, ryegrass, tall fescue, and timothy. Seed using conventional or no-till into a killed sod. It is not a good candidate for frost seeding or interseeding.

Kentucky bluegrass is the lowest yielding of the cool-season grasses commonly used for pasture. Productivity is greatest during spring and fall. Plants become dormant during the hot, dry months of summer. Kentucky bluegrass may be grazed to 1–2 inches. It requires relatively long rest periods to replenish carbohydrate reserves. Overgrazing will reduce forage yield. Grazing may be extended somewhat by including a legume. White clover is a good choice as both species are tolerant of close grazing. Nonetheless, a shortage of pasture will result if Kentucky bluegrass is the only forage source.

FIGURE 2.8 Kentucky bluegrass (A) seed heads, (B) tall membranous ligule, and (C) distinct boat-shaped leaf tip.

TABLE 2.5 Characteristics of Kentucky Bluegrass

Seed and Seedling Characteristics	Adult Vegetative Characteristics
Seeds less than 4 mm long	Dense sod; slender rhizomes
Shoot rolled, but slightly flattened	Narrow leaf blade
Narrow leaf blade	Leaf blade tip shaped like boat prow
	Oval-shaped leaf sheath; sides overlap
	Membranous ligule; smooth margin
	Auricles absent
	Relatively short plant height
	Panicle-type seed head; slender

Reed Canarygrass

Reed canarygrass is a sod-forming grass. It is the highest yielding cool-season grass when fertilized and an excellent choice in wet areas where it is difficult to grow other species. Reed canarygrass can be used to provide grazing during the "summer slump" of some other forage grasses.

Reed canarygrass reproduces from short, thick rhizomes. Aboveground shoots develop in early spring and late fall. Shoots that develop in spring only live for that year, while those that develop in fall overwinter and survive through the following year (Fig. 2.9) (Table 2.6). Carbohydrate storage occurs in rhizomes. The seedhead develops in spring and matures in July. New rhizomes form from buds on old rhizomes during the summer. Reed canarygrass has excellent winterhardiness and persistence. It is tolerant of wet soils but also does well on droughty soils due to a deep root system.

Reed canarygrass is more difficult to establish than other cool-season grasses, particularly by interseeding or frost seeding. Seed using conventional tillage or no-till into a killed sod. Seeding in late summer when there is reduced weed competition is often more successful than spring seedings.

Once established, reed canarygrass is a very aggressive species. Like orchardgrass, it forms tillers throughout the growing season. Reed canarygrass must be well managed to avoid overgrowth and subsequent low quality. An early grazing, before tillers form, will not harm plants. Following this period, wait until plants are 14−16 inches tall before

FIGURE 2.9 Reed canarygrass (A) seed heads and (B) membranous ligule with wide and flat leaf blade.

TABLE 2.6 Characteristics of Reed Canarygrass

Seed and Seedling Characteristics	Adult Vegetative Characteristics
Seeds less about 4 mm long	Dense sod; robust rhizomes
Seeds are grey-brown; shiny and slick	Leaves rolled in whorl; flat and wide
Rounded shoot	Sheath sides overlap near top
Slow to germinate and emerge	Prominent ligule
	Auricles absent
	Seed head is green and purple; turning tan
	Panicle-type head compacted into a cylinder after flowering

grazing again. Unlike other grasses, canarygrass will provide good quality forage up to 24 inches in height. Reed canarygrass may be established with a legume. However, the legume may disappear from the stand as the reed canarygrass develops a thick sod.

Reed canarygrass produces toxic alkaloids. Use only varieties that are certified as "low-alkaloid," "tryptamine-free," and "best-carboline-free." Four of these varieties are Palaton, Rival, Venture, and Bellevue.

Smooth Bromegrass

Smooth bromegrass is a high-yielding grass but requires longer recovery periods than other grasses. It is best adapted to well-drained soils and is an excellent choice for drought-prone areas.

Smooth bromegrass spreads by short rhizomes to form a dense sod. The plant stores most of the food needed for regrowth and overwintering in the rhizomes (Fig. 2.10) (Table 2.7). Smooth bromegrass is winter hardy, drought and heat tolerant, and is quite persistent.

Bromegrass has low seedling vigor and is more difficult to introduce into pastures by frost seeding or interseeding than orchardgrass, timothy, or ryegrass. Successful stands may be established by no-till seeding into killed sods or through conventional tillage methods.

Smooth bromegrass may be grazed before stems elongate, when plants are less than 6–8 inches tall. It is more sensitive than most other species to grazing while stems are elongating. For long-lasting stands and high-quality forage, wait until plants are at least 10 inches tall or until new basal tillers are visible before grazing. Graze no closer than 4 inches to avoid removing new shoots from the base of the plant. Smooch bromegrass requires long recovery periods, especially during the summer slump. Smooth bromegrass is most productive in spring. Subsequent production may be low, especially if nitrogen is limiting. Smooch bromegrass is very responsive to nitrogen; consequently, mid- and/or late summer applications will increase productivity. It can become sod-bound in pure stands if not well fertilized.

Quackgrass

Quackgrass is a wild-type sod-forming grass that has not been bred, but it is still prevalent throughout the cool-season areas of North America. It is present in many pastures and can help to provide excellent ground cover and grazing fodder for livestock. Management and growth characteristics are similar to smooth bromegrass. It spreads by long rhizomes that can be up to 3 m (10 feet) long. The tips of the rhizomes are so sharp that they can pierce and grow right through a tap root of alfalfa or dandelion (Fig. 2.11) (Table 2.8).

FIGURE 2.10 Smooth bromegrass (A) infloresence is a panicle rather than a spike, (B) leaf sheath is fused, (C) distinct "M" constriction on leaves, (D) rhizomes projecting horizontally, then upward.

TABLE 2.7 Characteristics of Smooth Bromegrass

Seed and Seedling Characteristics	Adult Vegetative Characteristics
Seeds brown, about 10–12 mm long	Dense sod
Tall and slender shoot	Numerous but short rhizomes
Slow to germinate and emerge	"M" constriction near middle of leaf blade
	Leaves rolled in whorl
	Sides of leaf sheath are fused
	Small ligule; ragged hairs
	Auricles absent
	Panicle-type seed head; long branches; often drooping

FIGURE 2.11 (A) Quackgrass "W" shaped leaf constriction near the leaf tip and (B) long clasping auricles.

TABLE 2.8 Characteristics of Quackgrass

Seed and Seedling Characteristics	Adult Vegetative Characteristics
Seeds about 10–12 mm long	Sod forming
Shoot often reddish at base	Long and numerous rhizomes; slender and white
	Flat leaf blade; constriction near leaf tip
	Sheath overlaps near top
	Short and membranous ligule
	Unique clasping auricles; very long
	Spike seed head; seeds in clusters of 4–6

Legumes

Legumes are an important component of pastures. They increase yield and quality of grass pastures and provide nitrogen to grasses through fixation of atmospheric nitrogen into organic compounds. A legume is defined as a plant with seeds in a pod that splits into two distinct halves. Some common examples are peas, beans, and peanuts. We rarely see seedpods on the plants, as they are usually harvested well before pods form. Many legumes have compound leaves (more than one leaflet per leaf). Many nonlegumes also possess this trait (Fig. 2.12).

The best time to identify seedlings is in the three- to four-leaf stage when vegetative characteristics are usually easily seen (Table 2.9). Forage legumes exhibit significant morphologic variation among populations of the same species and often grow in mixed stands of several species. Thus, it is best to examine several plant characteristics when identifying legumes.

Parts of a legume plant

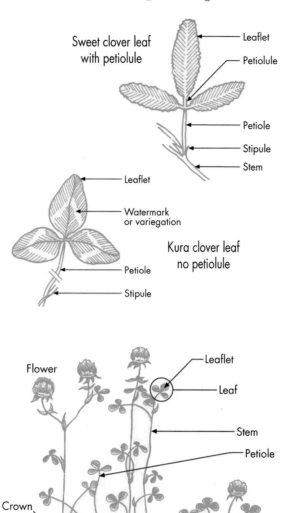

FIGURE 2.12 Structure and anatomy of legume plants showing the structure of a leaf with a petiolule (top), the structure of a leaf without a petiolule (middle), and the structure of a whole plant above and below the soil surface.

TABLE 2.9 Characteristics of Alfalfa Seeds and Seedlings

Seed Trait	Seedling Trait
Seed size	2.3–2.6 mm
Seed shape/color	Kidney shaped with small beak
Leaflets/leaf	3 (some varieties have 5–7)
Leaf margin	Serrations on upper third
Petiolule	Present
Pubescence	None
Variegation	None
Flower color	Purple, few yellow or cream
Flower type	Raceme
Growth habit	Erect

Glossary

- **Head:** a dense inflorescence of flowers without stems
- **Inflorescence:** the arrangement of flowers on the floral axis
- **Internode:** area of the stem between the nodes
- **Node:** the point on the stem where leaves are attached
- **Petiole:** the stalk of a leaf that connects the leaf to the stem
- **Petiolule:** extension of the petiole into the leaflets of a compound leaf
- **Pubescence:** small hairs on the surface of leaves and stems
- **Raceme:** an inflorescence in which flowers are mounted on short stems along a central axis
- **Rhizome:** an underground stem that is capable of producing new plants at the nodes
- **Stipule:** small, pointed, leaf-like structures at the base of the petiole
- **Stolon:** a prostrate aboveground stem that is capable of producing new plants at nodes
- **Tendril:** a slender modified leaflet used for support
- **Umbel:** an inflorescence in which flowers are mounted on short stems all arising from a common point
- **Variegation:** a pattern of lighter colored tissue on a leaf

Alfalfa

Alfalfa (*Medicago sativa* L.) is a widely grown forage legume. It is a productive, high-quality, long-lasting legume adapted to both hay and grazing. It is sensitive to low pH and soil drainage, and its use as a pasture legume is somewhat limited.

Growth Habit

Alfalfa is an erect growing plant with main stems arising from a large crown. Branches also occur from axillary buds on the stems (Fig. 2.13). Alfalfa has a strong, deep taproot that makes it well adapted to sandy soils. It has good winterhardiness, although it may winterkill in cold, open winters.

Alfalfa is easily established in conventional seedbeds or killed sods. It is not a good choice for frost seeding or interseeding into existing pastures. It does not reseed itself due to autotoxicity caused by older, established plants that produce chemicals that inhibit the growth of seedlings (Table 2.9).

Alfalfa does well in a cut forage system when three cuttings are made with 28- to 35-day intervals and root carbohydrate levels are allowed to replenish. Harvesting typically takes place more frequently in grazing systems, and root storage may not be adequate for regrowth after grazing. As a result, alfalfa should not be grazed closely, so some leaf area remains. Plants should be allowed to regrow to at least 10 inches tall between grazings. Quality is best when grazed prior to the flowering stage. This effect is most pronounced in spring. Quality drops less rapidly in late summer and fall. Allow a 6-week rest period prior to October 15 to reduce winter injury.

FIGURE 2.13   (A) Alfalfa plants with flowers, (B) an alfalfa seedling with three leaflets, no chevrons and serrations along the distal third of the leaflet margins and (C) a seedling with stem, petioles and leaflets.

Three cuttings of alfalfa prior to September 1 provide the greatest yield and quality of alfalfa. First cutting should be early and based on quality. Second cutting should be at bud stage and the third at 10%−25% bloom. This later cutting allows rebuilding of root reserves and better overwintering.

There are many alfalfa varieties available. Generally, good hay types will also be good grazing types. Consider yield, disease resistance, winterhardiness, and quality.

Alsike Clover

Alsike clover (*Trifolium hybridum* L.) is an annual or biennial clover best adapted to moist or poorly drained soils. It prefers cooler temperatures than red clover and so is well adapted to low-lying areas. It is not tolerant of drought or high temperatures. It will tolerate a soil pH as low as 5.0.

Alsike clover has erect stems like red clover, but as they are more fine, it is subject to lodging (Fig. 2.14). Flowers are borne along the entire length of the stem rather than at the tip as is red clover. Because the plant resembles red clover but the flower resembles white clover, alsike clover was once thought to be a hybrid between red and white clover. This has since been shown to not be the case (Table 2.10).

Alsike clover establishes best in a tilled seedbed or tilled sod. It is not a good candidate for frost or interseeding. Alsike clover forage is often of higher quality than alfalfa or red clover. It should be grazed at full bloom. There are no US varieties of alsike clover available. Most seed is common. Alsike clover is toxic to horses and should be eliminated from horse pastures.

Birdsfoot Trefoil

Birdsfoot trefoil (*Lotus corniculatus* L.) is a common pasture legume. It is well adapted to poorly drained, acidic soils. It is shallow rooted and therefore performs poorly on sandy soils. It is the only forage legume used in eastern North America not known to cause bloat. It can reseed itself, resulting in long-lived stands.

Three types of birdsfoot trefoil are available: prostrate, erect, and semierect. Prostrate types are low growing, more winter hardy, have less vigorous seedlings and recover more slowly than the more erect types. The prostrate types are more adapted to grazing. The erect types are best for haying. Semierect types are dual purpose and are the most commonly

FIGURE 2.14 Alsike clover (A) oval leaflets showing no serrations or chevrons, (B) infloresence dark pink at base, fading to nearly white at the top.

TABLE 2.10 Characteristics of Alsike Clover Seeds and Seedlings

Seed Trait	Seedling Trait
Seed size	1.0–1.3 mm
Seed shape/color	Mitten shaped with shallow notch dark color
Leaflets/leaf	3
Leaf margin	Smooth
Petiolule	Absent
Pubescence	None
Variegation	None
Flower color	Pinkish white
Flower type	Head
Other	Leaflets shiny underneath
Growth habit	Erect

grown. Birdsfoot trefoil has very fine stems and is prone to lodging. Many branches develop from the lodged stems and from the lower part of erect stems (Fig. 2.15).

Seedling vigor of birdsfoot trefoil is lower than other common legumes. It is readily established in tilled seedbeds or killed sods but is not a good candidate for frost seeding or interseeding in established pastures. Birdsfoot trefoil often has high levels of hard seed, so seed sown using these methods may germinate some time after planting (Table 2.11).

Birdsfoot trefoil utilizes root reserves to fuel early spring growth. Unlike most legumes, trefoil does not replenish root reserves again until late summer. Regrowth between grazings must come from photosynthesis. It is critical therefore to not

FIGURE 2.15 Birdsfoot trefoil (A) leaf with five apparent leaflets, (B) a seedling of an erect type, (C) bright yellow inflorescence and (D) seed pods.

TABLE 2.11 Characteristics of Birdsfoot Trefoil Seeds and Seedlings

Seed Trait	Seedling Trait
Seed size	1.2–1.7 mm
Seed shape/color	Round, brown
Leaflets/leaf	5
Leaf margin	Smooth
Petiolule	Absent
Pubescence	None
Variegation	None
Flower color	Yellow
Flower type	Umbel
Growth habit	Prostrate, semierect, or erect depending on variety

graze trefoil too closely, or regrowth will be slowed, and plant health will decline. A stubble height of at least 4 inches is recommended. Prostrate varieties of trefoil are preferred for grazing, as more leaf area remains following grazing. Trefoil will reseed itself, so allowing the plants to flower and set seed occasionally will help maintain long-lived stands. Trefoil tends to be of higher quality than alfalfa at similar growth stages, so it can be grazed into the flower stage and still provide good forage quality.

Trefoil is best adapted to grazing but can be used as a hay crop on poorly drained, acidic soils. Harvest trefoil for hay in the early flower stage. Leave a stubble height of 2–4 inches to allow regrowth. Trefoil may be difficult to harvest as a hay crop due to fine leaves that tend to shatter easily. Time of baling is more critical than for other hay crops.

Erect types of trefoil include Maitland and Viking. Prostrate types include AU-Dewey, Dawn, and Empire. Semierect types are Mackinaw and Norcen. While all are suited to grazing, prostrate and semierect types are best.

Crown Vetch

Crown vetch (*Coronilla varia* L.) is used primarily for land reclamation and roadside stabilization. It has not been used widely as a forage crop in the United States. It is tolerant of low soil pH and, once established, can tolerate prolonged dry periods.

Crown vetch has a prostrate growth habit. Much of the aboveground growth consists of long, pinnately compound leaves (Fig. 2.16). It spreads by way of a creeping rootstock. Crown vetch is slow to establish due to low seedling vigor. It is best planted with a faster growing grass species. Adequate stands of crown vetch may not be realized until the second year.

Crown vetch is a poor choice for forage because it contains compounds that reduce palatability and may be toxic to horses. Crown vetch regrows from axillary buds after haying or grazing, so a sufficient stubble (6 inches) should be left to ensure adequate stem area (Table 2.12). Grazing should take place 10–14 days after the first flowers appear, then grazed at 10–14 inches after that. Because of its prostrate growth habit and succulent nature, crown vetch is not a good choice for harvesting as hay. If necessary, harvest for hay 10–14 days after flowering begins. The most commonly used varieties of crown vetch are Chemung, Emrald and Penngift.

Hairy Vetch

Hairy vetch (*Vicia villosa Roth* L.) is an annual or winter annual legume. It is most often grown for soil improvement, roadside, or bank stabilization. It can also be grown as a pasture or hay crop but it can be toxic to horses, especially on high selenium soil. It grows well on a broad range of soils but is best adapted to sands or loamy sands.

Aboveground growth consists of viny, branched stems up to 6 feet long. Leaves are pinnately compound with a tendril at the tip (Fig. 2.17) (Table 2.13). Vetch is best seeded from July 25 to August 30. Winterhardiness may be a problem in open winters with no snow cover. When seeded the previous summer, hairy vetch can be grazed in May and June of the

FIGURE 2.16   Crown vetch (A) inflorescence, (B) leaf with fifteen leaflets, most in opposite pairs and (C) a seedling showing prostrate growth habit.

TABLE 2.12 Characteristics of Crown Vetch Seeds and Seedlings	
Seed Trait	**Seedling Trait**
Seed size	2.2−4.0 mm
Seed shape/color	Rod shaped
Leaflets/leaf	Many
Leaf margin	Smooth
Petiolule	Absent
Pubescence	None
Variegation	None
Flower color	Purple and white
Flower type	Umbel
Growth habit	Prostrate

following year. Following grazing, vetch may then be plowed down and a subsequent crop seeded for late summer pasture. Hairy vetch is typically seeded with a small grain companion crop when grown for hay. Winter rye is a common choice for this purpose. Hairy vetch should be harvested when the first pods are well developed. Earlier harvesting improves quality if rye is included in the mixture but reduces total yield. These combinations make for a tangled hay that is difficult to handle.

Madison is a cold-tolerant variety developed in Nebraska. Auburn, Oregon, and Lana are less cold tolerant and should only be grown in areas with mild winters.

FIGURE 2.17 Hairy vetch (A) Inflorescence, (B) leaf showing pairs of alternate leaflets, with a tendril at the tip and (C) a seedling.

TABLE 2.13 Characteristics of Hairy Vetch Seeds and Seedlings

Seed Trait	Seedling Trait
Seed size	2.7–4.9 mm
Seed shape/color	Large and round, dark color
Leaflets/leaf	Many
Leaf margin	Smooth
Petiolule	Absent
Pubescence	None
Variegation	None
Flower color	Purple
Flower type	Taceme
Other	Tendrils at end of leaf
Growth habit	Prostrate

Kura Clover

Kura clover (*Trifolium ambiguum* L.) is a rhizomatous clover well adapted to grazing. It is a high-yielding, persistent legume. Like most other clovers, it is well adapted to low pH soils. Kura clover is slow to establish, as most early growth is devoted to rhizome rather than top growth. Once established, however, kura clover will persist indefinitely.

Kura clover spreads by an extensive rhizome system (Fig. 2.18). Aboveground growth consists only of petioles and leaves arising from a crown. It tolerates poorly drained, acidic soils, but due to the rhizome system, it is also fairly drought tolerant. It is one of the most cold-tolerant legume species.

FIGURE 2.18 Kura clover (A) shows a stolon and some rhizomes from a vegetative plant, (B) shows the characteristic tapered leaflets bearing lighter green chevron-shaped markings, (C) is a closeup of an inflorescence with light pink color near its base, fading to white at the top and (D) is a seedling with leaflets that are more rounded than mature leaflets, and prominently elevated by petioles.

Kura clover has poor seedling vigor and is difficult to establish. Seedlings initially establish, then appear to stop growing, as most energy is devoted to belowground rhizome growth. This allows for weed encroachment during the seeding year. Kura clover should be established with some companion crop such as oats or other fast-growing forage species. Best results will be obtained with a clean, tilled seedbed. No-till establishment is possible if weeds are controlled. Interseeding into existing pastures will likely fail. Establishment of kura clover is enhanced by the use of the proper inoculum and by nitrogen applications of 50 lb/acre in the establishment year.

Due to underground rhizomes, kura clover is extremely tolerant of grazing. While it can be grazed in the seeding year, most forage production will take place in the second and subsequent years. As aboveground growth is all leaves, kura clover forage is of very high quality (Table 2.14). Kura is best grown with a grass species, as grazing pure stands can lead to bloat. Kura clover tolerates close grazing, but it also maintains high quality with less-intensive grazing systems, so graze the mixed stand according to the grazing needs of the grass species.

The succulence of the aboveground growth makes cutting and drying kura clover difficult. If grown for hay, it should be grown in mixed stands with forage grasses.

Three varieties of kura clover are currently available: Rhizo, Cossack, and Endura. Endura, the most recently released variety, has improved seedling vigor.

TABLE 2.14 Characteristics of Kura Clover Seeds and Seedlings

Seed Trait	Seedling Trait
Seed size	1.5–2.1 mm
Seed shape/color	Mitten shaped, brownish
Leaflets/leaf	3
Leaf margin	Smooth
Petiolule	Absent
Pubescence	None
Variegation	White "V" on leaf
Flower color	Whitish pink
Flower type	Head
Other	Rhizomes present
Growth habit	Erect

Red Clover

Red clover (*Trifolium pratense* L.) is widely used throughout eastern North America. It is adapted to a wide range of soil types and tolerates a pH as low as 5.5. It is a short-lived perennial that usually persists only 2 or 3 years due to susceptibility to a number of root diseases. Newer varieties may last longer than this.

Red clover has an erect growth habit similar to alfalfa, but the main stems originate lower on the plant. New shoots form from axillary buds at the crown. The crown of red clover is not as deep in the soil as alfalfa, making it more susceptible to winter injury. It has a shallow, highly branched root system, so it grows poorly on sandy soils without adequate rainfall (Fig. 2.19) (Table 2.15).

Red clover is one of the easiest of the clovers to establish. Successful stand establishment may be achieved by using a companion crop or by direct seeding with herbicides. Red clover is also an excellent candidate for frost seeding or interseeding for improvement of existing pastures.

Red clover provides high-quality forage throughout the grazing season. Red clover quality does not decline as rapidly as alfalfa. Ideally, red clover should be grazed between first flower and 20% bloom. However, in mixed pastures, grasses would be beyond the ideal grazing stage at this time. Graze red clover as close to flowering as the accompanying grass allows. If possible, allow the clover to flower once during the year. First cutting for hay should be made when the stand is

TABLE 2.15 Characteristics of Red Clover Seeds and Seedlings

Seed Trait	Seedling Trait
Seed size	1.5–2.1 mm
Seed shape/color	Mitten shaped, yellowish red
Leaflets/leaf	3
Leaf margin	Smooth
Petiolule	Absent
Pubescence	Present
Variegation	White "V" on leaf
Flower color	Reddish purple
Flower type	Head
Growth habit	Erect

FIGURE 2.19   Red clover (A) leaf with three tapered leaflets bearing light green chevron-markings and showing fine hairs on the petiole, (B) a darker-colored flower than on white or alsike clover and (C) a seedling.

between first flower and 20% bloom. One or two additional cuttings can be made at 5- to 7-week intervals. Allow 6 weeks for regrowth before a killing frost in fall. Red clover is difficult to dry for hay and should be used as haylage when possible.

Many good varieties of red clover are available. Look for varieties with resistance to anthracnose and powdery mildew. Avoid seed labeled as common or medium red clover.

White Clover

White clover (*Trifolium repens* L.) is common throughout humid regions of North America. It is found in lawns, athletic fields, and waste areas, as well as in pastures. It is very shallow rooted and has little drought tolerance. It is best adapted to areas with cool temperatures and adequate rainfall. It tolerates a soil pH as low as 5.5.

White clover is a short-lived perennial with prostrate growth habit. Leaves are borne on long petioles. There are no upright stems. Seedlings 6−8 weeks old begin forming stolons. These stolons spread and root at nodes along the stem (Fig. 2.20) (Table 2.16). After 1−2 years the original plant dies. White clover is less winter hardy than red clover, and

TABLE 2.16 Characteristics of White Clover Seeds and Seedlings	
Seed Trait	**Seedling Trait**
Seed size	0.9−1.2 mm
Seed shape/color	Heart shaped
Leaflets/leaf	3
Leaf margin	Smooth
Petiolule	Absent
Pubescence	None
Variegation	White "V" on leaf
Flower color	White
Flower type	Head
Other	Stolons present
Growth habit	Erect

FIGURE 2.20 While clover (A) leaf with three rounded leaflets, having white chevron markings and barely-visible serrations on the basal leaflet margins, (B) white flower heads, (C) a stolon held upside-down, revealing new petioles leading to leaves and (D) a seedling with several petioles.

plants may die without snow cover. It is a prolific seed producer, however, and while individual plants are not persistent, stands may persist almost indefinitely.

Pure stands of white clover are rare. It is included in many pasture mixes. It is very small seeded and should be seeded on the surface or no deeper than 1/4 to 1/2 inch (5−12 mm). Conventionally prepared seedbeds or killed sods are best for establishing white clover. It is not a good candidate for frost or interseeding.

White clover is well adapted to close grazing. Pastures should be grazed to about 2 inches and allowed to regrow to 8−10 inches. Close grazing allows light penetration to the low-growing clover and allows for better reseeding. Grazed forage of white clover is very high quality as animals graze mainly leaves, petioles, and flowers and very few stems. White clover should be seeded with grasses that tolerate close grazing such as perennial ryegrass or Kentucky bluegrass. Due its low, succulent growth and relatively low yields, white clover is rarely harvested as a hay or haylage crop.

Three types of white clover can be found: wild white clover, white Dutch clover, and ladino clover. Wild white clover and white Dutch clover are very prostrate, have small leaves, and are low yielding. They are widespread in lawns and continuously grazed pastures. Ladino white clover is larger and more productive. Recently, even larger, more robust white clover selections have been introduced from Holland and New Zealand.

REVIEW QUESTIONS

1. In pastures, flowering heads of grasses are almost never observed. What are the most important characteristics that can be used to identify grasses in the vegetative state?
2. In a new planting, how can you tell the difference between perennial grasses and annual grasses when the seedlings are in the one-leaf or two-leaf stage?
3. Of the eight grasses listed in this chapter, which are rhizomatous sod formers and which are bunch grasses?
4. What is the best stage to identify legume seedlings?
5. What is the difference between a stolon and rhizome?
6. Which of the legumes described are not perennials?

REFERENCE

Ball DM, Hoveland CS, Lacefield GD. Southern Forages. fifth ed. Norcross, Georgia: International Plant Nutrition Institute; 2015.

Chapter 3

Nutritional Value of Pasture Plants for Horses

Paul Sharpe

University of Guelph, Guelph, ON, Canada (retired)

INTRODUCTION

Plants have value for herbivorous animals because they supply chemical elements and compounds that are required for life, development, body maintenance, growth, movement, reproduction, lactation, and other bodily functions. These chemicals are collectively called **nutrients**. Within major categories of nutrients, some specific components are essential for normal animal life and metabolism, and some other components are not essential but can make up part of the major category. Animal nutritionists have summarized absolute quantities of nutrients that animals should consume in a day and ranges of concentrations of nutrients that should be in diets. The quantities of nutrients that animals eat in a day depend upon age, body weight, growth rate, other physiologic states such as stage of pregnancy and lactation, plus levels of exercise (work). Ambient temperature, humidity, wind speed, amount of sunlight experienced per day, coat thickness, and exposure to mud also influence ideal levels of intake of dietary energy. Plants vary in their concentration of nutrients due to many factors: plant species, variety, stage of maturity, specific parts of plants, levels of nutrients in the soil, availability of soil water, temperature, and amount of sunlight. Wild animals adapted mechanisms to make choices when eating that allow them to obtain enough essential nutrients to live and reproduce. Those wild animals that experience changes in supplies of available feedstuffs and nutrients learn to adapt or move away or they die. Domestic animals depend upon people to manage most of their nutrient supply. The more we understand about the nutrients, their concentrations in plants of different species and ages, and the nutrient requirements of the animals in our care, the better the level of welfare of those animals.

This chapter describes a little about the chemical structure of nutrients, leaving the details to textbooks on biology, organic chemistry, and biochemistry. The term, "organic" here means "produced by organisms," and in the realm of chemistry, organic molecules contain carbon, oxygen, and hydrogen. The emphasis of this chapter is on variations in concentrations of nutrients in plants, what causes them, how they can be assessed, and how they can be well-matched to the nutrient requirements of grazing horses.

For meat and milk livestock in modern production systems that are managed for high production levels, the concentrations of dietary nutrients need to be high. For horses that are not growing, pregnant, lactating, or doing significant exercise, feedstuffs can have slightly lower quality, which means lower concentrations of energy, protein, vitamins and minerals, since the production demands are not as high as for lactating dairy cows or feedlot beef cattle. Thus, pasture can meet the majority of nutrient requirements for many horses because pasture is an excellent source of nutrients. Wild and feral horses evolved and still thrive on wild pastures and rangelands that do not look very productive or very high quality. Thus, it appears that feral horses in at least some American herds are quite efficient at gleaning nutrients from range forage. Fig. 3.1 shows a band of feral horses in Utah that have ideal body condition and are obviously fertile, given the swollen bellies of some mares and the young foal present. These horses have the energy to run across the rangeland that provides all of their feed, but it would be very difficult to quantify the nutrient requirements and make fair comparisons with domestic horses. The plants are typical of those in several dozen photos taken during a study of these horses in 2010. Compared to tame pastures in areas of higher rainfall, this land has many weeds, a very low forage yield, and forage is probably of low quality. However, the horses are not fenced and have the freedom to choose forage plants that provide them with optimum concentrations of required nutrients and fiber.

Horse Pasture Management. https://doi.org/10.1016/B978-0-12-812919-7.00003-2

FIGURE 3.1   A band of fertile, feral horses in Utah, apparently in good condition on forage that is sparse and not of high visual quality. *Photograph courtesy of Helen MacGregor.*

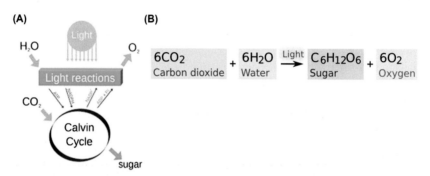

FIGURE 3.2   (A) **Chloroplasts** in green plants take in **carbon dioxide** (CO_2) and **water** (H_2O) within the light reactions and Calvin cycle; Author: Daniel Mayer. BNU Free Documentation License, Version 1.2, Free Software Foundation. (B) Energy from sunlight is used to turn these chemicals into sugars, which are used to build protective structures around cells (cell walls) and to store energy (starch). *(A) Source: https://commons.wikimedia.org/wiki/Filesimple_photosynthesis_overview.svg. (B)Source: https://commons.wikimedia.org. File: Photosynthesis_equation.svg; Author: ZooFari. (public domain).*

CARBOHYDRATES

One of the most important chemical processes that occurs in plants is photosynthesis. Within photosynthesis is a trio of processes called the Calvin cycle, wherein chlorophyll molecules in chloroplasts of green plants take in **carbon dioxide** (CO_2) and **water** (H_2O) and use enzymes to convert them to a three-carbon molecule called **glyceraldehyde-3 phosphate** (G3P) (Fig. 3.2A). The enzyme that uses the CO_2 and brings it into the Calvin cycle is called **rubisco**, and this enzyme action is a **carboxylase** action. Two molecules of G3P can combine to make one molecule of a simple sugar (Fig. 3.2B). This six-carbon sugar is called **glucose** ($C_6H_{12}O_6$) (Fig. 3.3).

Diversity in Carbohydrates

Chemists refer to glucose and similar simple sugars as **monosaccharides**. Some glucose molecules are modified to form a similar six-carbon sugar called **fructose**. Pairs of glucose and fructose molecules are bonded together to form the familiar **disaccharide** called **sucrose.** Humans and many other creatures recognize a property of disaccharides and some monosaccharides referred to as **sweetness**. This ability to recognize sweet molecules is possibly an evolutionary advantage to animals, since sugars and bigger molecules that contain sugars as building blocks provide a considerable amount of dietary energy.

Many glucose molecules can be joined together in chains called **starch**, and some of this starch is stored in the chloroplasts. Other starch can be stored in stems, fruits, seeds, and roots. Sucrose moves to other plant parts while

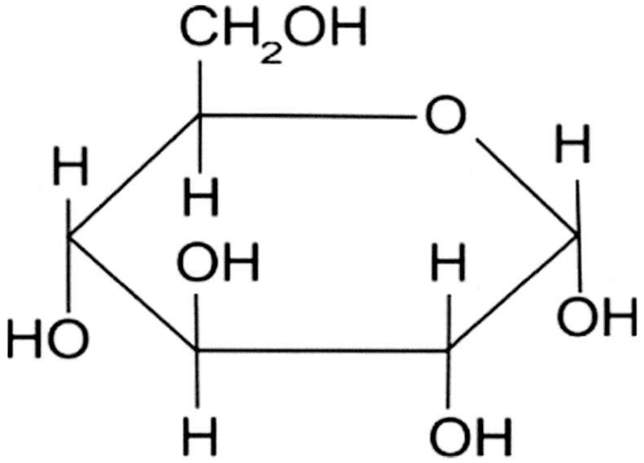

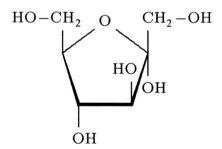

FIGURE 3.3 A molecule of α-D-glucose. Notice six carbon atoms; an oxygen atom in the hexagonal ring; —OH (hydroxyl) groups and hydrogen (H) atoms attached to carbon atoms. *Source: https://commons.wikimedia.org. Author: Lukas3 at Polish Wikipedia (public domain).*

dissolved in a solution called phloem sap, through tubes called phloem vessels. Once sucrose reaches other tissues, it can be split into its component monosaccharides, glucose and fructose, which are subsequently used to make other compounds, including amino acids, lipids, and building blocks of nucleic acid, needed for development and growth of the plant.

The six carbons of a **glucose** molecule can be arranged in a relatively straight line in a dry environment, but most of the time in plants, glucose is in an aqueous environment, which promotes formation of a ring structure (Fig. 3.3). In solution, D-glucose converts between its alpha form (with the first OH group clockwise from the oxygen in the ring, being below the plane of the ring) and its beta form (with the first OH group clockwise from the ring oxygen being above the plane). Most six-carbon sugars, called **hexoses**, have a six-sided shape, including an oxygen atom at one point of the hexagon, but **fructose** forms a five-sided ring with a branch containing a carbon atom (Fig. 3.4).

If you imagine looking at a glucose molecule in the ring form with the plane of the ring slightly tilted, so that the close edge is lower and the far edge is higher, you can discern a plane within the ring. Each carbon atom has hydrogen (H) atoms or hydroxyl (OH) groups attached to it. The carbon adjacent to and one position clockwise from the oxygen component of the ring is called carbon number 1. There are two possible orientations of the H and OH at carbon 1. If the H is above the plane and the OH is below the plane, this form is given the name "α-D-Glucose" and if the OH group is above the plane and the H is below the plane, this form is given the name "β-D-Glucose".

Linkages between monosaccharides and between a polysaccharide and a monosaccharide are catalyzed by enzymes and involve "condensation" or "dehydration" of a molecule of water as a by-product, since one hydroxyl group and one hydrogen atom will be cleaved off carbon 1 and one or the other off carbon 2 or 4. An atom of oxygen is left between the monosaccharides as part of a **glycosidic linkage**. Two monosaccharides bonded together form a disaccharide. A diversity of disaccharides exists because different monosaccharides can be paired. The position of a hydroxyl group below (α) or above (β) the plane of carbon number 1 determines whether the glycosidic linkage between monosaccharides is of the α or β type. For example, glucose and fructose are joined by a glycosidic linkage to form **sucrose**, which is also known as table sugar (Fig. 3.5). **Maltose** is a disaccharide formed where starch is broken down into two glucose units, for example, where

FIGURE 3.4 Monosaccharide sugars commonly have six carbons, but some have only five. A typical structure of a monosaccharide is a hexagon, containing five carbons and an oxygen, with the sixth carbon on a side chain. **Fructose** has five sides and two side chain carbons. Small differences, such as whether a hydroxyl group (-OH) is above or below the plane of the ring, cause different chemical properties. *Source: https://commons.wikimedia.org/wiki/File:Alpha-d-fructose.svg. Attribution: Rob Hooft. GNU Free Documentation License, Version 1.2, Free Software Foundation.*

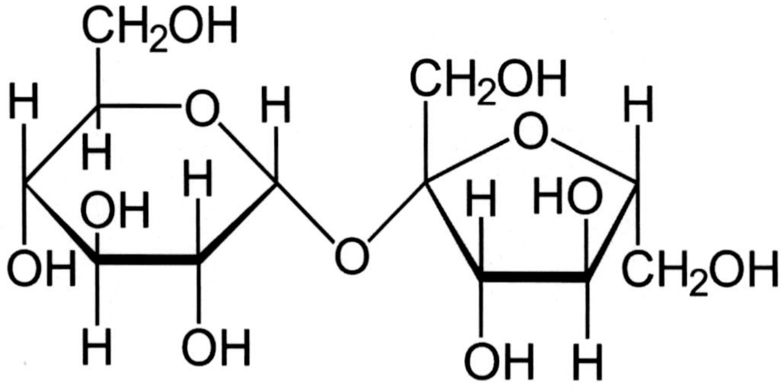

FIGURE 3.5 A diversity of disaccharides exists because different monosaccharides can be paired and different carbon numbers can be involved. The disaccharide **sucrose** is formed by an α-1,2 glycosidic linkage between α-D-glucose and fructose. *Source: https://commons.wikimimedia.org/wiki/File: Sucrose_structure__formula.png. Author: Bas. GNU Free Documentation License, Version 1.2, Free Software Foundation.*

yeast enzymes ferment starch, both in brewing and in the cecum. Animals can digest maltose to individual glucose molecules for energy metabolism. Maltose is sweet and tastes like caramel. **Cellobiose** is a disaccharide consisting of two β-D-glucose units, and it is formed where cellulose is being digested, for example in **cellulolytic bacteria** (Fig. 3.7).

Large polymers of sugars (e.g., starch and cellulose) are created by continuous additions of monosaccharide to an existing chain of them. The bonds or linkages between monosaccharides are either **α-1,4 glycosidic** linkages if the oxygen of the linkage is below the plane or **β-1,4 glycosidic** linkages if the oxygen is above the plane. This depends on whether the molecule that bonds its carbon #1 is α-D-glucose or β-D-glucose. Figs. 3.5 and 3.6 show how disaccharides are formed by glycosidic linkages. Fig. 3.7 shows part of a starch molecule with a linear arrangement of α-D-glucose units held together

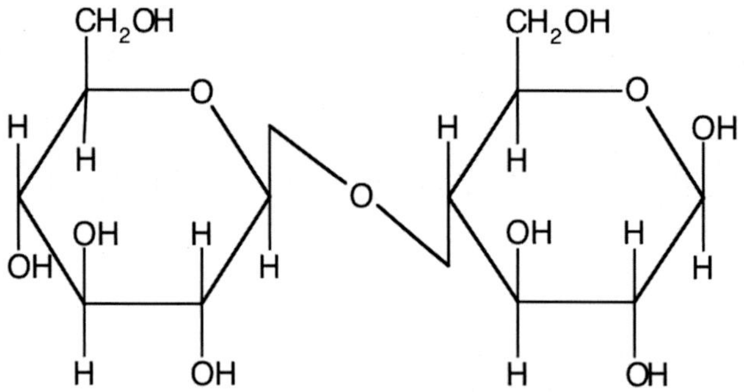

FIGURE 3.6 The disaccharide **cellobiose** has a β-1.4 glycosidic linkage between two β-D-glucose molecules. It would be rare to find accumulations of cellobiose in nature because it exists only briefly in the initiation of a cellulose polysaccharide and during cellulose digestion. Cellobiose can be formed in a lab by using the appropriate cellulase enzyme to split disaccharides from the ends of cellulose molecules. *Source: https://commons.wikimedia.org/wiki/File:Cellobiose.jpg. Public domain.*

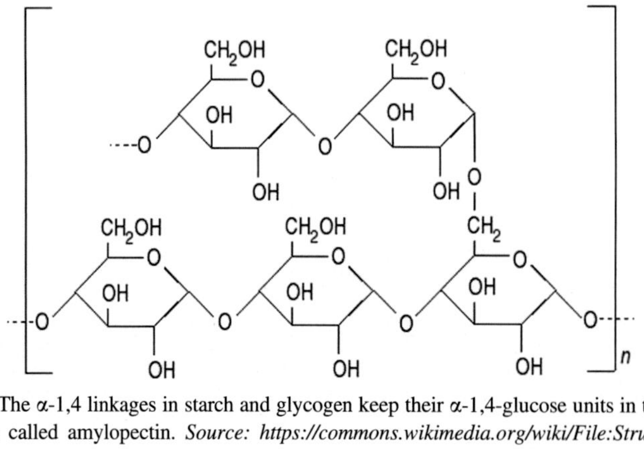

FIGURE 3.7 A molecule of **starch**. The α-1,4 linkages in starch and glycogen keep their α-1,4-glucose units in the same orientation. An α-1,6 linkage forms a branch in the type of starch called amylopectin. *Source: https://commons.wikimedia.org/wiki/File:Structure_de_l'amylopectine.jpg. Author: Laranounette. Creative Commons Attribution-Share Alike 3.0 Unported.*

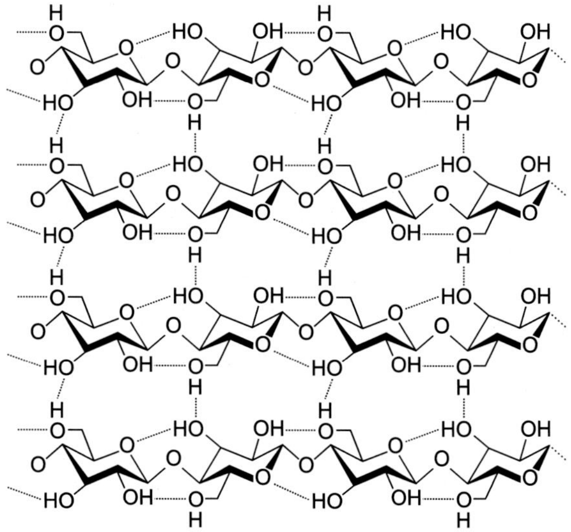

FIGURE 3.8 A molecule of cellulose, containing β-1,4 linkages in a polysaccharide made of glucose units. The β-1,4 linkages are resistant to digestion by mammals because mammals do not produce any enzymes that can break these bonds. Some microorganisms in the cecum and rumen of herbivores make cellulase enzymes that will break these β-1,4 bonds. *Source: https://commons.wikimedia.orgwiki/File:Cellulose_strand-es.jpg. Author: I. Laghi. Licensed under the Creative Commons Attribution-Share Alike 3.0 Unported license.*

by α-1,4 glycosidic linkages. This type of starch is called **amylose**. Adjacent α-D-glucose units in starch all have the same orientation. These linkages can be broken by α-1,4 amylase enzymes produced in the pancreas and intestines of mammals and birds, so starch is readily digested by them. Starch called **amylopectin** also has some branching at carbon 6 due to α-1,6 glycosidic linkages.

Fig. 3.8 shows a molecule of cellulose with β-D-glucose molecules joined by β-1,4 glycosidic linkages. Synthesis of cellulose starts with a β-1,4 glycosidic linkage that flips alternate glucose units over. Hydrogen bonding between adjacent polymer chains in cellulose causes formation of fibrils that help to provide strength to plant cell walls. Layers of cellulose fibrils in thick cell walls provide a third dimension of hydrogen bonding and provide greater strength.

The β-1,4 glycosidic linkages of cellulose can be represented as alternating above and below the plane. This minor structural difference between cellulose linkages and starch linkages is very important in animal nutrition because the enzyme called **amylase** that will break α-1,4 glycosidic linkages is produced by the pancreas of most animals and in some animals by tissues such as liver, small intestine mucosa, and salivary glands, whereas an enzyme that breaks β-1,4 glycosidic linkages (as found in cellulose) is not made by mammals at all! Herbivorous animals such as horses and ruminants have a symbiotic relationship with **cellulolytic bacteria** in their gastrointestinal tract. Cellulolytic bacteria (which are enteric microbes) produce a **cellulase** enzyme that will break β-1,4 glycosidic linkages. In ruminant animals (cow, sheep, goat, deer family, etc.), most of the cellulolytic bacteria are held in large expansions of the esophagus referred to as the rumen, reticulum, and omasum and a smaller amount is held in the cecum. In equids (horses, donkeys, zebras), there is no equivalent to the rumen, reticulum, and omasum, but there is a much-expanded cecum, and that is where the cellulolytic bacteria live and break down the cellulose and related plant fibrous molecules into metabolites that equids can digest and metabolize.

In addition to cellulose, plant fibers in cell walls can contain polysaccharides called **hemicelluloses**, each made of a mixture of diverse sugars, having β-1,4 glycosidic linkages and many branches. Hemicelluloses contain some nonglucose sugars, more branching, and some different linkages than cellulose. Hemicelluloses, along with **pectins**, adhere to cellulose by both hydrogen bonds and covalent bonds, forming a network of cross-linked fibers. Some hemicelluloses are soluble and some are not. Some hemicelluloses may act as antinutritive factors by binding minerals and other nutrients, blocking their availability to animals. Natural **hemicellulases** occur more frequently in gut fungi than gut bacteria, and the diversity among animal species is large. The distribution of *Piromyces*, the most abundant fungal genus known to produce hemicellulases, was low in the Equidae family and even absent in some horse samples compared to the cattle and sheep that were sampled (Liggenstoffer et al., 2010).

Polysaccharides that contain multiple fructose units are called **fructans**. These are part of the group of soluble carbohydrates found inside plant cells, but they have a lower degree of polymerization than starch or cellulose.

When feeding or providing feed for herbivores, it is important to remember that you are providing appropriate amounts of digestible nutrients and indigestible fiber for the animal and appropriate amounts of nutrients for the enteric microbes, since it is these organisms and their enzymes that will digest fiber and release energy from it, for use by the animal.

Structural Versus Nonstructural Carbohydrates

Plants do not have a skeleton in the way that animals do. The **structural strength** of plants that allows them to be tall and thin without falling over easily is due to **cell walls** that thicken as the plant matures and a combination of the following chemicals within the cell walls:

1. cellulose and hemicellulose fibers
2. pectins, which provide gelatin-like support
3. resin-like **lignin** incorporated in xylem and phloem tissues
4. cross-linking among plant fibers
5. small amounts of proteins, including collagen-like fibers and enzymes
6. hydrostatic pressure (**turgor**) pushing outward on the cell wall due to water and salts inside the cell.

When a new plant cell arises by mitosis, its cell (plasma) membrane, made of a lipid bilayer with protein inclusions, is surrounded by a **primary cell wall**, which is composed largely of cellulose, and the cellulose microtubules are supported by pectins. As the cell reaches its mature size, a **secondary cell wall** may be formed inside the primary cell wall with polysaccharide fibers and a matrix of tough, indigestible lignin. Different plant parts may differ in specific arrangement of cellulose, hemicellulose, pectin, and lignin within their cell walls. Within the cytoplasm and vacuoles of plant cells are water, ions, salts, organic acids, sugars, amino acids, proteins, and pigments. These can be referred to as **cell solubles**.

Due to their role in providing structural support, the polysaccharides, cellulose, and hemicellulose are referred to as **structural carbohydrates**. They are insoluble in water, and they require specific enzymes to be digested. Lignin is not chemically a carbohydrate, since it is made of a complex arrangement of phenyl propane monomers, rather than the monosaccharide monomers of starch and cellulose.

Nonstructural carbohydrates (NSC) are the sugars and starch in plants and glycogen in animals. NSCs are readily digestible in the small intestine of all animals, due to the secretion of pancreatic and intestinal amylases into the intestinal lumen. The term **amylase** means an enzyme that breaks bonds in **amylose**. Initial digestion of starch by α-amylase in animals is to oligosaccharides (composed of two to four monosaccharides). Maltase and sucrase enzymes from intestinal cells digest oligosaccharides to monosaccharides (glucose and fructose), which are absorbed into the bloodstream and used in metabolic processes. When large amounts of NSCs are ingested by horses, some of them are converted by particular microbe species to lactic acid, which can cause a drop in the pH of gut contents and of blood, sometimes contributing to development of **laminitis**.

The NSC fraction of a forage is the sum of percentages of all sugars + fructans + starch. Sugar contents of forages respond to the stimulus of sunlight on photosynthesis. Thus there is little or no sugar production during the dark hours and little sugar and starch in plants early in the morning. NSC and water-soluble carbohydrate (**WSC**, free sugars and fructans) concentrations rise in forages once sunlight strikes their leaves in the morning. In alfalfa, glucose and fructose increased from 6 a.m. to noon and sucrose concentration increased from 6 a.m. to 6 p.m. (Lechtenberg et al., 1971). Alfalfa leaf starch increased between 9 a.m. and 3 p.m., and the level of starch in stems did not vary over time. NSC accumulation occurs at different rates in different forage species. WSC ranged from 95 to 560 g/kg DM, while fructan was 32–439 g/kg (Longland and Byrd, 2006).

In warm environments, warm season grasses have lower NSC than cool season grasses. Meadow bromegrass produced lower levels of NSC than several other species on irrigated land in Utah. On dry land, the lowest NSC levels were in Sandberg bluegrass and tall wheatgrass (Jensen et al., 2014). Higher values of WSC and NSC occur at cooler temperatures probably because photosynthesis of sugars can continue under sunshine at cooler temperatures, but plant growth slows, so sugar concentration rises. After a frost, NSC levels can remain elevated for several days. Drought can cause fructan content to increase in stem bases. Factors that encourage growth, for example, application of N fertilizer and water, tend to decrease NSC content. WSC can more than double its concentration during daylight hours, but in reduced light and warm temperatures, WSC content only varies a little. Small amounts of starch are stored in leaves, to provide an energy source for plants through the night, but large amounts are stored in grass seed. Grazing horses have been seen stripping seed heads from grass plants, and if enough was consumed, it could potentially deliver a large dose of starch rapidly to the hindgut. NSC concentrations were lower in short (15 cm) than in tall (30–40 cm) tall fescue, and horses grazing tall forage had higher serum insulin concentrations, indicating a possible strategy for preventing insulin resistance in grazing horses (Siciliano et al., 2017).

What About Fructans?

Fructan concentrations in grasses have been reported as highest in May, lowest in August, and intermediate in October. Since about 2005, there have been many warnings in the media about high levels of fructans in cool season forages being the trigger for laminitis in horses. Fructan is a polysaccharide, rather than a monosaccharide like glucose or a disaccharide like sucrose. Fructan develops from a fructose molecule attaching to a sucrose and then more fructose units joining in sequence. Thus it is significantly bigger than a sugar (Baron, 2018). Fructan is not made of repeating glucose units like either starch or cellulose, and it cannot cause a spike of insulin from the pancreas, like sucrose and glucose can. Apparently the experiments in which horses were loaded with the fructan **inulin** at more than 8 pounds (3.6 kg) per dose via stomach tube may not have been a good model for a hypothesized intake of the more common fructan found in grasses, called **levan**, which is fermented in the hind gut more slowly than inulin. This dose was equivalent to a horse consuming 10 kg of forage DM having 37.5% fructan (Kellon, 2015). The range of total fructan reported by Longland and Byrd was 3.2% −44%, so the dose probably does not represent a normal pasture intake. Fructans are normal storage carbohydrates in cool season grasses, while starch serves this role in alfalfa and in grains. The time of year most associated with laminitis is in the spring with green grass, a time when horses may be allowed to graze too much too soon (Baron, 2018). Pasture grasses in spring are high in sugars and low in fiber, which are contributing factors to laminitis. Fat horses that get little or no exercise are often in a high-risk group for metabolic problems, including laminitis.

Horses graze for 12−17 h per day, so it is unlikely that laminitis on pasture is caused by a large bolus dose of NSC. Some horses might be chronically close to developing laminitis and are thus susceptible to slight increases in NSC intake (Longland and Byrd, 2006). To manage horses that seem to be susceptible to NSC in development of laminitis, the following suggestions might reduce the risk (Kellon, 2015; Baron, 2018):

- using C4 species if possible
- using C3 species that accumulate lower NSC concentrations
- maintaining short, leafy grass, with appropriate mowing, soil moisture, and fertility to keep grass growing
- restricting grazing to early morning
- grazing shaded pastures
- preventing grazing of pastures that have seed heads by mowing
- using grazing muzzles to slow forage intake
- testing pasture swards and hay supplies for laminitis-prone animals
- introducing high-risk horses to pasture gradually, starting with a couple of hours per day, and exercise them daily
- ensuring that in the spring, horses have or are attaining a moderate body condition score

LIPIDS

Lipids include the oils and fats. They are similar to carbohydrates in being composed primarily of the elements carbon, hydrogen, and oxygen. Some lipids have a phosphate group attached and are called **phospholipids**. The basic structure of a lipid is a **triglyceride**, which consists of a three-carbon chain called **glycerol** (Fig. 3.9A) with three **fatty acids** attached to it (Fig. 3.9B). The fatty acids are long chains of carbon atoms, with two hydrogen atoms attached to them until the end of the chain. This last carbon has three hydrogens attached to it and is called a **methyl group**. Thus arrangement of glycerol attached to three fatty acids provides the triglyceride structure (Fig. 3.9C). The bonds between a fatty acid and a carbon of glycerol include an oxygen atom between the two carbons and are called **ester linkages**. The length of the carbon chain in fatty acids can vary from two in **acetic acid** (which is called vinegar when in an aqueous solution) to about 20 (**arachidic**, **arachidonic**, and **eicosapentaenoic acids** (EPAs)). The fatty acids differ from each other in three other things: the number of double bonds between carbon atoms in the chain, whether a double bond creates a bend in the molecule, and the position of those double bonds, relative to the end of the chain. **Saturated fatty acids** (SFAs) are saturated with hydrogen atoms and have no double bonds, as shown in Fig. 3.9B. **Monounsaturated** fatty acids have a single double bond in the carbon chain, which results from two adjacent carbon atoms giving up their bond to one hydrogen and using the resulting free electrons to create the double bond. If the hydrogens attached to the doubly bonded carbons are on the same side, this is called the **cis** orientation, and it causes a bend in the molecule. If the hydrogens are on opposite sides, this is called the **trans** orientation, and it does not cause a bend. The bends prevent unsaturated fat molecules from packing as tightly as saturated fat molecules. Many such bends in a membrane lowers its melting point, making it fluid at lower temperatures. **Polyunsaturated fatty acids** (PUFAs) have two or more double bonds (Fig. 3.10).

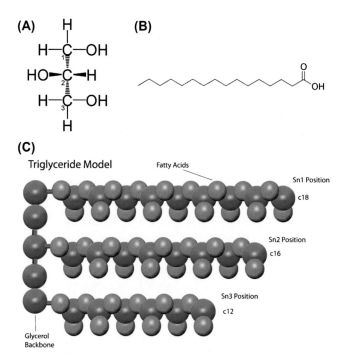

FIGURE 3.9 (A) A molecule of the three-carbon base of triglyceride lipids, glycerol. (B) A free **fatty acid** molecule, palmitic acid, containing 16 carbon atoms. The angles of the *zigzag line* represent carbon atoms and the *lines* represent bonds between the carbons. These carbon atoms are **saturated** with hydrogen atoms. The carbon at the right end has three hydrogens, forming a methyl group. Each carbon within the chain has two hydrogens attached to it. Carbon #16 has a double bond to an oxygen atom and a single bond to the oxygen of a hydroxyl (OH) group. (C) A triglyceride lipid molecule. Three FAs are linked to glycerol by ester linkages. *(A) Source: http://commons.wikimedia.org/wiki/File:SN-Glycerol.png. License: Public domain. (B) http:// commons.wikimedia.org/wiki/File:Palmitic_acid_structure.png. Author: Edgar181. Public domain. (C) http://en.wikipedia.org/wiki/File:Eie- TRIGLYCERIDE.jpg. Author: Roger Daniels, Bunge Oils Director of Development. GNU Free Documentation License, Version 1.2, Free Software Foundation.*

FIGURE 3.10 A polyunsaturated fatty acid (18:2, n-6) showing two *trans*-double bonds at carbon positions 6 and 9 away from the omega-1 (methyl) carbon. A *cis* double bond would create a bend in the molecule. *Source: http://commons.wikimedia.org/wiki/File:Linoelaidic_acid.png. Author: Edgar181. Public domain.*

Short descriptions of fatty acids indicate the number of carbon atoms, followed by the number of double bonds in them (e.g., 16:0 means 16 carbons and 0 double bonds, 18:2 means 18 carbons and 2 double bonds). Adding the letter "n" or the word "omega" after this number of double bonds indicates the position of a double bond relative to the final or end carbon of a fatty acid chain. This is the end with the carbon bearing three hydrogen atoms (-CH_3 or "methyl" end). An "omega-3 fatty acid" has a double bond three bonds away from the CH_3 or "methyl" end (Fig. 3.10).

In organisms, lipids store energy, provide thermal insulation, repel water on outer surfaces, and provide electrical insulation around nerves. Some modified lipids help plants capture light energy (carotenoids), regulate physiologic processes (vitamins, steroid hormones), and form the basic structure of cell membranes (**phospholipids**). In phospholipids, there are two fatty acid chains, and one chain is replaced by a phosphate-containing compound (Fig. 3.11). Choline, serine, and inositol are three small molecules that may replace the fatty acid. They bind to a **phosphate group** that has a negative electrical charge, causing that part of the molecule to be attracted to the partially positive parts of water molecules. Molecules that have relatively negative or positive sections are called **polar**. The negative charge near the oxygen atom of a water molecule and the partially positive charges around the hydrogen atoms are the cause of its polarity. Other polar molecules are attracted to the polarity of water because their positive poles are attracted to the negative poles of water molecules and the positive poles of water molecules are attracted to the negative charges of the other polar molecules. This attraction to water is called **hydrophilic**, meaning "water-loving." Nonpolar molecules are not attracted to water and are called **hydrophobic**, meaning "water-fearing." Thus other polar molecules are attracted to and readily soluble in water, and nonpolar molecules are repelled by and insoluble in water. The fatty acid carbon chains have no free electrical charges, so

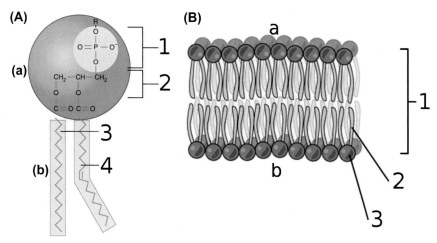

FIGURE 3.11 (A) A phospholipid, having a positively charged component (a), such as choline or serine, attached to a negatively charged phosphate (1), which is attached to the glycerol (2). Two fatty acids attached to glycerol form a nonpolar tail. (b) One fatty acid is saturated (3), and the other is monounsaturated with a *cis* double bond, nine carbons from the methyl end (4). Author: OpenStax. Creative Commons Attribution 3.0 Unported license (B) Two layers of phospholipids form a bilayer in cell membranes of organisms with hydrophobic tails inward (2) and hydrophilic heads (3) outward (1), in contact with water molecules of intracellular (a) and extracellular (b) fluids. Author: OpenStax. Creative Commons Attribution 4.0 International license. *(A) Source: https://commons.wikimedia.org/wiki/File:0301_Phospholipid_Structure_labeled.jpg. (B) http://commons.wikimedia.org/wiki/File:0302_Phospholipid_Bilayerlabeled.jpg.*

they are nonpolar and hydrophobic. In cell membranes, phospholipid molecules are stacked together in two layers with all the polar heads facing the same way within a layer (Fig. 3.11B). The polar heads face away from the polar heads of the other layer. Being hydrophilic, they are oriented toward the watery environment of the inside of a cell or toward the extracellular fluid surrounding cells. The nonpolar tails are hydrophobic, so they point toward the inside of the membrane.

Fatty Acids of Special Interest

Two fatty acids (linoleic (18:2,n-6) and α-linolenic (18:3, n-3)) are considered to be essential in the diets of some animals but not horses. Green leaves of forage plants and green algae, both of which receive plenty of sunlight, have higher concentrations of omega-3 fatty acids than grains (French et al., 2000). Mixtures of ryegrass, orchardgrass, and legumes are reported to contain about 63% linolenic and 16% linoleic acids (Wyss et al., 2005). Linseed and flaxseed oils are high in α-linolenic acid, relative to grains. Fatty oils extracted from cold-water fish such as salmon, menhaden, etc., are rich in omega-3 oils called eicosapentaenoic acid (EPA, 20:5n-3) and docosahexaenoic acid (DHA, 22:6n-3). The origin of these marine fatty acids is marine algae. To promote human health through PUFA-enriched meat, milk, or eggs, livestock are given diets enriched in DHA, EPA, and sometimes conjugated linoleic acid (CLA). Dairy cows on permanent pasture had 500% more CLA in their milk than cows fed a ration containing conserved forage and grain. Their dry hay did not contribute to milk CLA content (Dhiman et al., 1999). Where grazed grass had a concentration of alpha-linolenic acid (18:3n-3) 25 times higher than in concentrates, decreasing proportions of grains and hay while increasing proportions of grazed grass in diets for beef steers increased the ratio of PUFA to SFA in intramuscular beef fat (French et al., 2000). Studies in which horses were fed diets supplemented with omega-3 fatty acids from linseed, flaxseed, or fish have demonstrated modulation of inflammatory processes. Supplementation of hay and grain diets with n-3 fatty acids in horses may help manage chronic inflammatory conditions such as osteoarthritis, equine metabolic syndrome, laminitis, and inflammatory airway disease and thereby help improve longevity of sport horses (Brennan et al., 2017; Hess and Ross-Jones, 2014; Monteverde et al., 2016; Nogradi et al., 2015; Pritchett et al., 2015).

   A number of chemicals that are not exactly lipids but may be derived from them or are just soluble in lipids are called "fat-soluble vitamins." These vitamins are designated by the letters A, D, E, and K, and they promote normal animal cell metabolism and a variety of specific organ functions. Being soluble in lipids, these vitamins readily associate with or pass through the lipid bilayer of cell membranes, and some may be stored in fatty tissue.

Vitamin A

Vitamin A is derived from beta-carotene and other carotenoid pigments in plants. Enzymes in the intestines cleave beta-carotene into two molecules of vitamin A. Some of the beta-carotene is absorbed and is used in fat, skin, and the corpus luteum of the ovary, where it acts as an antioxidant. The plant sources with the highest concentrations of beta-carotene are green forage leaves and yellow vegetables. Growing forages can have more than 10 times the beta-carotene concentration of high-quality hay. As long as there is some green color in forage and it constitutes the majority of a horse's diet, it will proved enough beta-carotene to prevent deficiency signs. Consuming fresh green forage for 6 weeks will fill the capacity of the liver to store vitamin A for 3 to 6 months (Lewis, 2005). Horses use vitamin A in night vision, reproduction, embryo development, and to maintain immune responses to infections. Deficiency of vitamin A has often been recognized as night blindness, but riding and observing horses in the dark is not nearly as common in the 21st century as it was before the 20th century, so some opportunities to detect vitamin A deficiency may be missed. Low growth rates and reduced production of blood cells, along with impaired immune function, are also associated with vitamin A deficiency (NRC, 2007a,b,c,d,e,f,g).

Vitamin D

Dead, rather than live leaves of plants, stimulated by ultraviolet light, even on cloudy days, convert ergosterol to ergo-calciferol (vitamin D_2). In mammalian skin exposed to ultraviolet light, 7-dehydrocholesterol is converted to cholecal-ciferol (vitamin D_3). These two forms of vitamin D are converted to 25-hydroxy vitamin D in the liver, and when it circulates through blood vessels to the kidneys, it is converted to either its most active form, 1,25 dihydroxy vitamin D, or a less active form, 24, 25 dihydroxy vitamin D. These forms of vitamin D vary in their potency in helping to maintain plasma calcium concentration. This is accomplished through interactions with parathyroid hormone and calcitonin (primarily from the thyroid gland). A deficiency of vitamin D is very rare in horses, but in some other animals, it can result in broadening of bone growth places, lack of calcification of growing bone, and bowed legs. Experimental deficiency of vitamin D in foals caused low feed intake, slow growth, and low bone breaking strength but not the classic bowing of legs (Lewis, 2005).

Vitamin E

Vitamin E acts jointly with the mineral selenium in protecting cell membranes and some enzymes from oxidation damage. Oxidation is a normal process in the body. Free radicals are atoms or groups of atoms that have one or more unpaired electrons in their outer shells, making them highly reactive. Free radicals are attracted to other atoms, to share electrons and reach a more stable state. However, these reactions damage proteins, lipids, and DNA. Vitamin E blocks free radicals from attacking lipids. Selenium and vitamin E can substitute for each other in being oxidized so that other functional molecules are not. Vitamin E also enhances the immune system, promotes cellular respiration, and helps synthesis of DNA and another antioxidant, vitamin C. Other roles of vitamin E include protecting against harmful effects of lead, mercury, and silver, promoting vitamin A absorption and storage, and decreasing platelet aggregation. There are several forms of vitamin E, and the most active form in feeds is called d-alpha-tocopherol. Storage, ensiling, and processing feeds all cause loss of their vitamin E content. Green growing forage contains higher concentrations of vitamin E than horses in any physiologic condition require. Horses fed only dry hay and those getting significant exercise should receive supplemental vitamin E. Deficiencies of vitamin E and/or selenium can include capillary leakage, red blood cell damage, degeneration of muscles, nerve damage, other organ damage, reproductive disorders, adipose tissue inflammation, and yellowing of fat (Lewis, 2005).

Vitamin K

Natural forms of vitamin K are produced in green leafy plants, and some other forms are produced by bacteria in the cecum and colon. These various forms are converted to active hydroquinone in the liver. The role of vitamin K in the body is activating blood clotting factors and other proteins. Vitamin K deficiency in some animals prevents normal blood clotting, but horses eating good quality forages and with normal intestinal microbes rarely have this problem. The most commonly reported situation causing a lack of blood clotting in horses is a result of consuming moldy sweet clover hay or silage. A penicillium mold in sweet clover converts coumarin to dicoumarol, which interferes with the ability of vitamin K to promote blood clotting. The condition usually takes several weeks of consuming moldy sweet clover before any effects are noticed. Signs include persistent bleeding following trauma, surgery, or nosebleeds, or hematomas under the skin, pale mucous membranes, weakness, rapid heart and respiration rates, and difficulty breathing (Lewis, 2005).

PROTEINS

Proteins are composed of carbon, hydrogen, oxygen, and **nitrogen**, arranged in 20 different building blocks called **amino acids**. Two amino acids also contain sulfur. The central or α carbon is attached to four different chemical groups: a hydrogen, a carboxyl group (consisting of a carbon bonded to a hydroxyl group and a double bond to an oxygen), an amino group (consisting of a nitrogen bonded to three hydrogens), and an R group or side chain (Fig. 3.12). These first three groups are consistent in amino acids, whereas there are 20 distinct R groups that provide each amino acid with unique chemical properties. The simplest R group is a hydrogen atom in the amino acid "glycine." The other R groups contain from one to nine carbons. Some contain extra nitrogen or sulfur. Amino acids are joined by **peptide bonds** into chains, similar to the chains of monosaccharides in polysaccharides (Fig. 3.13). The chemical properties and electrical charges of the side chains determine solubility in water and other chemical properties. Short chains of amino acids are called **peptides**. Longer chains of amino acids (>20) are considered to be proteins, and these can be hundreds or thousands of amino acids long. Attractive forces between R groups determine places where long, flexible proteins form loops, helical shapes, and

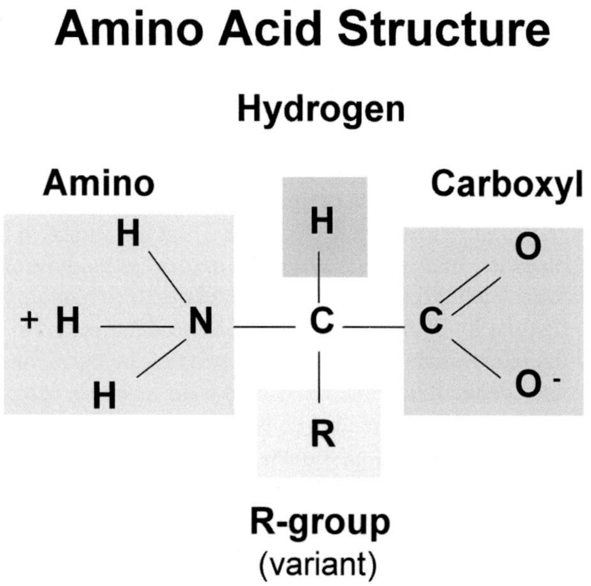

Amino Acid Structure

Hydrogen

Amino   **Carboxyl**

$$+ H - N - C - C$$

R-group
(variant)

FIGURE 3.12   General structure of amino acids. The "R" group has 20 various molecular structures in nature. *Source: http://commons.wikimedia.org/ wiki/File:Amino-acid-structure.jpg. Author: Johndoct. Creative Commons Attribution-Share Alike 4.0 International.*

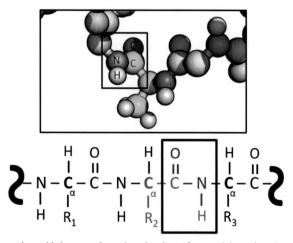

FIGURE 3.13   A peptide bond joins two amino acids between the carboxyl carbon of one and the amino nitrogen of the other. *Source: http://commons. wikimedia.org/wiki/File:PEPTIDE-BOND-FIGURE.png. Author: Chemistry-grad-student. Creative Commons Attribution-Share Alike 3.0 Unported license.*

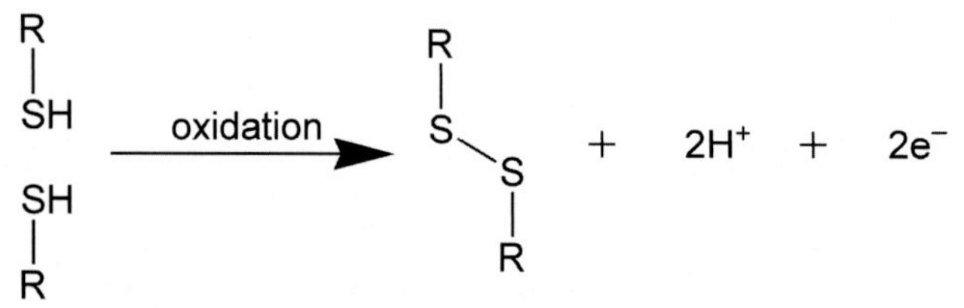

FIGURE 3.14 A disulfide bridge forms between the sulfur atoms of two different amino acids in the same amino acid chain, located a significant distance away from each other, effectively bonding two parts of the protein together and helping to establish tertiary protein structure. *Source: http:// commons.wikimedia.org/wiki/File:Disulfide-bond.png. Author: Benjah-bmm27 at English Wikipedia. Public domain.*

pleated sheets. In the amino acids called "cysteine" and "methionine" the sulfur atoms cause these amino acids to be attracted to each other and form strong bonds called **disulfide bridges**, which create specific molecular shapes of proteins (Fig. 3.14). Hydrogen bonds between +ve charges on hydrogens in NH_3 groups and −ve charges on double-bonded oxygens of nearby amino acids also contribute to shape and to physiologic properties of proteins.

The level of complexity of protein structure is described on four levels, which are illustrated well in biology and biochemistry textbooks. **Primary** structure refers to the order and sequence of the different amino acids. **Secondary** structure refers to a three-dimensional shape such as a **helix** (shaped like a ribbon wrapped around and along a pole) or a **pleated sheet**, both resulting from hydrogen bonds. The disulfide bonds are very strong and are common in keratin proteins in hair, hooves, and feathers. Some proteins are thousands of amino acids long. Opportunities for complexity in protein shape increase with number and diversity of amino acids in the chain. Bends and folds in a protein characterize its **tertiary structure**, causing some to be globular in shape (globins). Two or more proteins bonded together as subunits form **quaternary structures**, such as hemoglobin. The diversity of proteins and the attractive forces between them promote the following: adhesion of cells to each other, transfer of substances into cells, binding of reactants to enzymes for rapid biochemical reactions, and stimulation of cell processes when protein hormones bind to protein receptors on cell surfaces.

Dietary crude protein (CP) is digested by animals to its component amino acids. In the small intestine, these amino acids are absorbed and carried by the bloodstream to cells in most parts of the body for the synthesis of animal proteins. Some of the many uses of proteins include the following: receptors for hormones in cell membranes, transport proteins for ions and glucose in cell membranes, contractile fibers within cells to help change shape, aiding cell-to-cell communication, muscles for locomotion, gathering and ingesting food, reproduction, antibodies for fighting infections and foreign particles, enzymes, hormones, neurotransmitters, and carriers of other important molecules.

The amino acid **lysine** is said to be the "first limiting amino acid" in horses and some other livestock species. "First limiting" means that it is **essential** for a number of bodily functions and is in low enough quantities in most feedstuffs that it is most likely to be deficient. During protein synthesis in horse cells, if a DNA code calls for lysine and it is not present, the production of that protein stops until more lysine is supplied. **Methionine** is a sulfur-containing amino acid that is essential for making the protein **keratin**, which is a major structural component of hooves and hair. The **protein quality** of feedstuffs is a measure of how many essential amino acids the proteins contain and how close the balance of essential amino acids is to the requirements of the animals eating those feedstuffs. Good pasture tends to have high protein quality. In cases of amino acid deficiencies on pasture, soybean meal is usually a suitable supplement. Overfeeding protein to horses causes them to excrete more urea than normal and this can lead to a strong ammonia smell where urine accumulates.

WATER-SOLUBLE VITAMINS

A number of chemicals, collectively referred to as B vitamins, are soluble in water, generally produced by plant tissue and effective in some parts of animal metabolism, particularly of energy metabolism. Vitamin C is also water-soluble. All B vitamins are also produced by bacteria in the intestines, and an adequate amount of vitamin C is produced in the liver. Since they are not soluble in fats, most B vitamins are not stored for long periods in the horse, with the exception of vitamin B_{12}. Deficiency of some of the B vitamins can cause, in some animal species, decreased feed intake, poor growth rate, loss of weight, poor hair coat, plus weak and damaged hoof walls. However, horses seem to be more resistant than meat- and milk-type livestock to water-soluble vitamin deficiencies. Very few of the B vitamin deficiencies reported in these other types of livestock have been documented in horses. Presence of significant amounts of leafy green forage in the diets of horses and moderate ability to store B vitamins seem to work together to prevent these deficiencies. Vitamin C helps with development of bones and dentin, utilization of some B vitamins, intestinal absorption of iron, and synthesis of some amino acids and their derivatives. Collagen production is promoted by vitamin C. Along with vitamins A, D, and E, vitamin C acts as an antioxidant in animals, but horses and most other mammals do not require vitamin C in their diet (Lewis, 2005).

IMPORTANT MINERALS

The following roles and concentrations of minerals in plants and animals are summarized from Barker and Collins (2003) and Blevins and Barker (2007).

Boron is used by plants in amino acid synthesis, stability of membrane glycoproteins, and forming root nodules in legumes. Deficient plants turn yellow and grow poorly. Functions of boron in animals are not well-defined, but it is known to bind with several biomolecules. Normal concentration of boron in plants is 10−50 ppm, with deficiency possible below 10 ppm. There is not a known requirement for it in animals.

Calcium (Ca) tends to have a concentration of 0.1%−0.5% in grasses and 1.2%−1.5% in legumes. Horses are unlikely to become deficient in calcium because of its high concentrations in forages, relative to horse requirements, unless the soil is wet and acidic (Frape, 2010). The concentration of calcium in the soil contributes to soil structure and helps to moderate pH (Murphy, 1998). Calcium is readily extracted by plants and contributes to cellular regulation processes within them. The most common way of ensuring adequate plant-available calcium in soil is by testing for calcium and pH in soil samples and spreading appropriate amounts of lime.

Copper (Cu) aids in nitrate reduction and electron transfer during photosynthesis. Animals need copper for some enzymes, regulation of iron, and for immune function. A normal copper level in plants is only 5−15 ppm, and deficiency symptoms may occur below 5 ppm. Copper availability to animals is sometimes compromised by molybdenum.

Iron (Fe) has roles in chlorophyll production, nitrogen metabolism, and enzyme functions in plants, where its normal concentration is 50−1000 ppm, and deficiency for plants is <35 ppm. A typical sign of iron deficiency is chlorosis (lack of color) in young leaves. Iron is bound to sulfur-containing amino acids in plant proteins. In animals, iron is needed for production of hemoglobin, which carries oxygen in red blood cells. It is also used in normal functioning of the immune system.

Magnesium (Mg) concentration in forages is known to drop during rapid spring growth. Normal plant level of magnesium is 0.2%−0.8%, and deficiency is not seen until it drops to <0.05%. Plants require magnesium in ATP metabolism, in activating several enzymes, and as a component of chlorophyll, so a lack of healthy color between leaf veins might indicate a lack of magnesium. Sandy soils are more likely to be magnesium-deficient than heavier soils. Some commercial mineral preparations specifically have elevated magnesium levels for this purpose. A magnesium deficiency causes "grass tetany" in cattle more frequently than in horses, although lactating mares are more susceptible than less-productive animals. Binding of magnesium by potassium or calcium may reduce its availability (Frape, 2010).

Manganese (Mn) is involved in protein synthesis, maintaining chloroplast membranes, and enzyme systems in plants. Animals need manganese for bone matrix formation. Typical foliage concentrations of manganese are 30−300 ppm, and deficiencies can develop in plants at 20 ppm.

Molybdenum (Mo) is involved in nitrogen fixation, so it is primarily important to legumes, which eventually leave nitrogen in the soil for grasses. Normal plant level is 1−100 ppm, and they do not suffer deficiency until it drops to <0.2 ppm. Molybdenum is more available in alkaline than acidic soils, and adding lime will thus improve molybdenum availability (Murphy, 1998). Since amounts of molybdenum recommended per acre are only fractions of a pound, molybdenum should be blended into other fertilizer components by the fertilizer dealer.

Nitrogen (N) is needed by plants to make amino acids, which are metabolized to proteins, neurotransmitters, and other necessary molecules, including the purines and pyrimidines in the nucleic acids. Maintaining legumes in the range of 10%−25% can provide all of the nitrogen that a pasture needs for a reasonable growth rate. When seeding legumes, the appropriate variety of **Rhizobia** bacteria, sold as an **inoculant**, should be purchased, stored, and applied as directed to ensure a high level of nitrogen fixation.

Phosphorous (P) concentrations in forages vary from about 0.2% to 0.4%. Phosphorus is used in energy metabolism, as part of phospholipids in cell membranes, to activate many molecules (e.g., ADP and ATP), as a component of nucleic acids, and as a structural component of bone. Once phosphorus concentration is high enough for good forage health and growth, horses will return most of the phosphorus they eat through their manure. So, if most manure is harrowed and spread evenly and there are not many pasture-raised horses that leave the farm, the recycling of phosphorus will continue without interference. Once minimum requirements of calcium and phosphorus have been met, the total dietary ratio of calcium:phosphorus for horses should vary from about 1.1:1 up to about 6:1 (Lewis, 2005).

Potassium (K) is a major component of some soil types, and where this occurs, there is much more potassium than horses normally require (Frape, 2010). Several aspects of plant physiology require potassium, including enzyme activation and synthesis of starch and proteins. While 2.0%−2.5% potassium is common in plants, some grasses can have 5%. Clay soils tend to be high in potassium. Much of the potassium consumed by herbivores is returned to the land through urine.

Selenium (Se) is deficient in the soils of much of North America and is present at high levels in other areas. It should be at least 0.2 ppm in forages and 0.3 ppm in total diets (Murphy, 1998). While a role for selenium in plants is questionable, it is required by animals for glutathione peroxidase function, integrity of cell membranes, and muscle development. Selenium and vitamin E work together as antioxidants. Raising soil pH helps to make selenium more available. Soils and the forages and grains that grow in them are relatively deficient in selenium through most of the eastern third of North America, extending through the subarctic and arctic regions and down the west coast to northern California. An irregular wedge of land containing adequate soil selenium concentration extends from Alabama, west to southern California, and north to the mid latitudes of Alberta, Saskatchewan, and Manitoba. There are exceptional pockets of variable selenium content throughout this high selenium zone. Green forages have relatively high levels of selenium compared to grains. Clinical deficiencies of selenium and/or vitamin E are more common in foals than adult horses and include inadequate immune response to pathogens, myopathy (lack of muscle development), and fat inflammation (Lewis, 2005).

Sodium (Na) and chlorine (Cl) requirements of animals are generally low in forages but easily met by a forage diet supplemented with free-choice loose NaCl salt. Sodium works with potassium in osmotic balance across cell membranes in plants and animals. Chlorine has roles in photosynthesis and maintaining osmotic pressure. Plants function with 100−200 ppm sodium and 500−10,000 ppm chlorine, but these two minerals are not required. Blocks of salt are not consumed at as high a rate as loose salt, possibly because they require more effort and could cause discomfort. Sources of salt and minerals should be moved around on pastures to prevent hoof traffic from trampling too much forage. Placing salt and mineral feeders in weedy patches can help to weaken and kill some weeds from trampling. Adult horses need about five pounds of salt per month (Murphy, 1998).

Sulfur (S) concentration in forages ranges from about 0.13% to 0.34%. Two amino acids, cysteine and methionine, need sulfur as a component. Organic sources of sulfur in soil must become mineralized before they are available to plants. Superphosphate fertilizer and gypsum (calcium sulphate) are reasonable sources of sulfur for soil, but triple superphosphate has no sulfur in it (Murphy, 1998).

Zinc (Zn) concentration is normally 10−100 ppm in forages and is considered deficient at <10 ppm. Its main function is enzyme activation in plants and animals. The plant enzymes promote starch formation and seed maturation. Zinc deficiency can cause leaves to be white or striped.

General improvement of pasture in Australia increases uptake of Cu, Zn, Mn, P, Ca, and Mg (Frape, 2010). The parent material of soils will have a large influence on both macro-and micromineral pools available to plants. Soil pH influences the rates of minerals dissolving from soil into groundwater. Low amounts of rainfall will prevent adequate levels of some minerals from reaching plant roots, and excess rainfall can cause dilution of minerals in soil water. Agricultural extension agents in North America have maps of soil types and information on nutrient concentrations of soils in their region. Collecting and analyzing soil samples from your property can reveal more precisely the concentrations of various nutrients in the rooting zone. Trends in concentrations of cobalt, copper, manganese, and selenium, for example, are known to exist in different parts of a continent. Applying trace minerals to the soil is more efficient for some minerals than others. Foliar sprays (onto foliage) of mineral solutions can provide more successful supplementation to forage plants. Intraruminal boluses have been used to slowly release deficient minerals in sheep in Australia, but such boluses would not likely remain in the stomach of a horse. Intramuscular injection of selenium provides a reasonable supply for up to a few months. Provision of a powdered multimineral supplement, separate from but fed next to loose salt, is a reasonable way to supply minerals. It allows a pasture manager to monitor the rate of consumption of both the salt and the mineral products, especially if the products are dispensed from a rain- and snow-resistant feeder. Growth rates of plants are very responsive to nitrogen and mineral levels in soil. In addition to monitoring growth rates, nutrient analyses of forage samples provide indications of which nutrients may be in short supply.

Plant concentrations and deficiency conditions in forages and animals were summarized for several minerals by Barker and Collins (2003). This could be a useful resource for a pasture manager who needs to research a troublesome mineral imbalance.

NUTRIENT ANALYSIS OF FEEDS

The plant material used as feed for horses, whether it is pasture, hay, or grain, is composed of a combination of organic and inorganic matter. The organic component is created by organisms and consists mostly of carbohydrates, lipids, and

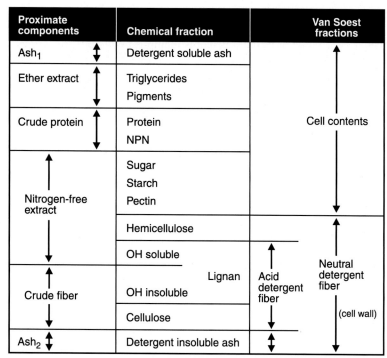

Proximate components	Chemical fraction		Van Soest fractions	
Ash₁ ↕	Detergent soluble ash			
Ether extract ↕	Triglycerides / Pigments			
Crude protein ↕	Protein / NPN		Cell contents	
Nitrogen-free extract ↕	Sugar / Starch / Pectin			
	Hemicellulose			
	OH soluble			Neutral detergent fiber
Crude fiber ↕	OH insoluble — Lignan	Acid detergent fiber ↕		
	Cellulose			(cell wall)
Ash₂ ↕	Detergent insoluble ash	↕		

FIGURE 3.15 Components and fractions of forage dry matter. *Reproduced with permission from Schroeder, J.W., 2013. Forage Nutrition for Ruminants. North Dakota State University Extension Publication AS-1250. http://www.ag.ndsu.edu/pubs/ansci/dairy/as1250w.htm. North Dakota State University Extension Service (NDSUES).*

proteins. Chemists and nutritionists developed methods over several decades to analyze both major and minor components of feedstuffs, so nutritional quality could be compared and discussed. Fig. 3.15 shows three columns of chemical fractions of forages that have been of interest as analysis techniques have continued to be developed. A modern forage analysis will include ether extract (representing lipids) and CP from the first column. The crude fiber analysis was one of the first chemical tests developed, but it was incomplete, failing to capture all of the lignin and none of the hemicellulose. Thus determining the nitrogen-free extract by subtraction was uncertain. Specific analyses from the second column are not often carried out, but some of these fractions can be determined by subtraction. The third, "Van Soest fractions" column is relied upon to determine cell wall components (including acid detergent fiber and neutral detergent fiber), which relate to digestibility and voluntary intake of the feed, plus cell contents (sometimes called cell solubles). The procedures used in modern feed-testing labs are chosen for practical reasons such as ease of analysis, reliability, cost, and relevance to nutrition.

The first division of the feed is into moisture versus dry matter. For air-dried feeds in most of North America, 10% −15% moisture is common. In Death Valley, where humidity is very low, expect 10% or less, and on the west coast of Washington, expect up to 15%, due to high humidity. The dry matter fraction includes all of the organic compounds that have been discussed so far (carbohydrates, lipids, and proteins) plus some found in much smaller quantities (DNA, RNA, vitamins, organic acids, pectins, β-glucans, etc.) and all the minerals, expressed as Ash in Fig. 3.15. The inorganic or ash fractions (soluble and insoluble) contain elemental minerals, such as calcium, iron, magnesium, and their salts. Each mineral can be determined individually.

The only way that two or more feedstuffs can be fairly compared for their nutrient concentrations is on a dry matter basis. The 10%−15% moisture of air dry feeds means that the dry matter content of these feeds is $100 - 15 = 85\%$ to $100 - 10 = 90\%$. If we have a need to compare the protein concentration of a cool season, mature grass hay that is 85% dry matter and 11.9% protein on an "as is" or "as fed" basis against the protein concentration of a ryegrass/white clover pasture that is 20% dry matter and 3.2% protein, we need to do some arithmetic. The concentration of the nutrient is expressed on a dry matter basis by dividing its concentration by the percentage moisture in that feed (expressed as a decimal). On a 100% dry matter basis, the grass hay has $11.9\%/0.85 = 14\%$ protein. The ryegrass/white clover pasture has a protein concentration of $3.2\%/0.2 = 16\%$ on a dry matter basis.

Some of the organic matter contains proteins and other compounds that contain nitrogen. Chemists developed a technique to measure the nitrogen concentration in organic materials. Since the average protein has 0.16 g of nitrogen per gram of protein, the nitrogen concentration of feed dry matter is multiplied by $(1.0/0.16) = 6.25$ to give the percentage of CP on a dry matter basis. For example, 2.56% nitrogen $\times$ 6.25 = 16% CP. The word "crude" is used to describe the protein because there are amino acids, nucleotides, enzymes, and a few other nitrogenous compounds in small amounts that contribute to the nitrogen concentration. If nutritionists need to examine protein content in more detail, they can analyze concentrations of individual amino acids. One amino acid, "lysine" is considered the first limiting amino acid in horse nutrition, which is the first amino acid to become deficient in horses. Thus, it is essential to supply at least a minimum required amount in horse diets.

The carbohydrates can be broken down by several techniques to a variety of fractions. The nitrogen-free extract (NFE) term refers to sugars, starches, some water-soluble vitamins, and small amounts of other cell wall components such as hemicellulose and lignin. NFE is rarely used in modern horse nutrition.

Crude fiber (CF) is a term commonly used to describe human foods and some commercial concentrate feed products. It is not used much in nutrition of herbivores because it does not precisely measure one chemical fraction. CF mostly consists of cellulose plus variable proportions of other polysaccharides, lignin, and insoluble ash (Jung, 1997; Schroeder, 2013).

The cell contents and cell wall components that can be analyzed by the detergent system are shown in Fig. 3.16. A detailed discussion of the relevance of these plant fractions to horse nutrition is available in NRC (2007b).

DIGESTIBILITY OF FEEDS

The proportion of a feed that animals ingest and that is absorbed from the digestive tract is **digestibility**. Trials to determine digestibility involve analyzing the nutrients in a feed sample, weighing an amount of feed that is then eaten by an animal, collecting the manure, and analyzing the nutrients in the manure. The difference between amounts of organic matter or nutrients fed and amounts detected in manure represents the apparent digestibility of the organic matter or nutrients.

For example, to determine the percentage digestibility of the organic matter (OM) in a feed, use the following formula:

$$((\text{OM in feed} - \text{OM in manure}) \div (\text{OM in feed})) \times 100\% = \text{OM digestibility \%}$$

If there were 10 kg of OM in a ration fed to a horse, and the resulting manure contained 2 kg of OM, the digestibility of OM is $((10-2)/10) \times 100\% = 80\%$. This is an **apparent digestibility** because some of the OM or nutrient of interest that is found in the manure may be contributed by the horse's digestive tract, perhaps consisting of old cellular material.

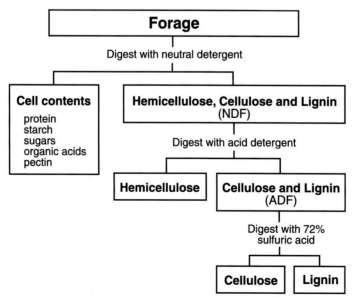

FIGURE 3.16 A schematic diagram of the detergent system of forage analysis. *Reproduced with permission from Schroeder, J.W., 2013. Forage Nutrition for Ruminants. North Dakota State University Extension Publication AS-1250. http://www.ag.ndsu.edu/pubs/ansci/dairy/as1250w.htm. North Dakota State University Extension Service.*

Link Between Digestibility and Energy Evaluation of Feeds

Feed energy is not a chemical like carbohydrates, vitamins, or minerals. Energy is the ability to do work. Carbohydrates, lipids, and proteins can all be used to supply energy to animals that consume them, mostly from breaking carbon—carbon bonds. The greatest mass of organic material that supplies energy to herbivores is supplied by carbohydrates. Lipids have greater energy density (by 2.25 times) than carbohydrates, but their mass in horses' diets is much smaller. The digestion of proteins provides valuable amino acids for animal metabolism but very little energy. Movement, thermoregulation, digestion, and stimulating reactions are all body processes that need energy from feed.

A historic expression of the energy contained in feeds and needed by animals is **total digestible nutrients** (TDN), which can be expressed as pounds, kg, or percentages. TDN is determined by measuring concentration and digestibility of the major nutrients and placing them in a formula. The TDN formula is TDN = (digestible crude protein) + (digestible crude fiber) + (digestible nitrogen-free extract) + (digestible crude fat × 2.25). Notice the 2.25 × fat to account for its higher energy density than other organic compounds. Determining TDN on a novel feed sample is a great deal of work in a lab and in digestibility trials. Over time, faster, more practical measures of feed energy were developed.

USEFUL ENERGY TERMINOLOGY

Samples of grass, legumes, grains, and any other plant material can be place in a device called a bomb calorimeter (Fig. 3.17) to determine their **gross energy** concentration. This is the amount of energy released when a sample of about 1 g of dry OM is placed in a small bomb chamber containing oxygen and ignition wires. The sample is ignited, and the heat produced by the explosion increases the temperature of about 2 L of water surrounding the bomb chamber. The water is stirred, and the rise in temperature is recorded, then used in a formula to determine the number of calories of heat energy produced.

One calorie is the amount of heat energy required to raise 1 g (1 mL or cubic centimeter) of water one degree Celsius from 19.5 to 20.5°C. One kilocalorie (kcal) equals 1000 calories (a kilogram calorie), and one megacalorie (mcal) equals

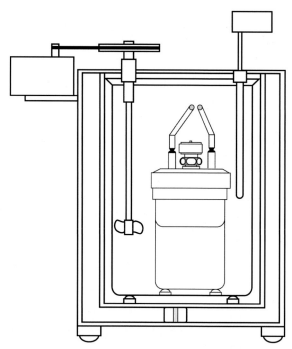

FIGURE 3.17 A **bomb calorimeter**, for determination of gross energy of organic materials. The outer wall has two layers plus some insulation and the inner layer contains water. A motor on the left spins a pulley and rotates a shaft containing a propeller to mix the water to an even temperature. The inner vessel is a combustion chamber in which a dry sample is placed and pure oxygen is added. Ignition wires (appearing cut off) carry an electric current to explode the sample. The change in water temperature is determined by a thermometer (on the right) and the temperature change correlates to the number of calories of gross energy. *Source: https://commons.wikimedia.org/wiki/File:Bomb_Calorimeter_scheme.png. Author: Ichwarsnur. Creative Commons Attribution-Share Alike 4.0 International license.*

one million calories. Energy values of human food are frequently expressed as "Calories" (equivalent to Kcal) as opposed to "calories." The calorie as defined here is equivalent to 4.182 J.

Animals cannot use all of the gross energy in a feed because it is not all digestible. Stomach acid, enzymes from the pancreas and small intestine, and intestinal microbes all contribute to digestion, but some manure leaves the animal, and its gross energy concentration can be determined in a bomb calorimeter. Likewise, the energy lost in urine, gases, and heat can be measured. If the energy lost in feces is subtracted from the gross energy of a feed, what is left is called **digestible energy (DE)**. This is a more useful term for nutritionists and the feed industry than gross energy and is the standard energy term for horses in North America. If the energy lost in urine and gases is subtracted from DE, what is left is called **Metabolizable Energy (ME)**, and this is commonly used in analyzing feed for pigs and designing their diets. If energy lost in the form of heat is subtracted from ME, the remainder is called **Net Energy (NE)**. NE is further subdivided into the energy required for maintenance of the body (NE_m) and energy required for production, either NE_L for lactation or NE_G for gain of weight. The use of ME and NE systems allows more precision in formulating diets, but it also requires more information.

Table 3.1 lists daily requirements for DE and CP in a mature 500-kg mare depending on her physiologic status and level of activity. The lowest requirements for DE and CP are just to maintain the mare's body weight and condition score. For this table, an average maintenance level was chosen. A minimum maintenance can be chosen for "horses with a sedentary lifestyle, due either to confinement or to a docile temperament." An elevated maintenance requirement can be chosen for "horses with nervous temperaments or high levels of voluntary activity" (NRC, 2007e). Numbers in parentheses in Table 3.1 are percentages above the daily maintenance requirements. For example, light, moderate, and heavy levels of exercise raise the DE daily requirement by 20%, 39%, and 59%, respectively.

The analysis of feeds and design of rations (diets) for beef and dairy cattle in Western countries depends on the NE system. The high nutrient demands of beef feedlots and lactating dairy cows justify this level of complexity and precision. For the horse industry in North America, there is not yet enough information available from research to convert from a DE to a NE system. However, in France, the NE system is used for horses. NE values of common feeds in France are related to the NE value of a standard, well-accepted feed (barley grain), allowing users to compare and substitute feeds based on barley. When enough information becomes available, more countries might also adopt NE systems for horse nutrition.

CHEMICAL ANALYSIS OF PLANTS

Horse trainers and nutritionists who formulate and analyze horse rations learn to collect representative samples of all feeds available to them, then send the samples to feed analysis laboratories for nutrient analyses. A typical nutrient analysis for horse diets can include over 30 components. The following basic analyses should be requested:

- % dry matter
- % crude protein
- % acid detergent fiber

TABLE 3.1 Daily Digestible Energy and Crude Protein Requirements of a 500-kg Mare, Depending on Physiologic Status and Level of Activity (NRC, 2007e)

Physiologic State and Level of Activity	Daily Digestible Energy Requirement (Mcal) and (Percent Above Maintenance)	Daily Crude Protein Requirement (Grams) and (Percent Above Maintenance)
Maintenance, no work, average maintenance requirement[a]	16.7	630
Light exercise	20.0 (20)	699 (11)
Moderate exercise	23.3 (39)	768 (22)
Heavy exercise	26.6 (59)	862 (37)
Pregnancy, fifth month	16.7	630
Pregnancy, 11th month	21.4 (28)	893 (42)
Lactation, first month	31.7 (90)	1535 (144)

[a]*Alert temperament and moderate voluntary activity.*

- % neutral detergent fiber
- % water-soluble carbohydrates or % ethanol soluble carbohydrates
- % crude fat
- % ash
- % calcium
- % phosphorus
- % magnesium (samples collected in spring may be deficient)

The following should be requested as a calculated value:

- horse DE (Mcal/lb) or horse TDN %

The development of equations to determine DE concentrations in feeds is discussed in NRC (2007e).

In geographic areas where certain minerals are known to be deficient or excessive in the soil, it is worth having these analyzed in forages and grains. Vitamin deficiencies are rare if the bulk of a ration is well-managed forage, but if any signs of deficiencies are noticed or are common in the region, it would be advisable to test for vitamin levels. If it is probable that excess nitrate from fertilizer, manure, or a septic system entered the soil where your forages grow or if there was a sudden frost or drought soon after nitrogen fertilizer was applied to forages, they should be tested for nitrate concentration and the result discussed with a nutritionist or veterinarian. Owners of horses that are prone to equine metabolic syndrome and/or laminitis are advised to have all three of the soluble carbohydrates tested: water soluble, ethanol soluble, and starch. Before choosing a feed testing lab, it is worth learning which ones will calculate DE for horses and all the analyses that your nutritionist or veterinarian wants to see. The analyses suggested before can be used by a nutritionist, along with accurate information on a horse, such as its lifestyle [sedentary: minimum voluntary activity; average: alert temperament and moderate voluntary activity; elevated: nervous temperament or high level of voluntary activity], weight, age, physiologic condition (idle, light, moderate, heavy, or very heavy exercise), month of pregnancy, month of lactation, and rate of growth. The daily nutrient requirements of individual horses can be determined from the aforementioned information and tables supplied from a source such as the National Research Council's "Nutrient Requirements of Horses," 2007, or in computer programs containing this data. The amounts of various available feeds, for which you have nutrient concentrations, can be matched with the nutrient requirements to determine how much of each feed ingredient a horse should receive in a day. This process is called formulating a ration or diet.

Some feed testing laboratories are connected to universities or extension services, and some are independent commercial businesses. An example of a commercial feed testing lab that specializes in analyzing feed for horses is Equi-Analytical Laboratories (http://equi-analytical.com/). A variety of different nutrient analysis packages are available, with the analysis of individual components available at extra cost (mostly $5 to $10). Forage analysis packages can vary in the range of US $18 to $80.

In attempts to quantify all of the components of plant cells, there is a distinction between cell contents and cell walls. Cell contents include proteins, starches, sugars, lipids, some minerals, organic acids, and some pectins. Cell walls contain hemicellulose, cellulose, lignin, some pectin, and some minerals. Unfortunately, analyzing cellulose, hemicellulose, lignins, pectins, and sugars is not straightforward.

A system of analyzing plant cell wall components was developed by Van Soest in the 1960s, and it is still in common use today. The first step in the Van Soest method is to dissolve a dried and ground plant sample in a neutral (neither acidic nor basic) detergent solution. The cell contents are largely dissolved in the neutral detergent solution and are called **neutral detergent solubles**. The solution and remaining plant material are passed through a filter to separate the solubles from **neutral detergent fiber (NDF)**, which is largely cell wall materials. The NDF fraction contains cellulose, lignin, and most of the hemicellulose. The NDF fraction is dried and weighed to determine the %NDF. A subsample of plant NDF is digested in an **acid detergent solution**, and residue is collected on a filter, and then weighed. The residue is called **acid detergent fiber (ADF)**, which includes lignin, cellulose, silica, insoluble forms of nitrogen, and some variable amounts of minor components, but not hemicellulose. Estimated hemicellulose concentration is determined by subtracting ADF from NDF (Figure 3.16). Lignin is not digestible by animals or by bacteria commonly found in digestive systems of herbivores, but hemicelluloses and cellulose are partially fermented by some herbivore gut microorganisms (Jung, 1997; Schroeder, 2013; NRC, 2007c; AAFCO, 2017). One diagram summarizing fractionation of plant carbohydrates and related compounds can be found in NRC (2007b). NDF analysis has replaced CF analysis for forages, but CF is still in common use for grains and for human food, even though it recovers only a part of the polysaccharides and lignin and thus severely underestimates the total plant cell wall content (Jung, 1997).

Uses and Relative Advantages of NDF and ADF Measurements

NDF is the best available estimate of cell wall components, but it underestimates cell wall concentration because several minor components (fructans, gums, mucilages, β-glucans, and most pectin) are solubilized and not captured. Recovery rates of lignin, cellulose, and the ADF component in NDF were 91%–95% in the 1960s (Colburn and Evans, 1967). These components are not digested by mammalian digestive enzymes but are rapidly fermented by intestinal microbes. This is a problem for pectin-rich legumes. NDF can be used in equations to calculate digestibility, TDN, and NE. Voluntary consumption of forages declines as NDF levels rise, so it is a good predictor of DM intake.

ADF represents the least digestible fiber portion of forage. The higher the ADF concentration is, the lower the energy concentration. DE concentration of forages is determined by using %ADF and %CP in a formula, such as this one stated in NRC (2007d): DE (Mcal/kg) = $4.22 - 0.11 \times (\%ADF) + 0.0332 \times (\%CP) + 0.00112 \times (\%ADF^2)$. NRC uses a different formula for determining DE concentration in energy feeds and protein supplements, without inclusion of CP%. Commercial laboratories may use different formulas based upon recent research and availability of other analyses such as % hemicellulose, % acid ether extract, % CF, and % NFE.

Different formulas for DE have been suggested by Pagan (1998) and Zeyner and Kienzle (2002), and these are described in detail in NRC (2007e). These include more nutrient components than just ADF and CP. However, Pagan's formula requires analysis of hemicellulose, fat, NSC, and ash, while the formula developed by Zyner and Kienzle requires analysis of CP, acid ether extract, CF, and NFE. These formulas do not have great advantages over the DE formula used by NRC since 1989, and performing these extra analyses adds time and expense to the determination of energy concentration.

The Van Soest detergent methods were compared to more modern enzymatic-chemical methods for analyzing cellulose and nonsoluble hemicellulose in forages, and the results were considered equivalent (Hindrichsen et al., 2006).

Further Uses of ADF and NDF

Agronomists and dairy nutritionists use a term called **relative feed value (RFV)** to express intake potential and digestibility of a forage in one number. RFV uses ADF to calculate the amount of **digestible dry matter (DDM)** in a forage plus the dry matter intake potential as a percent of body weight from NDF.

DDM = digestible dry matter = $88.9 - (0.779 \times \%ADF)$

DMI = dry matter intake (% of body weight (BW)) = $120/(\%NDF)$

RFV = $(DDM \times DMI)/1.29$

While RFV was developed in studies of ruminant nutrition and is not in common use for developing horse diets, recognizing what it represents can be useful when searching listings of batches of hay for sale. Animals fed forages of identical RFV can sometimes perform differently, perhaps because of differences in digestibility of the NDF. To account for differences in digestibility, the **relative forage quality (RFQ)** index was developed.

RFQ = (DMI as a % of BW) × (TDN as a % of DM)/1.23

DMI is determined from a formula using NDF and NDF digestibility.

Use of TDN is less common lately, due to requirements to determine digestibility of the fiber, protein, lipid, and carbohydrate components of feedstuffs and the fact that animal performance can be predicted more accurately using NE than TDN. However, average digestibilities for nonfiber carbohydrates (NFC), CP, fatty acids (FAs), and NDF have been developed, and formulas were developed for predicting TDN from these measures. In the following formula for TDN, "d" refers to "digestible."

TDN = (dNFC + dCP + dFA*2.25 + dNDF-7)/1.23.

Further details of these calculations are available from Moore and Undersander (2002) and Garcia (2017).

Data from hay auctions reveals that the premium for higher quality forage is worth $0.90/ton as RFQ increases from one value to another. The range of RFQ values for forages includes Utility grade at RFQ < 90, Standard grade at 90–110, Choice at 140–160, and Supreme at >185 (Saha et al.,). A two-variable model of equine forage DM digestibility was developed, using NDF and CP concentrations on a limited selection of forages, and it will likely be further developed to cover a broader range of forages (Hansen and Lawrence, 2017).

Ruminant nutritionists are also working with variability in the digestibility of NDF in their attempts to refine energy calculations. During passage through a ruminant digestive system, some of the NDF will not be digested. This can be called undigested NDF (uNDF), and it represents undigested residue after fermentation for a given length of time, from 24 up to 240 h. Another refinement in use is to express uNDF on an OM basis, since variable amounts of ash (soil minerals) are contained in feed samples. These refinements have not yet been developed for nonruminants.

In summary, as plants mature, their cell walls thicken with buildup of cellulose, hemicellulose, pectins, and lignins, and these can be measured indirectly by the ADF and NDF analyses of plant fiber developed by Van Soest. ADF and NDF values can be used in formulas to estimate energy concentration and digestibility of forages and to rank their voluntary intake by herbivores. Some associations that organize hay auctions will provide nutrient analyses (usually CP, ADF, and NDF) of large lots of forage for sale.

Alternative Methods of Analysis of Forage Components

Near-infrared reflectance spectroscopy (NIRS) involves subjecting dried, ground forage samples to near-infrared light. The process is quicker, less costly, safe, and accurate compared to wet chemistry methods. NIRS is used by many commercial labs to measure protein, ADF, and NDF. It is not suitable for minerals, since they do not absorb light energy in the near-infrared spectrum.

Factors Influencing Forage Dry Matter Intake

A range of estimates of voluntary dry matter intake (VDMI) is provided in NRC (2007a,b,c,d,e,f,g), and it is 1.5%−3.1% of BW. Age, weight, body condition score, and physiologic demands such as body maintenance versus growth, pregnancy, lactation, and work affect a horse's appetite. An older horse that is just maintaining itself, not pregnant, lactating, or working, only needs 1.5% of its BW in FDM per day. Lactating mares consumed about 2.8% of their BW per day in some studies. A collection of studies reviewed by NRC (2007d) indicated that if pasture DE concentration is at least 2.4 Mcal/kg and CP is 10.5% or more, lactating mares will eat about 2.4% of the BW per day and will not require supplementation of either energy or protein. Irrigated pastures have been known to have such a high water content that they could support nonproductive mares but not lactating mares. Later, when the same pastures were not irrigated, they could also support lactating mares (NRC, 2007e).

Aspects of forage quality also influence intake. Generally the higher the forage quality is, the higher the intake. Chapter 10 discusses behavioral aspects of feed intake. Forage quality measures that are reported to increase VDMI include sugar content, higher CP, and reduced cell wall content. These trends make forages more attractive to eat, but they might also mean that horses reach a point of satiety sooner when eating higher quality forages. Growing horses may have positive or negative VDMI responses to increasing forage quality, and shifts in the response may be attempts to maintain constant DE intakes (NRC, 2007e).

Cool season grasses tend to have higher CP and DE percentages than warm season grasses, such as Bermudagrass. This is illustrated in Table 3.2. Warm season grasses have more-lignified cell walls than cool season grasses, and this causes the warm season grasses to be less digestible (NRC, 2007e). As bluegrass/alfalfa swards grew from 11 cm (4.3 in) to 47 cm (18.5 in), digestibility of DM decreased from 73% to 52%. Seminatural grassland pastures had 61% digestibility in May and 53% in July (NRC, 2007e). DE and digestible protein levels of forages also decline as plants mature. Dry matter digestibilities of hay were reported as low as 30% for mature threshed grass hay by mature ponies and as high as 63% for alfalfa fed to yearling quarter horses (NRC, 2007e).

Along with the factors mentioned before, appropriate stocking rates and forage DM yields interact to influence the carrying capacity of pastures. FDM yield of horse pastures has varied from as low as 1 ton/ha and in reseeding and fertilizing trials from 4 to 14 tons/ha (Hopkins et al., 1990; NRC, 2007e). Fields that produce increased FDM yields after receiving nitrogen fertilizer can probably supply the nutrients required for horses that are lactating or rapidly growing, at higher stocking densities (more horses per ac or ha), or for more grazing days per season.

CONCENTRATIONS OF NUTRIENTS IN TYPICAL PASTURE PLANTS

North American horse nutritionists depend on the National Research Council's 2007 publication, "Nutrient Requirements of Horses, Sixth Revised Edition," 2007, for much of the information they use when formulating diets for horses. NRC (2007g) provides a table of "Nutrient Composition of Selected Feedstuffs," and it is also available through the NAP website at https://www.nap.edu/read/11653/chapter/18#304. Feedstuffs are grouped as Concentrates, Forages, and Fats and Oils. For each individual feedstuff, the following nutrient concentrations are given in percentages: dry matter, CP, lysine, fat, NDF, ADF, Ash, Ca, P, Mg, Cl, K, Na, and S. DE is listed as Mcal/kg of dry matter. Several micromineral concentrations are listed in mg/kg. Within the forages section, the only pasture that is mentioned specifically is "Legume Forage Pasture, Vegetative." The "immature" and "vegetative" categories of grass, legume, and mixed forages provide reasonable approximations of the concentrations in pasture plants as they mature. While the concentrations of PUFAs are

TABLE 3.2 Concentrations of Nutrients in Forages (Based Upon Stored Forages) (NRC, 2007a,b,c,d,e,f,g)

Feeds	DM %	DE Mcal/kg DM	CP % of DM	Lysine % of DM	Fat % of DM	NDF % of DM	ADF % of DM	Ca % of DM	P % of DM
Legume hay, immature	84.2	2.62	20.5	1.05	2.1	36.3	28.6	1.56	0.31
Legume hay, mature	83.8	2.21	17.8	0.89	1.6	50.9	39.5	1.22	0.28
Mostly legume hay, mid maturity	84.2	2.35	19.1	0.9	2	47.2	35.4	1.17	0.3
Mostly grass hay, mid maturity	87.3	2.19	17.4	0.68	2.6	55.1	36.4	0.88	0.36
Grass hay, cool season, mature	84.4	2.04	10.8	0.38	2	69.1	41.6	0.47	0.26
Grass hay, cool season, immature	84	2.36	18	0.63	3.3	49.6	31.4	0.72	0.34
Grass pasture, cool season, vegetative	20.1	2.39	26.5	0.92	2.7	45.8	25	0.56	0.44
Bermudagrass hay, coastal	87.1	1.87	10.4	0.36	2.7	73.3	36.8	0.49	0.27

expected to be lower in the dry hays than in growing pasture, the nutrient concentrations presented are reasonable approximations of what could be expected of pasture plants at the maturity levels given. Table 3.2 contains data from the NRC source on grass, legume, and mixed forages and reveals the following trends in nutrient concentrations:

- As plants mature, DE concentration decreases slightly, CP and lysine decrease significantly, NDF and ADF increase, Ca and P generally decrease.
- Legumes, relative to grasses, have slightly higher DE, significantly higher CP, lysine, and calcium, and lower NDF, ADF, and phosphorus.
- The popular warm season coastal Bermudagrass, compared to cool season grasses, has lower DE, CP, lysine, calcium, and phosphorus but higher NDF.

It is also worth looking at graphic representations of these concentrations, along with relative proportions of leaves and stems in forage plants. Fig. 3.18 shows how the weight of dry matter and other components accumulate in a grass plant as it matures from a young, leafy stage to a bloom stage. Dry matter increases in a sigmoid pattern, then peaks at the bloom stage, and declines as leaves dry out and drop off after flowers have turned into seed heads. Cellulose and other fibrous components have a less steep increase than dry matter and reach a plateau shortly after the bloom stage. Nonstructural carbohydrates, which are largely inside plant cells, reach their peak weight between the boot and heading stages. Protein content has a relatively shallow slope of accumulation until a plateau between the heading and bloom stages. It starts to decline after the NSC and shortly before dry matter. The one component that continues to accumulate through all four of these stages and until the plant becomes dormant is lignin, which is the tough, indigestible binder within secondary cell walls.

A slightly different pattern is seen when plant components are expressed as percentages of the total dry matter weight (Fig. 3.19). Leaves are initially at their highest proportion of the plant and protein concentration is maximal at the same time, since leaves have a higher protein concentration than other shoot parts. Enzymes, including rubisco, are major components of leaves and plant protein. Relative amounts of leaves and CP decline at the same rate with growth stage advancement. This is a good reminder that leafy grass plants tend to have the highest CP concentrations. Mineral concentrations follow a similar pattern to leaves and CP. Stems are required for structural strength as well as for conducting water and dissolved substances within plants. Thus, the cell walls of stem tissue accumulate the largest percentages of fiber and lignin as a plant ages.

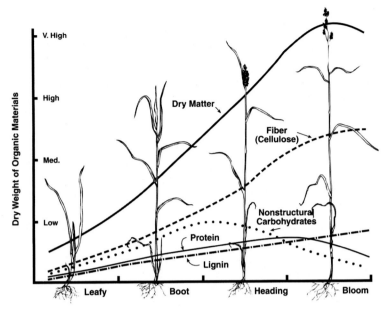

FIGURE 3.18 Changes in plant composition of growing grass plants from leafy through bloom stages (White and Wolf, 2009). The weights of nonstructural carbohydrates and proteins peak in mid-maturity, then decline as the plants mature, while the weights of fiber, lignin, and total dry matter increase to the bloom stage. *Reproduced with permission of Virginia Cooperative Extension.*

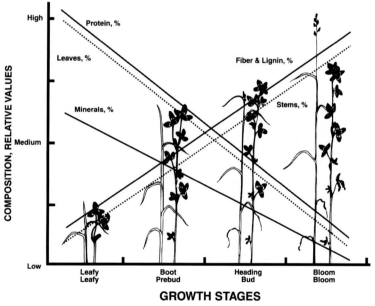

FIGURE 3.19 The relative composition of chemical protein and minerals declines with decreased leaf percentage, and the composition of fiber and lignin increases with an increase in stem percentage as perennial grass and legume plants mature (White and Wolf, 2009). *Reproduced with permission of Virginia Cooperative Extension.*

DIGESTION OF PLANT NUTRIENTS BY HORSES

Rather than a complete explanation of the anatomy of the digestive system and the physiology of digestion in horses, this section highlights digestion of carbohydrates and focuses mainly on the cecum. Horses are herbivores, which are required to digest structural carbohydrates (primarily fibrous material) from plant cell walls, while not being able to make the enzymes to do the job. Horses evolved a system with a small stomach and a large postgastric region, containing a wide variety of microbes that do produce the enzymes for digesting cellulose, hemicellulose, and the other components of plant cell walls, with the exception of lignin.

Carbohydrate Digestion in the Stomach and Small Intestine

The stomach of a horse occupies about 10% of the gastrointestinal tract. When this fact is considered in light of the many hours (up to 17) that a horse will graze in a day, it makes sense that most feed moves through the stomach rapidly. However, some digesta can remain in the stomach for 2—6 h (Frape, 2010). Major secretions of the stomach into its lumen include hydrochloric acid and pepsin, and the hormone gastrin is secreted into the bloodstream. Fermentation by *Lactobacilli* and *Streptococci* yields lactic acid in the esophageal and fundic parts of the stomach. Bacterial populations increase over a period of at least 3.5 h following a single meal. Other microbes use some of the lactic acid to produce volatile FAs, of which acetate is the majority (Varloud et al., 2007). The pH of chime (the liquid phase of stomach contents) falls after a meal and in the direction of the pylorus. Layering of stomach contents may contribute to the change in pH from esophageal and fundic to pyloric regions. Pepsin activity is much greater toward the pylorus than upstream. Smaller particles of forages pass through the stomach faster than larger particles.

The small intestine of a 450-kg horse is about 21—25 m long, and some digesta can pass all the way through it, to reach the cecum in 30—45 min (Frape, 2010). Starches from grass seeds, grains, and leaves are digested in the small intestine by amylases from the pancreas and α-glucosidase from the intestinal mucosa. Horses evolved on diets that are relatively low in starch, so the concentration of α-amylase in pancreatic juice is only about 5% of what is in the pig. Alpha-amylase cleaves α-1,4 glycosidic bonds in starches, resulting in products containing two or three glucose units or five to six glucose units if a branch in the starch molecule is involved. Alpha-glucosidases remove individual glucose units from these two or three glucose fragments and oligosaccharides (starch fragments with 4—20 glucose units). One glucosidase is sucrase, which splits sucrose into glucose and fructose. Another is maltase, which breaks maltose into α-D glucose and β-D glucose. Material leaving the small intestine for the cecum contains fibrous feed, starch, protein, microbes, cell debris, and intestinal secretions.

Carbohydrate Digestion in the Cecum and Colon

Microbes in the cecum, like those in the rumen of ruminants, produce enzymes that convert cellulose, hemicellulose, and other nonlignin components of fiber initially to monosaccharides and then to **volatile fatty acids** (VFA). Some microbe species, for example Bacteroidetes, end their processing with the two-carbon **acetic acid**. Other species process fiber to the three-carbon **propionic acid** and still others, for example, Firmicutes, to the four-carbon **butyric acid**. Some microbes called archaea are producers of methane (CH_4). These diverse microbes interact and have symbiotic relationships with the host animal (den Besten et al., 2013). Each VFA has a methyl (-CH_3) group at one end and a carboxyl (-COOH) group at the other end. The carboxyl group easily gives up the hydrogen ion from its hydroxyl (-OH) portion, thus making it an acid. After some modification, each of these VFAs can be inserted into metabolic pathways that require simple organic inputs of two, three, or four carbons. Acetic acid (known as acetate or vinegar in aqueous solution) is made in the largest proportion of the three, and the higher the percentage of fiber in the diet, the higher the proportion of acetate produced. All of the VFAs are quickly absorbed across the cecum or colon wall into the bloodstream. A rise in VFA concentration in cecal fluid increases its osmolarity, and this may stimulate a satiety signal when a threshold concentration is reached (Geor, 2013). As VFAs are absorbed by the colon, bicarbonate is secreted into the lumen, reducing the acidity (den Besten et al., 2013).

Acetate is metabolized rapidly, having a half-life of only 1.5—4 min. Several animal tissues use acetate in their energy metabolism and convert both its carbon atoms to carbon dioxide (CO_2) (Sabine and Johnson, 1964). Acetate is used by the liver, for building fat in adipose and mammary cells, and for energy in digestive organs and skeletal muscles.

Propionate is produced in increasing quantities as the proportion of dietary starch increases. Liver and other digestive organs use propionate. Two molecules of propionate are readily combined into glucose. Infusion of propionate into ponies has caused plasma glucose concentration to rise (Geor, 2013).

Of the three VFAs, butyrate is produced in the smallest amounts by cecal microbes. A relatively small amount of butyrate is used by the liver (to produce β-hydroxybutyrate), but much of it is utilized by cells of the intestines (Geor, 2013).

VFAs are important to omnivores, as well as their essential role in metabolism of herbivores. Each individual horse and person has their own unique microbiome or microbe population distribution, which changes frequently due to relative concentrations of protein, fiber, NSCs, and microbes ingested, plus changes of pH and any physiologic condition that can be favorable to one species of microbe and unfavorable to another (den Besten et al., 2013). Even in humans, up to 10% of daily caloric requirements can by met by VFAs (Bergman, 1990).

High-starch diets stimulate growth of anaerobic and lactic acid–utilizing bacteria and suppress growth of cellulolytic bacteria. Starch is also associated with greater production of lactic acid and lower pH in the cecum. Higher fiber concentration in the diet and additional yeast tend to moderate these effects of starch (Medina et al., 2002).

After a change in protein concentration of the diet, relative concentrations of various bacteria can change within a day, and this can result in changes in the proportions of VFAs produced. If the new diet persists, the change in relative microbe populations can persist (Muhonen et al., 2008).

In microbes, β-glucosidase hydrolyzes β-1,4 glycosidic bonds in the disaccharide cellobiose, which is cleaved off cellulose by enzymes earlier on a cellulase pathway.

Digestibility of grass hay in the cecum was lower in horses fed a higher rate of timothy hay than a low rate (Miyaji et al., 2014).

A diet containing adequate amounts of cell wall fiber is necessary for good horse health. In addition to cellulose and hemicellulose, these less fibrous components, galactans, fructans, gums, mucilages, and pectin, are all fermentable by gastrointestinal microbes to VFAs, without causing the buildup of lactic acid and the drop of pH caused by starch and sugar. Horse nutritionists and veterinarians have learned that many metabolic disorders in horses can be managed nutritionally by ensuring that caloric intake is not too high and that energy sources are appropriate (Pagan, 2012).

Fat Digestion

The composition of dietary fat is reflected in the composition of body fat in horses, suggesting that intact FAs are absorbed through the small intestine (Frape, 2010). Most of the fat in grasses and legumes includes the phospholipids of cell membranes and any oils stored in vacuoles. An addition of vegetable oil to horse diets provides a highly digestible source of energy without adding starch. Medium-chain FAs (6–12 carbons) are metabolized to ketones in the liver.

Nitrogen Digestion

Not all of the nitrogen in plant material is in the form of amino acids and proteins. Some nonprotein nitrogen (NPN) is in the form of urea. When large amounts of protein are ingested, some of the N is converted to ammonia (NH_3).

NUTRIENT REQUIREMENTS OF HORSES

Discovering appropriate amounts of nutrients that horses should eat per day is an ongoing effort. Observations of horse feeding behavior, body fatness, resistance to sickness or lameness, and abilities to do work (pulling, running, jumping, herding) have always provided horse managers with ideas that they should feed a little more or a little less of something. Nutritionists and veterinarians added experiments to these everyday observations and, through sharing reports of their work, were able to develop recommendations about how much horses need to eat to develop, grow, work, and reproduce to the satisfaction of their owners. The Sixth edition of a summary of this equine nutrition research was produced as "Nutrient Requirements of Horses, Sixth Revised Edition" (2007a-g) by a Committee on Nutrient Requirements of Horses, working under the authority of the National Research Council of the National Academies of the United States. The daily requirements for DE, CP, the amino acid "lysine, 14 minerals, and five vitamins are included in tables for horses of the following BWs: 200, 400, 500, 600, and 900 kg. A list of daily DE and CP requirements for a 500-kg (1102 lb) mare in various physiologic states, extracted from NRC (2007f), is presented in Table 3.1.

Free software associated with the Sixth Edition is available to allow nutritionists and other trained people to determine the nutrient requirements of individual horses based upon their physiologic needs to maintain their bodies, relative to their metabolic rate, their BW, their level of exercise, whether they are breeding or not, their month of pregnancy, their month of lactation, and their age while growing. The software also accesses Table 16.6 Nutrient Composition of Selected Feedstuffs, which was described earlier (see Table 3.2). Knowing what feedstuffs are available and their nutrient contents, the nutritionist can formulate a ration or diet to match the nutritional requirements of horses. The nutrient concentrations of feeds in the program can be adjusted, in line with reports from feed analysis labs. Assessing BW and body condition score and keeping records of them, along with records of what has been fed, will assist horse managers in making diet adjustments to ensure that horses receive optimum levels of nutrients.

FEED INTAKE

Reports over several years indicated that grazing horses ate about 1.4%–3.0% of their BW per day as feed dry matter, based on access to pasture for 24 h per day. In many situations, horse farm managers find it advantageous to have horses grazing for intervals ranging from about 3 h to about 18 h per day. Horses grazed at higher dry matter intake (DMI) rates when they were subjected to restricted turnout times (Glunk et al., 2013), making prediction of pasture DMI difficult when grazing for less than 24 h per day. A mathematical model was developed to predict pasture DMI as a function of hours of pasture access (Siciliano, 2012). The model was this:

$$\text{Pasture DMI} = 5.12\sqrt{\text{Pasture access hours}} - 2.86 \ (R^2 = 0.7; \ P < .001)$$

The equation can also be used to calculate the hours of pasture access required for horses to obtain a target DE intake and the amount of hay a horse needs to meet its daily energy requirement when pasture access is available for only a part of a day. Several other factors that can affect the rate of pasture DMI include herbage mass available, physiologic status of the horse, ambient temperature, forage height, chemical composition, maturity, and species. Further research is anticipated to quantify the degree of influence each of these factors has and refine the equation to make more precise predictions of pasture DMI (Siciliano, 2012). The ability to calculate hay needs, along with knowledge of hay prices, allows managers to calculate costs of hay fed. When a cost of using and managing pasture is determined per horse per day, one can calculate the savings of feeding pasture versus hay.

REVIEW QUESTIONS

1. Explain two differences between each of the following pairs of chemicals:
   a. carbohydrates and lipids
   b. carbohydrates and proteins
   c. fatty acids and amino acids
   d. disaccharides and polysaccharides
2. Why is the difference between alpha-1,4 bonding and beta-1,4 bonding in sugars important in herbivore nutrition?
3. What use are hydrogen bonds and disulfide bonds in the three-dimensional structure of proteins?
4. List four components of plant cell wall fiber.
5. What device is used to determine the gross energy concentration of an organic substance, and how does it work?
6. What is the difference between the gross energy and digestible energy concentrations of a feed sample?
7. In the detergent system of fiber analysis of feeds, how is the NDF percentage determined, what does neutral detergent fiber (NDF) represent, and what use is this measurement?
8. How is the acid detergent fiber (ADF) value of a feed determined, what does it represent, and what use is this measurement?
9. Which of the following increase in their absolute dry weight within a forage plant as it matures to the heading stage: lignin, protein, nonstructural carbohydrates, cellulose, and dry matter?
10. Which of the following decrease as relative values (percentages) while a forage plant matures: minerals, leafiness, percentage of stem material, crude protein percentage, digestible energy concentration, and NDF percentage?
11. Which of these three feeds should be most valuable per ton? Why?
    a. CP 8%, ADF 42%, NDF 70%
    b. CP 22%, ADF 25%, NDF 45%
    c. CP 17%, ADF 36%, NDF 55%

REFERENCES

AAFCO (Association of American Feed Control Officials). Critical Factors in Determining Fiber in Feeds and Forages. 2017. https://www.aafco.org/Portals/0/SiteContent/Laboratory/Fiber_Best_Practices_Working_Group/Fiber-Critical-Conditions-Final.pdf.

Barker DJ, Collins M. Forage fertilization and nutrient management. In: Barnes RF, Nelson CJ, Collins M, Moore KJ, editors. Forages: An Introduction to Grassland Agriculture. sixth ed. Ames Iowa: Blackwell; 2003. p. 263–94.

Baron, V. 2018. Don't Freak-Out About Fructans. Alberta Horse Industry. https://www.albertahorseindustry.ca/index.php?option=com_content&view=article&id=485:2012-hboc-speaker-spotlight-dr-vern-baron&catid=142:2011-news-archive&Itemid=353.

Bergman EN. Energy contributions of volatile fatty acids from the gastrointestinal tract in various species. Physiol. Rev. 1990;70:567−90.

Blevins DG, Barker DJ. Nutrients and water in forage crops. In: Barnes RF, Nelson CJ, Moore KJ, Collins M, editors. Forages: The Science of Grassland Agriculture. sixth ed. Ames Iowa: Blackwell; 2007. p. 67−82.

Brennan KM, Whorf C, Harris LE, Adam E. The effect of dietary microalgae on American Association of Equine Practitioners lameness scores and whole blood cytokine gene expression following a lipopolysaccharide challenge in mature horses. J. Anim. Sci. 2017;95(Suppl. 4):166.

Colburn MW, Evans JL. Chemical composition of the cell-wall constituent and acid detergent fiber fractions of forages. J. Dairy Sci. 1967;50(7):1130−5.

den Besten G, van Eunen K, Groen AK, Venema K, Reijngoud D-J, Bakker BM. The role of short-chain fatty acids in the interplay between diet, gut microbiota, and host energy metabolism. J. Lipid Res. 2013;54:2325−40.

Dhiman TR, Anand GR, Satter LD, Pariza MW. Conjugated linoleic acid content of milk from cows fed different diets. J. Dairy Sci. 1999;82:2146−56.

Frape D. Chemical Composition of Feedstuffs Used for Horses, Appendix C. In: Equine Nutrition and Feeding. fourth edition. Chichester, United Kingdom: Wiley-Blackwell; 2010. p. 371−8.

French P, Stanton C, Lawless F, O'Riordan EG, Monahan FJ, Caffrey PJ, Moloney AP. Fatty acid composition, including conjugated linoleic acid, of intramuscular fat from steers offered grazed grass, grass silage, or concentrate-based diets. J. Anim. Sci. 2000;78:2849−55.

Garcia A. Hay Analysis: Understanding Relative Fed Value and Relative Feed Quality. South Dakota State University Department of Extension; 2017. https://igrow.org/up/resources/02-2006-2017.pdf.

Geor RJ. Endocrine and metabolic physiology. Chapter 2. In: Equine Applied and Clinical Nutrition: Health, Welfare and Performance. Elsevier, Ltd; 2013. p. 33−63.

Glunk EC, Pratt-Phillips SE, Siciliano PD. Effect of restricted pasture access on pasture dry matter intake rate, dietary energy intake, and fecal pH in horses. J. Equine Vet. Sci. 2013;33(6):421−6.

Hansen TL, Lawrence LM. Composition factors predicting forage digestibility by horses. J. Equine Vet. Sci. 2017;58:97−102.

Hess T, Ross-Jones T. Omega-3 fatty acid supplementation in horses. R. Bras. Zootec. 2014;43(12):1−10. http://www.scielo.br/scielo.php?script=sci_arttext&pid=S1516-35982014001200677.

Hindrichsen IK, Kreuzer M, Madsen J, Bach Knudsen KE. Fiber and lignin analysis in concentrate, forage and feces: detergent versus enzymatic-chemical method. J. Dairy Sci. 2006;89:2168−76.

Hopkins A, Gilbey J, Dibb C, Bowling PJ, Murray PJ. Response of permanent and reseeded grassland to fertilizer nitrogen. 1. Herbage production and herbage quality. Grass Forage Sci. 1990;45:43−55.

Jensen KB, Harrison P, Chatterton NJ, Bushman BS, Creech JE. Seasonal trends in nonstructural carbohydrates in cool-and warm-season grasses. Crop Science 2014;54:2328−40.

Jung H-JG. Analysis of forage fiber and cell walls in ruminant nutrition. In: 37th Annual Ruminant Nutrition Conference "New Developments in Forage Science Contributing to Enhanced Fiber Utilization by Ruminants". at: Experimental Biology 96, April 14, 1996. American Society for Nutritional Sciences; 1997. 810S−3S.

Kellon E. It's Time to Put the Fructan Theory to Rest. 2015. Dr. K's Horse Sense. June 22, 2015, https://drkhorsesense.wordpress.com/2015/.../its-time-to-put-the-fructan-theory-to-rest.

Lechtenberg VL, Holt DA, Youngberg HW. Diurnal variation in nonstructural carbohydrates, in vitro digestibility and leaf to stem ratio of alfalfa. Agron. J. 1971;63:719−24.

Lewis LD. Feeding and Care of the Horse. second ed. Ames, Iowa: Blackwell Publishing; 2005. p. 42−61.

Liggenstoffer AS, Youssef NH, Couger MB, Elshahed MS. Phylogenetic diversity and community structure of anaerobic gut fungi (phylum Neocallimastigomycota) in ruminant and non-ruminant herbivores. ISME J. 2010;4:1225−35. https://www.nature.com/articles/ismej201049#t1.

Longland AC, Byrd BM. Pasture nonstructural carbohydrates and equine laminitis. In: The Waltham International Nutritional Sciences Symposia, American Society for Nutrition; 2006. 2099S−102S. https://academic.oup.com/jn/article-abstract/136/7/2099S/4664903.

Medina B, Girard ID, Jacotot E, Julliand V. Effect of a Preparation of Saccharomyces cerevisiae on Microbial Profiles and Fermentation Patterns in the Large Intestine of Horses Fed a High Fiber or a High Starch Diet. 2002.

Miyaji M, Ueda K, Hata H, Kondo S. Effect of grass hay intake on fiber digestion and digesta retention time in the hindgut of horses. J. Anim. Sci. 2014;92(4):1574−81. OPTIONAL.

Monteverde V, Congiu F, Vazzana I, Dara S, Pietro S, Piccione G. Serum lipid profile modification related to polyunsaturated fatty acid supplementation in thoroughbred horses. J. Appl. Anim. Res. 2016;45(1):615−8.

Moore JE, Undersander DJ. Relative Forage Quality: A Proposal for Replacement for Relative Feed Value. Proceedings National Forage Testing Association; 2002.

Muhonen S, Connysson M, Lindbert JE, Julliand V, Bertilsson J, Jansson A. Effects of crude protein intake from grass silage-only diets on the equine colon ecosystem after an abrupt feed change. J. Anim. Sci. 2008;86(12):3465−72.

Murphy B. Pasture nutrition. In: Greener pastures on your side of the fence. 4th ed. Colchester, Vermont: Arriba Publishing; 1998. p. 71−96.

Nogradi N, Couetil LL, Messick J, Stochelski MA, Burgess JR. Omega-3 fatty acid supplementation provides an additional benefit to a low-dust diet in the management of horses with chronic lower airway inflammatory disease. J. Vet. Intern. Med. 2015;29:299−306.

NRC. Nutrient Requirements of Horses. Sixth Revised Edition. National Research Council of the National Academies. Washington, D.C: The National Academies Press; 2007a. p. 109−27.

NRC. Nutrient Requirements of Horses. Sixth Revised Edition. Washington, D.C: National Research Council of the National Academies. The National Academies Press; 2007b. p. 203—6.

NRC. Nutrient Requirements of Horses. Sixth Revised Edition. Washington, D.C: National Research Council of the National Academies. The National Academies Press; 2007c. p. 34—9.

NRC. Nutrient Requirements of Horses. Sixth Revised Edition. Washington, D.C: National Research Council of the National Academies. The National Academies Press; 2007d. p. 1—33.

NRC. Nutrient Requirements of Horses. Sixth Revised Edition. Washington, D.C: National Research Council of the National Academies. The National Academies Press; 2007e. p. 144—56.

NRC. Nutrient Requirements of Horses. Table 16-3. Sixth Revised Edition. Washington, D.C: National Research Council of the National Academies. The National Academies Press; 2007f. p. 298—9.

NRC. Nutrient Requirements of Horses. Table 16-3. Sixth Revised Edition. Washington, D.C: National Research Council of the National Academies. The National Academies Press; 2007g. p. 304—7.

Pagan JD. Measuring the digestible energy content of horse feeds. In: Pagan JD, editor. Advance in Equine Nutrition. Nottingham, UK: Nottingham University Press; 1998. p. 71—6.

Pagan J. In: Past, Future Advances in Equine Nutrition: In the Past Quarter-century, Several Advances Have Been Made in Equine Nutrition, and Five of the Most Important Are Reviewed Forthwith. Feedstuffs 84.54 (Dec 31, 2012); 2012. p. 10.

Pritchett KB, Leatherwood JL, Vandergrift B, Anderson JJ, Stutts JJ, Beverly MM, Kelley SF. Influence of omega-3 polyunsaturated fatty acid supplementation on plasma prostaglandin E2 production in young exercising horses. J. Equine Vet. Sci. 2015;35:416.

Sabine JR, Johnson BC. Acetate metabolism in the ruminant. J. Biol. Chem. 1964;239(1):89—93.

Saha, U., Hancock, D. and Kissel, D, How Do We Calculate Relative Forage Quality in Georgia? Agricultural and Environmental Services Laboratories. Cooperative Extension Service, College of Agricultural and Environmental Sciences, University of Georgia. 2010. http://aesl.ces.uga.edu/publications/Feeds/RFQ_Calc_Circ.pdf.

Schroeder JW. Forage Nutrition for Ruminants. North Dakota State University Extension Publication AS-1250. 2013. http://www.ag.ndsu.edu/pubs/ansci/dairy/as1250w.htm.

Siciliano PD, Gill JC, Bowman MA. Effect of sward height on pasture nonstructural carbohydrate concentrations and blood glucose/insulin profiles in grazing horses. J. Equine Vet. Sci. 2017;57:29—34.

Siciliano PD. Estimation of pasture dry matter intake and its practical application in grazing management for horses. In: Proc. 10th Atlan. Nutr. Conf.; 2012. Timonium, MA, N.G. Zimmermann, editors, 9-12.

Varloud M, Fonty G, Roussel A, Guyonvarch A, Julliand V. Postprandial kinetics of some biotic and abiotic characteristics of the gastric ecosystem of horses fed a pelleted concentrate meal. J. Anim. Sci. 2007;85(10):2508—16.

White HE, Wolf DD. Controlled Grazing of Virginia's Pastures. Virginia Cooperative Extension factsheet 2009. 418-012, http://pubs.ext.vt.edu/418/418-012/418-012.html.

Wyss U, Morel I, Collomb M. Fatty acid content of three grass/clover mixtures. Grassl. Sci. Eur. 2005;11:348—50. www.seepastos.es/docs%20auxiliares/Actas%20Reuniones%20escaneadas/.../2.348.pdf.

Zeyner A, Kienzle E. A method to estimate digestible energy in horse feed. J. Nutr. 2002;132:1771S—3S.

Chapter 4

Soils for Horse Pasture Management

Paul Voroney

School of Environmental Sciences, University of Guelph, Guelph, Ontario, Canada

INTRODUCTION

Folklore says that big trees are a good indicator of prime farm land suitable for raising healthy and strong-boned horses (Fig. 4.1). This is because the soils in these areas are inherently fertile and grow grasses that are nutritious pastures for horses. There are several farming areas in North America that historically have produced great horses of many breeds, attributed to their soils, which have developed from limestone rocks rich in phosphorus. Additionally, it is well known that several factors besides the rocks and minerals contribute to soil formation and are the foundation of fertile healthy soils: the climate, vegetation, and water. This chapter describes the nature and properties of soils. With this knowledge as a foundation, you should be able to manage soils sustainably for horse pasture production. Download *The Global Soil Biodiversity Atlas* (https://globalsoilbiodiversity.org/node/271); it contains excellent photos and descriptions of soil features and is available for free.

WHAT IS SOIL?

There are many perspectives on soil, depending on whether you are managing it for plant growth or use in construction of the farm. If you are raising horses and want nutritious pastures for them to graze on, then understanding soil fertility and water management is critical. Nutritious pastures require good healthy soil, so the quality of the plant growth provides an indicator of soil conditions. Likewise, if you are growing hayfields and other forage crops for feed, then knowledge of soil management is important. Uniquely, soils provide the conditions necessary for plants to grow. However, soil materials are also used in construction of the stable yards, corrals, walkways, paddocks, exercise yards, riding arenas, berms, and dams.

Soils (the pedosphere) occur at the interface where organisms (biosphere) interact with rocks and minerals (lithosphere), water (hydrosphere), and air (atmosphere). Here, climate and human activity control the intensity of the processes responsible for soil formation (Fig. 4.2). Components of each of these spheres make up the soil and provide a habitat for

FIGURE 4.1 Mare and foal grazing pastures on the fertile soils in Kentucky. *Photo by Jimmy Henning, University of Kentucky.*

Horse Pasture Management. https://doi.org/10.1016/B978-0-12-812919-7.00004-4

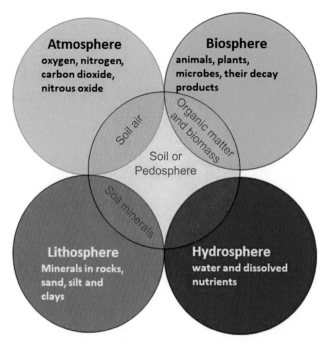

FIGURE 4.2 The soil or pedosphere is where the rocks or lithosphere, air or atmosphere, water or hydrosphere, and plants and animals or biosphere interface and interact.

plants to grow. Soils anchor plants and allow their roots to explore a stable pore space containing water, nutrients, and air ("Soil Stories, the Whole Story," https://www.youtube.com/watch?v=Ego6LI-IjbY&feature=youtu.be).

The sun's energy is captured by plants during photosynthesis to form the organic compounds making up their tissues, and it is the biologic decomposition of plant litter that is the driving force for many chemical and biologic processes in the soil. Physical and chemical weathering of rocks into finer particles accompany the release of plant available nutrients and initiate the soil-forming process (https://youtube.com/watch?v=bTzslvAD1Es).

Soil formation depends on five interacting factors: climate, parent material, topography, living organisms, and time (Fig. 4.3). The relative influences of these soil-forming factors differ with location on the earth's surface, depending on climate and parent material, and they act in unique combinations in each environment ("Classes of Soil Parent Material," https://www.youtube.com/watch?v=BYBBE6KzHLM). Soils in most of North America have formed since retreat of the last glaciers 10,000–30,000 years ago ("The Five Factors of Soil Formation," https://www.youtube.com/watch?v=bTzslvAD1Es).

During soil formation a unique profile extending downward from the surface develops, having horizontal layers or horizons that appear different from one another. The horizons vary in thickness, color, and composition, depending on the intensity of the soil-forming factors (see *Web Soil Survey* https://www.nrcs.usda.gov/wps/portal/nrcs/site/soils/home).

Soils derived from weathered rocks and minerals are referred to as mineral soils. These soils typically contain from 1% to 10% organic matter. When land areas are submerged in water for long periods of the year, decay of plant residue is slow and organic matter accumulates. These areas become organic soils and include peats and muck lands. Their organic matter content is much greater than that in mineral soils and can be near 100%. The percentage of the total soil volume that is occupied by solid particles in mineral soils is about 45%–65%, whereas for organic soils, it is 10%–20%.

Soil varies in thickness from less than an inch (a few millimeters) to several feet (1 foot = 0.305 m). The minimum depth for it to be classified as soil is 4 inches (10 cm).

COMPOSITION OF MINERAL SOILS

Soils are made up of solids, the minerals and organic matter, and pore space, containing air and water (Fig. 4.4). The nature and relative proportions of these components vary depending on the soil's location in the landscape and on the parent material of the region. The proportions of water and air filling soil pores vary with local climatic conditions and are affected by irrigation and drainage practices.

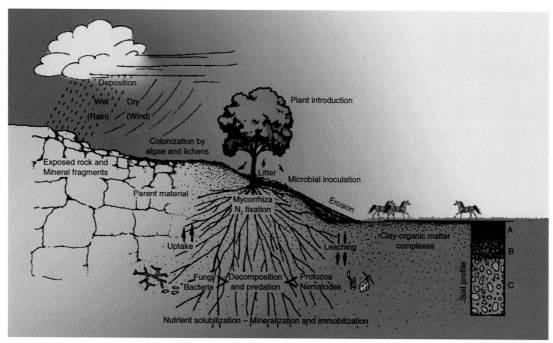

FIGURE 4.3 Soils and soil profiles are formed by the five interacting factors of climate, parent material, topography, organisms, and time. *Adapted from Eldor A. Paul (Ed.), Soil Microbiology, Ecology and Biochemistry, fourth ed. Academic Press.*

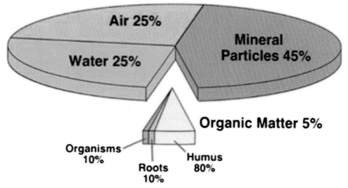

FIGURE 4.4 Volumetric composition of mineral soils: solids ($\sim 50\%$) comprised of mineral particles and organic matter, and pore spaces ($\sim 50\%$) containing air and water.

The mineral particles are derived from ground-up and weathered rock and mineral fragments and constitute more than 90% of the soil's dry weight. Large mineral particles, including boulders, stones, and gravels, affect the physical attributes of a soil; however, only those individual mineral particles less than 0.078 inch (2 mm) in diameter are considered soil. This group of mineral particles is known as the fine earth fraction. In preparation for analysis of soil physical and chemical properties, the coarser mineral particles are discarded by passing the soil through a 10 mesh sieve having 0.0787 inch openings.

The mineral particles of the fine earth fraction range in size over four orders of magnitude: diameters from 2.0 mm to smaller than 0.002 mm, and it is common for this fraction to be divided into separates of decreasing size. Sand-size particles (0.05−2 mm) can be seen by the naked eye and feel rough and grainy, whereas silt-sized particles (0.05−0.002 mm) are microscopic and feel slippery and smooth. The smallest of the mineral particles, the clay-sized particles (<0.002 mm), are also microscopic, and when wet, they feel sticky and can be molded into shapes that hold together.

The proportions of sand, silt, and clay in the fine earth fraction determines the soil's texture. When investigating a field site, considerable insight into the properties of the soil can be inferred from its texture (e.g., soil water characteristics,

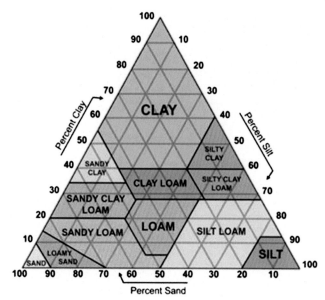

FIGURE 4.5 Soil textural triangle.

infiltration, drainage, and storage, nutrient retention, susceptibility to compaction, ease of tillage), thus it is often one of the first properties to be measured. Soils with a high proportion of sand are referred to as light soils because they are relatively easy to work with farm equipment and do not retain much water. Those rich in clay are called heavy soils, as they require much more effort to till due to their stickiness and high capacity to retain water.

If the textural analysis is measured by a soil testing laboratory, the soil can be assigned to a specific textural class using a textural triangle (Fig. 4.5) ("Soil Textural Triangle Tutorial," https://www.youtube.com/watch?v=4hW59WZ0EQI).

With practice and patience a soil's textural class can be determined adequately by feel (rubbing the soil between thumb and fingers) of the molded, moistened soil ("How to Determine Soil Texture by Feel," https://www.youtube.com/watch?v=GWZwbVJCNec&t=122s).

There are 12 textural classes, and soils within a textural class have similar properties with respect to water storage and tilth, which is the ease of working soil with tillage and planting equipment. Other general names commonly used to describe soil texture include coarse for sandy soils, medium for loamy soils, and fine for clayey soils. Because the soil's textural class has such an important role in defining its physical and chemical properties, it is used in naming the soil. Examples of soils in well-known horse farming areas include Maury silt loam, McAfee silt loam, and Guelph loam. These medium-textured soils have optimal water retention and aeration properties, and they provide an ideal rooting pore space for growing lush, healthy pastures. Also, the consistence of these soils is usually soft, making them comfortable yet secure for horses to walk on. Maps of soils located throughout the United States and descriptions of their compositions are available at this website: https://casoilresource.lawr.ucdavis.edu/gmap/.

ORGANIC MATTER AND SOIL HUMUS

Surface soil is enriched in organic matter, which accounts for 1%−10% of the total soil mass (Fig. 4.6), the exact amount depending on farm management practices, soil tillage, and cropping system. A typical well-managed pasture soil contains about 100 tons/acre (200 Mg/ha) of organic matter in total to a depth of 4 feet (1 m). The soil organic matter contains 55 tons/acre (110 Mg/ha) of organic carbon, 5.5 tons/acre (11 Mg/ha) of organic nitrogen, and 1.1 tons/acre (2.2 Mg/ha) of organic phosphorus: a very significant organic carbon sink and rich nutrient pool indeed!

The organic matter consists of plant, faunal, and microbial residues in various states of decay, from fresh litter to humus. Grasses growing on these soils provide inputs of this organic material, and the litter produced is subsequently consumed and decomposed by the organisms living in the soil. The living component of soil organic matter consists of plant roots (5%−15%) and soil organisms (85%−95%), of which microorganisms account for 80%−90% and fauna 10%−20% (see section on Soil Biology).

Because it is difficult to separate the components of living organisms from the by-products of their decomposition or to isolate organic matter separately from the mineral particles (sand, silt, and clay), the precise chemical nature of soil organic

FIGURE 4.6   Pasture soils are typically rich in organic matter. *Photo by Peter Nowell.*

matter is far from being completely understood. Indeed, soil organic matter is considered by soil scientists to be the most chemically complex organic material on earth.

Typically, during sampling and preparation of soils for analyses, visible plant roots, larger organisms, and plant litter are discarded from the fine earth fraction during sieving, leaving the <0.0787-inch (2-mm) particulate fragments of recognizable plant and faunal tissue and microscopic remains of biologic decay that are intimately associated onto the surfaces of minerals in the fine earth fraction. The chemistry of this nonliving component of soil organic matter, which ranges from large organic compounds such as carbohydrates, proteins, and lipids to simple biomolecules such as sugars, amino acids, and fatty acids, is well understood. It accounts for ~50%–60% of the total soil organic matter.

Over a period of months to a few years and provided that soils have adequate aeration, eventually all of the organic residues from dead plants and organisms decompose. Over a much longer term, a brown to black-colored, colloidal organic matter is formed from the products of organic matter decay by a unique chemical process called humification (Fig. 4.7).

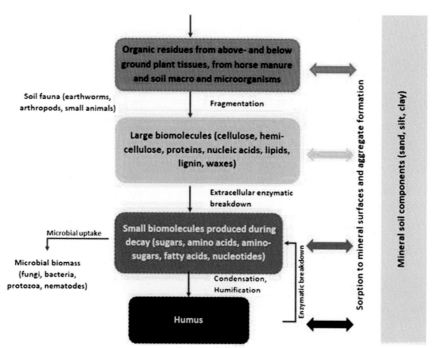

FIGURE 4.7   Decomposition of grass residues and animal manures to form humus.

This component of soil organic matter is known as humus and accounts for 40%—50% of total soil organic matter. However, the precise chemical structure of the remaining organic matter, the dark brown—black colored humus formed in soil from the products of decay, is not known, even though it has been studied for hundreds of years. Together, these organic matter constituents benefit soil physical, chemical, and biologic properties. Though difficult for soil organisms to decompose because they are chemically so complex, humus does slowly decay at an annual rate of 2%—3%.

AGGREGATION

Most of the organic matter in surface soils becomes intimately associated with the fine earth fraction as the particulate fragments undergo decay and become glued by microbial biofilms to the surfaces of the finer mineral particles, the clays and silts. When individual mineral particles become bound together by these organic constituents and by inorganic cements (calcium and magnesium carbonates, iron and aluminum oxides, silicates), they form larger stable units known as peds or aggregates (Fig. 4.8). Together with soil texture, the size, nature, and arrangement of soil aggregates are of particular importance to the structure of the soil because of their influence on soil pore size distribution, ease of tillage, resistance to compaction, and resistance to erosion.

Soil organic matter and clay minerals together play an especially important role in initiating aggregate formation due to their colloidal properties, specifically due to their small size, high surface area, and electrostatic charge. Plants roots and fungal hyphae help physically to bind together the mineral particles contained in larger aggregates. Usually, loamy or clayey soils are strongly aggregated, whereas sandy and silty soils are weakly aggregated.

Two size classes are commonly used to describe soil aggregation based on their resistance to break down and their associations with soil organic matter. Macroaggregates are >1/100 inch (250 μm) in diameter, and microaggregates are ≤1/100 inch (53—250 μm) in diameter. Macroaggregates are formed during decomposition of plant roots and residues from the production of microbial binding agents that glue and hold together both microaggregates and unaggregated mineral particles.

Macroaggregates that are 1—2 mm diameter give the soil a crumb-like structure that is ideal for plant root growth. They also make it easier to till the soil and to prepare a friable seed bed for planting. In fine-textured, clayey, and loamy-textured soils, the formation of macroaggregates can result in the beneficial pore space characteristics of a sandy soil, allowing rapid water infiltration and drainage, and promoting soil aeration. Macroaggregation in sandy soils can make them more resilient to compaction. Soil structural characteristics can be evaluated visually in the field by examination of the size and appearance of the aggregates (*Soil Structure Score Chart for Evaluation of Aggregation in the Field*, https://www.sruc.ac.uk/vess).

Macroaggregates usually remain intact as long as the soil is left undisturbed and when the soil surface is covered with plant vegetation and litter, such as fields remaining in pasture or forage production. However, intense rainfall, intensive soil tillage, and soil compaction can break down macroaggregates. They are especially vulnerable to breaking apart and

FIGURE 4.8 Soil aggregation is promoted by grass roots and decaying organic residues. Note aggregates of different size. Macroaggregates are greater than 1/100 inch (250 μm) in diameter and contain microaggregates and particulate organic matter. Macroaggregates 1/2—1 inch (1—2.5 cm) in diameter give soils a crumb-like structure that is ideal for plant root growth. *Photo by Peter Nowell.*

disintegrating when soils are wet. Therefore, it is important that soils not be cultivated or driven and walked on when they are wet. Microaggregates tend to be much more resistant to break down from the impact of rainfall, from freezing/thawing and drying/rewetting events, and from soil tillage. The following soil and plant management practices are useful for promoting aggregation and for preventing aggregate destruction:

1. Minimize tillage events that leave the soil bare, exposed to the sun and wind, and that dry out the soil.
2. Protect surface soil aggregates from intense rainfall by maximizing continuous coverage with growing plants (pastures, forages, fall planted crops, cover crops). Plants and root growth also help to reduce soil erosion.
3. Protect the surface soil by the application of mulches (crop residues or composts).
4. Maximize soil amendments with readily decomposable organic matter to promote soil microbial activity (animal manure bedding, green manure crops).
5. Restrict soil tillage to periods when the soil is sufficiently dry and friable to minimize smearing and formation of clods.

NATURE OF SOIL PORE SPACE

Soil pore space is defined as the percentage of the total pore volume occupied by soil pores:

$$\text{Soil pore space } (\%) = [\text{pore volume/soil volume}] \times 100$$

Total soil pore space can be estimated from data on soil bulk density, which is relatively easy to do. To measure it, a soil core sampler of known volume is pushed into the soil, taking care to avoid compaction. The soil in the core is dried (105°C) and the mass recorded. The formula to calculate soil bulk density (D_b) is:

$$D_b \text{ (g/cm}^3) = \text{soil mass (g)/soil bulk volume (cm}^3)$$

Soil pore space (%) is calculated using this measurement of soil bulk density and assuming that the particle density (D_p) of mineral soils is 2.65 g/cm^3.

$$\text{Soil pore space } (\%) = 100 - [(D_b/D_p) \times 100]$$

Total soil pore space can be divided into two size classes, macropores and micropores, based on the interaction of soil mineral and organic matter particles with water and air (Table 4.1). Macropores are greater than ~10 μm in diameter and include the pore space where plant roots and larger fragments of plant litter are found. Micropores are less than 10 μm in diameter and are largely contained within microaggregates. Because of their small size, they are important for governing soil water storage and availability for plant growth. Ultimately though, it is both the size and interconnection of the pores that determine how water moves and is retained in soil.

SOIL WATER

Soil water content affects the moisture and amount of nutrients available to plants and soil aeration status. Soil water content can be measured on a mass or volume basis. Gravimetric soil water content is the mass of water in the soil, measured as the difference between the moist soil and the soil dried at 105°C, known as the oven-dry weight. Note that gravimetric soil water content is expressed per unit mass of oven-dried soil.

Gravimetric soil water content (%) = [mass of moist soil (g) − mass of oven-dried soil (g)/mass of oven-dried soil (g)] × 100

TABLE 4.1 Soil Macropore and Micropore Distribution Across a Textural Gradient (% of Total Soil Porosity)

Pore Category	Pore Function	Soil Pore Diameter (μm)	Sandy Loam	Loam	Clay Loam
Macropores	Aeration/water infiltration, and drainage	>10	58	35	32
Micropores	Water retention	<10	42	65	68

Measurements of soil gravimetric water content are considered destructive (oven-drying), so the soil sample should not be used for further chemical analysis.

Volumetric soil water content is the volume of water per unit volume of soil.

Volumetric soil water content (%) = [volume of water (cm^3)/volume of soil (cm^3)] × 100

This equation can be rewritten to calculate the depth equivalent of water in a soil, which makes it easy to relate volumetric soil water content to units commonly used to describe amounts of water added to soils from rainfall or irrigation, which are expressed in inches or cm.

Volumetric soil water content (%) = [depth of water in inches (cm)/depth of soil in inches (cm)] × 100

Depth of water as rainfall or irrigation in inches (cm) = volumetric soil water content (%) × depth of soil in inches (cm)

Rugged, low-cost sensors are commercially available that allow direct, nondestructive measurements of soil volumetric water content. These measurements of soil volumetric water content are essential for assessing the status of plant available water in soil ("Determining Soil Moisture Content," https://www.youtube.com/watch?v=jzYCuspFhwo) and for scheduling irrigation events ("Assessing Soil Moisture Content for Irrigation," https://www.youtube.com/watch?v=XCddABhV3bg).

Water is held in the soil by its adhesion to the surfaces of mineral and organic particles and by cohesion or attraction to itself, the latter being responsible for its surface tension. These attractive forces combined with the force of gravity affect the energy status of the water in soil. How tightly water is held in the soil and in which direction it moves depend on its energy status, more specifically the potential energy of the water in soil. The difference in the energy status of pure water (at a standard reference state) and soil water is termed the soil water potential. The units commonly used to express soil water potential are pascals (Pa) or kilopascals (kPa). Other units, bars, for example, are often used, and their equivalency is shown in Table 4.2.

When the total soil pore space becomes filled with water, such as after an extensive rainfall or an irrigation event, the soil is said to be saturated with water. The water potential when soils are saturated is near 0 kPa, that is, essentially the same potential as that of pure water and free to flow under the influence of gravity. The volumetric soil water content at the saturation point is equivalent to the total soil pore space.

Water on the soil surface flows overland following a downward slope, and that contained in soil macropores drains downward, both due to gravitational forces. After a rainfall or irrigation event that has saturated the soil and has stopped, soil water content drops quite rapidly as water drains from the macropores. It may take 1−3 days for all of the macropore water in surface soil to drain, depending on the soil's texture and structure, and further drying is much slower.

The water retained after drainage has stopped is held within soil micropores. At this point the soil is said to be at field capacity, as this represents its maximum water-holding capacity ("Demonstration of Soil Field Capacity and Permanent Wilting Point," https://www.youtube.com/watch?v=m7DAej5-d6w). The soil water potential at field capacity ranges from −10 kPa (sandy soils) to −33 kPa (loam and clay loam soils). Field capacity is a useful term because it represents the maximum amount of water the soil can store. However, plants are able to use just a portion of the water stored in the micropores.

Water moves relatively slowly within soil micropores in any direction from a region of high water potential to a region of low water potential. As an example, water uptake by plant roots lowers the nearby soil water potential. If the water potential of the surrounding soil is higher, perhaps deeper in the soil or between plants, water moves toward the roots upwards and sideways.

TABLE 4.2 Soil Water Status and Units Commonly Used to Express Soil Water Potential

Soil Water Status	Soil Water Potential (bars)	Soil Water Potential (kPa)
Saturation	0	0
Field capacity	−1/3	−33
Permanent wilting point	−15	−1500
Air-dry soil	−31	−3100

PLANT AVAILABLE WATER

Without rainfall or irrigation to recharge the soil's water content, plants over time begin to wilt. Water flow within micropores is slow, so plants may wilt during the daytime when transpiration rates are high but recover overnight when leaf stomates are closed. However, eventually the plants stay wilted because the soil holds onto the water too tightly for plants to take it up. At this point the soil is said to have reached permanent wilting point and a soil water potential of -1500 kPa. The soil appears dry at this point, but it still contains water, though plants are not able to take it up from the soil. The difference in soil water content at field capacity and permanent wilting point represents the plant available water, and this varies with soil textural class (Table 4.3).

The plant available water of a clay loam soil is less than that of a loam soil, despite having higher water contents at field capacity. ("How Soil Properties Affect Soil Water Storage and Movement," https://www.youtube.com/watch?v=ego2FkuQwxc).

Soil water characteristics can be estimated based on soil texture and organic matter content, which are commonly measured physical and chemical soil properties (*Soil Water Characteristics Explained*, https://www.nrcs.usda.gov/wps/portal/nrcs/detailfull/national/water/manage/drainage/?cid=stelprdb1045310). At this website, a computer program (SPAW) developed by Solomon and Rauls can be downloaded to estimate soil water characteristics important for water management. Knowledge of the soil's moisture content at field capacity and its plant available water content provide a basis for understanding water management issues in the field. A 1-inch (2.5-cm) rainfall or irrigation event would recharge the plant available water in the top foot (30 cm) of a sandy loam soil, whereas a 2.5-inch (6.3-cm) event would be required for a loam soil and 2-inch (5.4 cm) event in a clay loam. In these same soils, plants would wilt after 4 d growing in a sandy loam, after 8.4 d in a loam, and after 7.2 d in a clay loam, assuming a rooting depth of 12 inches (30 cm) and a typical evapotranspiration rate of ~ 0.03 inch/d (~ 0.75 mm/d). To ensure that plant growth will not be limited by water availability, it is recommended that plant available water content should not be depleted by more than 50%. All of this information can be used to schedule irrigation frequency and depth of water application required.

WATER DRAINAGE

Soils that remain saturated with water for extended periods of the year can cause difficulties to working the fields with farm equipment and to plant growth. Most agricultural crops and grasses grow best in well-drained soils because under these conditions, soils are well-aerated, rooting depth is not restricted, and these soils warm up more quickly in the spring time. Therefore, drainage systems are installed to remove excess surface water and to lower the soil water table, the depth to where the soil is saturated ("Drainage Water Management," https://www.youtube.com/watch?v=j4mYch4RFsY).

There are two types of drainage systems, a surface system that hastens water runoff and a subsurface system that allows macropore water from within the soil to drain. Excess water moves most rapidly overland, so grading fields and construction of ditches to allow for controlled water runoff should be a first consideration when dealing with excess water conditions. Installation of subsurface drainage, either as deep ditches or as perforated plastic pipes, promotes excess water flow when the soil is saturated, thereby lowering the water table. Thus, surface soils become drier and better aerated, whereas soil below the ditches or drainage pipes can remain saturated.

TABLE 4.3 Soil Water Potential (kPa) and Volumetric Soil Water Content (%) (100 × Volume of Water/Volume of Soil) Across a Textural Gradient

Soil Water Characteristic	Soil Water Potential (kPa)	Sandy Loam	Loam	Clay Loam
Saturation	0	44	49	53
Field capacity	−33	18	32	36
Permanent wilting point	−1500	8	11	18
Air-dry soil	−3100	4	7	14
Plant-available water		10	21	18
Plant-available water[a] inch (cm)		1 (2.5)	2.5 (6.3)	2 (5.4)

[a]*Surface soil to 1 foot (30 cm).*

SOIL AERATION

Soil aeration is a critical factor affecting plant growth, as plants require oxygen (O_2) for metabolism and leguminous plants require nitrogen (N_2) for nitrogen fixation. The movement of gases in soil is by diffusion, from a region of high concentration to a region of low concentration, so oxygen and nitrogen move from the atmosphere to the soil, and carbon dioxide (released from plant roots and soil microorganisms) diffuses from the soil to the atmosphere. Diffusion through the soil's air-filled pores—the macropores—allows relatively rapid gaseous exchange between the atmosphere and the soil. However, diffusion of gases in water is ~1/10,000 that in air, so management practices to promote and maintain macroporosity are critical for ensuring adequate soil aeration. Soil compaction should be avoided because it reduces total porosity, crushes macropores, and increases the proportion of smaller micropores, thereby reducing plant available water. Ideally the soil should have a macropore:micropore ratio of between 0.5 and 0.8.

SOIL CHEMISTRY

The soil's chemical properties are inherited from the processes of soil formation, during weathering and transport of the parent material from which the soil has formed. Thus the chemical nature of the rocks and minerals and the intensity of the weathering processes are fundamental in determining the chemical properties of the soil.

Weathering of rocks and minerals containing high proportions of silicates (SiO_2), such as quartz and feldspars, gives rise to soils that are acidic and deficient in plant nutrients, whereas those high in magnesium and iron, such as olivines, pyroxenes, and amphiboles, give rise to soils that are basic and richer in plant nutrients. Limestones are calcium- and magnesium-rich carbonates that are relatively easily weathered, and their presence in soil significantly affects the soil's chemical properties. Limestone-containing parent materials give rise to basic soils because of their abundance of basic cations. Also, because these soils have an abundance of plant-available calcium, the forage produced on these pastures promotes growth of strong animal bones. The farmlands surrounding Lexington, Kentucky, Ocala, Florida, and Shelbyville, Tennessee, are areas of phosphorus-rich limestone that historically have produced strong-boned horses of many breeds.

Thus, there are two main types of parent materials: those rich in limestone and referred to as calcareous, and those deficient in limestone and referred to a noncalcareous. The presence of limestone in the parent material buffers the naturally occurring acidification processes and reactions in soils, thereby slowing them from becoming acidic.

Parent material also determines the quantity and nature of the clay minerals present in the soil. Because of their unique chemical structure, clay minerals have a special role in determining the chemical properties of the soil. Both clay minerals and soil organic matter are known as soil colloids because of their extremely small sizes and high surface areas, which make them very reactive with water and with plant nutrients dissolved in soil solution. In addition, both colloids have negatively charged surfaces, and this common property attracts and retains important nutrient cations (positively charged ions) for plant uptake ("Origin of Charges and Higher Concentrations of Cations at Colloid Surfaces," https://www.youtube.com/watch?v=IHHE3rHGrqo). The nutrient cations adsorbed to these colloids include macronutrients such as ammonium (NH_4^+), calcium (Ca^{++}), potassium (K^+), magnesium (Mg^{++}), and iron (Fe^{+++}) and micronutrients such as copper (Cu^{++}), Zinc (Zn^{++}), and manganese (Mn^{++}). The strength of adsorption of cations to colloid surfaces from most tightly held to more weakly held is as follows:

$$Al^{+++} > Ca^{++} > Mg^{++} > K^+ = NH_4^+ > Na^+$$

Cations attracted and adsorbed onto colloidal surfaces are referred to as exchangeable cations, as they are readily exchangeable with other cations dissolved in the soil solution and available for uptake by plant roots and by soil organisms. By an exchange reaction with adsorbed cations, H^+ released from plant root hairs and from soil organisms force adsorbed cations off into the soil solution, where they are taken up.

The total amount of exchangeable cations that a soil can adsorb is referred to as its cation exchange capacity (CEC). Measurements of soil CEC are expressed as the number of centimoles of positive charge per kilogram soil ($cmol_c$/kg). Another unit commonly reported that has the same numeric value is milliequivalents per 100 g soil (me 100 g^{-1}): $cmol_c$/kg = 1 me 100 g^{-1}. The CEC of soils ranges from 3 to 50 $cmol_c$/kg soil.

Soil cation exchange capacity is an important soil parameter to measure because it represents a reservoir of plant-available nutrients. It also plays an important role as a buffer to resist changes in soil acidity. Because the concentration of cations adsorbed to soil colloids is 10−100× higher than the concentration of cations in the soil solution, leaching of exchangeable cations from surface soils in drainage waters is greatly reduced. Thus nutrient cations are readily leached from sandy soils but not so much from loamy and clayey soils that have high cation exchange capacities and retain the cations.

SOIL ACIDITY AND ALKALINITY

There are several reasons why soils can become acidic. Many of the processes contributing to soil acidification are naturally occurring. For example, excessive rainfall or irrigation can leach away the basic cations from surface soil, which are less tightly held by the soil's CEC than the two acidic cations (H^+, Al^{+++}), thereby the soil becomes acidic. Plant roots and soil organisms produce carbon dioxide, which reacts with water to form carbonic acid and slowly acidifies the soil. Also, plant roots and soil microorganisms release other acidifying compounds during metabolism and nutrient uptake (organic acids) that contribute to soil acidity. Other soil microbial processes, such as nitrification that oxidizes ammonium in soil solution to form nitrate, acidify the soil. The main source of the ammonium can be from microbial decomposition of soil organic matter, though it is also present in animal manures and composts and produced during decay of cover crop residues that are rich in organic nitrogen.

The concentration of H^+ in soil solution is a measure of soil acidity, and it is expressed in units of pH on a logarithmic scale that ranges from pH 0 to pH 14, with pH 7 being the midpoint and by definition neutral. Soil pH typically varies from pH 8 to pH 4, though it may be lower or higher under extreme conditions (Table 4.4). Soils that have a pH within the range from >6.5 to <7.4 are referred to as neutral pH soils. Those soils that have a pH > 7.3 are considered alkaline, and they contain an abundance of basic cations such as Ca^{++}, Mg^{++}, K^+, and Na^+ that resist acidification. Soils that are deficient in basic cations have a lower pH, and those with a pH < 6.6 are considered acidic. Soils in areas of high rainfall tend to become acidic, especially if they are sandy textured with low water-holding capacity and have a low CEC. Soils in drier regions (semiarid and arid) with less leaching tend to retain sufficient basic cations to prevent them from becoming acidic and maintain a neutral pH. Soils that have developed from a limestone-rich parent material usually have a neutral to slightly alkaline pH.

The nutrients needed for plant growth are most readily available in soils that have a pH in the neutral range. In addition, many of the plants commonly grown for pasture, forage, and cereal production grow best in this pH range, which is why this soil pH range is a recommended target. Because soil pH is so critical to plant growth, it is routinely measured in laboratory soil tests. Recommendations for correcting soil acidity by lime additions are available from county agricultural extension specialists (https://www.youtube.com/watch?v=9bk9q5lzCxM).

SOIL FERTILITY

Besides carbon, hydrogen, and oxygen, which plants obtain from air and water, plants take up most of their other nutrient requirements from the soil (Table 4.5). The nutrient elements include nitrogen, phosphorus, potassium, calcium, magnesium, sulfur, iron, manganese, zinc, copper, molybdenum, boron, and chlorine. Selenium is not usually considered an essential plant nutrient, but because it is such a critical element for animal health and nutrition, deficient pasture soils can be fortified with selenium. Some nutrients such as nitrogen and sulfur can be deposited onto soils and plants as dusts and as atmospheric contaminants with rainfall.

TABLE 4.4 Soil Description Based on pH

Degree of Acidity or Alkalinity	pH Range	Typical pH Range in Soils
Extremely acid	Below 4.5	pH in acid sulfate soils
Very strongly acid	4.5–5.0	
Strongly acid	5.1–5.5	
Moderately acid	5.6–6.0	pH range in humid regions
Slightly acid	6.1–6.5	
Neutral	6.6–7.3	
Slightly alkaline	7.4–7.8	
Moderately alkaline	7.9–8.4	pH range in arid regions
Strongly alkaline	8.5–9.0	
Very strongly alkaline	Above 9.0	

TABLE 4.5 Forms of Plant Nutrients Obtained From Soil

Nutrient Category	Nutrient	Chemical Symbol	Forms Absorbed From Soil
Primary or macro nutrients	Nitrogen	N	NH_4^+ (ammonium) and NO_3^- (nitrate)
	Phosphorus	P	$H_2PO_4^-$, HPO_4^{-2} (orthophosphate)
	Potassium	K	K^+
Intermediate or secondary nutrients	Sulfur	S	SO_4^{-2} (sulfate)
	Calcium	Ca	Ca^{+2}
	Magnesium	Mg	Mg^{+2}
Micronutrients	Iron	Fe	Fe^{+2} (ferrous), Fe^{+3} (ferric)
	Zinc	Zn	Zn^{+2}
	Manganese	Mn	Mn^{+2}
	Molybdenum	Mo	MoO_4^{-2} (molybdate)
	Copper	Cu	Cu^{+2}
	Boron	B	H_3BO_3 (boric acid), $H_2BO_3^-$ (borate)
	Chlorine	Cl	Cl^-
	Nickel	Ni	Ni^{+2}
	Cobalt	Co	Co^{+2}
	Selenium	Se	Se^{+4} (selenite), Se^{+6} (selenate)

Nitrogen is typically required by plants in the largest amounts, as it is a component of many important plant constituents such as proteins and chlorophyll. Legumes grown in pastures and for forage, together with the help of rhizobia bacteria associated with their roots, are able to convert atmospheric N_2 into forms that the plants can use. In this way, legumes are able to supply much of their own nitrogen requirements and also contribute to overall soil nitrogen fertility. Nevertheless, a good yield of grass hay or alfalfa can remove up to ~ 270 lbs N/ac (~ 300 kg N/ha) in the annual harvest, so it is critical that all sources of plant available nitrogen (nitrogen fixation, amendments with animal manures and composts, and crop residue return) are included in the farm's nitrogen fertility management program.

Most plants have mycorrhizal fungi intimately associated with their roots that assist with nutrient uptake. The fungal hyphae act as extensions of the plant's root system, thereby allowing the plant to explore further out into the soil. For nutrients such as phosphorus, which is held tightly by soil colloids, and micronutrients, which are present in the soil in low amounts, the mycorrhizae can transport these nutrients directly from the soil solution into the plant.

Soils often have adequate supplies of some of the nutrients required for plant growth but are deficient in others. For example, soils may have adequate amounts of calcium and magnesium but be deficient in phosphorus and nitrogen. These nutrient deficiencies can critically limit plant growth, even when other nutrients are present in sufficient amounts. Careful soil sampling of the fields and soil test analyses performed by an accredited laboratory are essential for assessing the field's fertility status ("Soil Testing: a Must for Proper Pasture Management," https://www.youtube.com/watch?v=mn2YbNFFSwQ). The soil test will indicate which nutrients are sufficient and those that are deficient. Typically the soil test also considers previous cropping history and amendment applications in making recommendations for amounts of nutrients to apply. Therefore, the best nutrient management practices can be followed to deal with these deficiencies. Animal manures and composts can be applied to supply phosphorus requirements, for example, while planting of legumes that fix atmospheric nitrogen can be used to manage a nitrogen deficiency.

SOIL BIOLOGY

The nature and extent of diversity of organisms in soils is impressive (reminder to download a copy of the *Global Soil Biodiversity Atlas*). All three domains of life, bacteria, archaea, and eukarya, live in the soil and surface litter, the most abundant of which are the microorganisms, bacteria, archaea, and fungi. The huge and diverse populations of the different organisms range in size from microscopic (1—2 μm) microflora and microfauna, up to macro- and megafauna such as earthworms that can be up to 20 inches (0.5 m) in length. A single teaspoon of soil contains billions of bacteria, hundreds of thousands of archaea and fungal hyphae, extending up to a mile (1600 m) in length ("Living Soil," https://www.youtube.com/watch?v=MlREaT9hFCw). Astonishingly, the biomass of organisms living within the surface foot of topsoil (0—30 cm) of an acre (0.5 ha) is equivalent in biomass to ~25 horses!

The entire population of organisms exists as a complex food web responsible for the transformation of organic matter and nutrients in soil. Their interactions with each other, their feeding habits, and their effect on the physical structure of the soil are all at play in a healthy soil ("Science of Soil Health," https://www.youtube.com/watch?v=IHOF6NfLm7M&list=PL4J8PxoprpGa3wFYSXFu-BW_mMatleIt0).

Most, especially the smallest microflora, are responsible for decomposing soil organic matter and plant litter, turning these sources of organic carbon and nutrients back into forms plants can reuse, CO_2 and inorganic nutrients. Microfaunal predators, the nematodes and protozoans, graze on the microflora, thereby enhancing the flow of energy and nutrients to larger organisms and to plants (https://www.youtube.com/watch?v=EyKfpOso8q8&list=PL4J8PxoprpGa3wFYSXFu-BW_mMatleIt0&index=3).

Larger soil animals, mainly mites and collembolans, tear apart the decomposing litter and in this manner speed up decomposition by the microorganisms. The largest of organisms, including earthworms, ants, or termites, and beetles typically drag plant litter and animal manures down into the soil, which provides a better environment than the soil surface for microbial decomposition. They are also the soil engineers in that they alter the physical structure of the soil by creating soil biopores, building mounds, and excreting fecal pellets and casts. For example, the biomass of earthworms, whose populations are highest in fertile pasture soils, can reach 2—3 million worms per acre (0.405 ha), weighing 1000 lb (500 kg). Earthworms annually "plow up" soil in their casts equivalent to 100 tons (~50 Mg) per acre (0.405 ha) ("Amazing Earthworms," https://www.youtube.com/watch?v=9ZHTerOJYMA).

SECRETS TO SOIL MANAGEMENT FOR SUSTAINABLE PASTURE PRODUCTION

The secrets to best management practices for sustainable pasture production have been known for a long time (Fig. 4.9). The management goal should always be to protect the quality of the soil resources while minimizing the impact of pasture production on the surrounding natural environment.

The quality of pasture soils can be maintained for future generations by following these best management practices:

1. Maximize the period of pasture growth to protect the soil against erosion. Keep the surface soil covered and never leave soil bare. Nutrient-rich clay particles are carried away during erosion events that pollute streams, rivers, and lakes. Plants shield surface soils from the impact of rainfall, and their root systems bind soil particles together. Additionally, organic matter—rich surface soils are better able to store plant-available water than soils with less organic matter. Remember, it takes a century to build an inch of topsoil and only a few minutes for erosion to remove topsoil or form ugly gullies.

FIGURE 4.9 Late fall pasture grazing in southern Ontario. *Photo by Maren Oelbermann, University of Waterloo.*

2. Maximize the period of pasture growth to minimize losses of soil nutrients and soil amendments (synthetic fertilizers, animal manures, and composts). Nutrients, particularly nitrogen and phosphorus, are converted into forms that can contaminate surface and ground waters. Actively growing plants take up these nutrients as they are released, so they do not accumulate and become susceptible to losses.
3. Maximize the period of pasture growth to minimize emissions of greenhouse gases. Carbon dioxide, nitrous oxide, methane, and ammonium are potent greenhouse gases emitted from soils, but pasture production has the potential to minimize these emissions. Growing plants take up carbon dioxide from the atmosphere and inorganic nitrogen from the soil, and their residues add organic carbon and organic nitrogen to the soil. The net result is pasture soils become a reservoir of organic carbon and nitrogen, not a source of carbon dioxide and nitrous oxide.
4. Promote plant biodiversity in pasture swards to enhance soil fertility. Inclusion of legumes in the pasture, for example, can supplement soil nitrogen fertility, thereby reducing the need for nitrogen fertilizers. Additionally, plants with deep rooting systems can take up nutrients from the subsoil, and with time, this enriches the surface soil.
5. Minimize disruption of soils during tillage operations to protect the habitat of soil organisms. While soil tillage may be necessary for seedbed preparation and to manage compaction problems, conservation tillage should be practiced. A healthy soil is a biodiverse soil.

REVIEW QUESTIONS:

1. Describe the process to determine the classification of the soils in your horse pasture.
2. What observation about the vegetation in a region can indicate the quality of the soil (fertility, plant-available water retention, and drainage) for its potential to promote growth of strong-boned horses?
3. In which textural class do the following soils belong?

Textural class	Percent clay	Percent silt	Percent sand
	20	40	40
	10	25	65
	20	60	20
	70	10	20

4. a. Calculate the mass of soil in an acre-furrow-slice, given that it corresponds to 1 acre in area and to a depth of 0.5 ft (6 inches). Assume a soil bulk density of 81.2 lbs/ft3. b. Similarly calculate the mass of soil in a hectare-furrow-slice, given that it corresponds to 1 hectare in area (10,000m^2) and to a depth of 0.15 m. Assume a soil bulk density of 1300 kg/m^3.
5. What chemical and physical properties would long-term pasture soils likely have?
6. In which size pores is water retained when the soil is at field capacity?
7. Calculate the plant-available water in the surface soil to 1 foot depth (30 cm) given the information shown in the table: Soil water potential (kPa) and volumetric soil water content (%) (100 x volume of water/volume of soil) across a textural gradient

Soil water characteristic	Soil water potential (kPa)	Sandy loam	Loam	Clay loam
Saturation	0	47	50	53
Field capacity	−33	16	29	36
Permanent wilting point	−1500	8	13	18
Air-dry soil	−3,100	4	7	14
Plant available water (%)				
Plant available water* inch (cm)				

8. Why does installation of a drainage system by the pasture manager, in areas where the water table is high, benefit the soil and plant growth?
9. What is a benefit of a high CEC to pasture soil and to plant growth?
10. How can the acidity of a soil be made more neutral?

11. What is the name of the soil microorganisms that convert atmospheric nitrogen into forms that plants can use, and where are they located?
12. What is a benefit of mycorrhizal fungi to forage plants?
13. What are some of the important activities of soil microflora?
14. How does maximizing the period of pasture growth enhance soil quality and minimize detrimental effects to the environment?

Chapter 5

Introduction to Pasture Ecology

Edward B. Rayburn[1] and Paul Sharpe[2]

[1]*West Virginia University Extension Service, Morgantown, WV, United States;* [2]*University of Guelph, Guelph, ON, Canada (retired)*

Pasture-based horse production is a near miraculous process. After an egg from a mare is fertilized by a sperm cell from a stallion, the fertilized egg develops into a foal. Plants gather sunlight, fix solar energy, and take up soil minerals to make food for the grazing horses. The horses graze the pasture; microbes in each horse's cecum and colon digest the forage and are then digested by the animal to provide energy, protein and minerals for maintenance, milk production, movement, and growth. Treaded plant tops, dead roots, manure, and urine provide energy and protein to soil organisms that help maintain soil structure, water infiltration, and water-holding capacity of the soil. The majority of minerals are cycled back to the soil in dead plant material and manure or urine from the horses. These nutrients are then used to grow another flush of pasture. If all goes well, at the end of the season, we have a new, well-grown foal from each mare that was mated. Likewise, each new plant is produced from two cells, carbon dioxide, solar energy, and minerals harvested from a healthy soil by the seedling plant in the pasture.

This interplay of sunlight, plants, soil, and animals are the parts of pasture ecology. The horse farm manager who understands plant, soil, and animal ecology is prepared to be an excellent pasture manager, to understand how to adapt to changes in weather, and interpret how research and farmer experience from other areas apply to their farm. For our introduction to pasture ecology, we will discuss the four living components of the pasture ecosystem: plants, grazing animals, soil community, and human managers. But first let us discuss some overarching ecologic principles.

OPTIMAL ENVIRONMENT VERSUS LIMITING FACTORS

Each plant, animal, bacteria, protozoa, and fungus has its niche or place in its ecosystem. Each has an optimum physical and chemical environment and habitat. The habitat provides adequate food and cover, allowing the species to reproduce and maintain itself. The environment is determined by the climate, time of year, soil texture, position in the landscape, and human management.

The flow of energy from the sun follows the food web, which is composed of primary producers that fix solar energy and other organisms that eat plants and each other to get their part of the fixed solar energy. Each organism in the food web fills one of these roles: primary producer, primary consumers, higher level consumers (predators and omnivores), parasites, or detritus feeders (decomposers).

All organisms grow and reproduce best when they live in an environment that provides ideal levels of all their needed resources (temperature, water, nutrients, solar radiation, and protection from predators and parasites). If any one of those needed resources is below a lower threshold (e.g., low soil fertility) or above an upper threshold (e.g., excessive temperature), that resource will become a limiting factor to the health and productivity of the organism. If a second organism is better adapted to this level of limiting resource, it will be healthier and more productive and will outcompete the first organism for space and use of other resources. When managing pasture plants or animals, it is imperative that we know the optimal resource conditions needed by the plants and animals and provide them if economically possible. When not economical, we should then look at alternative plants and animals that do well within these limiting resource constraints.

In economics, there is a law of diminishing returns (Fig. 5.1). This principle applies to organisms within ecosystems for all of their resource needs. If the availability of a resource falls below a critical value the health and productivity of an organism decreases. For example, clover plants need a soil pH at or above 6.0 and soil test phosphorus and potassium in the high/optimum range, or their health and productivity decreases. If another plant, say tall fescue, is more tolerant of low

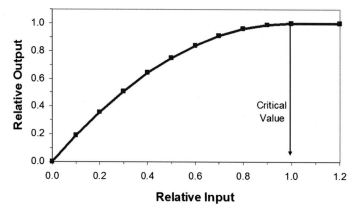

FIGURE 5.1 Law of diminishing returns. In some cases, if the relative input is too high, there is an upper critical value that causes production to decrease.

pH or medium soil test potassium, it will have a competitive advantage and will increase in the stand at the expense of the clover.

When there is an excess of a resource, there may be an upper critical value where health and productivity decreases. Looking at temperature as a resource, cool-season grasses have a lower critical temperature near 50°F. When average daily temperature is below that, their growth rate is limited by temperature. Likewise, cool-season grasses have an upper critical temperature near 70°F. As daily mean temperature increases from 70 to 90°F, growth of cool-season grasses decreases from near maximum at 70 degrees to zero at 90 degrees. If a warm-season grass is in the stand, its growth rate goes from slow at 70 degrees to near maximum at 90°F. That is one reason Bermuda grass or crabgrass can be dominant in a pasture in August, but fescue is dominant in October.

PLANTS

Plants are the primary producers in the pasture ecosystem. They intercept sunlight through photosynthesis, using its energy to fix carbon dioxide and water to make carbohydrates, some of which is also converted to other plant components. Leguminous plants like clover and lespedeza use *Rhizobia* bacteria in root nodules to fix nitrogen from the air that is used by plants for making proteins. Plants take up mineral nutrients from the soil to make enzymes, coenzymes, DNA, RNA, and cell structures within the plants. These minerals are used for similar purposes in the livestock that eat and digest the plants.

There are many different species of plants that live in a naturalized pasture. They differ in preferred thermal and radiant environment, growth habit, life history, forage quality, and antiquality components. Pasture ecology is complex, but it is not difficult. There are a number of basic principles that apply to all plants. Some plants like it hot; some like it cool. Some need full sun; others tolerate shade. Some need long days to reproduce; others need shorter days. Some plants grow upright, some along the surface of the soil, and some in between. Plants differ in their cell wall thickness and cell contents due to species, growth stage, and time of year. Some plants make toxins to discourage animals from eating them; other plants may have antitoxins to the toxins. There are grasses, legumes, forbs, shrubs, trees, tall plants, short plants, plants that have rhizomes, and those that grow as a bunch of tillers. Some plants have roots that are fibrous; some have taproots. Some plants reproduce vegetatively, some sexually every year, others their first year then die. Some plants live for 2 years, and produce seed, then die. Knowing these characteristics for the plants in the pasture helps in understanding how to manage them.

RESPIRATION

Plants also use respiration to maintain life and to grow. In a simplistic sense, respiration is the conversion of carbohydrates to carbon dioxide and water to release energy needed to perform metabolic functions. The release of carbon dioxide and water by plants is part of the global carbon and water cycles.

During the day, both photosynthesis and respiration occur with photosynthesis dominating, so there is a net increase in carbohydrates. At night, respiration continues, but photosynthesis ends when the sun sets. So during the night, there is a net loss of carbohydrates. This can be seen in the cycle of sugar in plant leaves. The sugar content in a leaf is lowest at sunrise. During the day, sugar content increases, and sugars are used for leaf and root growth and are stored for reserve energy.

After sunset, respiration uses up leaf sugar, bringing it back to a low point in the morning. Both photosynthesis and respiration are affected by air temperature, with different plant species having different responses to temperature.

In cool-season plants, there is a second type of respiration called photorespiration. These plants are called C3 plants because their metabolic pathway uses 3-carbon sugar. Warm-season plants are called C4 plants because their metabolic pathway uses 4-carbon sugar, which does not allow photorespiration. Photorespiration increases with air temperature more quickly than conventional respiration. The C3 plants grow well when average daily temperatures are cool (50 to 70°F) but poorly when temperatures are hot (over 80°F), thus their name of cool-season plants. The C4 plants do well at high air temperatures (80 to 90°F), thus being called warm-season plants. The C4 plants have a semitropical origin and are usually sensitive to cool weather and light frosts, while many C3 plants tolerate light frosts and some hard freezes.

LIGHT INTERCEPTION: CANOPY HEIGHT AND TIME OF YEAR

Light interception is determined by how tall and thick the pasture canopy is and how high the sun is in the sky (Fig. 5.2). A tall, thick pasture intercepts more light than a short, thin pasture. The light that is not intercepted by the leaves is lost to the plant for making sugar. The sunlight hitting the ground will heat the soil and air, which may increase respiration in the plants. On cold spring days, this could be good. On hot summer days, this could be bad. In the fall, C3 plants benefit from having some sunlight at the base of the canopy since this is needed to stimulate the formation of tiller buds that will produce tillers the following year.

When the June sun is high in the sky, rays of sunlight shine deeper into the canopy than do rays from the October sun, which is low in the sky. In June a pasture height of 8 inches is needed to capture 90% of the sunlight (midday in WV), while in October a height of about 4.5 inches captures 90% of the sunlight (Fig. 5.2). More light penetrates the canopy in May than June. When pastures start going to head in May, grass tillers elongate, raising some of the leaves above others, allowing more light to penetrate into the canopy. Once the reproductive tillers are removed and the pasture is again dominated by vegetative growth, light interception increases.

More information on reasons for the seasons and the angle of light shining on the earth is available from *Understanding Astronomy: The Sun and the Seasons*.

As primary producers, terrestrial plants can capture enough energy from photosynthesis to produce an average of several kilograms of dry plant material per square meter per year. Ecologists measure primary productivity biomass as the *dry* mass of organisms because their water content, which fluctuates with water uptake or loss, has no energetic value. An ecosystem's productivity is the *rate* at which the standing crop produces *new* biomass. The standing crop biomass is the total dry mass of plants present at a given time.

ENERGY RESERVES CYCLE WITH GROWTH

In many plants, there is an energy reserve cycle during periods of growth after harvest. Consider tall fescue and white clover as examples (Fig. 5.3). When leaves of these plants are grazed off, new leaves start to grow. When most of the leaves are removed, energy for new growth comes from stored energy reserves. The site of stored energy reserves is

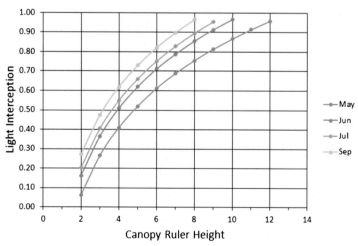

FIGURE 5.2 Light interception by a pasture is determined by the sun's height in the sky and the pasture's height, which is an estimate of the pasture's forage mass and leaf area.

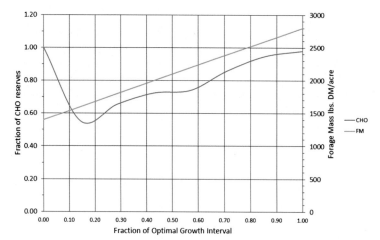

FIGURE 5.3 When a plant's leaves are grazed off, new growth is powered by carbohydrate reserves and photosynthesis in remaining leaves. As new leaves grow, they use reserve carbohydrate, and the reserves decline. Later, when new leaves produce carbohydrates in excess to plant growth needs, carbohydrate reserves increase.

species-specific. Tall fescue stores energy reserves in the lower leaf sheath and underground stems called rhizomes. White clover stores energy reserves in the aboveground stem called a stolon.

Over a period of days the energy reserves decrease as new leaves grow. As leaf area increases, more sunlight is intercepted, and photosynthesis increases, providing more energy for growth. At some point, photosynthesis is great enough to produce more sugar than is needed for growth, resulting in sugar moving into storage with an increase in energy reserves in the plant. As leaf area increases further, energy reserves return to the high level needed to sustain the next growth cycle. As leaves grow beyond maximum light interception, they shade one another, and older leaves become less efficient and die. Beyond this point, net growth is reduced and forage quality declines.

GROWTH UNDER ROTATIONAL GRAZING

When all photosynthetically active leaves are removed from a plant, new growth is dependent entirely on stored energy reserves. When some leaf area is left, plant regrowth may be faster since it is powered by photosynthesis and stored energy reserves. This is one reason we do not want to overgraze pastures. For example, a grass-clover pasture in West Virginia was allowed to grow to a dry matter (DM) forage mass of 3000 lbs. DM/acre. Cattle then grazed the pasture to 1500 lbs. DM/acre. After the animals were removed, growth was linear during all growth intervals (Fig. 5.4). In September, growth plateaued at just over 2000 lbs. DM/acre due to dry weather. Throughout the grazing season, residual forage mass was maintained above 1200 lbs. DM/acre since animal forage intake is limited and animal performance decreases at lower residual forage mass. If grazing had continued at each rotation so that residual forage DM was less than 1200 lbs. DM/acre,

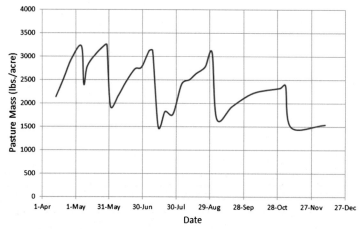

FIGURE 5.4 Weekly measurements of pasture forage mass in a rotationally grazed grass-clover pasture showing the maintenance of linear growth powered by carbohydrate reserves and residual leaf area.

there would have been less leaf area to resume a new growing cycle, and it would have taken longer to reach the target grazing forage mass of 3000 lbs. DM/acre.

ROOT GROWTH

Root growth determines the ability of a plant to take up water and nutrients. Root growth is controlled by actively photosynthesizing leaves. When energy is in short supply, it is used by plant tissue nearest to the site of photosynthesis. Therefore, roots receive energy when more energy is produced by photosynthesis than is being used by top growth. (Grass roots are not used for reserve carbohydrate storage, while the tap root of legumes and forbs are used for storage. In grass plants, carbohydrates are stored in stem and tiller base tissue.)

Some cool-season grasses tolerate close grazing (2-inch height) if adequately long rest intervals (4 weeks) are provided between grazing events (Fig. 5.5). However, upright-growing, native, warm-season grasses do not tolerate close grazing since little leaf area remains and leaf sheaths that may contain reserve carbohydrates are removed (Fig. 5.6).

Optimal yield depends on harvesting leaves without adversely affecting top or root growth. This is based on the proper timing and intensity of grazing relative to energy reserves and residual leaf area.

GROWTH HAS TWO PHASES

Pasture growth is observed as the plants increase in height (Fig. 5.4). However, within the plant, growth is composed of two phases: cell division then cell enlargement. Both phases require adequate energy, protein, mineral nutrients, and water.

Plant cells differ from animal cells in that they have a cell wall. A young cell has a thin cell wall and a high proportion of cell contents. As the cell matures the cell wall thickens, and the proportion of cell contents decreases (Fig. 5.7). Young cells need to have thin walls, so they can stretch during expansion. Mature cells in stems need thick cell walls to support the stem and seed head.

The growth pattern from young to mature cell affects forage quality as well as production. Cell contents are about 98% digestible, while the cell wall material can be 70% digestible in young forage but only 30% digestible in mature forage. The cell wall becomes less digestible as more lignin is deposited around the structural fiber. In cool-season forages, this is most pronounced in hot weather.

GROWING POINTS

Cell division occurs in specialized cells located at growing points of the plant. There are four main types of growing points in forage plants: terminal, axillary, intercalary, and root tip.

Terminal growing points are on the end of a shoot. Axillary growing points are in the axis of the leaf, the angle between the leaf and stem it is attached to. Intercalary growing points are found in grass leaves and provide growth on each side of

FIGURE 5.5 When grazed to a 2-inch residual height tall fescue root growth responds to rest interval: left to right root growth for 7-day, 14-day, and 28 day rest intervals. *Photo courtesy Dr. James Green, North Carolina State University.*

ROOT DEVELOPMENT IN RELATION TO TOP REMOVAL

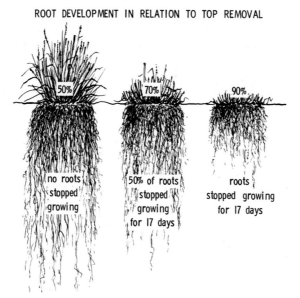

FIGURE 5.6 In native warm-season grasses that grow upright, root growth after defoliation is proportional to the intensity of removal of top growth (United States Department of Agriculture, Soil Conservation Service (USDA/SCS), Ag. Handbook No. 389, 100 Native Forage Grasses in 11 Southern States).

the region of cell division. Root tip growing points are at the tip end of the root, as would be expected. Growing point locations are characteristic of each plant species and determine what the plant looks like (its morphology).

CELL WALL CONTENT CHANGES WITH SEASON AND PLANT TYPE

Forage cell wall content is estimated using neutral detergent fiber (NDF) analysis. The NDF content of pasture changes over the growing season due to growth stage and temperature affecting the balance between photosynthesis and respiration. The NDF content is low in the cool weather of the spring when forage is vegetative and high in nonstructural carbohydrates, increases in the warm weather of summer, and then decreases in the cool weather of the fall as nonstructural carbohydrates increase. Pasture NDF is also affected by species composition since grasses have higher NDF than forbs (a.k.a. broad leaf weeds) and legumes have the lowest NDF content.

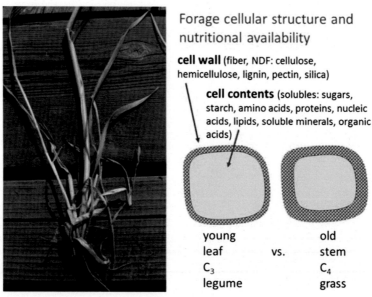

Forage cellular structure and nutritional availability

cell wall (fiber, NDF: cellulose, hemicellulose, lignin, pectin, silica)

cell contents (solubles: sugars, starch, amino acids, proteins, nucleic acids, lipids, soluble minerals, organic acids)

young leaf C_3 legume   vs.   old stem C_4 grass

FIGURE 5.7 A young plant cell is relatively thin and can stretch as the cell expands to grow. As the cell matures the cell wall thickens. The cell wall is a relatively larger part of the cell in old versus young forage, in C3 grasses versus C3 legumes, in C4 grasses versus C3 grasses, and in stems versus leaves. *Courtesy Dr. Tom Griggs, WVU.*

FORAGE QUALITY, ANTIQUALITY, AND PALATABILITY

Plants differ in forage quality and palatability. The primary determinant of forage quality is age of the plant material. Young growth will be higher quality than older growth. Another determinant is that legumes have lower NDF and higher cell soluble content than grasses (Figs. 5.8 and 5.9). Some plants have major antiquality components such as wild endophyte in KY-31 tall fescue.

COMPETITION BETWEEN PLANTS

In natural ecosystems, the proportion of different plant species changes as part of a process called succession. Drivers of succession can be plant life history and adaptation to changes in soil (pH, depth, organic matter content, and mineral content), water availability, and climate. Environmental disturbances such as fire, storms, or human activity can cause relatively sudden, dramatic effects on an ecosystem. Some organisms make changes that help others (facilitation), and some species negatively affect others (inhibition). For example, horseweed (*Conyza Canadensis*) prevents the growth of asters because horseweed shades asters and releases substances from its roots that inhibit aster seedling growth. The fact that ecosystems can contain many different species of plants is related to the tolerance of one species for another. Species that cannot tolerate the presence of other species or changes in the environment or disturbances such as switching from a hay crop management to pasture management will contribute a smaller percentage of the total number of plants and may die out.

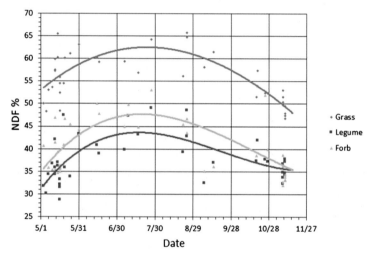

FIGURE 5.8 When grown in the same climatic and soil environment and harvested at the same regrowth age grasses are high in neutral detergent fiber (NDF), legumes are low in NDF with forbs being slightly higher in NDF than legumes. The NDF content of all forages tend to be lower in the cool weather of spring and fall compared to summer.

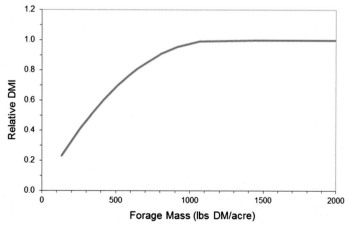

FIGURE 5.9 Forage mass or dry matter (DM) yield of a pasture affects the relative DM intake of livestock on the pasture (relative intake of 1 is 100% of potential intake). When forage mass drops below about 1200 lbs. DM/acre, intake decreases due to lack of forage availability.

Plants compete with each other for space, sunlight, water, and nutrients. When allowed, tall-growing plants will overtop short plants, allowing them access to more sunlight. Deep-rooting plants will be more competitive for water than shallow-rooted plants. However, in a dry year with light showers that only wet the upper soil, plants with dense, fibrous, shallow roots will capture the water before it penetrates deep enough for the deep-rooted plants.

Grasses are more competitive than legumes when nitrogen is available and when potassium is low. When legumes fix nitrogen for a few years, the nitrogen status of the soil will increase, which stimulates the growth of grasses. This vigorous growth can crowd out the legumes for a while until the excess nitrogen is used up, allowing the legumes to come back into the stand. Grass roots take up potassium more readily than legume roots. Therefore, when managing for legumes as the nitrogen source in pastures, maintaining the soil test potassium in the high/optimum range is essential to keep potassium from limiting legume health and vigor.

PLANT DIVERSITY, MORPHOLOGY, AND TOLERANCE TO GRAZING

Most forage plants can be grazed successfully when the timing and intensity of grazing is controlled within the bounds tolerated by the plant. Some plants tolerate long regrowth periods, while others do best when defoliated at shorter intervals. A main factor is lack of tolerance to shading or self-shading, which reduces tiller production or growth. Lack of tolerance to self-shading often occurs in species considered pasture species (Table 5.1). Hay species tend to be more tolerant and do well under longer growth periods. When species differing in tolerance or requirement for rest interval are present in a pasture the rest interval imposed will determine the dominant species in the pasture (Table 5.2).

THE GRAZING ANIMALS

Grazing animals are primary consumers. Their nutritional requirement is determined by age, production state, and production rate. An adult mare doing no work, not pregnant or lactating, and having minimum requirements to maintain her body weight and condition has the lowest daily nutrient requirements among horses (Table 5.3). A pregnant mare's nutrient

TABLE 5.1 Plant Species Differ in Tolerance to Long Growth Periods and Shading

Tolerant (Hay Type)	Intolerant (Pasture Type)
Smooth bromegrass	Tall fescue
Timothy	Orchardgrass
Reed canarygrass	Perennial ryegrass
Quackgrass	Kentucky bluegrass
Red clover (±)	White clover
Alfalfa	
Birdsfoot trefoil	
Native WS grasses	

TABLE 5.2 Plant Growth Habit is Related to the Plant's Tolerance to Harvest Frequency

Weeks rest	2	3	4	5
Avg. no. cuts/yr.	16	11	8.5	7
Species	Avg. lbs. forage/harvest			
Ladino clover	699	1209	1725	2286
Broadleaf trefoil	455	831	1231	1977
Alfalfa	465	884	1715	2943
3-way mix	661	1138	1772	2809

At frequent harvest intervals Ladino clover was most productive and yield of the 3-way mix reflected Ladino clover yield. At a lax harvest interval alfalfa was the most productive and yield of the 3-way mix reflected alfalfa yield.
Adapted from Peterson, M.L., Hagan., R.M., 1953. Production and quality of irrigated pasture mixes as influenced by clipping frequency. Agron. J. 45, 283–287.

TABLE 5.3 Daily Nutrient Requirements of Horses With a Mature Body wt. of 1100 lbs. (500 kg)[a]

Type of Horse	Body wt. (lbs.)	Average Daily Gain/ Milk[b]	Dry Matter Intake (lbs.)[c] (Approx.)	Digestible Energy (Mcal)[d]	Crude Protein (grams)
Adult, no work, minimum	1100		22	15.2	540
Pregnant mare, 5 months	1109	0.31	25	17.1	685
Pregnant mare, 8 months	1150	0.70	26	18.5	759
Pregnant mare, 11 months	1245	1.4	28	21.4	893
Lactating mare, 1 month	1100	36	33	31.7	1535
Lactating mare, 3 months	1100	33	33	30.6	1468
Lactating mare, 6 months	1100	24	27.5	27.2	1265
Growing horse, 4 months	370	1.85	9.25	13.3	669
Growing horse, 6 months	475	1.6	11.9	15.5	676
Growing horse, 12 months	706	1.0	17.6	18.8	846
Growing horse, 18 months	851	0.64	21.3	19.2	799
24 months, moderate exercise	944	0.4	21.2	24.8	888
24 months, very heavy exercise	944	0.4	23.6	32.5	1091

[a]Nutrient Requirements of Horses, sixth revised edition, National Research Council (2007).
[b]Gain in body weight or production of milk in lbs./day.
[c]2% of body wt. for no work; 2.25% of body wt. for pregnancy and moderate exercise; 2.5% for growing horses and very heavy exercise; 3% for 1 month through 3 months into lactation; 2.5% for 6 months into lactation.
[d]Megacalories or 1,000,000 × the amount of energy needed to raise the temperature of 1 mL of water from 20 to 21°C.

requirements increase gradually toward foaling; then they increase rapidly at foaling to provide nutrients for peak milk production. These requirements decline as milk output declines. As a foal's body grows, its total daily nutrient requirements rise, until it reaches mature size. The rate of growth begins to level off around 24 months of age (with variation among and within breeds). After that, daily nutrient requirements vary with the level of work (Table 5.3).

The choices animals make in selecting food can be described by the mathematical models in optimal foraging theory, which predicts that an animal's diet is a compromise between the costs and benefits associated with different types of food. Animals eat to maximize the energy intake in a meal, with minimal costs in terms of time and energy to find, capture, and consume their food. The benefits are the nutrients and energy obtained from the food. Horse feeding behavior is described in more detail in Chapter 8.

FORAGE MASS AND DRY MATTER INTAKE

Timing and intensity of grazing determine animal as well as plant performance. Forage age impacts forage quality and forage mass, which affects selective grazing and intake. Horses, cattle, and sheep are able to eat the most forage when pasture is young, thick, and tall. When forage mass drops below about 1200 lbs. DM/acre, there is not adequate forage for an animal to eat as much as the digestive tract can hold every day (Fig. 5.9). In a mixed grass-clover pasture, 1200 lbs. forage DM/acre is approximately represented by a 3- to 4-inch ruler height.

Estimates of voluntary dry matter intake (VDMI) for grazing horses range from about 1.5% to 3.1% of body weight, according to studies summarized by the National Research Council, in *Nutrient Requirements of Horses, sixth edition*, 2007. Many reports on horse VDMI list 2.0%−2.5% of body weight per day. An increase of pasture quality may result in an increase of VDMI if a horse previously had an insufficient digestible energy (DE) intake, whereas horses with previously adequate DE intake might reduce their VDMI to maintain a constant DE intake (NRC, 2007).

SELECTIVE GRAZING

Livestock have the ability to selectively graze. Selective grazing occurs when animals select and consume forage of a higher quality than the average forage quality in the pasture (Rayburn et al., 2008). The degree of selectivity is determined by how intensively the pasture is grazed (Table 5.4). The quality of the diet consumed is determined by the average pregrazing forage quality (reflected in Table 5.4 by Whole Pasture Total Digestible Nutrients Percent) and the degree of utilization. As TDN percent increases or utilization percent decreases, the apparent diet quality increased. As pasture utilization percent increases, due to either a higher stocking rate or a longer time on the pasture, there is more competition for immature, high quality parts of plants and a higher percentage of mature, lower quality plant parts will be eaten.

Foraging preference by animals is partly a species behavior and partly learned behavior. Grazing animals differ in their ability to be selective grazers based partly on muzzle width (deer vs. horse vs. cow). Animals also differ in their ability to consume coarse, high-fiber forage based on stomach size and flow rate. For example, deer have a relatively small rumen with high flow rate and need to be selective foragers of rapidly digestible plant parts. A beef cow has a relatively large

TABLE 5.4 Selective Grazing Allows Cattle to Eat a Higher Quality Diet Than the Average Forage Quality in the Pasture if They are Not Required to Graze the Pasture Too Close. As Pasture Utilization Increases, Selective Grazing Decreases

Whole Pasture TDN Percent[a]	Pasture Utilization Percent[b]			
	20	40	60	80
	Apparent Diet Quality TDN Percent			
60	68	65	63	60
65	73	71	68	66
70	79	77	74	72

[a]Total Digestible Nutrients (TDN) can be used to reflect the energy content of feeds, either as a weight or a percentage. TDN is the sum of digestible carbohydrates, plus digestible protein, plus digestible fats times 2.25, because fats provide 2.25 times more energy per gram than a gram of carbohydrate or protein.
[b]Utilization increases with stocking intensity and time that a paddock is grazed.

rumen and does well with high-fiber forage that resides in the rumen for some time to be digested. The horse is a monogastric herbivore with a relatively small stomach that seems to function best when horses eat many small meals per day, sometimes giving the impression of grazing through most of the day. A collection of reports on grazing time revealed that horses may graze for as short as 10 h or as long as 17 h per day (NRC, 2007).

It is difficult to find research reports that help to generalize about horses' preferences for different forage plants, but NRC (2007) summarized the following: pastures of mixed species were preferred to monocultures, and grasses were preferred to legumes and herbs. Wild horses thrive on wild growing plants in many parts of North America, where there is a mixture of grasses and forbs. It is logical that wild horses evolved in environments where similar mixtures prevailed.

It would be useful to know which forage heights are preferred by horses, but reports are conflicting. Timothy swards were preferred at 5 cm rather than 20 cm tall (Hayakawa, 1991), but perennial ryegrass was preferred at 15 cm rather than 4.5 cm tall (Naujeck et al., 2005). Fresh herbage intake is probably governed by the interaction of a number of factors, including plant maturity, herbage quality, and sward characteristics that dictate herbage bite mass, rather than sward height alone (NRC, 2007). Where herbage conditions allow a large bite mass, the rate of biting might decrease, so intake rate stays constant (NRC, 2007). The bite rate of horses on pasture has been reported in the range of 8−18 bites per minute, depending on the forage species (NRC, 2007).

REVIEW QUESTIONS

1. What are the four parts of pasture ecology?
2. What class of organisms are the classical primary producer in any ecosystem?
3. The balance between ____ and ____ determine plant growth and carbohydrate content or reserves?
4. What are three environmental resources that neighboring plants compete for?
5. In a pasture, what management factors have a great impact on plant and animal health and nutrition?
6. What is the goal of good pasture management?
7. What factors determine the nutrient requirement of an animal?

REFERENCES

Understanding Astronomy: The Sun and the Seasons. http://physics.weber.edu/schroeder/ua/SunAndSeasons.html.

Hayakawa U. Grazing management of yearling racehorses. 2. Sward canopy height in set grazing. J. Jpn. Soc. Grassl. Sci. 1991;31:337−42.

Naujeck A, Hill J, Gibb MJ. Influence of sward height on diet selection by horses. Appl. Anim. Behav. Sci. 2005;90:49−63.

NRC (National Research Council). Nutrient Requirements of Horses. Sixth Revised Edition. National Research Council of the National Academies. Washington, DC: National Academy Press; 2007.

Peterson ML, Hagan RM. Production and quality of irrigated pasture mixes as influenced by clipping frequency. Agron. J. 1953;45:283−7.

Rayburn EB, Whetsell MS, Lozier JD, Smith BD, L Shockey W, Seymour DA. Initial Nutritive Value and Utilization Affect Apparent Diet Quality of Grazed Forage. Online. Forage and Grazinglands; 2008. https://doi.org/10.1094/FG-2008-0903-01-RS. http://www.plantmanagementnetwork.org/fg/.

FURTHER READING

Campbell NA, Reece JB. Biology Seventh Edition. San Franciso, CA: Pearson Education, Benjamin Cummings; 2005.

Hodgson J. The influence of grazing pressure and stocking rate on herbage intake and animal performance. In: Hodgson J, Jackson DK, editors. Pasture Utilization by the Grazing Animal. British Grassland Society; 1975. p. 93−103. Occasional Symposium 8.

Russell PJ, Hertz PE, McMillan B, Brock Fenton M, Addy H, Maxwell D, Haffie T, Milsom B. Biology: Exploring the Diversity of Life. Second Canadian Edition. Toronto, Ontario, Canada: Nelson Education; 2013.

Sadava D, Heller CH, Orians GH, Purves WK, Hillis DM. Life: The Science of Biology. eighth ed. 2008.

Chapter 6

Pasture Plant Establishment and Management

S. Ray Smith and Krista L. Lea

Department of Plant and Soil Sciences, College of Agriculture, Food and Environment, University of Kentucky, Science Center North, Lexington, KY, United States

CHARACTERISTICS OF A HEALTHY PASTURE

There are over 10 million horses in the United States and Canada, and the majority of their lives are spent on pasture. Pastures supply nutrition, provide for exercise, control erosion, and add to the aesthetic value of horse farms. The ability to establish and manage horse pastures is therefore important to horse owners. Good pasture management on equine operations is essential to maintaining the productivity and value of the pasture by providing the proper nutrients and environment to grow a healthy horse.

Horses graze closer than cattle and tend to repeatedly graze the same areas of a pasture, so desirable forage plants in a pasture can be reduced or eliminated. Hooves can also damage pastures, even with grasses that form tight sods. Areas around gates, fence lines, waterers, and hay feeders endure the most traffic and are the hardest to maintain.

Good establishment and management principles must be employed to maximize the value of pasture forages, and in a broader sense, the overall value of pastures to the horse. Establishment principles include proper fertilization, species and variety selection, seeding date and rates, seeding method, and control of weed competition, which are discussed in detail in this chapter. Management principles include grazing plans or rotations, paddock design and fencing, parasite control, and regular fertilization and weed management.

Establishment

The following recommendations will increase the chance of success whether seeding all or part of a pasture.

Ensure proper soil fertility. A current soil test will indicate the amount of lime, phosphorus, potassium, and other nutrients (except for nitrogen) needed for the species to be seeded. Non-nitrogen soil amendments can be applied any time of the year that weather conditions allow equipment on the field.

Use high-quality seed of an improved variety. Many varieties of commonly established grasses, such as Kentucky bluegrass, orchardgrass, tall fescue, perennial ryegrass, and Bermudagrass, are available for pasture plantings in North America. It is recommended to seed grass varieties that have been proven to be top performers under the environmental conditions of the region. For Kentucky and adjoining states, the University of Kentucky has an extensive forage variety testing program. To review these reports, simply download at http://forages.ca.uky.edu/variety_trials. Many states and provinces have similar variety trials that can be easily accessed online.

High-quality seed has high rates of germination and zero or minimal contamination from other crop or weed seed. Look for this information on the seed tag; in Canada and the United States a blue certified seed tag is a guarantee of seed quality and purity. One of the largest determiners of seed price is whether the seed has gone through the certification process, and certification is an important guarantee of quality. In addition, state and national seed laws around the world require that all bags of agricultural seed contain a white tag that shows the germination percentage, inert matter, weed seed, and the date these parameters tested. Sometimes, this information is "stamped" directly on the bag. Use this information to ensure that the seed purchased has a high germination rate, low weed seed count, and has been tested within the last year (Figs. 6.1 and 6.2).

FIGURE 6.1 Blue certified seed tag is a guarantee of seed quality and purity.

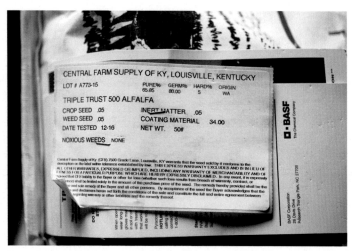

FIGURE 6.2 White seed tags show germination percentage, inert matter, weed seed, and other parameters of the particular seed lot.

When buying tall fescue or perennial ryegrass seed for pasture used by pregnant mares, make sure that it also contains a green tag that clearly states that this variety is endophyte-free or low endophyte (usually less than 5% infection). If this information is not clearly stated, assume the seed is infected with a fungal endophyte that increases stand longevity but which also produces toxic ergot alkaloids. Do not use seed that is infected with endophyte in pastures to be grazed by pregnant mares. Never plant turf-type tall fescue or perennial ryegrass varieties because they contain high endophyte levels. Fortunately, there are now varieties of both species that contain a novel or beneficial endophyte that do not produce ergot alkaloids and are safe for pregnant mares. These varieties show longer term survival than endophyte-free varieties because of the beneficial effect of the endophyte.

Plant enough seed at the right time. When seeding a new pasture, it is essential to use the recommended seeding rate to ensure good ground cover and competition with weeds (Table 6.1).

The upper seeding rate listed for each species increases seed cost but also increases the chances for rapid establishment. When sowing a mixture of plant species, less seed of each component is used than when sown alone. Grasses and clovers can be seeded in either spring or fall. However, cool season grasses (Kentucky bluegrass, orchardgrass, tall fescue, and perennial ryegrass) are most easily established in the late summer and early fall, except in the northcentral United States and Central Canada, where early spring seeding is usually required. Bermudagrass should only be planted in early summer since it is a warm season grass and will not tolerate below-freezing temperatures at a seedling stage. Seedings of white clover are usually best seeded in early spring but may also be planted in late summer.

TABLE 6.1 Common Seeding Rates and Optimum Seeding Dates for Pasture Plant Species

Species	Rate lb/acre (kg/ha) (Seeded Alone)	Rate lb/acre (kg/ha) (In Mixtures)	Optimum Seeding Dates
Endophyte-free or novel endophyte tall fescue	20–40 (22–45)	10–20 (11–22)	Late summer/early fall
Orchardgrass	15–30 (17–37)	10–15 (11–17)	Late summer/early fall
Kentucky bluegrass	15–30 (17–37)	10–15 (11–17)	Late summer/early fall
Endophyte-free perennial ryegrass	20–40	5–15 (6–17)[a]	Late summer/early fall
Bermudagrass	5-8 (6–9)	N/A	Early summer
White clover	N/A	1–3 (1–4)	Winter[b], early spring

[a]When seeding a grass mixture, never seed more than 20% perennial ryegrass. Ideally, only 10%–15% is needed to provide quick cover without out-competing the other perennial grasses.
[b]Winter seeding of clovers, often called frost seeding, can be made on closely grazed or mown pastures 4–6 weeks before the average last date for below zero temperatures.

Use the best seeding method available. In general, planting into a conventionally tilled seedbed is more effective than no-till seedings, especially when a cultipacker or a modified roller is used to firm the seedbed before and after seeding. Brillion seeders combine packing and seeding into one operation. No-till seeding is preferred on sloping land and when overseeding existing pastures. This practice does not disturb the existing pasture sod and therefore is most common on horse pastures. A common, but less accurate method is to harrow or disk a pasture, broadcast the seed, then harrow or drag the pasture to cover the seed. With this method, a roller or cultipacker is recommended after seeding to achieve good soil-to-seed contact. All three methods can be successful if the seed is placed in firm contact with the soil at a depth from which the seedling can emerge. The forage crops listed in Table 6.1 should all be sown at a depth of ¼ to ½ inch (0.6–1.3 cm), with the exception of Bermudagrass, which should be seeded at ⅛ to ¼ inch (0.3–0.6 cm).

Frost seeding refers to the practice of broadcasting seed on top of the ground during the winter and relying on the freeze/thaw cycle of late winter to work the seed into the soil. Frost seeding is recommended only for clover seed sown onto very closely grazed or clipped sod, 4–6 weeks before the average last date of below-freezing temperatures (mid-February in the transition zone of the United States). Seeding grasses via frost seeding is not recommended for some grass species and in some regions since the success rate for grasses varies greatly depending on environmental conditions (Fig. 6.3).

In addition to providing the nutrition needed by the horse, pasture should be established with long-term sustainability in mind. Establishment of pastures with multiple species not only provides horses with multiple options for forages but also creates a more sustainable grassland ecosystem (Soder et al., 2007). A pasture established with a mixture of grasses and legumes will be more sustainable because legumes effectively self-fertilize through nitrogen fixation (Graham and Vance, 2003). Many forages perform well in some areas due to appropriate temperature and rainfall parameters but perform poorly

FIGURE 6.3 The transition zone represents areas that both cool season and warm season forages can survive and be utilized in designing rotational grazing plans.

in other regions. Pastures should be established with species that will perform well within specific regional climatic parameters.

Control competition. Weeds and existing forages can outcompete young seedlings by shading and taking up soil moisture. Control competition by using a combination of heavy grazing, close clipping, or herbicides. Herbicides often have a post-application waiting period before seeding. Read and consider all label instructions before selecting a herbicide for weed control. Do not let weeds or grasses go to seed the year of establishment. If allowed, these unwanted seeds will also have an opportunity to germinate when desirable grasses are seeded and will contaminate an otherwise clean pasture.

Common Forage Species for Horse Pastures

Choose the forage species best adapted to your climatic zone and soil conditions. For example, Bermudagrass is ideally adapted for horse pastures across the southern United States, and certain winter hardy varieties will survive in the transition zone, but Bermudagrass is not adapted to the northern United States or Canada. Conversely, Kentucky bluegrass is uniquely adapted for horse pastures across Canada, the northern United States, and much of the transition zone but will not survive the hot summers of the southern United States. Perennial ryegrass survives across the northern United States and areas of southern Ontario, but it provides only short-term pasture in the transition zone and southern United States.

Common Grass Species for Horse Pastures

- Kentucky bluegrass is a long-lived, cool season grass species with a spreading or rhizomatous root system. This sod-forming ability along with high palatability, and vigorous spring and fall growth make it the most desirable grass species for horse pastures where it is adapted (transition zone and northward across North America). It shows excellent cold tolerance but goes dormant during summer drought conditions, and forage yield is lower than most other cool season grasses.
- Orchardgrass is a productive cool season grass species with a bunch-type growth habit. It is known as a quality forage grass with high yield potential and is ideally suited for hay production. It does not tolerate close continuous grazing, and even under managed grazing, stand survival is usually 4–5 years. It is adapted to much of North America with the exception of the US Plains states, western Canada, and the southern United States.
- Endophyte-free or novel endophyte tall fescues are ideal cool season grasses for horse pastures and are adapted to a slightly larger zone than orchardgrass. In contrast to orchardgrass, tall fescue is adapted to more frequent grazing. This species has the added benefit of being tolerant to drought and waterlogged soil conditions, but it will go dormant under high temperatures in the summer. Tall fescue has a rough leaf texture that is not as palatable as other cool season grasses. Endophyte-free varieties usually survive 4–5 years, while novel endophyte varieties typically survive 10+ years if not overgrazed. Avoid planting turf-type tall fescue because of high levels of toxic endophyte in these varieties, and the variety KY-31 should not be used for pastures grazed by pregnant mares because it contains high levels of the toxic endophyte.
- Perennial ryegrass is a short-lived, cool season grass that has exceptionally high seedling vigor and can be used for a 2-year solution or to thicken up troublesome areas around the farm in the transition zone or a 3- to 4-year pasture in the northern United States and Southern Ontario. Insist on endophyte-free, forage-type perennial ryegrasses. The turf-type perennial ryegrasses are almost always highly infected with a toxic fungal endophyte, while forage types are not. It is a high-quality and high-yielding forage crop, but high concentration of soluble sugars can be an issue for laminitis-prone or diabetic horses.
- Bermudagrass makes an excellent horse pasture due to its grazing tolerance, heat and drought tolerance, and ability to carry high stocking rates during the summer. This species requires high levels of nitrogen and potassium to maintain productive stands. Bermudagrass does not survive north of the transition zone, and even here, it is essential to only plant varieties that have proven winter survival. It spreads rapidly with underground rhizomes and aboveground stolons, forming an excellent sod, but this aggressive spreading ability can make it a weed in non-pasture situations.
- Timothy is an ideal grass species for hay production and poorly drained soils, but it is limited by a short stand life and poor survival under close grazing. It is best suited for the northern United States and Canada with marginal adaptation in the transition zone.
- Many forage grass species are found in horse pastures, but those listed before are the most common and typically the best adapted. For example, species like smooth bromegrass and reed canarygrass perform well for hay in the northern United States and Canada, but they show poor stand survival under frequent or continuous grazing. In contrast, crabgrass shows good grazing tolerance, but is a warm season annual and has to emerge from seed each summer.

Common Legume Species for Horse Pastures

- White clover is the best adapted legume for horse pastures. It has a low growth habit and ability to spread by above-ground stolons, allowing it to tolerate frequent and close grazing. As a legume, it is self-sufficient for nitrogen and provides nitrogen to surrounding grasses through N-fixation. It is lower yielding than legumes like red clover and alfalfa, and it goes dormant under high temperatures and dry conditions during the summer. It is similar to orchardgrass and tall fescue in range of adaptation.

- Alfalfa and red clover are both known as high-quality, high-yielding species for hay, but they are rarely used in horse pastures in the United States and Canada because of poor tolerance to frequent close grazing. Interestingly, it is common for horses to graze alfalfa in Argentina and Australia, but in these countries, well-designed rotational grazing systems are utilized. In addition, red clover can occasionally cause excessive salivation mid to late summer, often called "the slobbers." This is caused by a mycotoxin called slaframine, a by-product of the fungus *Rhizoctonia leguminicola*, which can infect red clover and other legumes (Arnold and Smith, 2015).

Management Following Establishment

Soil Sampling and Fertilizing

Once pastures have been fully established (12–18 months after planting, so a solid sod has formed), sound pasture management is critical for maintaining a healthy and vigorous stand that benefits the horse, the owner, and the environment. Once soil fertility reaches recommended levels, pastures should be soil tested every 3 years to track changes in pH and fertility. Hay fields should be tested every year because hay harvests remove large amounts of soil nutrients. Soil pH can dramatically affect nutrient availability and plant growth. Maintain pH between 6.2 and 6.5 for grass-legume pastures by applying lime according to the soil test. Grasses will survive lower pH levels, but they are most productive when pH is maintained above 6.0. Phosphorous and potassium should be maintained in the high range as determined by soil testing; in certain regions of North America, adding micronutrients such as copper, zinc, and selenium may be recommended.

Maintaining proper fertilization will encourage continued development of pasture grasses and is often the most overlooked aspect of pasture management. For typical hay harvest situations, a spring nitrogen fertilization plan is recommended to increase yield at first cutting. However, in pasture management, the goal is sustained growth and health of the stand over time. For this reason, horse farm managers with predominately cool season grass pastures are encouraged to apply one to two applications of nitrogen each fall (typically, 30–60 lb/acre [34–67 kg/ha]) in early fall and a second application in late fall/early winter (Smith et al., 2010). Fall fertilization encourages tillering and promotes thicker stands, lengthens growing season by 2–4 weeks, and promotes early green-up in the spring. If only one nitrogen application can be made, it should consist of 60 lb/acre (67 kg/ha) in late fall, as long as the vegetative material is still green. At high stocking densities, nitrogen should be applied at low application rates throughout the grazing season. This may consist of 30–40 lb/acre (34–45 kg/ha) at green-up and again late spring and in the fall. It is important to remember that fertilizer recommendations are typically based on the level of the actual nutrient and not what is in the bag of fertilizer. For example, to apply actual nitrogen at 40 lb/acre (45 kg/ha) would require 87 lb/acre (96 kg/ha) of urea since urea is only 46% nitrogen. Warm season grasses like Bermudagrass should be fertilized with N at green-up and perform best when additional N applications are provided monthly or bimonthly until midsummer. (Fig. 6.4).

Weed Control

The effects of weeds in a pasture are often underestimated. Weeds greatly reduce the productivity and carrying capacity of a pasture. Weeds often become dominant when the grasses are abused by overgrazing, mowing too close, or when soil fertility is low. When proper fertility is maintained, pasture grasses will typically outcompete weeds. In addition, most weeds are not toxic to horses and moderate amounts can be tolerated and can even provide useful nutrition to grazing horses.

Many new seedings fail due to competition from weeds. Broadleaf weeds can be controlled with herbicides on newly established stands after the forage seedlings are well established. Otherwise, herbicide damage to the new seedlings is possible. Spraying broadleaf weeds before seeding is an effective option, but it is essential to follow label directions for the recommended waiting period before seeding. Mowing is effective on erect, upright weeds taller than the seedling grasses, such as ragweed. Low-growing weeds, such as dandelions and plantain, are not controlled effectively by mowing.

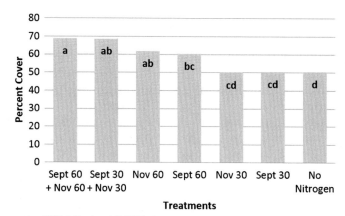

FIGURE 6.4 Percent forage cover spring 2007 following fall 2006 nitrogen applications in Lexington, Kentucky. This illustrates the advantage of two fall nitrogen applications to spring forage cover.

Herbicide applications can be useful to control problem weeds. Knowing the weed species is essential when selecting a herbicide and deciding the application timing. Always follow herbicide label recommendations, but there is little to no information on withholding periods for horses after treating pastures with herbicides. Pesticide manufacturers rarely undertake grazing/residue studies with horses to develop their labels as they do with cattle or sheep. Most categorize horses as a "non- food" animal and specify that no withholding period is needed.

Many horse pasture specialists have developed some of their own policies over the years. Ross Watson (personal communication) in Australia advises farm managers to follow the recommended withholding period that is specified for cattle (normally 7—14 days) for dry mares, yearlings, nonpregnant, and/or nonbreeding horses. For pregnant mares or mares with foals, breeding stallions, and particularly high value animals, he recommends doubling the labeled withholding period before re-entry of these horses (e.g., 7 days on label becomes 14 days).

Clipping/Grazing

Many studies have revealed grazing to significantly improve overall plant species diversity in pastures and grasslands. One such study in southern Europe suggested that this relationship between grazing intensity and species richness approximates a bell curve: both ungrazed and overgrazed areas experience lower species richness, while the maximum species richness correlates with a moderate grazing intensity somewhere in between (Hadjigeorgiou et al., 2005). Loucougaray et al. (2004) found that grazing horses alone, and even more so in conjunction with cattle, led to improved species richness in coastal grasslands in France. They found that the horses in their study tended to avoid grazing their defecation areas, which became highly localized patches of low species diversity dominated by as few as two species. However, in combination trials, cattle grazed the horse latrine areas, increasing the species richness in those areas (Loucougaray et al., 2004).

Managing Endophyte-Infected Tall Fescue

The majority of tall fescue in North America is the variety KY-31. This variety is infected with a fungal endophyte that produces toxic ergot alkaloids and is a concern to the equine industry. These ergot alkaloids lead to complications in pregnancy and delivery when consumed in the third trimester (Smith et al., 2009). Last trimester pregnant mares should only be placed in paddocks that contain little or no toxic tall fescue. KY-31 tall fescue has become a naturalized grass species in the United States and now occurs on over 14 million ha (34.6 million ac). Fig. 6.5 illustrates the simple way that one Thoroughbred farm designated pastures that have been carefully evaluated to contain "no tall fescue" and therefore are safe for late-term pregnant mares (Fig. 6.5).

Endophyte-free and novel endophyte fescue varieties have been produced by plant breeding programs to reduce or eliminate ergot alkaloids by targeting the fungal endophyte that produces them. Another option is to eradicate tall fescue altogether and plant instead a different species, such as orchardgrass, KY bluegrass, perennial ryegrass, or a combination of these species. Careful grazing management of infected fescue can reduce risk of toxicosis. Rotating horses to a different pasture when alkaloid content is particularly high in tall fescue (i.e., late spring), supplementing forage with a warm season alternative, seeding a legume into a fescue stand, and clipping highly toxic seedheads are all possible alternatives (Roberts and Andrae, 2004).

FIGURE 6.5 One horse farm in central Kentucky uses "No Fescue" signs at the gates to indicate which pastures and paddocks are safe for late-term mares.

Pasture Renovation

Renovate means to renew and improve. A pasture or hay field that has become less productive can become more productive through renovation or by "renewing" it. In the transition zone of the United States, this usually means adding lime and fertilizer, controlling weeds, and planting an adapted legume such as white clover. The primary benefits of renovation come as a result of getting a legume established in grass-dominated fields.

Renovating pastures with clover increases forage quality, adds nitrogen to the system, and is desirable in horse pastures. White clover is preferred on horse pastures in most areas of the world, over red and other upright growing clover species, as it tolerates closer grazing. Adding legumes to hay and pasture fields brings at least four benefits: higher yields, improved quality, nitrogen fixation, and more summer growth.

Most of the growth of cool season grasses occurs during the spring and fall. Legumes such as red clover, alfalfa, and lespedeza make more growth during the summer months than cool season grasses. Growing grasses and legumes together improves the seasonal distribution of forages and provides more growth during the summer. Fig. 6.6 shows the seasonal distribution of common forages grown in the transition zone of the United States.

Competition from existing vegetation will reduce the success of renovating or overseeding an existing pasture. When seeding into an existing sod, remove as much top growth as possible. This can be accomplished by hay harvest, intense grazing, clipping close, or a controlled burn. When a complete reestablishment is required, one or more cycles of crop harvest followed by a fallow spray of glyphosate is recommended. When a crop is not grown in the field, at least two applications of glyphosate should be used 4—6 weeks apart to ensure that the existing stand is completely killed out. If the goal is to remove an unwanted species such as tall fescue, it is essential to not let the field go to seed the year of establishment. Finally, the seedbed should be prepared based on what species are being planted. Proper preparation of the field is essential in establishing high-quality horse pasture.

Improving or reestablishing grass cover in high-traffic areas around fences, gates, and barns is probably the most common pasture problem facing horse owners. On the assumption that it is not possible to keep horses off such areas until vegetative cover is attained, the question becomes how best to get vegetative cover as fast as possible and maintain it as long as possible.

The weedy pasture (Fig. 6.7A) and the bare pasture (Fig. 6.7B) are likely a result of overgrazing and poor soil fertility. Taking a close look at a pasture by walking across it can help farm owners and managers get a better understanding of their pasture composition and the management steps needed to improve forage production. Generally, pastures that are less than 75% desirable forages could benefit from weed control and overseeding. Pastures less than 50% desirable are candidates for complete reestablishment. The pasture in Fig. 6.7C could also be a candidate for complete reestablishment if grazed by pregnant mares. It contains a significant portion of endophyte-infected tall fescue with few other forages to dilute the ergot alkaloids produced by the toxic endophyte. A healthy pasture is dominated by a mixture of desirable, non-toxic grasses, such as Kentucky bluegrass, orchardgrass, and endophyte-free (or novel endophyte) tall fescue (Fig. 6.7D). The following references provide more practical guidelines for when and how to renovate horse pastures (Teutsch and Fike, 2009; Johnson and Russel, 2007) (Fig. 6.7A—D).

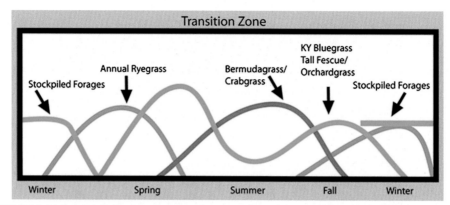

FIGURE 6.6 Seasonal distribution of common forage types grown in the transition zone of the United States.

FIGURE 6.7 (A) A weedy pasture. (B) A pasture with significant bare soil. (C) Broodmare pasture dominated by toxic tall fescue. (D) A healthy pasture contains a mixture of desirable grasses, in this case, Kentucky bluegrass, orchardgrass, and endophyte-free tall fescue.

RESTING PASTURES

Allowing seedlings to mature before being grazed is essential to the success of pasture establishment. Whether overseeding an existing pasture or starting with a new pasture, 6 months or more of growth are needed before seedlings are established. Rest is a key factor in the success or failure of a new pasture seeding. The adjacent pastures seen in Fig. 6.8A and B were

FIGURE 6.8 (A) Fall seeded pasture rested over winter is productive the following spring. (B) Fall seeded pasture heavily grazed all winter is a total loss and will require seeding again the next fall.

managed in a similar fashion in the fall, with both being sprayed and reestablished with a cool season grass mix. Both pastures showed excellent forage stands 6 weeks after seeding. One pasture (A) was rested all winter and the following spring was thick, lush and productive. However, the other pasture (B) was not rested, but heavily grazed all winter. Virtually no seedlings survived, resulting in a failed establishment. This pasture will have to be reseeded again in the fall.

The length of a rest period between grazings is influenced by obvious factors such as ambient temperature and precipitation. An important management factor is the height of grazing, for shorter grazing of plant shoots causes a proportional shortening of roots and a decrease in photosynthetic leaf area. Following a very short grazing, there is a lag phase before growth rate and leaf area increase enough to promote a rapid increase in photosynthesis and carbon fixation (Humphries, 1997). As plants grow, some parts begin senescence and decomposition, having a negative effect on forage production. So rest periods should be short enough that forage production does not decline significantly. Furthermore, as plant height increases, there will be an initial increase in forage intake, followed by a leveling off at a maximum for animals of a certain size and physiologic state.

During the resting period, carbohydrates produced by photosynthesis will be stored in roots, stems, rhizomes, or stolons. Plant species with plenty of leaf close to the ground and below the grazing height will regrow faster than plants with very little leaf area close to the ground. Increasing the number of paddocks can increase the amount of time available for each paddock to regrow. Being able to vary the rest period allows grazing managers to provide paddocks at ideal stages of yield and quality to meet the changing physiologic needs of horses and for optimum plant longevity. During periods of rapid plant growth an ideal rest period could be around 17 days, and during slow plant growth, it could be more than 35 days.

Two overseeding trials at the University of Kentucky provided unexpected results and showed the regenerative properties of cool season grass pastures that are rested. In these trials, heavily grazed pastures were overseeded using

FIGURE 6.9 Regrowth with just 6 weeks of rest; animals still graze beyond the electric fence.

different seeding methods and grass species in a randomized design. In both locations, over both years, the control strip (not seeded) performed as well as the seeded strips because horses were removed and the existing perennial grass species tillered (produced new basal shoots) and resumed competitive growth. While these studies did not provide the overseeding data that was expected, they do provide a good demonstration of the importance of resting overgrazed pastures. In Fig. 6.9, little difference can be seen in the seeded strips, but the entire plot area was significantly more productive than the area just beyond the electric fence, where animals still graze. This photo was taken just 6 weeks after the experimental area was fenced off. It is important to note that overseeding occurred in the spring when environmental conditions were ideal for cool season grass growth (Fig. 6.9).

REVIEW QUESTIONS

1. List five establishment principles to increase the chances of successful pasture establishment.
2. Describe what colored seed tags tell the end user.
3. Three seeding methods are described in this chapter: seeding into a prepared seed bed, no-till drilling, and drag/broadcast/drag. What principles are required for all of these methods to be successful?
4. Complete the following chart:

Species	Season	Positive Attribute	Negative Attribute
KY bluegrass	Cool	Sod-forming, high palatability, vigorous growth	
Orchardgrass		Quality forage, high yielding, good for hay production	
E− or novel tall fescue			
Perennial ryegrass			Short lived, high sugar content
Bermudagrass			Requires high levels of nitrogen and potassium, does not survive north of the transition zone
Timothy	Cool	Good hay production, adapted to poorly drained soils	
White clover	Cool		
Alfalfa and red clover	Cool		Poor tolerance to frequent close grazing

5. How often should pastures and hayfields be soil sampled and why are the recommendations different?
6. Based on the text, when should pastures be considered a candidate for complete reestablishment?

REFERENCES

Arnold M, Smith SR. Slaframine Toxicosis or "Slobbers" in Cattle and Horses. Univ. of Kentucky Agric. Exper. Station Publication; 2015. ID-110, https://uknowledge.uky.edu/anr_reports/110.

Graham PH, Vance CP. Legumes: importance and constraints to greater use. Plant Physiol. 2003;131:872−7.

Hadjigeorgiou I, Osoro K, Fragoso de Almeida JP, Molle G. Southern European grazing lands: production, environmental and landscape management aspects. Livest. Prod. Sci. 2005;96:51−9.

Humphries LR. Evolving Sciences of Grassland Improvement. Cambridge University Press; 1997.

Johnson KD, Russell MA. Maximizing the Value of Pastures for Horses. Purdue University Agronomy Extension Publication; 2007. ID-167, https://www.agry.purdue.edu/ext/forages/publications/ID-167.htm.

Loucougaray G, Bonis A, Bouzille J. Effects of grazing by horses and/or cattle on the diversity of coastal grasslands in western France. Biol. Conserv. 2004;116:59−71.

Roberts C, Andrae J. Tall fescue toxicosis and management. Onling Crop Manage. 2004. https://doi.org/10.1094/CM-2004-0427-01-MG.

Smith SR, Schwer L, Keene TC. Tall fescue toxicity for horses: literature review and Kentucky's successful pasture evaluation program. Online Forage Grazinglands 2009. https://doi.org/10.1094/FG-2009-1102-02-RV.

Smith SR, Lacefield G, Schwer L, Witt W, Coleman R, Lawrence L. Establishing Horse Pastures. Univ. of Kentucky Agric. Exper. Station Publication; 2010. ID-147, http://www2.ca.uky.edu/agcomm/pubs/id/id147/id147.pdf.

Soder KJ, Rook AJ, Sanderson MA, Goslee SC. Interaction of plant species diversity on grazing behavior and performance of livestock grazing temperate region pastures. Crop Sci. 2007;47:416−25.

Teutsch CD, Fike JH. Virginia's Horse Pastures: Forage Establishment. Virginia Cooperative Extension. Publication; 2009. 418−103, https://pubs.ext.vt.edu/418/418-103/418-103.html.

Chapter 7

Forage Yield and Its Determination

Paul Sharpe[1] and Edward B. Rayburn[2]

[1]University of Guelph, Guelph, ON, Canada (retired); [2]West Virginia University Extension Service, Morgantown, WV, United States

INTRODUCTION AND QUESTIONS ABOUT FORAGE YIELD

If you want to manage something, measure it first.

Ensure that you know what to measure and how to measure it accurately.

The amount of forage dry matter (FDM) that each grazing horse eats on pasture in a day is within a range of 1.5%–3.2% of the horse's body weight. Since feeds vary in their dry matter percentage and nutrient requirements are expressed on a dry matter basis, amounts of forage produced and consumed are considered on a dry matter basis. Live, growing forage plants consist of about 80% water and 20% dry matter. Hay varies around 85%–90% dry matter. Determining how much FDM a horse needs over a period such as a week involves calculating the amount needed per day (e.g., 1000 pound (lb) horse × 2% = 20 lbs) and multiplying by the number of days (20 × 7 = 140 lbs for a week).

This chapter describes methods of FDM yield determination for pastures of known size. Since plant height is proportional to yield of FDM and height should be quicker to measure than weight, it is practical to use a method of height measurement that is accurate and highly repeatable.

New horse owners do not always appreciate the need to measure and analyze what they are measuring. Imagine that a horse owner bought a hobby farm with three acres (ac) of pasture for her four horses. In spring, the four horses were turned out to the whole three acres, where the horses grazed as much as they liked for the whole grazing season. The horses were in good condition when they were first turned onto the pasture.

After a few weeks it was apparent that the horses liked to graze in some areas of the pasture but not in others and they liked to eat some types of plants more than others. By late summer, the ribs, hips, and withers were showing on the horses (indicating that the horses were not getting enough to eat) and much of the pasture was eaten down to a height of about half an inch (in) (1.27 cm) (indicating that the horses were left on their pasture for too long and some plants had been grazed repeatedly).

Eventually the horse owner started to wonder whether her horses had enough to eat. Then she realized it was time to buy some hay to feed the horses through the winter. A neighbor who had plenty of hay to sell asked, "How much do you want?" A friend wanted to help her answer this question and inquired, "Well, how much did the horses eat since you bought them?" Now the owner realized that there were some important numbers she needed to learn about. After an in-depth discussion, she came up with a list of questions:

1. How much does a horse eat in a day?
2. Does the amount that a horse eats in a day change if the horses are eating pasture versus hay?
3. How much hay or how much pasture will feed a horse for a day or a month or a grazing season?
4. How much land is needed to supply the pasture and the hay for each horse for a year?

The owner checked her favorite social media sites for answers to these questions. She received plenty of replies and noticed to her frustration that there were big differences in the answers. So she decided to go to a presentation on how to manage pastures, put on by her local horse club, with presentations by extension people and visits to horse pastures. Later, she bought a book on pasture management.

Horse Pasture Management. https://doi.org/10.1016/B978-0-12-812919-7.00007-X

ANSWERING THE HORSE OWNER'S FIRST THREE QUESTIONS

1. How much does a horse eat in a day?

The National Research Council report "Nutrient Requirements of Horses, Sixth Revised Edition" (2007) summarized many studies on how much feed horses need per day. A horse's voluntary dry matter intake (VDMI) of feed varies between about 1.5% and 3.2% of body weight per day. Just to maintain the adult body weight of a nonpregnant, non-lactating horse, about 1.5% of body weight is required per day. As pregnancy advances, the VDMI increases. As the level of daily physical exercise increases, VDMI increases, so that a typical racing thoroughbred needs about 2.2% per day. A mare in the peak of lactation needs about 3.2% of her body weight per day as FDM. A convenient average VDMI for most horses is 2% of body weight per day.

2. Does the amount that a horse eats in a day change if the horses are eating pasture versus hay?

The **water** content of forages varies considerably, from about 80% for fresh, growing grasses and legumes (thus, 20% **dry matter**) to about 10%–15% for most types of hay (85% - 90% dry matter). Consequently the mass of feed that a horse eats in a day depends upon the form of the feed and its dry matter percentage (% DM). For consistency, feed intake is reported as feed dry matter per day (FDM/d) or dry matter intake as a percentage of body weight. Feed maturity, palatability, and digestibility may also influence VDMI. Palatability is a level of acceptance that results from a combination of experiences and flavors. Several factors that can affect voluntary intake and choices of alternative forages are discussed in Chapter 8. For example, if spoilage or mold growth occurs in any form of feed, the VDMI will likely decrease. However, whether the form of the feed is fresh versus hay versus silage (a wet, fermented feed) has little effect on the VDMI.

3. How much hay or how much pasture will feed a horse for a day or a month or a grazing season?

A horse weighing 1000 lbs (454 kg) and eating 2% of this needs $0.02 \times 1000 = 20$ lbs (9 kg) of dry matter per day. If all of this feed is forage (hay, silage, or pasture), then this represents 20 lbs (9 kg) of FDM. Using simple arithmetic, you can calculate the approximate FDM requirement of any number of horses for a given number of days. For example, to feed four horses with an average weight of 1100 lbs (498 kg) for 30 days, you need $4 \times 1100 \times 0.02 \times 30 = 2640$ lbs (1197 kg) of FDM.

If the forage dry matter is in the form of pasture, which has 20% DM, dividing the 2640 lbs of FDM by 0.20 reveals 13,200 lbs (5987 kg) of fresh live pasture needed for the four horses over 30 days. Since the dry matter portion contains all of the nonwater nutrients, it is best to do all of your calculations on a dry matter basis.

If the forage dry matter is in the form of hay, which has 90% DM, divide the 2640 lbs of FDM by 0.9 to obtain 2933 lbs (1330 kg) of hay on an "as fed" basis. This is enough for these four horses for 30 days.

Good pasture management includes a recommendation to "take half and leave half" of the FDM in a pasture. Therefore, the 2640 lbs of FDM can be divided by 0.5 to give 5280 lbs (2395 kg) of FDM that should be available as pasture to the four horses over the 30-day period.

PREPARING TO ANSWER THE FOURTH QUESTION

To answer the horse owner's fourth question (How much land is needed to supply pasture and hay for a horse for a year?), she needs to determine the forage yield from the pasture and hay land.

The simplest way to estimate the weight of FDM from pasture land is by measuring forage height and using that number in a formula for calculation of forage yield.

METHODS AND TOOLS FOR DETERMINING FDM YIELD

One of the simplest ways to estimate the FDM yield in a pasture is to measure the height of the plants and compare that to information about the relationship between plant height and FDM yield. The population of plants in a pasture is referred to as a "sward." A simple ruler can be used to measure approximate plant heights in or in centimeters (cm). Scientific reports use metric measures such as cm or millimeters (mm). Books and reports written outside the United States also use the metric system of measurement, whereas most American extension and trade publications use the Imperial system, including in and feet (ft). See Appendix 1 for information on sizes of measurements and conversions.

Direct Measurement of Forage Height and Yield

The simplest tool for plant height measurement is a ruler. Some people may put the measurements of a ruler onto their boot or onto some other implement that they will always have in the pasture. As a manager of horses becomes more concerned with accuracy and feed costs, he or she could try working with more sophisticated tools for measuring pasture height and yield. The most direct measure of forage yield in a pasture would be to cut the entire pasture, weigh the cut plants, and determine their dry matter percentage. Since this is not practical, a reliable estimate can be made by measuring height of about 20–30 representative sample squares, cutting them, weighing them, drying them, and calculating the mass in a unit of area, usually as lbs of FDM per ac or kilograms per ha (kg/ha). The method is described in detail in Appendix 2. Basically, you use a small sampling square of known area (perhaps 1, 2, or 3 square ft) (0.09–0.28 sq m) and throw it randomly 20 times throughout the whole pasture. You measure the forage height by applying a ruler next to standing plants and estimating average height. Then using clippers (electric or hand-powered) you harvest the forage at ground level. Weigh the forage immediately, then dry it either in a microwave or a 220°F (104°C) convection oven overnight and weigh it again (Appendix 2).

The weight of FDM per square foot can be expanded to the weight per acre by multiplying by 43,560 square ft/ac. (Weight of FDM per square meter (sq m) can be expanded to the weight per ha by multiplying by 10,000 sq m per ha.) Yields of 1000–4000 lbs of FDM/ac are common (1121–3363 kg/ha). You can also calculate how many lbs of FDM are represented by each inch of forage height when you divide the FDM mass per ac by the number of in of height. A figure of about 200–300 lbs of FDM per ac per in of height is common (88.25–132.38 kg/ha/cm). This method is time-consuming, very accurate, but not very expensive if you have access to an oven. In some extension projects, large lots of forage are air-dried and then a subsample is used to determine dry matter percentage.

If someone tries drying forage samples in the kitchen oven that is shared with a partner, there will be a very unhappy partner, due to the odor generated. Forage drying is best done in an old or cheap oven outside the house. Convection ovens and microwave ovens are faster than those with only a heat source. Having used a low-temperature heat source once or twice, a manager will likely want a faster method and be willing to pay a little more money for equipment and be willing to sacrifice a small amount of accuracy. The relationship between sward height and sward yield varies among different plant species, mixtures, and even among seasons.

Indirect Measurement to Estimate Yield

Indirect measures of FDM yield can be obtained by using tools that have been calibrated to obtain yield estimates quickly. Visual estimates with no tools can be used to roughly determine when to move horses into and out of a paddock, although accuracy and repeatability are low. Using the tools described subsequently should be more reliable than just comparing forage plant heights to the height of your boots.

Pasture Rulers and Pasture Sticks

If you take a ruler that is 1, 1.5, or 3 ft long (or a meter stick) into a pasture or a lawn, you can try to find an approximate height of the majority of grass and legume plants, but you soon realize that there are problems with this method. Very few plants are absolutely straight and pointing straight upward. Most plant parts extend from the ground at some angle other than 90 degrees, and as stems and leaves lengthen, they curl downward. The uppermost parts of plants are at many different heights. Some form of compensating for these facts would be useful. Some rulers designed for forage height measurements have a sliding tab that can be moved downward to touch the first plant in its path. The height can be measured there.

Many extension services and the US Department of Agriculture-Natural Resources Conservation Service (NRCS, 2017) provide at no cost, special pasture rulers or "pasture sticks" that are embossed with information on managing different forage species based on the height of the plant canopy (Fig. 7.1). These pasture sticks are designed with a principle in mind: as plant canopy height increases, forage yield increases. Instructions for use are printed on the sticks, and more detail on their use is available from NRCS and other agricultural extension publications. The NRCS pasture stick has a grid for estimating sward density. The stick should be laid on the ground within the area of forage to be sampled. From directly above, observe the grid and count the dots within it. The number of dots that you can see determines whether the forage density is thick, moderate, or thin. Next, hold the stick vertically and estimate the average height of the plants. Finally, look at the list of forage species and species mixtures plus three columns of numbers in a table on the pasture stick and read the estimated range of pounds of dry matter/acre per inch of forage height. For example, if the pasture was a

FIGURE 7.1 A pasture ruler or grazing stick. It bears markings for forage heights, recommended grazing heights, and information to help the user estimate forage yield. Measure pasture height as the height of the tallest leaf within a closed hand-span (about 4.5 inches) of the ruler.

TABLE 7.1 Estimated Pounds of Forage per Inch of Sward Height From a West Virginia Pasture Ruler According to Forage Species Mix and Sward Density (Rayburn, 2016)

	Sward Density		
	Fair	Good	Excellent
Forage Species	(lb FDM/acre/inch of height)		
Tall fescue + legume	100–200	200–300	300–400
Red clover or alfalfa	150–200	200–250	250–300
Orchardgrass + legume	100–200	200–300	300–400
Mixed pasture	150–250	250–350	350–450

mixture of orchardgrass and clover and considered thin, the yield would be 150–200 lbs FDM/acre/inch, and if the pasture was perennial ryegrass fertilized with nitrogen and considered thick, the yield would be 350–400 lbs of FDM/acre/inch.

The University of Kentucky publication AGR-191 includes the instruction to "Spread your hand and lower it onto the canopy," before taking the forage height measurement with a pasture stick (Smith et al., 2010). The greater the number of measurements that you take and then use in calculations of average density and average height, the more reliable is the estimate of yield. Tables are available that provide information on the mass of forage per inch of height and sward conditions, to help in converting a plant canopy height to lbs of FDM per ac (Table 7.1). A University of Vermont Extension factsheet lists several sources of pasture sticks in different states and provides information on contacting their manufacturers (Heleba, 2012).

Falling Plate Meters

Plate meters (Rayburn and Rayburn, 1998) contain a ruler plus a plate of standard area and mass (weight) that will slightly and uniformly compress the forage below it. Measuring the compressed height of forages reduces some of the variability in

FIGURE 7.2 A falling plate meter, which uniformly compresses forage to allow fair comparisons of forage height, density, and mass. Compressed forage height is correlated to forage dry matter yield.

measurement with a ruler and provides a more accurate estimate of FDM yield than a ruler. A falling plate meter can be made simply and cheaply at home (Rayburn and Lozier, 2003b), and it is suitable for occasional use (Figs. 7.2 and 7.3). The "falling" plate is lowered onto the forage canopy and a measuring stick inserted into a hole in the center of the plate is used to measure a standardized forage height. Rayburn and Rayburn (1998) originally described the cost of making a falling plate meter as $12.00 in 1996. A "purchasing power calculator," using the consumer price index increase, estimated the real value and contemporary opportunity cost of making a falling plate meter at $19.00 in 2014 (Williamson, 2016).

Rising Plate Meters

Rising plate meters (Fig. 7.4) are commercial products produced in New Zealand and available as a manual version for $320 USD to $472 USD plus shipping. A notched center post is placed like the end of a cane on the ground while the operator walks a zigzag pattern on a pasture. A metal plate of known area and weight rises up the post by an amount depending on the height and density of the forage. As the plate rises it triggers a counter. Both manual and electronic models record the height in 0.5 cm increments and provide formulas for calculation of yield by the operator. An electronic model uses the same formulas to quickly convert the average of 30 or more height measurements to a calculated FDM yield. Manufacturers recommend a general formula but encourage users to select or develop improved formulas for local conditions. Dairy Farmers of America Farm Supplies sells three models of Jenquip rising plate meters for prices between $435 and $825 (February 11, 2016).

Capacitance Probes

Capacitance probes (Fig. 7.5) look like a straight walking stick with an electronic meter on the top and a metal prong on the bottom. The bottom end of the probe on the ground sets up an electric field. An electrical capacitance change is directly proportional to the moisture content of the surrounding plant material. The probe is placed vertically on the pasture every two to six paces, usually 20 to 40 times per area sampled. One New Zealand model is available for $1036 USD, plus shipping (January 2016). Some studies have revealed high correlations of capacitance data with yields based on clipping, but in one report the device had problems in heavy dew and drought conditions (Serrano et al., 2014).

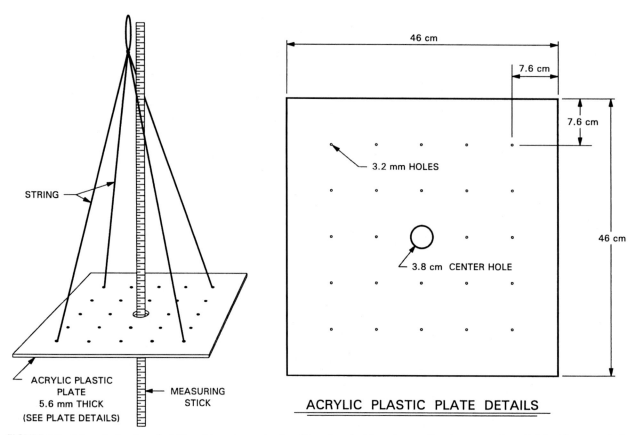

FIGURE 7.3 Schematic drawing of a falling plate meter, made of acrylic, and used with a ruler to uniformly measure forage height. More recently the only 3.2 mm holes in the plate are the four at the corners. This has not significantly affected the plate mass.

FIGURE 7.4 A rising plate meter. As the handle is pushed down through forage to the ground, the aluminum plate rises in 0.5-cm increments. Gears and a counter determine the height of the plate above ground. A calculator within the box converts the height to a forage dry matter yield per acre or per hectare.

FIGURE 7.5 A capacitance probe for estimating forage yield. Electrical capacitance, due to water content in plants, varies with plant height and density.

Comparisons of Pasture Sticks and Meters

Plate meters, rulers, and capacitance meters were compared and cross-calibration values developed to enable people to convert estimates of forage mass obtained with one tool to estimates obtained with other tools when used on similar forage stands (Rayburn et al., 2007).

The FDM density of any plant varies from the bottom to the top of the plant. Table 7.1, which is an extract from Rayburn (2006) shows estimated lbs of forage per in of sward height in four different forage species, from use of a pasture stick. There was no apparent effect of the mixture of forage species, but differences were apparent due to classification of sward density as Fair (<75%), Good (75%−90%), or Excellent (>90% of ground covered by forage). As sward density improved, the median FDM yield per in of forage height increased from about 170 to about 350 lbs of forage per in.

Rayburn and Lozier (2003a) produced a factsheet called "Estimating Pasture Forage Mass From Pasture Height" in which they showed relationships among plant height, forage density (as indicated by numbers of "tillers" or stems with leaves), and forage mass. It included a calibration table, based on their research, which provided measurements by ruler, falling plate meter, and a rising plate meter in swards with thin, average, and thick **pasture tiller density** (in pounds of FDM per ac per in of falling plate height). For each **ruler height** and pasture **tiller density** (thin, average, or thick), there is a corresponding calibrated forage density (in lbs of FDM per ac per in of height) and a calibrated forage mass (in lbs of FDM per ac). Based upon research between 2003 and 2015, Rayburn (2016) produced a revised data set, summarized here as Table 7.2. To help you understand solutions to a horse owner's problems, it is worth studying this table in detail. The forage or sward densities of Table 7.2 are considerably greater than those of Table 7.1, which emphasizes the fact that plant density varies considerably from place to place and time to time. Thus, measuring what the actual plant density and forage mass are on the fields that you manage will be a valuable aid in predicting how much feed you can grow. Table 7.3 shows the effect of growth in height of a very dense species, perennial ryegrass, on forage mass per acre. Compared to the data for multi-species of forage in Table 7.2, mass of perennial ryegrass does not increase much in mass/acre from 1.8 to 3.5 in high but it has a relatively steep yield increase between 4.1 and 5.4 in high.

Once you have acquired a pasture stick, plate meter, or capacitance meter, it is useful to spend time working with it in a variety of forage conditions and calibrating it against your own clipping and weighing data. The difference between initial height at turning animals into a paddock and residual height after removing them represents the amount of grazeable forage.

In Table 7.2, plant height measurements by falling and rising plate meters are highly correlated to measurements by ruler (>0.99). The falling and rising plate measurements are lower than the ruler measurements due to the uniform

TABLE 7.2 General Calibrations for Pasture Forage Density and Forage Mass at Different Mean Pasture Heights as Measured With a Ruler, a Falling Plate Meter, and a Rising Plate Meter (Rayburn, 2016)

Measurement Method			Forage Density (DM lbs/ac/ in of Falling Plate ht)			Forage Mass (DM lbs/ac)		
			Pasture Tiller Density					
Ruler Height (in)	Falling Plate (in)	Rising Plate (in)	Thin	Average	Thick	Thin	Average	Thick
2	1.0	0.8	340	664	988	350	683	1018
3	1.7	1.3	345	643	942	576	1073	1572
4	2.3	1.8	350	622	895	809	1437	2066
5	3.0	2.4	356	601	848	1049	1774	2500
6	3.6	2.9	361	580	801	1295	2084	2874
7	4.2	3.4	366	560	754	1549	2367	3188
8	4.9	3.9	371	539	707	1809	2624	3442
9	5.5	4.4	377	518	660	2076	2855	3636
10	6.2	4.9	382	497	613	2350	3059	3769
11	6.8	5.4	387	477	566	2630	3236	3843
12	7.4	5.9	393	456	519	2918	3386	3856

TABLE 7.3 Sward Height Measured by Ruler and Forage Mass in Perennial Ryegrass Swards (Naujeck et al., 2005)

Sward Height in Perennial Ryegrass, Measured by Ruler in Inches (cm)	Forage Mass (Pounds FDM/Acre)
2.8 (7.0)	553
3.5 (8.9)	651
4.1 (10.5)	1936
5.4 (13.6)	2293

compression of the plates on the plants. The Forage Density columns of Table 7.2 show how management, timing, and forage species influence density along the length or height of plants. The descriptors "Thin," "Average," and "Thick" were evaluated subjectively. At thin density, represented by young swards or "aftermath meadow" (recently grazed), the density has the least range (340–393 lbs/ac/inch) and is the lowest. This makes sense since the youngest, least developed plants have the least degree of tillering or branching and leaf production. Average swards were mixed cool season grasses and clovers with moderate density between 456 and 664 lbs of FDM per ac per in of forage height (about twice as dense as in Table 7.1). Thick swards were mostly mature, fertilized tall fescue. The slope of decline in density was similar between thin and average swards. Thick, tall fescue grass was denser at any height than the average sward at any height except for the tallest samples. The fact that densities of average and thick swards declined in density with increasing height illustrates that as plants reached their mature height, their higher shoots became less leafy. This is consistent with the production of tall stems that bear flowers and seeds but is not necessarily the case with all grass species.

It is worthwhile to become familiar with and keep records of typical forage yields in fields that you manage and in those that you compare to your own. The "Forage Mass" columns indicate the yields of feed available as hay or as pasture. For all three tiller densities, the forage mass increased as plant height increased. The median FDM masses for thin, average, and thick stands were approximately 1550, 2370, and 3190 lbs/ac, respectively. At heights of plants that are very leafy, growing fast, and very nutritious (8–12 in by ruler), thin and average swards would provide about 1800–3400 lbs of

FDM/ac and would be suitable for horses that are growing, lactating, or exercising a great deal. The thick, tall fescue data provided higher forage mass at 8–12 in ruler height (about 3400–3900 lbs of FDM/ac) than the thin and average swards. Such high yields of leafy tall fescue are highly desired for rapid weight gains in beef cattle, which will be sold by the lb. For horses, this crop is less desirable, especially as a pure forage, and this is explained in detail in other chapters. Appendix 3 shows the data from Table 7.2 as graphs and provides further explanations.

As a comparison to the table from Rayburn (2016) the following grass height and yield data are extracted from an experiment on perennial ryegrass swards in England, following cutting with a lawn mower, a light experimental grazing, then 1 week of regrowth. Experimental patches with sward heights measured by ruler at 2.8, 3.5, 4.1, and 5.4 in (7.0, 8.9, 10.5, and 13.6 cm) produced 553, 651, 1936, and 2293 lbs of FDM/acre, respectively (Naujeck et al., 2005). Using similar ruler heights to those in Table 7.2 and looking across to the Forage Mass columns, you can see that most of the perennial ryegrass tiller densities would have been considered either average or thick (Table 7.3).

Research trials from New Zealand and the United States have compared various plate and capacitance meters against the standard of clipping, weighing samples then drying and reweighing them to determine FDM yield in the areas sampled. The plate and capacitance meters are quicker to use than the clipping method, but they have not been big sellers in North America. A manager in charge of the production and quality of feed for a large horse farm would be wise to evaluate the cost of one of these tools relative to the benefit of easily and quickly determining the feed supply in pastures and hay fields. A major value of determining forage yields several times per grazing season is in weekly or biweekly feed budgeting. Calculating yield following a change in weather can reveal a change in forage growth rate. Then alternative feeding plans can be made to avoid running out of grass.

The USDA and United States Forest Service (USFS) (2000) compared several herbaceous measurement devices against the direct measurement method, including clipping, weighing, and oven drying. A capacitance meter was slightly better than a rising plate meter in its correlation with the standard oven method for determining forage yield.

THE RELATIONSHIP BETWEEN FORAGE HEIGHT AND FORAGE YIELD IN PASTURE

Fig. 7.6 illustrates how the yield of components of a grass plant increase as the plant grows in height until it has produced blooms. Then yield or dry weight decreases as leaves die off, especially near the bottom of the plant (White and Wolf, 2009). Cellulose (a structural carbohydrate that readily digests in the gut of the horse) accumulates in plant cell walls until this bloom stage. Highly digestible nonstructural carbohydrates (NSC) such as starches and sugars increase in quantity until the flower head develops, then decrease. The total content of proteins peaks later in the heading stage. The content of lignin, an indigestible binder of other plant chemicals, continues to accumulate through all stages of plant maturity until growth stops. Lignin binding to other components prevents them from being digested by microbes in the digestive system.

Fig. 7.7 shows differences in chemical composition among common forage grasses and legumes from a young, leafy stage to a full bloom stage (White and Wolf, 2009). Concentrations of protein, leaves, and minerals decline, while concentrations of fiber, lignin, and stems increase. Notice that Fig. 7.7 showed protein percentage decreased with maturity, while the actual weight of protein increased in Fig. 7.6. Conversely, both the fiber dry weight in Fig. 7.6 and the fiber percentage in Fig. 7.7 increased.

Grass species vary in the structure of their shoots and roots and in their growth habit and maximum height (Fig. 7.8). This illustration of four grasses represents aftermath (post-grazing) growth, although the tall fescue should have more young tillers and rhizomes. Bluegrass is usually much shorter and has finer leaves than orchardgrass, tall fescue, and Bermudagrass at similar stages of development. Bluegrass and orchardgrass shoots consist of tillers and leaves, whereas tall fescue and Bermudagrass have definite stems from which leaves project. Nonstructural carbohydrates are stored in rhizomes and tillers. Note the changes in structure of these different grass species at any height of the canopy from 2 to 8 in (White and Wolf, 2009). If all these grasses were growing in a uniform mixture in a pasture, a horse grazing from 8 down to 5 in would obtain no bluegrass, significant amounts of leaf from orchardgrass and tall fescue, and a mixture of seed heads, stems, and leaves of Bermudagrass. Another horse grazing from 5 in down to 2 in would obtain a few leaf tips of bluegrass, and plenty of leafy material from the other species, with variable amounts of stem. This heterogeneity complicates the estimation of FDM yield and nutrient intake at various plant heights.

Different forage legumes are represented at prebloom stage in Fig. 7.9, showing big differences in typical height of white and ladino white clovers compared to red clover and alfalfa (White and Wolf, 2009). White and ladino white clover, which are short and tall members, respectively, of the same species, depend on relatively thin petioles to push their leaves upward, whereas red clover and alfalfa have thicker, stiffer stems that allow them to grow much taller. Thus, the species composition of a sward can influence its yield and density. Plants of short stature can be useful components of a slightly

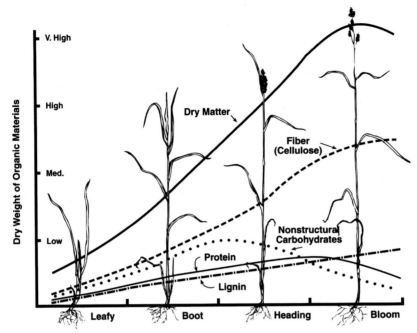

FIGURE 7.6 Changes in plant composition of growing grass plants from leafy through bloom stages (White and Wolf, 2009). The weights of nonstructural carbohydrates and proteins peak in mid-maturity, then decline as the plants mature, while the weights of fiber, lignin, and total dry matter increase to the bloom stage. *Courtesy of Office of Communication and Marketing, College of Agriculture and Life Sciences, Virginia Tech, 127B Smyth hall (0904), 185 Quad Lane, Blacksburg, VA 24061. Original artwork commissioned by Roy Blaser.*

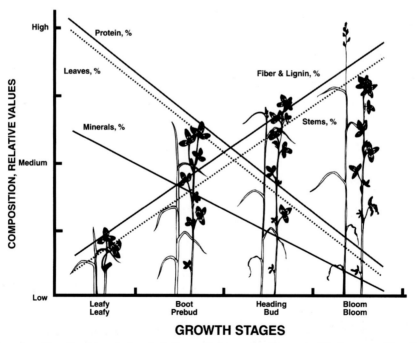

FIGURE 7.7 The relative composition of protein and minerals declines with decreased leaf percent, while the composition of fiber and lignin increases with an increase in stem percent as perennial grass and legume plants mature (White and Wolf, 2009). *Courtesy of Office of Communications and Marketing College of Agriculture and Life Sciences, Virginia Tech, 127B Smyth Hall (0904) 185 Ag Quad Lane, Blacksburg, VA 24061. Original artwork commissioned by Roy Blaser.*

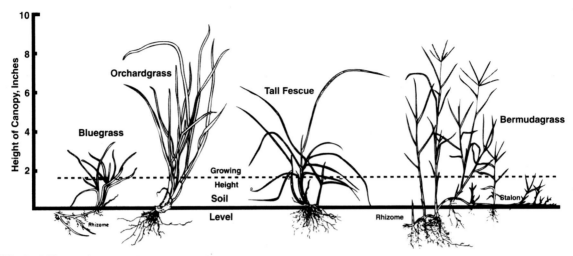

FIGURE 7.8 Differences in shoot and root structure of bluegrass (with a low growth habit) and three other forage grasses with tall growth habits (White and Wolf, 2009). New shoots and roots can arise from rhizomes and stolons. At this stage, there are no obvious stems in bluegrass and orchardgrass, but there are in tall fescue and Bermudagrass. *Courtesy of Office of Communications and Marketing College of Agriculture and Life Sciences, Virginia Tech, 127B Smyth Hall (0904) 185 Ag Quad Lane, Blacksburg, VA 24061. Original artwork commissioned by Roy Blaser.*

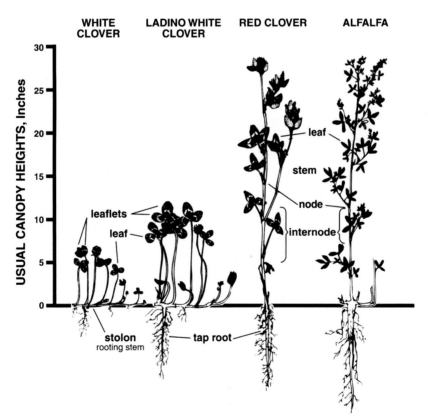

FIGURE 7.9 Canopy heights and structures of four common forage legumes, highlighting tap roots and tall stems. The shorter legumes are well-suited to grazing and the taller legumes are better-suited to use as hay (White and Wolf, 2009). *Courtesy of Office of Communications and Marketing College of Agriculture and Life Sciences, Virginia Tech, 127B Smyth Hall (0904) 185 Ag Quad Lane, Blacksburg, VA 24061. Original artwork commissioned by Roy Blaser.*

mature forage mixture to provide soil cover plus density at lower plant heights where the taller plants have mostly stems and few leaves. A sward that has been well-managed for several years will have a mixture of plant species, with no visible bare soil when viewed from above.

USING NUMBERS FROM SAMPLE MEASUREMENTS TO ESTIMATE FORAGE YIELD FOR AN ACRE AND A WHOLE FIELD

Consider a clipping, drying, and weighing estimate of FDM yield. Imagine that 30 randomly chosen sites in one paddock were sampled with a ruler and a 1-sq-ft quadrat (a sampling square). The average height of forage plants was 10 in and the average dry weight of all the forage samples that were clipped to ground level and dried in an oven was 0.025 lb. To calculate the number of lbs of FDM/acre, multiply the number of lbs of FDM in 1 square (sq) ft times the number of sq ft in 1 ac (43,560). Thus 0.025 lb/sq ft × 43,560 sq ft/ac = 1089 lbs FDM/ac.

Good grazing management practice includes the suggestion, "Take half and leave half," so you leave about half of the starting forage height to encourage regrowth and allow some organic matter to be trampled into the soil, thus contributing to soil nutrients, organic matter, and water-holding capacity. So a good grazing manager is interested in how many lbs of FDM are available per ac in each in of forage height that will be eaten. Dividing the yield per ac by the plant height provides the forage density in lbs/acre/inch, thus 1089 lbs/ac/10 in = 108.9 lbs FDM/ac/in.

Since the starting forage height was 10 in and half of this is 5 in, we should calculate how much forage we expect to be eaten in a short grazing period before animals are moved to another paddock. Multiply the forage density by the number of inches of forage that will be grazed, 108.9 lbs/ac/in × 5 in = 544.5 lbs/ac. If the paddock being sampled is 2 ac in size, the amount of FDM we offer for grazing is 544.5 lbs/ac × 2 ac = 1089 lbs FDM. If you had eight horses in a group and each needed 22 lbs of FDM per day, the daily need is 8 × 22 = 176 lbs. A 2-ac paddock with this forage yield could feed these horses for 1089 lbs/176 lbs/day = 6.2 days. This is approaching the maximum of 7 days for a grazing period under good management-intensive grazing. These horses could be moved to another paddock sooner, so they are not repeatedly grazing their most preferred plants and weakening them.

If you had just four horses with an average weight of 1100 lbs and wanted to feed them 2% of their body weight per day as in the section on horse owner's third question, the calculations are similar. 4 × 1100 lbs × 0.02 = 88 lbs of FDM needed per day. The 2-ac paddock containing 1089 lbs FDM can provide 1089 lbs/88 lbs/day = 12.4 days of feed. Good rotational grazing practice includes grazing periods up to 7 days, so there is a considerable surplus of feed for these four horses on these two acres, at this time.

While it is very common to see managers leaving four horses on such a 2-ac paddock for 12 days or longer, such a low stocking rate creates an **undergrazing** scenario. All plants in the paddock would grow considerably during this time, and the FDM yield would probably increase since there are not enough horses to reduce it. The horses would tend to graze their favorite plants repeatedly and avoid their less-favored plants, which would have a chance to produce flowers, then seeds, thus reproducing. The favorite plants would not have a chance to reproduce. Therefore, over time, less-desirable plant species would predominate. Keeping horses in a paddock long enough for these changes and for animals to re-graze plants causes overgrazing.

Appropriate rotational grazing management in this case involves subdividing paddocks with temporary electric fence, for example, in thirds. Then the available forage in one grazing paddock would be 1089/3 = 363 lbs. The four horses would be supported for 363 lbs/88 lbs/day = 4.125 days before being moved onto another third of this paddock. While some horse owners are reluctant to use electric fence for rotational grazing of horses, they would be wise to research the use of this valuable tool. Many horse managers use it successfully. Fencing and Watering systems are described in Chapter 15.

CONVERTING FORAGE YIELDS BETWEEN POUNDS PER ACRE AND KILOGRAMS PER HECTARE

If you wish to work in the metric system, you would be measuring plant height in cm, land area in ha, and FDM mass in kg. The 10-in plant height equals 25.4 cm. Many websites, factsheets, and books (See Appendix 1) contain methods to convert between these metric and the standard US or Imperial measures. The 2 ac equals 0.81 ha. The 1089 lbs equals 494 kg. One lb per ac equals about 1.121 kg/ha. So the 1089 lbs FDM/acre calculated before is equivalent to 1089 × 1.121 = 1220.8 kg/ha.

ANSWERING THE HORSE OWNER'S FOURTH QUESTION

The fourth question the horse owner had was, "How much land is needed to supply the pasture and the hay for each horse for a year?" Going back to the FDM requirement of 88 lbs/day for the four horses, multiply it by 365 days in a year to obtain 32,120 lbs of FDM. If the horses were to be confined in a barn and fed hay all year, then this is how much hay FDM would be required for a year. Since hay in humid environments can be about 85% dry matter, the amount of required hay on an "as fed" basis would be 32,120/0.85 = 37,788 lbs of hay. If hay bales weighed 45 lbs on average, this would mean about 37,788/45 = 840 bales. If the horses need 32,120 lbs FDM/year and the land produced, for example, 4000 lbs FDM/ac, then 32,120/4000 = 8 ac of hay land are needed. Depending on soil fertility, precipitation, latitude, average and extreme temperatures and stand thickness, the total annual forage yield could result from one to four cuts or harvests of hay.

If the horses are on pasture all year, a "take half and leave half" system is being practiced, and if an average yearly FDM yield is 2000 lbs FDM/ac, then multiply the 2000 lbs FDM/ac by 0.5 because only half the yield is grazed, to reveal the number of lbs of FDM available for grazing per acre 2000 × 0.5 = 1000 lbs FDM/ac. How much land is needed to supply the pasture for the four horses for a year? 32,120 lbs/year/1000 lbs/ac = 32.12 acres. For one horse, this would be 32.12/4 = 8.03 ac (the answer to the fourth question).

These answers should raise another question: If I need 32.12 acres to graze horses and only 16.06 acres to grow hay for them, why would I bother managing pasture and grazing? The first reason is that grazing management requires less labor, less machinery use, and less fuel use than harvesting, raking, baling, hauling, stacking, unstacking, and feeding hay. Then there is the issue of manure. You can scoop it up, move it to a pile, and later put it in a manure spreader and pull that around with a tractor while the manure is spread or let the horses spread it as they walk and a couple of times per year, break it up with a harrow.

If you wondered about the recommendation to take half and leave half of the forage, since that means half the feed is left behind, think about normal plant and soil activities. The taller the residual, ungrazed forage, the larger the surface area for photosynthesis and the less shrinkage of roots, thus encouraging rapid regrowth. The lowest portion of a forage plant has the lowest leaf/stem ratio, thus it is of relative low feed quality. Due to splashing of soil during rainfall, the lowest portions of forage plants have the most soil on them. Some plants will be trampled or bent by falling manure, into contact with the soil, before they can be grazed. This plant material is not wasted. It is exposed to soil organisms and moisture, so that it is naturally composted, adding organic matter and recycling nutrients to the soil.

Next consider that in an all hay system, nutrients and organic matter are being taken off the land with every load of hay. Most of those nutrients and organic matter are not stored in the horse, so they are excreted in manure and urine. A portion of the nutrients and organic matter might be returned to the land, if the manure is spread onto it.

Further consequences of taking more than half of the available forage are that forage palatability, digestibility and crude protein content decrease towards the lower parts of the plants, so intake of desired nutrients per hour of grazing time is reduced. Furthermore, some grasses accumulate sugars in stem bases and some horses that are prone to obesity may develop laminitis because of an increased sugar intake.

Taking less than half the available forage results in paddocks that are unevenly grazed, horses leave more good quality feed behind, the remaining ungrazed forage matures and becomes less palatable, some of the more mature plants may develop flowers, then seeds to reseed the sward and paddocks will have suitable yields for re-grazing after a slightly shorter rest period.

In livestock operations, the costs of acquiring and distributing feed plus costs of handling manure in an all hay system are greater than the costs of an all grazing system. Since there are advantages and disadvantages to both systems and in much of North America the climate is not suitable for grazing all year long, many farms feed hay in the colder months and depend mainly on grazing for feed in the warmer months.

The length of the growing and grazing seasons in different areas will determine which proportions of a horse's yearly diet will be met with hay and pasture. The amount of hay wasted will also have a significant effect on how much hay is needed to support a given number of horses. A study led by Martinson et al. (2011) at the University of Minnesota compared several commercial round bale hay feeders versus a no feeder control treatment. The percentage of offered hay that was wasted varied from 5% to 33% for the commercial bale feeders. All commercial feeders reduced waste compared to the 57% for the no feeder control. FDM intakes for horses using bale feeders were 2.0%–2.4%, and no groups of horses using bale feeders lost weight. However, the horse group fed round bale hay without a feeder had a dry matter intake of only 1.3%, and they lost body weight. The payback period for feeders varied from about 1 month to 20 months. Thus it would make sense to purchase efficient hay bale feeders, learn how to manage them, and refer to the Martinson study to learn what level of hay waste to use when calculating your hay needs.

REVIEW QUESTIONS

1. Imagine that you need to provide feed for six horses weighing 1200 lbs on average and needing 2.2% of their body weight/day, due to active training. Assume that all feed will be forage and calculate the amount of feed that:
   a. The group of six horses needs to be fed per day.
   b. Needs to be supplied by hay feeding for 8 months of the year and by pasture if grazing occurs for 4 months of the year.
   c. Needs to be supplied by hay feeding for 2 months of the year and by pasture if grazing occurs for 10 months of the year.
2. Your pastures and hay fields have the ability under your management and average rainfall to produce 3000 lbs of grazeable FDM/ac over a whole growing season. How many ac of pasture land do you need for 5 months of grazing the six horses mentioned in question 1 above?
3. If you planned to start measuring your forage production and budgeting your horse feed, what tool would you plan to use first in measuring forage production? Why?
4. If you become well known in your community for your excellent management of pasture, hay, and horses and neighbors ask you to measure their forage production (for a fee), what tool would you probably use in measuring forage? Why?
5. According to Table 7.2, if you measured average forage height in a paddock at 4.9 in with a falling plate meter and the forage was thin, what is the forage density and what is the forage mass? Provide the correct units with your answers.
6. Look at Fig. 7.6. Describe the increases of forage height and forage dry matter yield as plants develop, then mature to the bloom stage.
7. Look at Figs. 7.6 and 7.7. (a.) What happens to the weight of cellulose, nonstructural carbohydrates, protein, and lignin as the plant grows from the leafy stage through the bloom stage? (b.) What happens to the percentage concentration of fiber and lignin, stems, protein, leaves, and minerals as the plant grows from the leafy stage through the bloom stage?
8. If you can increase the degree of tillering of pasture grasses, what will this likely do to the yield of FDM per ac?
9. What are the dimensions of each side of
   a. 1 sqft?
   b. 1 square yard?
   c. 16 sq ft?
   d. 1 sq ac?
10. If you read an article that described forage yields as 2000 kg/ha, what would this be in lbs/ac?

REFERENCES

Heleba D. Finding a Pasture Stick in Your Area for Your Organic Dairy Farm. 2012. Factsheet, http://articles.extension.org/pages/28873/finding-a-pasture-stick-in-your-area-for-your-organic-dairy-farm.

Martinson K, Wilson J, Cleary K, Lazarus W, Thomas W, Hathaway M. Selecting a Round-Bale Feeder for Use during Horse Feeding. University of Minnesota Extension; 2011. http://www.extension.umn.edu/agriculture/horse/nutrition/selecting-a-round-bale-feeder/.

Naujeck A, Hill J, Gibb MJ. Influence of sward height on diet selection in horses. Appl. Anim. Behav. Sci. 2005;90:49–63.

National Research Council. Nutrient Requirements of Horses, 6th Revised Edition. Washington, DC: The National Academies Press; 2007.

NRCS. MA NRCS Pasture Stick Instructions. United States Department of Agriculture; 2017. https://efotg.sc.egov.usda.gov/references/public/MA/MA_NRCS_Pasture_Stick_Instructions.pdf.

Rayburn EB, Rayburn SB. A standardized plate meter for estimating pasture mass in on-farm research trials. Agron. J. 1998;90:238–40.

Rayburn EB, editor. Forage Utilization for Pasture-Based Livestock Production. Chapter 2 Assessing Forage Mass and Forage Budgeting. Ithaca, New York: Natural Resource, Agriculture and Engineering Service Cooperative Extension NRAES-173; 2006. p. 20–42.

Rayburn E, John L. Estimating Pasture Forage Mass from Pasture Height. West Virginia University Co-operative Extension Service Fact Sheet; 2003a. http://www.wvu.edu/~agexten/forglvst/passmass.pdf.

Rayburn E, John L. A Falling Plate Meter for Estimating Pasture Forage Mass. West Virginia University Co-operative Extension Service Fact Sheet; 2003b. https://www.wvu.edu/~agexten/forglvst/fallplate.pdf.

Rayburn EB, Lozier JD, Sanderson MA, Smith BD, Shockey WL, Seymore DA, Fultz SW. Alternative methods of estimating forage height and sward capacitance in pastures can be cross calibrated (Online). Forage and Grazinglands; 2007. https://doi.org/10.1094/FG-2007-0614-01-RS.

Rayburn EB. Personal Communication. 2016.

Serrano J, Shahidian S, da Silva JM. Validation of GRASSMASTER II for measurement and mapping of dry matter yield in Mediterranean pastures. In: International Conference of Agricultural Engineering. Ref: C0096. Zurich. July 6–10, 2014; 2014.

Smith R, Panciera M, Probst A. Using a Grazing Stick for Pasture Management. University of Kentucky Cooperative Extension Service Publication AGR-191; 2010.

United States Forest Service. Herbaceous Measurement Device Comparison. 2000. http://www.fs.fed.us/td/programs/im/herbaceous_veg/hmdcs.shtml.

Williamson SH. Purchasing Power of Money in the United States from 1774 to Present, Measuring Worth. 2016. www.measuringworth.com/ppowerus/.

White HE, Wolf DD. Controlled Grazing of Virginia's Pastures. Virginia Cooperative Extension Factsheet 418-012; 2009. http://pubs.ext.vt.edu/418/418-012/418-012.html.

FURTHER READING

Lemus R, Parish J. Assessing Needs and Feed Sources: How Much Forage Do I Have?. Publication 2458 Mississippi State University Extension Service; 2016. http://msucares.com/livestock/beef/forageavailability.pdf.

Mills AM, Smith MC, Moot DJ. Relationships between dry matter yield and height of rotationally grazed dryland Lucerne. J. N.Z. Grassl. 2016;78:185−96. https://www.grassland.org.nz/publications/nzgrassland_publication_2832.pdf.

Chapter 8

Grazing Behavior, Feed Intake, and Feed Choices

Paul Sharpe[1] and Laura B. Kenny[2]

[1]University of Guelph, Guelph, ON, Canada (retired); [2]Extension Division of the College of Agricultural Sciences, Penn State University, University Park, PA, United States

INTRODUCTION

Do you have a favorite food? Do you prefer sweet or salty snacks? Even though most people need similar levels of nutrients to maintain our health, individuals have their own preferences. Grazing animals similarly, have preferences, both on a species and individual level. Discerning these preferences for an entire species is not an easy task. Some horses love mints as treats, others refuse them. In addition to individual preferences, numerous other factors may affect grazing behavior, such as climate, turnout time, plants available, feeding experiences as youngstock, and the quality of feed offered. Therefore, it is not prudent to make sweeping generalizations about the horse's grazing behavior as a whole. This chapter will summarize grazing research from around the world to illustrate what has been learned in this developing field of study.

GRAZING BY HORSES COMPARED TO OTHER HERBIVORES

Horses have dexterous lips that can gather forages and bring them close to the teeth. Upper and lower incisor teeth meet and are capable of clipping forage close to the ground (Matches, 1992; Singer et al., 1999). Tactile hairs and touch receptors on the lips and muzzle enable a horse to determine how far its nose is from the ground and to obtain physical feedback about vegetation structure as it grazes (Avery, 2003). Show horses often have these hairs clipped, decreasing their sensory abilities. Ruminant animals, such as cattle and sheep, depend less on their lips and more on their tongues to bring food into the mouth. They also have only one set of incisor teeth (lower), which reduces their ability to clip plants. Goats might appear more like sheep than horses, but their mobile lips are similar in action to horse lips. During grazing, horses move forward almost constantly, sampling forages frequently and lingering on patches that they seem to prefer (Archer, 1971, 1973). The timing of grazing episodes is influenced by the quality of available forages (Tyler, 1972).

Horses are more active on pasture and their single (noncloven) hooves cause greater damage to soil than the cloven hooves of cattle, sheep, and goats (McClaran and Cole, 1993; OMAFRA, 2000). Thus, horses should be allowed 2 to 3 times more space per livestock unit (1 livestock unit equals 1000 pounds of animal) (OMAFRA, 2000). Horses are more selective grazers than cattle, preferring to consume grasses over forbs and shrubs (Archer, 1973; Olson-Rutz et al., 1996). Smelling odors from manure informs horses about the location of manure on a pasture and whether the manure is from horses (Avery, 2003). Horses and cattle avoid grazing forages growing through or adjacent to manure of their own species (Fig. 8.1). This probably helps them reduce the likelihood of ingesting parasite larvae that specifically infest their own species.

When compared to cows, ewes, and goat does, horses spent more time grazing per day than the ruminants (Ferreira et al., 2013). This may be due to the smaller size of the equine stomach compared to the multicompartmentalized rumen (Field and Taylor, 2012); horses tend to eat many small meals throughout the day. Horses also spent more time grazing on improved pasture than on native, low-quality pasture. Goats tended to browse the unimproved pasture, suggesting that they would be a useful species to graze a rough pasture either along with or in rotation with horses (Ferreira et al., 2013). Horses cannot regurgitate and do not chew cud like ruminants.

Year 1

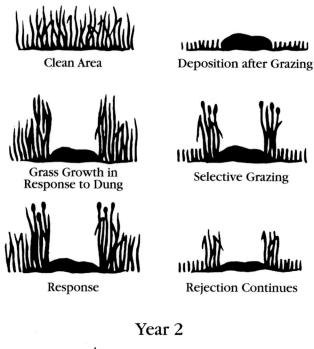

Clean Area

Deposition after Grazing

Grass Growth in
Response to Dung

Selective Grazing

Response

Rejection Continues

Year 2

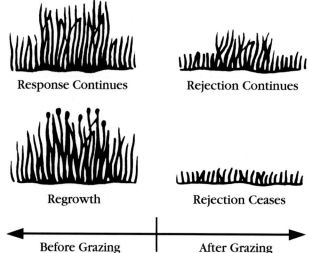

Response Continues

Rejection Continues

Regrowth

Rejection Ceases

Before Grazing | After Grazing

FIGURE 8.1 The effect of manure on grazing. Animals will avoid grazing areas contaminated with manure for a long period of time. *From Fig. 3.1. OMAFRA, 2000. Pasture Production. Publication 19. Ontario Ministry of Agriculture, Food and Rural Affairs. pp. 21−31. http://www.omafra.gov.on.ca/ english/crops/pub19/pub19toc.htm. © Queen's Printer for Ontario, 2015. Reproduced with permission.*

In addition to the physiologic differences between horses and other livestock, we must also consider the objectives in grazing management. Horses are raised for athleticism, hardiness, and longevity, while cattle and other production animals are managed for goals like maximal weight gain or milk yields. Horses must maintain a moderate weight; feeding very high-quality feed in large amounts like one would to a feedlot steer results in equine obesity and potential metabolic problems. In fact, even foals should not be managed for rapid growth, as it can cause developmental orthopedic disease that can affect soundness later in life (Archer, 1973; Davidson and Harris, 2007; Hoskin and Gee, 2004). Thus, diets for horses that are fully grown and idle or only lightly worked do not need to have high concentrations of nutrients, compared to the diets of livestock raised for meat, milk, or fiber. Exceptions to this generalization are horses in early lactation and horses that are exercised vigorously and frequently, such as those used for racing and other athletic competitions.

FACTORS REGULATING FEED INTAKE

Horses and ponies utilize digestive and metabolic capabilities common to carnivores, omnivores, and ruminants (Ralston and Baile, 1983). Pregastric cues (appearance, taste, smell, and texture) and stimuli from passage of feed influence feed choices and the onset and lengths of meals in horses and ponies (Ralston and Baile, 1983). Passage of feed through the mouth and pharynx of horses triggers the onset of normal satiety behavior (Ralston, 1982).

Blood concentrations of glucose, volatile fatty acids, and fats stimulate onset of satiety, delaying the onset of the next meal in horses and ponies with free choice access to feed (Ralston and Baile, 1983). Once a horse starts eating, the size and duration of the meal are also influenced by how hungry the horse was at the start of the meal. Mechanical distension of the digestive organs does not seem to signal the end of feeding in horses, as it does in several carnivores and omnivores (Ralston and Baile, 1983).

Physical form of a feed can also affect the rate of its intake. Horses consumed feed more quickly when it was pelleted than whole, despite having the same ingredients (Hintz and Loy, 1966), and horses consumed more alfalfa in a day when it was in wafer or pellet form compared to loose hay (Haenlein et al., 1966).

THE DIETS OF FERAL HORSES

Study of feral horses has useful lessons for horse nutrition and grazing management. Watts (2007) reported levels of nonstructural carbohydrate (NSC; sugars, starches, and fructans) to range from less than 9% to 12% in native and naturalized grass of the intermountain western United States. By comparison, average pasture grass was 15%−18% NSC, and improved varieties of grass grown in the same region frequently contained higher than 20% NSC. Native range in this region was said to have lower NSC per acre than cultivated pastures, yet feral horses manage to survive and thrive under these conditions.

Feral horses normally have foals in spring and early summer, closely followed by breeding. This means that the period of highest daily nutrient requirements for lactating mares coincides with the highest NSC concentrations for some species of rangeland forage. Watts (2007) also pointed out that feral horses need to exercise to eat, requiring more walking activity to acquire an average diet than most domestic horses in confinement. It is normal for feral horses to gain weight during months that forages are high in NSC, protein, and other required nutrients and lose some weight during winter, when standing forages are mature and have declining quality. Feral horses in the United States can have good body condition scores in mid-summer when much of the available forage looks mature and lacks green color. (See Fig. 8.2 for wild horses in Utah and Ransom and Cade (2009) for a collection of photos of wild horses in Colorado, Wyoming, and Montana). Apparently, feral horses can thrive in environments where the supply of highly nutrient-dense forage is seasonal, and much of the year, the forage and browse are of lower quality. In this chapter, we will learn how these horses maintain good condition by selecting diets that are of higher quality than the sward average.

TIMING OF GRAZING PATTERNS IN FERAL AND CAPTIVE HORSES

Partly due to the small size of the equine stomach, horses naturally consume many small meals throughout the day. Several researchers have observed both feral and confined horses grazing approximately 14 h per day, eating predominantly before dawn and after dusk (Edouard et al., 2009; Fleurance et al., 2001; Mayes and Duncan, 1986). It has been suggested that

FIGURE 8.2   Wild mares and foal in Utah during July are in good body condition when forage in their environment appears sparse and of moderate quality. *Photo courtesy of Helen MacGregor.*

inactive adult horses requiring only a maintenance diet can consume their nutritional requirements in 8—10 h of grazing (Dowler et al., 2012). Therefore, care must be taken to avoid obesity and related problems in captive "easy keeper" type horses on improved pastures.

Feral Horses

Studies in feral and wild horses revealed that the amount of time spent grazing varies with daylength. In these studies, horses grazed between 43% and 62% of daylight hours (King, 2002; Random and Cade, 2009). The percentage of time that Camargue horses spent grazing was greater during daylight than during night in autumn, winter, and spring but equal during summer due to a decrease in summer grazing around noon (Mayes and Duncan, 1986). Feral ponies and horses observed overnight grazed from 1900 to 2200, reduced grazing for a few hours, and then resumed grazing around 0400 (Keiper and Keenan, 1980; Mayes and Duncan, 1986).

These studies give us a glimpse into the natural, evolutionary grazing behavior of horses. Season and weather appear to affect grazing activity, with horses spending less time grazing during the peak activity of flies (Mayes and Duncan, 1986; Singer et al., 1999).

Captive Horses

Presently, most horses are kept in captivity, so numerous researchers have observed pastured and stalled horses to characterize their grazing and eating patterns. Horses turned out for 24 h a day grazed approximately 45% of the time (King et al., 2013), much like feral and wild horses. However, horses on limited turnout display some interesting grazing behaviors. Those with 1—6 h of turnout each day spent an average of 82% of their time grazing, likely as compensation for spending most of their time in stalls (Ödberg and Francis-Smith, 1976). Building on this finding, Dowler et al. (2012) found that horses on 8-h turnout consumed twice as much dry matter from pasture in the first 4 h than the second 4 h. However, this only occurred in October, but not February and May, suggesting that season plays a role in determining grazing behavior (Dowler et al., 2012). Another researcher also observed that horses confined for 12 h before being released onto pasture then grazed for 3 h continuously upon turnout. Horses that were stalled for 24 h with free choice hay consumed the most feed during daytime hours (King et al., 2013). The authors suggested that disruption in the behaviors of the natural time budget with 12 or 24 h confinement may be related to increased incidence of digestive disorders and stereotypies documented in confined horses (King et al., 2013).

Some of these studies also measured the amount of time horses engaged in locomotion and resting. Wild/feral horses rested 20%—38% and traveled 7%—16% of the time (Boyd et al., 1988; King, 2002; Ransom and Cade, 2009). The turnout-restricted horses spent only 1% of time resting and 7% of time in locomotion (Ödberg and Francis-Smith, 1976), while the horses out for 24 h rested for 21% of the time and moved for 24% of the time (King et al., 2013). This suggests that horses with limited turnout time will waste little time with moving and resting in favor of grazing, while horses with unrestricted pasture time will spend more time seeking out preferred grazing sites.

MEASURING THE AMOUNT AND RATE OF FEED INTAKE

Researchers have several tools and methods to estimate feed intake of grazing animals. These include sampling and measuring forage yield before and after grazing, weighing animals before and after grazing, and using inert marker methods. Being able to estimate feed **dry matter intake** (DMI) helps researchers to compare animals' preferences for different forage characteristics.

Forage DMI rate is calculated as bite rate × bite size, and it increases with body size (ponies 8—22 g/min; heavy mares 22—103 g/min), so it is often reported per unit of body weight (Fleurance et al., 2009). As the amount of available forage increases, horses can take larger bites and increase their intake rate, especially in larger horses (Fleurance et al., 2009). DMI has been expressed as 14—32 g per kg of live weight per day (Edouard et al., 2009), which corresponds to 1.4%—3.2% of body weight per day, in line with typical horse feeding recommendations for maintenance of body weight (2.0%) and lactation (2.5%) (NRC, 2007).

Factors Influencing Bite Rate

Observed bite rates in horses range from 8 to 57 bites per minute (Duren et al., 1989; Edouard et al., 2010). The mass of forage dry matter per bite can be less than 0.5 g, especially for short forages, and it can rise to 3 or even 5 g on tall, poor-quality forage (Duren et al., 1989; Edouard et al., 2009). As bite sizes increase, bite rates decrease, and time to chew and swallow increase (Fleurance et al., 2009).

The rate of feed intake of grazing animals reflects the quantity and quality of swards and the amount of time they are turned out. Duren et al. (1989) suggested that horses compensate for time in stalls by rapidly consuming forage immediately upon turnout because the experiment showed that average biting rates in the first 20 min of grazing were higher than for the rest of a limited 3-h grazing session. This study also made an interesting observation that exercised horses had lower biting rates and more contented behavior than nonexercised horses, which displayed more restless grazing behavior, but bite size was the same. Fleurance et al. (2009) concluded that handling time (processing, chewing, and swallowing) limits DMI rate, and it increases with bite size but is not affected by the fiber content of feeds. Larger horses processed forage faster than smaller horses, and smaller horses had a lower intake rate than larger horses when bite size increased, presumably due to their smaller mouths.

Horses Prefer Fresh Pasture to Dried Hay

When given the choice between fresh, tall fescue pasture and hay cut from that same pasture the year before, horses consumed more DM on pasture. The pasture had higher digestibility than the hay, which may explain the preference for forage as pasture, although the stage of maturity of both the hay and pasture when consumed could play a role as well (Chavez et al., 2014). Muller and Uden (2007) also found that the drier a forage was, the less horses consumed. Another study estimated DMI and rate of DMI of horses grazed for 3, 6, 9, or 24 h. Horses with shorter turnout times (hay was provided when not turned out) had lower *total pasture intake*, but higher *rates of intake* than horses with longer turnout, suggesting that they preferred to consume forage as pasture rather than hay (Glunk et al., 2013).

Forage Quality Affects Intake Rates

As forage quality decreases, horses (and other herbivores) tend to consume less of it (Ungar and Noy-Meir, 1988; van Langevelde et al., 2008). This has been illustrated in goats, where they increased bite rate and bite size when feed had a higher leaf to stem ratio (an indicator of lower maturity and higher nutritional quality) (Illius et al., 1995). Recall from Chapters 3 and 7 that as forage maturity increases, quality decreases as the plant becomes stemmier and less leafy by weight. In cattle, more energy must be expended to gather nutritious leaves when forage contains a high density of stems versus a low stem density (Drescher et al., 2006). This has implications for analyzing different types of forages (browse vs. grasses) and different maturities of forages.

To further elucidate the role of forage quality on intake rate, Edouard et al. (2010) offered horses pasture swards of varying heights and found that as long as protein was adequate, horses selected the sward that allowed higher DMI (tall height/poor quality). However, if protein was low, then horses selected swards with higher nutritional quality (lower height/high quality). The horses always made use of both swards, balancing protein and energy needs with fiber needs.

HOW YOUNG ANIMALS LEARN TO MAKE FEED CHOICES

The mother plays an important role in teaching youngstock about feed choices. Mothers that avoid poisonous plants teach their offspring to avoid those plants (Mirza and Provenza, 1990; Provenza et al., 1992) and to consume alternatives (Mirza and Provenza, 1992). Lambs that observed their dams eating wheat grain consumed more wheat than lambs that were exposed to wheat grain in the absence of their dams (Green et al., 1984).

Feed neophobia is a phenomenon in animals in which they are cautious when sampling novel foods (Launchbaugh and Provenza, 1991). In young animals, this phenomenon can be minimized by exposing them to many feeds and flavors at a young age. This concept has been observed in lambs and calves, and it has been suggested that flavors introduced in utero or through mother's milk may also influence flavor preferences later in life (Green et al., 1984; Nolte and Provenza, 1991; Smotherman, 1982). For example, calves exposed to straw consumed more straw at age 5 than cows that were not exposed to straw at a young age (Wiedmeier et al., 2002).

Understanding the concept of feed neophobia can be useful when managing young animals to ensure that they are willing to eat a more diverse diet when they mature. Many times, if animals are only exposed to a small number of different feeds while young, they are reluctant to try new feeds. It can be a disadvantage when the animals' diets must change, perhaps due to transport to a novel environment or when all of an original feed source is depleted. Launchbaugh et al. (1997) determined that feed neophobia could be overcome by observing that lambs offered four novel feeds in different orders consumed more of the fourth novel feed than the first one. These results support the idea that repeated exposure to novel feeds may increase future intake of novel feeds (Launchbaugh et al., 1997).

Some research has suggested that feed preferences can begin as early as gestation and may have a genetic component. It is known that foraging behavior is affected by the diversity and quantities of chemical compounds in plants and by an herbivore's experiences with each feed (Villaba et al., 2015). Nutritional experiences of fetuses and young animals influence neural, morphologic, and physiologic changes that influence food preferences and intake much later in life (Provenza et al., 2015; Villaba et al., 2015). These interactions of ingested substances with animal genomes likely influence gene expression and behavioral responses, which, over generations, facilitate animal populations becoming locally adapted to landscapes (Provenza, 2008). Support for these ideas comes from reports of fetal lambs exposed to saltbush shrubs (which their dams ate). The exposed lambs grew faster and handled a salt load better than lambs from dams on grass pastures (Chadwick et al., 2009).

These studies show that animals that learn to eat a wide variety of foods early in life have advantages when they are subjected to changes in the mixture of feeds available to them, being more adaptable and resilient than their contemporaries with narrower assortments of feeds that they willingly eat. This would be a fascinating topic to study in horses, especially since horses are considered to be picky eaters. While the studies discussed in this section involve sheep and cattle, which have significant differences in their digestive systems and feed preferences from horses, there is great value in considering their findings, since ruminants often use grazing environments similar to those used by horses.

Horses add to their behavioral predispositions and physical abilities by accumulating knowledge about habitat quality and refining their foraging skills. The earliest learning probably occurs as they watch their mothers sort and make feed choices. Throughout a foal's young life, foraging decisions are made daily and are likely dependent on their geographic location, dam's feed choices, and the variety of feeds to which they have access; however, equine-specific research could shed more light on these factors. Decisions that young animals learn to make include what, where, and how much to eat. These decisions affect the individual animal, other animals in the same environment, and the composition and productivity of the forage resource (Launchbaugh and Dougherty, 2007). Herd social behaviors influence individual grazing behavior, just as human social behaviors influence individual adolescent food and drink choices.

ANIMALS MAKE FEEDING DECISIONS BASED ON FEEDBACK FROM PAST CHOICES

Foraging decisions are likely to result in nutritional benefits with low probability of toxic or poor-quality plant material (Launchbaugh and Dougherty, 2007). But why? How do grazing animals know which plants contain the most optimum collection of nutrients and which are toxic?

Fig. 8.3 shows external factors (such as social interactions with peers and plant attributes) plus internal factors (animal attributes that influence ingestion, digestion and metabolism) that direct foraging decisions. As animals mature, they sample various foods and develop flavor aversions and preferences. Flavors and presumably odors of plants are associated with **postingestive feedback** (Launchbaugh and Dougherty, 2007).

Studies on people and lambs have revealed that an individual's preference for food depends on its nutritional needs, particularly for energy and minerals, and preference declines when foods are eaten to a level that satisfies the nutritional needs (Provenza, 1995b). The ruminant preference for diets including several diverse species is consistent with a theory that herbivores attempt to avoid toxins and acquire nutrients due to aversive stimuli combined with postingestive feedback

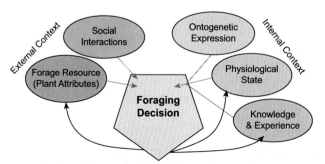

FIGURE 8.3 Foraging decisions are directed by many factors and conditions that are external to the grazing animal, such as social interactions with other herbivores or plant quality and antiquality factors. The environment inside the animal also directs foraging decisions through development of physical and physiologic capabilities (ontogeny), current physiologic state, and acquired knowledge and experience. *From Fig. 44.2 Launchbaugh, K.L., Dougherty, C.T., 2007. Grazing animal behavior. In: Forages. The Science of Grassland Agriculture, vol. II. sixth ed. Blackwell Publishing, Oxford, England, p. 677.*

unique to each feed (Provenza, 1995b). Aversions can be caused by high concentrations of toxins, high concentrations of rapidly digestible nutrients, deficiencies of certain nutrients, and eating nutritionally adequate feeds in excess, causing satiety and surfeit (filled to a level causing nausea). Aversions encourage animals to eat a balanced diet, reduce toxin intake, and sample diverse feeds (Provenza, 1995b). The encouragement to sample many feeds is lost if pastures are monocultures of nutritious species, and the risk of satiety and surfeit may result. Sheep and cattle, if given a chance, will switch to grazing different types of feed several times per day.

Animals probably cannot directly taste or smell nutrients and toxins in plants, but intake in ruminants can be suppressed by elevated levels of toxins and either deficits or excesses of nutrients. If animals can experience positive feedback from nutritious, nontoxic feeds and experience negative feedback from nutritionally unbalanced or toxic feeds, these sensations provide an evolutionary advantage, enabling them to survive changing conditions and reach the age of reproduction (Provenza, 1995a).

Numerous studies have been performed on several species to test and illustrate this concept. Aversions have been observed based on malaise caused by high concentrations of fermentation by-products (propionate, acetate, ammonia nitrogen) and presence of toxins, as well as animals simply no longer preferring certain flavors (Provenza et al., 1994; Provenza, 1995b; Wang and Provenza, 1997).

Factors that improved feed preference included low doses of fermentation by-products, feeds that were high in nutrients and low in toxins, and flavors associated with experimental recovery from threonine deficiency (Gietzen, 1993; Provenza, 1995b; Wang and Provenza, 1997). One interesting experiment started with the assumption that high-grain diets can create malaise in ruminants by increasing the acidity of the gut contents. Lambs drank more sodium bicarbonate solution (a neutralizer of acid) when eating a high-grain rather than a low-grain diet. Normally lambs would prefer plain water to sodium bicarbonate solution (Provenza et al., 1994; Provenza, 1995b).

Noticeably, none of these studies involve horses. While aversions from negative feedback are certainly a natural behavior, the physiology and mechanisms of equine aversions may differ from those of ruminants and warrant study.

A wide variety of grazing animal species, including horses, select for net energy, protein, and some forms of carbohydrate. Grazing animals appear to minimize their level of discomfort from tastes, textures, and physical and chemical properties of feeds eaten, plus physiologic responses to eating, such as hormones released, mobilization of body reserves, and inhibition or stimulation of parts of the central nervous system through feed selection. This selection effort may help to explain why intake declines as pasture is grazed down below the leafier portion to the stemmier portion with lower levels of digestible nutrients and higher levels of indigestible components (Poppi, 2011). This idea is based on the theory of **minimal total discomfort** in which animals experiment with their intake of various feeds until they strike a balance between feedback signals of excess and deficiency (Forbes, 2001, 2007). According to this theory, animals can adjust their diet to their nutrient requirements (Forbes, 2007). The theory is based mostly on research in ruminants, and it provides an interesting starting point for future research into feed choices made by horses.

MAKING CHOICES WHILE GRAZING

Through evolution, animals making feeding choices that result in avoidance of poisoning and improved nutrient intake are likely to survive and reproduce. **Foraging theory** aids in understanding the survival value of feeding choices (Sadava et al., 2008). The benefits of foraging (seeking and selecting feeds) are the nutrients and energy obtained by the animal for its maintenance, growth, and reproduction. The costs of foraging are the energy expended, the time lost for other activities that could enhance fitness, and the risk of increased exposure to predators. More valuable feed types yield more energy per unit of time or energy expended. As the most valuable food types become scarce, energy-maximizing animals add progressively less valuable food types to their diets. Horse herd dynamics, involving complex hierarchies and "guarding resources," can influence which horses access the most desirable forage plants in a range or pasture (Lesté-Lasserre, 2016).

Foraging theory does not apply perfectly to many domesticated horses. Confined horses on improved, well-managed pasture at a low stocking rate do not have to contend with the costs of foraging despite being evolutionarily equipped to do so (Fig. 8.4). In fact, individuals within many hardy horse breeds need to have grazing restricted to avoid obesity and related metabolic problems. However, if forage is allowed to become scarce due to overgrazing, then there arises a need to seek the highest quality plants among what remains.

Fig. 8.5 shows a progression of levels at which grazing animals make decisions about which plants and plant parts they will eat (Launchbaugh and Dougherty, 2007). At each of the three major decision levels, there are selection cues or attributes of the plants and their environment that guide animals in deciding what and where to eat. The Bite and Plant level

FIGURE 8.4 A horse on lush pasture. This horse does not need to expend a great deal of energy to locate the highest quality feeds in its pasture. *Contributed by Carey Williams.*

reflects the smallest space and the most immediate feeding decisions. The Patch and Feeding Site level reflects slightly larger areas where animals choose to be when grazing. The Camp and Home Range level refers to the largest areas in which animals choose to spend relatively long periods of time. When grazing animals are making feed choices at these levels, they can choose forages more nutritious and less toxic than the average forage because of conditioned flavor aversions and preferences (Launchbaugh and Dougherty, 2007).

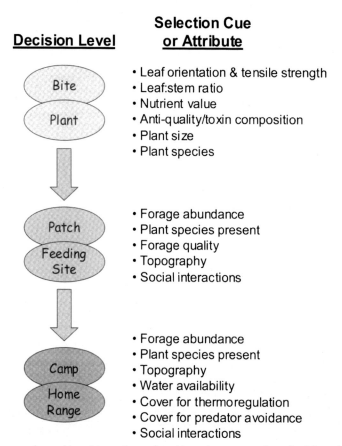

FIGURE 8.5 Foraging decisions occur along a hierarchic continuum from bite to home range. At each of these decision levels, animals respond to varying selection cues related to plant and habitat attributes. *From Fig. 44.1 Launchbaugh, K.L., Dougherty, C.T., 2007. Grazing animal behavior. In: Forages. The Science of Grassland Agriculture, vol. II. sixth ed. Blackwell Publishing, Oxford, England, p. 676.*

Factors Affecting the Bite and Plant Level

Plant Species Preference

At the bite level, horses are adept at using their lips to sort through forages and select their preferred plants. Horses generally prefer grasses to weeds and shrubs (Archer, 1973). When confined to a small area by picketing, horses only consumed forbs after depleting the supply of grasses (Olson-Rutz et al., 1996). A review of diet composition reports in wild horses, livestock, and wild ungulates revealed that diet composition averages across the four seasons for horses were 77%–89% for grasses, 4%–15% for forbs, and 3%–10% for browse, with no significant seasonal effects (Scasta et al., 2016). Horses appear to have preferences for some grass species over others. Commonly seeded cool-season grasses Kentucky bluegrass, tall fescue, perennial ryegrass, timothy, and meadow fescue have been reported to be highly preferred by horses, while tall fescue and orchardgrass were less preferred in other studies (Allen et al., 2013; Archer, 1971, 1973; Hayes et al., 2009; Martinson et al., 2016).

There are a number of variables that can affect horse preference of the same species between studies, and even experimental design can play a role: preference can only be determined based on the species presented to the animals. Marten (1978) suggested that species preference is dependent on factors relating to the animal, the plant, and the environment. Animal factors include breed or species, senses, individual variation, past experiences (i.e., negative feedback from toxins), and physiologic condition (an increasingly hungry horse may sample plants that it normally would not). Plant variables include species, variation within the species (i.e., cultivar), chemical composition (nutrient concentrations or toxins), morphology, maturity, availability, and effects of management (like mowing or fertilization). Lastly, environmental variables that can affect species preference include plant diseases, soil fertility, presence of feces, supplemental feed provided, climate, and seasonal/diurnal variation. Allen et al. (2013) found that the preference by horses for some grass species was markedly different between seasons in the same year, and between years in the same season, illustrating the complexity of species preference in horses.

Forage Nutritional Quality

There is general acceptance that "higher quality" forage (younger plants with high NSC) is more likely to be consumed than "lower quality" forage (more mature plants with higher fiber) (Hoskin and Gee, 2004). Several equine studies have illustrated this using various indicators of forage quality, including DM, NSC, and water-soluble carbohydrates (a component of NSC) (Allen et al., 2013; Dowler et al., 2012; Longland and Byrd, 2006). Interestingly, researchers have been able to prompt cattle consumption of previously rejected forage by adding sugary substances to it, suggesting that taste plays a large role in forage selection (Marten and Donker, 1964; Marten, 1978).

As a result of this preference for forages with high sugar content, the diets selected by animals are of higher nutrient quality than the average sward. Availability of green leaves is an important component of intake, and tough stems are deterrents to foraging (Benvenutti et al., 2006; Drescher et al., 2006). Thus, pasture maturity and reproductive stage have important effects on grazing behavior and intake (Prache and Delagarde, 2011). These effects will be at both the bite and the patch levels.

Familiar Versus Novel Forages

Feed neophobia has been observed in horses. In one study, horses showed a greater preference for familiar forages compared to novel forages regardless of their hunger levels at the time (van den Berg et al., 2016). In some cases, horses rejected novel forages without sampling them at all. Some other horses sampled novel forages on day 1 but would not eat as much of them on the next 2 days.

If familiar, nutritious forage is sparse, horses may choose to eat weeds and toxic plants (Offord, 2006). Horses grazing a pasture dominated by the toxic legume spotted locoweed (*Astragalus lentiginosus*) in Arizona began eating some locoweed on the second day and increased consumption of it over time. Horses appeared to prefer green grass (which was sparse) and reject abundant dry grass. Pfister et al. (2003) suggested that the horses' propensity to eat green forage influenced their locoweed consumption.

Individual Variation

Farm managers might wonder whether there are real differences in grazing behaviors among horses. Horses can be classified as efficient, semiefficient, or inefficient grazers, depending on their relative test scores in choosing, sorting, and adaptive behavior (Marinier and Alexander, 1991). In this study, efficient horses scored highly at all three behaviors, so the authors inferred that they may safely graze on pastures infected with toxic plants. The horses differed significantly in the magnitude of their ability to choose and sort different feeds. A second study revealed that laboratory-type tests can be used to predict grazing efficiency in the field and can thus be used to decide which horses are likely to be inefficient and thus vulnerable to poisoning while grazing (Marinier and Alexander, 1992).

Factors Affecting the Patch and Feeding Site Level

Sward Height

The relative height of forages can serve as a rough estimation of maturity and has been tested as a predictor of grazing sites in horses. Results have been mixed, indicating that height alone does not play a large role in grazing location selection. Some other factors tied to forage height include contamination with soil, sensory stimuli of mown versus natural forage surfaces, maturity and nutritional quality, and proximity to manure (Naujeck et al., 2005; Fleurance et al., 2010).

When forage quality was equal among different sward heights, horses chose the tallest sward (17 cm) that maximizes intake and grazing efficiency (Edouard et al., 2009). This finding was supported by another study that observed horses grazing longest on the tallest of four swards (15 cm) when swards had similar maturity (Naujeck et al., 2005). However, when forage height and maturity varied naturally from 1 to 56 cm, horses selected mostly intermediate forage heights (6–7 cm), likely because they strike a balance between abundance and quality (Fleurance et al., 2010).

Distance to Water

A study of feral horses on Sable Island in Canada determined that water sources affect horses' feeding sites (Rozen-Rechels et al., 2015). Different groups of horses may drink at ponds, where several horses can drink at once, or at self-excavated water holes, where they line up to drink and drinking takes longer than at ponds, or at both. Herds using both ponds and water holes exhibited higher stocking rates than herds using only water holes, and these stocking rates declined as the horses moved farther from water. Forage near the water sources was poor-quality heathland, but farther from water, horses could graze higher quality grasslands. The herds using ponds were able to drink more efficiently and then had time to travel farther for high-quality forage. This study demonstrated how foraging can be constrained by important central places (drinking water source) and by forage density (Rozen-Rechels et al., 2015).

Presence of Manure: Lawn and Rough Grazing

Equines, like many other animals, have been observed to follow a specific set of behaviors when eliminating (Fig. 8.1). For example, stallions tend to defecate onto the fecal piles of other stallions. Mares approach and smell the stallion manure piles and without turning, defecate a body-length away (Avery, 2003).

When horses are confined to relatively small paddocks and allowed to graze continuously, three distinct areas develop. "**Lawns**" are areas of short-cropped grass. "**Roughs**" are tall and overgrown areas of group defecation where horses typically avoid grazing. **Bare areas** are devoid of forages, likely due to overgrazing and/or heavy hoof traffic. One study described a horse pasture consisting of 48% lawns, 31% roughs, and 21% bare areas (Ödberg and Francis-Smith, 1976), meaning that less than half of the fenced area was actually used for grazing if horses truly do not graze the roughs. Pastures that exhibit these characteristics are often referred to as "**horse-sick**" (Avery, 2003). Fig. 8.6 shows a horse pasture in autumn that displays the horse-sick appearance of long-grass roughs, short lawns, and bare areas where overgrazing and heavy traffic have killed pasture plants. With low stocking rates but long grazing periods, pasture losses because of fouling by manure can be as high as 45%. Shorter grazing periods, even with higher stocking rates, can reduce this loss (OMAFRA, 2000).

It has been suggested that horses graze in the lawn and rough pattern to avoid ingesting gastrointestinal parasite larvae in the feces, and research results have been mixed. It does appear that feces, not urine, is the cause of rejection, as ponies rejected feeding sites even when feces was hand-placed and then removed from the site (Archer, 1978). Another study revealed that horses continuously grazing at a moderate stocking rate (1.9 acres per horse) avoided tall grass (>16 cm) and only selected short grass when it was more than 1 m from a pile of feces (Fleurance et al., 2007). The parasite load of individual horses had no apparent effect on where they grazed. However, when horses were grazed at a higher stocking density (<0.5 acres per horse) during winter and spring, they defecated in both grazing areas and lawns (Medica et al., 1996). The distinction of roughs as areas of only defecation did not hold true under high stocking rates in the nongrowing season. Additionally, the lawn and rough behavior is not observed in feral horses with plenty of space to roam (Lamoot et al., 2004). The differences in stocking rates between these studies certainly help to illustrate the variation in conditions

FIGURE 8.6   The appearance of a horse-sick pasture with lawns, roughs, and bare areas.

under which equids are managed. It reminds us not to generalize about horse grazing and elimination behavior without clear definitions of management conditions, and to consider management conditions when comparing study results.

Forage Nutritional Quality

While we know that concentrations of DM and NSC have been shown to influence horses' local bite choices, they can also influence patch choice if the grazing land is particularly heterogeneous. In horse-sick pastures, this may tie in with the lawn and rough grazing pattern, which produces patches of mature, nutritionally poor forage. On unmanaged rangeland where animals must travel to find feed, it would be reasonable to expect that animals select feeding patches that provide the highest energy intake.

However, research in beef cattle has shown that nutritional quality is not the sole predictor of patch selection. Cattle grazing mountainous range avoided grazing locations with steep slopes despite the exceptionally high nutrient profiles of forages on those slopes. On flatter rangeland, the cattle did select grazing locations with higher than average nutrition (Ganskopp and Bohnert, 2009). This study indicated that models relying solely on forage quality indicators are poor predictors of grazing cattle distribution and suggested that horizontal distance to stock water and degree of slope were significant predictors of grazing livestock distribution. Given the number of variable factors that affect equine species preference, it is likely that a similarly complex process determines their selection of feeding sites.

Terrain and Slope

Feral horses in a forest reserve in Canada spent most of their time in lowland grasslands, but in the spring and summer, they also grazed shrublands, and in the winter, they also spent time in forests (Girard et al., 2013). This suggests that they selected habitat based on vegetation type, distance to forest, and solar radiation. Work in beef cattle has illustrated that spend less time at a single site on gentle terrain and stay longer (up to 10 days) at a grazing a site on rugged terrain, as a result of the energy costs associated with moving to a new feeding site (Bailey et al., 2015). This energy may also play a role in grazing patterns on sloping terrain. Sheep in New Zealand showed higher grazing slope than medium and high slope areas (Lopez et al., 2003). Cattle in Montane rangelands of Alberta reas on a landscape scale primarily by slope but also distance to water (Kaufmann et al., 2013). that horses attempt to maximize their feed digestible energy intake relative to the energy and time tain the feed, there will be a practical limit to the steepness of slopes that horses are willing to owledge of the supply of forage up the slope.

-Term Changes

d to use elevation to increase comfort (King, 2002). A pattern of habitat use uring the morning and evening then moved to high places around mid-day nd provided some relief from heat and flies.

Factors Affecting the Camp and Home Range Level

Foraging decisions by several species on the broadest level (camp and home range) are impacted most by geography. The equine work at this level has been performed by observing wild or feral horses, but the concepts may also apply to tame horses grazing on large rangeland pastures in the central and western United States. However, horses grazing on smaller improved pastures generally do not get to select a home range because humans manage their grazing.

Altitude and elevation may play a role in feral or rangeland horses' home range selection. Feral horses living at high elevations in the Andes Mountains have changes in genes affecting hypoxia adaptation and their nervous system compared to their Spanish ancestors (Hendrickson, 2013). The author suggested that the unique vegetation in high altitude environments and the natural history of the animal species may also influence selection of gene pathways.

Horses may select a home range based also on the availability of cover for thermoregulation. Horses without shade on a hot day show more pronounced signs of heat stress compared to horses with shade (Holcomb et al., 2013). They also tend to use shade in short, frequent bouts throughout the day (Holcomb et al., 2014, 2015; Holcomb and Stull, 2016), suggesting that proximity to shade in hot weather would be an important factor when selecting a camp location. Holcomb (2017) concluded that direct sun in hot weather is a mild challenge to horses, and they can choose relief from it if they have access to shade or shelter.

Shelter from cold weather is also physiologically important, as metabolic rates of severely cold-stressed horses can increase by 70% to provide energy for warmth (Cymbaluk, 1994). Horse farm managers may provide three-sided shelters, which improve heat conservation (Cymbaluk, 1994). Feral or rangeland horses would need to select a home range with adequate shelter available or increase their DE intake in cold weather, which would be difficult in the nongrowing season.

EFFECTS OF GRAZING BEHAVIOR ON THE ENVIRONMENT

Horses are particularly harsh on their grazing sites compared with other livestock. For more information, this topic is also covered in Chapter 14.

Animal Waste

Horse waste is rich in nutrients, which can be good for plants and soil health but detrimental for water quality. Some nitrogen and water from urine volatilizes and evaporates, respectively, into the atmosphere and is lost from the pasture system. The physical structure of the manure is broken into progressively smaller particles by drying, horse hooves, intestinal parasites, soil fauna, freeze-thaw cycles, and pasture management tasks. Rainwater carries manure particles deeper into the crowns of plants and pores of the soil, where microorganisms can further break them down into nutrients useable by plant roots. Manure contributes organic matter and thus water-absorbing capacity to soil. Some nitrogen forms in manure are more stable and released more slowly than the nitrogen in commercial fertilizers. Nitrogen from manure and urine can encourage leafy grass growth, but, as anyone with a dog and a lawn knows, urine-soaked plants may be scorched and killed (OMAFRA, 2000).

In addition to natural and largely beneficial actions, there can be negative consequences from manure deposition on pastures. Nitrogen and phosphorus from manure can contaminate surface water and ground water, threatening the health of the waterway. Animal manure contains bioactive compounds such as hormone metabolites, antibiotics, and pathogens, which can also enter water bodies.

Trampling

The physical damage to turf by horse hooves is significant, resulting in damaged plants, some death of plants, and mixing of topsoil. When confined in a small pasture, horses may beat a path around the perimeter, thus reducing the area for growth of edible forage. Through ideal grazing management, the length of grazing time, the length of pasture resting time, and the stocking rate, horse traffic patterns can be adjusted to minimize turf damage. Another consequence of trampling is soil compaction, which plugs and crushes soil pores, reducing availability of water and oxygen for roots (OMAFRA, 2000). In contrast, trampling of long or partly grazed forage puts the plant matter in contact with soil organisms, which gradually break the forage down into simpler organic components, thus contributing to soil organic matter. Much trampling damage is repaired by freeze/thaw cycles and activity of worms and other soil organisms.

Overgrazing

If grazing animals remain in one area for **too long** and graze regrowth before grazed plants recover, they are overgrazing (Murphy, 1998a; Voisin, 1959). The severity of defoliation of plants by grazing animals depends on how much plant material is removed and how often. Confined animals tend to overgraze areas of forage, continuing to graze forages down to a point where the plant physiology, reproduction, and growth are compromised (Vallentine, 2001). Continued overgrazing generally results in reducing pasture condition and a deterioration of botanical composition of the plant community (Vallentine, 2001). Remaining plants tend to be stemmy, indigestible, and coated with soil, and animals cannot efficiently graze the short stubble.

Overgrazing is a particular concern with confined horses, as they have strong plant preferences when grazing. They seek out preferred plants first (spot grazing), graze other plants, then return to the preferred plants as soon as regrowth starts. These plants do not have a chance to recover, regrow photosynthetic tissue, and replenish energy stores before they are defoliated again. Thus, preferred plants stay short, become weakened, and may die, to be replaced by potentially less palatable but more aggressive plants that can survive the overgrazed condition. Unpalatable plants are not grazed and are able to reproduce, eventually predominating in the pasture (Murphy, 1998b). Overgrazing can be minimized by grazing management strategies such as rotational grazing.

Positive Effects of Grazing on Plants

Grazing animals can have positive effects on forages. Grazing a small amount of a plant (less than 50%) makes more light available to young leaves at the base of the plant, especially in grasses. This can increase the rate of photosynthesis in the remaining part of the plant. Carbon supply and nutrient absorption are initially reduced following grazing, so plants respond to restore growth and carbohydrate supply. Cells in active shoot meristems multiply and expand. Newly photosynthesized carbohydrate is allocated to growing shoots (Richards, 1993; Sollenberger and Newman, 2007).

Gardeners remove apical buds of young plants when they "pinch back" tips of shoots. The result is increased branching from axillary buds lower in the plant, resulting in a denser, bushier plant. The same thing can happen in legume forages. When wild mule deer and elk graze young single stalks of scarlet gilia (*Ipomopsis* aggregate) the plant grows four replacement stems (Sadava et al., 2008). Removing apical buds of some trees and shrubs by grazing can extend the growing season of the plant, presumably since the new parts resulting from axillary buds have as long a life expectancy as the original plant material (barring a killing frost). Some plants benefit from grazing because herbivores spread their pollen or seeds.

MANAGING GRAZING BEHAVIOR

Based on the concepts presented in this chapter, we can recommend some practical management solutions to encourage or discourage horses from grazing, improve uniform pasture utilization, and minimize the negative effects of grazing on pastures.

Rotational Grazing

Rotational grazing or management-intensive grazing practices were developed to encourage grazing of forages at optimal stages of height, maturity, nutrient concentration, and digestibility (Voisin, 1959). Rotational grazing involves a number of paddocks (generally 4—30) within a pasture, which are only grazed one at a time (Gerrish, 2004; Hodgson, 1990; Hopkins, 2000). Based on grazing management of livestock, we recommend that a group of horses is introduced to the first paddock in the rotation when the average plant condition is optimal, as described earlier. When about half of the forage height has been removed and there is another paddock in the optimal condition, the horses are moved to the second paddock in the rotation, and the first paddock rests and regrows until it is again optimal for grazing or making hay (Murphy, 1998a). Use rotational grazing with short grazing times (<5 days) and long resting times (15—40 days) or as indicated by forage height and maturity.

Rotational grazing is covered in more detail in Chapter 9, and it has many advantages. By grazing smaller paddocks for shorter periods of time, horses do not have a chance to regraze new growth of their preferred forages, which both weakens the plant and contributes to the formation of lawns and bare spots. Horses may still designate certain areas for defecation and create roughs, but moving the horses from paddock to paddock spreads out the manure more evenly over the entire pasture area. Horses rotated frequently do not have enough time in any one paddock to create high-traffic loafing areas and kill vegetation, thus preserving the forage while still being allowed to exercise. Free-ranging horses do not form roughs or latrines since they defecate wherever they graze (Lamoot et al., 2004).

Multispecies Grazing

While we have learned that horses (and other species) avoid grazing near their own manure, presumably to avoid infection by gastrointestinal parasites, this behavior does not extend to the manure of other species. Horses are affected by four common parasite species that do not affect sheep or cattle and two species that do affect sheep and cattle (Fox, 2016a,b; Lane, 2016), so alternate grazing or co-grazing of cattle, goats, or sheep with horses can help to break parasite life cycles and thus reduce parasite loads in these herbivores. Multispecies grazing may also reduce the area of a pasture that is fouled by manure and thus not grazed because these herbivores tend to be repelled by manure of their own species but not others (Archer, 1978; Aoyama et al., 1994; Daniels et al., 2001; Dohi et al., 1991; Forbes and Hodgson, 1985; Teixeira et al., 2012).

The different grazing choices made by horses and cattle can combine to influence the plant species composition of pastures. A 6-year study of three different plant communities revealed that mixed grazing by horses and cattle produced the most species-rich and structurally diverse swards due to additive effects of the two herbivore species. Mixed grazing by cattle and horses reduced numbers of two very competitive plant species from horse latrine areas, thus showing that each animal species compensated for the predominance of a grass caused by the other animal species. The authors suggested that this effect could be used to manage plant diversity in grassland ecosystems (Loucougaray et al., 2004) (see Chapter 10). As each grazing animal species expresses its unique set of preferences, it offsets the effect of the other species of animals in the same environment. Thus, fewer plant species are suppressed, and biodiversity increases.

Manure and Parasites

The most obvious way to reduce parasite infection is to remove all manure from pastures; however, this is an arduous task on all but the smallest of farms. It is often recommended to drag or harrow pastures on hot, sunny days to break up fecal balls and dry out manure, although infective small strongyle larvae can survive for several weeks at warm temperatures and even longer at cooler temperatures or when desiccated (Lane, 2016; Neilson et al., 2007). This is one advantage to using long resting times when practicing rotational grazing.

Keep internal parasites under control through good communication with an equine veterinarian who stays current on recent advances in parasite control. Cooperation between a veterinarian and horse owner/manager is important due to the large variety of parasite organisms, the diverse efficacy of the different drugs approved for use against parasites, and the many factors that influence response of parasite populations to control measures (Lane, 2016). Modern recommendations for parasite control in horses include diagnosis of active parasite species in horses of different ages, choosing the most effective drugs, treating horses at the right times, and performing about 12 other nonchemical parasite management tasks (Lane, 2016).

Maximizing Pasture Intake and Nutrition

Using what we have learned about equine grazing preferences, we can manage the characteristics of our pastures to promote grazing. The following suggestions will help to optimize intake of DM and nutrients. Select forage species that have been shown to be preferred by horses in your environment, tolerant of grazing and trampling, and well-suited to the location, and keep them at an optimal height for grazing (Martinson et al., 2016). Plants should remain in the highly nutritious vegetative stage of growth at the time of grazing (provided the horses have fairly high nutrient requirements, due to growth, exercise, pregnancy, or lactation), and those plants that are not grazed can be mowed back to prevent the start of reproductive growth. Forage should be around 6–10 in (15–25 cm) tall at the onset of grazing, with high leaf:stem ratio and high digestible protein concentration (Edouard et al., 2009, 2010). Horses with lower nutritional needs can eat taller, more mature forage, but remember that they will seek out the most nutritious plants to consume a diet of higher quality than the sward average. If possible, select forage varieties with a range of days from first growth to maturity. This helps to ensure that there are frequently some plants at the ideal stage of maturity throughout a growing season. These efforts should optimize the digestible organic matter and digestible protein concentrations of the grazed forages.

Provide plenty of fresh, palatable drinking water at all times and close to the grazing area to cut down on travel time. More detailed information on grazing management and use of watering systems can be found in Chapters 9 and 15, respectively. Manage pastures to avoid overgrazing so that forage is abundant. Do not allow plants to be grazed shorter than about 50% of their pregrazing height or, in the case of tropical plants, not shorter than their growth points so that regrowth is not impeded. Poppi (2011) suggested using 70% of forage that is grown per year in temperate pastures and only 20% of the forage grown on improved tropical range. Use of forage in wetter tropical areas could be intermediate between these values (Poppi, 2011).

To minimize neophobia, introduce young horses to many forages, including those that they will probably encounter throughout their life. Watch horses for enthusiasm and reluctance to graze. Investigate reasons for reluctance.

Limiting Dry Matter and NSC Intake

While some horses have increased caloric needs due to reproduction or exercise, many others are considered "easy keepers" and maintain or even gain weight on low to moderate quality forage. The slightly different physiology and persistent grazing behavior of these easy-keeping horses can cause health problems if they are free-fed. These horses need their caloric and NSC intake controlled to avoid obesity and, if underlying metabolic conditions exist, laminitis and founder.

To restrict intake of DM and NSC, pasture managers can implement one or more of the following:

1. Sow grass species that are known to be only moderately preferred by horses or species that may be lower in NSC such as smooth bromegrass, tall wheatgrass, meadow bromegrass, and Bermudagrass (Longland and Byrd, 2006; Undersander, 2013).
2. Avoid grazing after several hours of sunshine, followed by a sudden drop in temperature (Bowden et al., 1967; Holt and Hilst, 1969; Lechtenberg et al., 1971).
3. Reduce sward height by mowing (Siciliano et al., 2017).
4. Use grazing muzzles on ponies and horses that tend to put on weight and/or are prone to laminitis (Glunk et al., 2014; Longland et al., 2016a,b).

CONCLUSIONS

There is still much to be learned about equine grazing behavior. A great deal of research exists for ruminants and other grazing species, but horses' different physiology and preferences make it difficult to extrapolate results from these studies to horses. As horse sports grow in popularity and economic impact, there are more opportunities to attract funding for research in this field.

A review of impacts of equine grazing on pasture quality and environmental conditions (Bott et al., 2013), reflecting views of authors from eight American states, summarized current knowledge and identified needs for equine-specific research on the effects of grazing animals on the environment.

REVIEW QUESTIONS

1. Describe the typical percentage of their outdoor time that horses graze and explain factors that tend to decrease or increase the length of time horses spend grazing each day.
2. Explain factors that increase and decrease the rate of feed intake.
3. Describe factors affecting choices that horses make while grazing at the level of bites, patches, and home ranges.
4. Explain how presence of horse manure in a pasture affects grazing of plants by horses.
5. Explain factors that affect horses in their earliest learning to make food choices.
6. Explain factors that affect adult horses in their continued learning to make food choices.
7. Explain effects that grazing horses have on pasture plants and their environment.
8. Describe a set of factors that will discourage a horse from grazing a plant.
9. In "Measuring the Amount and Rate of Feed Intake," tall fescue appears in both the "highly preferred" and "less preferred" categories in different studies. What are some factors that could explain this seeming discrepancy?
10. Define the term "horse-sick pasture" and describe how to avoid the development of such pastures.

REFERENCES

Allen E, Sheaffer C, Martinson K. Forage nutritive value and preference of cool-season grasses under horse grazing. Agron. J. 2013;105:679—84.
Archer M. Preliminary studies on the palatability of grasses, legumes, and herbs to horses. Vet. Rec. 1971;89:236—40.
Archer M. The species preferences of grazing horses. J. Br. Grassl. Soc. 1973;28:123—8.

Archer M. Studies on producing and maintaining balanced pastures for studs. Equine Vet. J. 1978;10:54—9.

Avery A. AG1054 Grazing Management for Horses. 2003. Agriculture Victoria factsheet, http://agriculture.vic.gov.au/agriculture/livestock/horses/management-for-horse-owners/feed-requirements-of-horses/grazing-and-feeding/grazing-management-for-horses.

Aoyama M, Dohi H, Shioya S, Takeuchi Y, Mori Y, Okubo T. Feeding-deterrent substance in cattle feces: its effects on ingestive behavior in goats. Appl. Anim. Behav. Sci. 1994;40(3—4):253—62.

Bailey DW, Stephenson MB, Pittarello M. Effect of terrain heterogeneity on feeding site selection and livestock movement patterns. Anim. Prod. Sci. 2015;55(3):298—308.

Benvenutti MA, Gordon IJ, Poppi DP. The effect of the density and physical properties of grass stems on the foraging behaviour and instantaneous intake rate by cattle grazing an artificial reproductive tropical sward. Grass Forage Sci. 2006;61(3):272—81.

Bott RC, Greene EA, Koch K, Martinson KL, Siciliano PD, Williams C, Trottier NL, Burk A, Swinker A. Production and environmental implications of equine grazing. J. Equine Vet. Sci. 2013;33:1031—43.

Bowden DM, Taylor DK, Davis WEP. Water-soluble carbohydrates in orchardgrass and mixed forages. Can. J. Plant Sci. 1967;48:9—15.

Boyd LE, Carbonaro DA, Houpt KA. The 24-hour time budget of Przewalski horses. Appl. Anim. Behav. Sci. 1988;2(1—2):5—17.

Chadwick MA, Vercoe PV, Williams IH, Revell DK. Programming sheep production on saltbrush: adaptations of offspring from ewes that consumed high amounts of salt during pregnancy and early lactation. Anim. Prod. Sci. 2009;49:311—7.

Chavez SJ, Siciliano PD, Huntington GB. Intake estimation of horses grazing tall fescue (*Lolium arundinaceum*) or fed tall fescue hay. J. Anim. Sci. 2014;92:2304—8.

Cymbaluk NF. Thermoregulation of horses in cold, winter weather: a review. Livest. Prod. Sci. 1994;40:65—71.

Daniels MJ, Ball N, Hutchings MR, Greig A. The grazing response of cattle to pasture contaminated with rabbit faeces and the implications for the transmission of paratuberculosis. Vet. J. 2001;161(3):306—13.

Davidson N, Harris P. Nutrition and welfare. In: Waran N, editor. The Welfare of Horses. Dordrecht, The Netherlands: Springer; 2007. p. 45—76.

Dohi J, Yamada A, Enstu S. Cattle feeding deterrents emitted from cattle feces. J. Chem. Ecol. 1991;17(6):1197—203.

Dowler LE, Siciliano PD, Pratt-Phillips SE, Poore M. Determination of pasture dry matter intake rates in different seasons and their application in grazing management. J. Equine Vet. Sci. 2012;32:85—92.

Drescher M, Heitkonig IMA, Raats JG, Prins HHT. The role of grass stems as structural foraging deterrents and their effects on the foraging behaviour of cattle. Appl. Anim. Behav. Sci. 2006;101:10—26.

Duren SE, Dougherty CT, Jackson SG, Baker JP. Modification of ingestive behaviour due to exercise in yearling horses grazing orchardgrass. Appl. Anim. Behav. Sci. 1989;22:335—45.

Edouard N, Fleurance G, Dumont B, Baumont R, Duncan P. Does sward height affect feeding patch choice and voluntary intake in horses? Appl. Anim. Behav. Sci. 2009;119:219—28.

Edouard N, Duncan P, Dumont B, Baumont R, Fleurance G. Foraging in a heterogeneous environment-an experimental study of the trade-off between intake rate and diet quality. Appl. Anim. Behav. Sci. 2010;126:27—36.

Ferreira LMM, Celaya R, Benavides R, Jáuregui BM, García U, Santos AS, García RR, Rodrigues MAM, Osoro K. Foraging behaviour of domestic herbivore species grazing on heathlands associated with improved pasture areas. Livest. Sci. 2013;155:373—83.

Field TG, Taylor RE. Chapter 16 digestion and absorption of feed. In: Scientific Farm Animal Production Tenth. Upper Saddle River, NJ: Prentice Hall; 2012. p. 264—7.

Fleurance G, Duncan P, Mallevaud B. Daily intake and the selection of feeding sites by horses in heterogeneous wet grasslands. Anim. Res. 2001;50:149—56.

Fleurance G, Duncan P, Fritz H, Cabaret J, Cortet J, Gordon IJ. Selection of feeding sites by horses at pasture: testing the anti-parasite theory. Appl. Anim. Behav. Sci. 2007;108:288—301.

Fleurance G, Fritz H, Duncan P, Gordon IJ, Edouard N, Vial C. Instantaneous intake rate in horses of different body sizes: influence of sward biomass and fibrousness. Appl. Anim. Behav. Sci. 2009;117:84—92.

Fleurance G, Duncan P, Fritz H, Gordon IJ, Grenier-Loustalot M-F. Influence of sward structure on daily intake and foraging behaviour by horses. Animal 2010;4:480—5.

Forbes JM. Consequences of feeding for future feeding. Comp. Biochem. Physiol. Part A 2001;128:463—70.

Forbes JM. A personal view of how ruminant animals control their intake and choice of food: minimal total discomfort. Nutr. Res. Rev. 2007;20:132—46.

Forbes TDA, Hodgson J. The reaction of grazing sheep and cattle to the presence of dung from the same or the other species. Grass Forage Sci. 1985;40:177—82.

Fox MT. Gastrointestinal parasites of cattle. In: Merck Veterinary Manual; 2016. http://www.merckvetmanual.com/digestive-system/gastrointestinal-parasites-of-ruminants/gastrointestinal-parasites-of-cattle.

Fox MT. Gastrointestinal parasites of sheep and goats. In: Merck Veterinary Manual; 2016. http://www.merckvetmanual.com/digestive-system/gastrointestinal-parasites-of-ruminants/gastrointestinal-parasites-of-sheep-and-goats.

Ganskopp DC, Bohnert DW. Landscape nutritional patterns and cattle distribution in rangeland pastures. Appl. Anim. Behav. Sci. 2009;116:110—9.

Gerrish J. Management-intensive Grazing, the Grassroots of Grass Farming. Ridgeland, Mississippi: Green Park Press, Mississippi Valley Publishing Corp; 2004. p. 56—9.

Gietzen DW. Neural mechanisms in the responses to amino acid deficiency. J. Nutr. 1993;123:610—25.

Girard TL, Bork EW, Nielsen SE, Alexander MJ. Seasonal variation in habitat selection by free-ranging feral horses within Alberta's forest reserve. Rangeland Ecol. Manag. 2013;66:428—37.

Glunk EC, Pratt-Phillips SE, Siciliano PD. Effect of restricted pasture access on pasture dry matter intake rate, dietary energy intake and fecal pH in horses. J. Equine Vet. Sci. 2013;33:421−6.

Glunk EC, Sheaffer CC, Hathaway MR, Martinson KL. Interaction of grazing muzzle use and grass species on forage intake of horses. J Equine Vet. Sci. 2014;34(7):930−3.

Green GC, Elwin RL, Mottershead BE, Lynch JJ. Long-term effects of early experience to supplementary feeding in sheep. Proc. Aust. Soc. Anim. Prod. 1984;15:373−5.

Haenlein GFW, Holdren RD, Yoon YM. Comparative response of horses and sheep to different physical forms of alfalfa hay. J. Anim. Sci. 1966;25:740−3.

Hayes SH, Smith SR, Olson GL, Lawrence L. Relationship of plant grazing tolerance to equine grazing preferences. J. Equine Vet. Sci. 2009;29:429−30 (Abstr.), https://doi.org/10.1016/j.jevs.2009.04.126.

Hendrickson SL. A genome wide study of genetic adaptation to high altitude in feral Andean horses of the páramo. BMC Evol. Biol. 2013;13(1):1−21.

Hintz HF, Loy RC. Effects of pelleting on the nutritive value of horse rations. J. Anim. Sci. 1966;25:1059−62.

Hodgson J. Grazing Management, Science into Practice. Harlow, Essex, England: Longman Scientific and Technical; 1990. p. 157−79.

Holcomb KE, Tucker CB, Stull CL. Physiological, behavioral, and serological responses of horses to shaded or unshaded pens in a hot, sunny environment. J. Anim. Sci. 2013;91:5926−36.

Holcomb KE, Tucker CB, Stull CL. Preference of domestic horses for shade in a hot, sunny environment. J. Anim. Sci. 2014;92:1708−17.

Holcomb KE, Tucker CB, Stull CL. Shade use by small groups of domestic horses in a hot, sunny environment. J. Anim. Sci. 2015;92:5455−6.

Holcomb KE, Stull CL. Effect of time and weather on preference, frequency, and duration of shade use by horses. J. Anim. Sci 2016;94:1653−61.

Holcomb KE. Is shade for horses a comfort resource or a minimum requirement? J. Anim. Sci. 2017;95:4206−12.

Holt DA, Hilst AR. Daily variation in carbohydrate content of selected forage crops. Agron. J. 1969;61:239−42.

Hopkins A. Grass, its Production & Utilization. third ed. London, England: Blackwell Science Ltd.; 2000. p. 252−9.

Hoskin SO, Gee EK. Feeding value of pastures for horses. N. Z. Vet. J. 2004;52:332−41.

Illius AW, Gordon IJ, Milne JD, Wright W. Costs and benefits of foraging on grasses varying in canopy structure and resistance to defoliation. Funct. Ecol. 1995;9:894−903.

Kaufmann J, Bork EW, Blenis PV, Alexander MJ. Cattle habitat selection and associated habitat characteristics under free-range grazing within heterogeneous Montane rangelands of Alberta. Appl. Anim. Behav. Sci. 2013;146(1−4):1−10.

Keiper RR, Keenan MA. Nocturnal activity patterns of feral ponies. J. Mammal. 1980;61:113−6.

King SRB. Home range and habitat use of free-ranging Przewalski horses at Hustai National park, Mongolia. Appl. Anim. Behav. Sci. 2002;28(2−4):103−13.

King SS, Jones KL, Schwarm M, Oberhaus EL. Daily horse behavior patterns depend on management. J. Equine Vet. Sci. 2013;33:365 (Abstr.).

Lamoot I, Callebaut J, Degezelle T, Demeulenaere E, Laquiere J, Vandenberge C, Hoffmann M. Eliminative behaviour of free-ranging horses: do they show latrine behaviour or do they defecate where they graze? Appl. Anim. Behav. Sci. 2004;86:105−21.

Lane TJ. Parasite control in horses. In: Merck Veterinary Manual; 2016. http://www.merckvetmanual.com/management-and-nutrition/health-management-interaction-horses/parasite-control-in-horses.

Launchbaugh KL, Provenza FD. Learning and memory in grazing livestock application to diet selection. Rangelands 1991;13(5):242−4.

Launchbaugh KL, Provenza FD, Werkmeister MJ. Overcoming food neophobia in domestic ruminants through addition of a familiar flavor and repeated exposure to novel foods. Appl. Anim. Behav. Sci. 1997;554:327−34.

Launchbaugh KL, Dougherty CT. Grazing animal behavior. In: Forages. The Science of Grassland Agriculture. sixth ed., vol. II. Oxford, England: Blackwell Publishing; 2007.

Lechtenberg VL, Holt DA, Youngberg HW. Diurnal variation in nonstructural carbohydrates, *in vitro* digestibility, and leaf to stem ratio of alfalfa. Agron. J. 1971;63:719−24.

Lesté-Lasserre C. Understanding Herd Dynamics, TheHorse.com, July 3, 2016. http://www.thehorse.com/articles/35555/understanding-herd-dynamics.

Longland AC, Byrd BM. Pasture nonstructural carbohydrates and equine laminitis. J. Nutr. 2006;136:2099S−102S.

Longland AC, Barfoot C, Harris PA. Efficacy of wearing grazing muzzles for 10 hours per day on controlling bodyweight in pastured ponies. J. Equine Vet. Sci. 2016a;45:22−7.

Longland AC, Barfoot C, Harris PA. Effects of grazing muzzles on intakes of dry matter and water-soluble carbohydrates by ponies grazing spring, summer, and autumn swards, as well as autumn swards of different heights. J. Equine Vet. Sci. 2016b;40:26−33.

Lopez IF, Hodgson J, Hedderley DI, Valentine I, Lambert MG. Selective defoliation by sheep according to slope and plant species in the hill country of New Zealand. Grass Forage Sci. 2003;58(4):339−49.

Loucougaray G, Bonis A, Bouzille J-B. Effects of grazing by horses and/or cattle on the diversity of coastal grasslands in western France. Biol. Conserv. 2004;116:59−71.

Marinier SL, Alexander AJ. Selective grazing behaviour in horses. Appl. Anim. Behav. Sci. 1991;30:203−21.

Marinier SL, Alexander AJ. Use of field observations to measure individual grazing ability in horses. Appl. Anim. Behav. Sci. 1992;33(1):1−10.

Marten GC. The animal-plant complex in forage palatability phenomena. J. Anim. Sci. 1978;46:1470−7.

Marten GC, Donker JD. Selective grazing induced by animal excreta. I. Evidence of occurrence and superficial remedy. J. Dairy Sci. 1964;47:773−6.

Martinson KL, Wells MS, Sheaffer CC. Horse preference, forage yield and species persistence of twelve perennial cool-season grass mixtures under horse grazing. J. Equine Vet. Sci. 2016;36:19−25.

Matches AG. Plant response to grazing: a review. J. Prod. Agric. 1992,5.1−7.

Mayes E, Duncan P. Temporal patterns of feeding behaviour in free-ranging horses. Behaviour 1986;96:106−29.

McClaran MP, Cole DN. Packstock in wilderness: impacts, monitoring, management and research, U.S. Department of Agriculture, Forest Service. Ft. Collins, CO: Intermountain Forest and Range Experiment Station; 1993.

Medica DL, Hanaway MJ, Ralston SL, Sukhdeo MVK. Grazing behavior of horses on pasture: predisposition to strongylid infection? J. Equine Vet. Sci. 1996;16(10):421−7.

Mirza SN, Provenza FD. Preference of the mother affects selection and avoidance of foods by lambs differing in age. Appl. Anim. Behav. Sci. 1990;28:255−63.

Mirza SN, Provenza FD. Effects of age and conditions of exposure on maternally mediated food selection by lambs. Appl. Anim. Behav. Sci. 1992;33(1):35−42.

Muller CE, Uden P. Preference of horses for grass conserved as hay, haylage or silage. Anim. Feed Sci. Technol. 2007;132:66−78.

Murphy B. Voisin management intensive grazing. In: Greener Pastures on Your Side of the Fence. fourth ed. Colchester, Vermont: Arriba Publishing; 1998a. p. 117−57.

Murphy B. Grazing animals: effects on pastures & vice versa. In: Greener Pastures on Your Side of the Fence. fourth ed. Colchester, Vermont: Arriba Publishing; 1998b. p. 97−115.

Naujeck A, Hill J, Gibb JJ. Influence of sward height on diet selection by horses. Appl. Anim. Behav. Sci. 2005;90:49−63.

Neilson MK, Kaplan RM, Thamsborg SM, Monrad J, Olsen SN. Climatic influences on development and survival or free-living stages of equine strongyles: implications for worm control strategies and managing anthelmintic resistance. Vet. J. 2007;174:23−32.

Nolte DL, Provenza FD. Food preference in lambs after exposure to flavors in milk. Appl. Anim. Behav. Sci. 1991;32(4):381−9.

NRC (National Research Council). Nutrient Requirements of Horses. Sixth Revised Edition. Washington, DC: National Research Council of the National Academies; 2007.

Ödberg FO, Francis-Smith K. A study on eliminative and grazing behaviour − the use of the field by captive horses. Equine Vet. J. 1976;8:147−9.

Offord M. Plants Poisonous to Horses. An Australian Field Guide. Australian Government Rural Industries Research and Development Corporation; 2006. Publication no 06/048 Project no. OFF-1A, https://rirdc.infoservices.com.au/items/06-048.

Olson-Rutz KM, Marlow CB, Hansen K, Gagnon LC, Rossi RJ. Packhorse grazing behavior and immediate impact on a timberline meadow. J. Range Manag. 1996;49:546−50.

OMAFRA. Pasture Production. Publication 19. Ontario Ministry of Agriculture, Food and Rural Affairs; 2000. p. 21−31. http://www.omafra.gov.on.ca/english/crops/pub19/pub19toc.htm.

Pfister JA, Stegelmeier B, Gardner DR, James LF. Grazing of spotted locoweed (*Astragalus lentiginosus*) by cattle and horses in Arizona. J. Anim. Sci. 2003;81(9):2285−93.

Poppi DP. Nutritional constraints for grazing animals and the importance of selective grazing behaviour. In: Lemaire G, Hodgson J, Chabbi A, editors. Grassland Productivity and Ecosystem Services. CAB International; 2011. p. 19−26.

Prache S, Delagarde R. The influence of vegetation characteristics on foraging strategy and ingestive behaviour. In: Lemaire G, Hodgson J, Chabbi A, editors. Grassland Productivity and Ecosystem Services; 2011. p. 27−36.

Provenza FD. Postingestive feedback as an elementary determinant of food preference and intake in ruminants. J. Range Manag. 1995a;48(1):2−17.

Provenza FD. Origins of food preference in herbivores. In: National Wildlife Research Center Repellents Conference 1995. Paper 29; 1995. http://digitalcommons.unl.edu/nwrcrepellants.

Provenza FD. What does it mean to be locally adapted and who cares anyway? J. Anim. Sci. 2008;86(E.Suppl.):E271−84.

Provenza FD, Ortega-Reyes L, Scott CB, Lynch JJ. Antiemetic drugs attenuate food aversions in sheep. J. Anim. Sci. 1994;72:1989−94.

Provenza FD, Pfister JA, Cheney CD. Mechanism of learning in diet selection with reference to phytotoxicosis in herbivores. J. Range Manag. 1992;45(1):36−45.

Provenza FD, Gregorini P, Carvalho PCF. Synthesis: foraging decisions link plants, herbivores and human beings. Anim. Prod. Sci. 2015;55:411−25.

Ralston SL. Factors in the Control of Feed Intake of Ponies (Ph.D. thesis). University of Pennsylvania; 1982.

Ralston SL, Baile CA. Factors in the control of feed intake of horses and ponies. Neurosci. Biobehav. Rev. 1983;7:465−70.

Ransom JI, Cade BS. Quantifying Equid Behavior-A Research Ethogram for Free-roaming Feral Horses. U.S. Geological Survey Techniques and Methods 2−A9. 2009. 23p.

Richards JH. Physiology of plants recovering from defoliation. In: Baker MJ, editor. Grasslands for Our World. Wellington, New Zealand: SIR Publishing; 1993. p. 46−54.

Rozen-Rechels D, van Beest FM, Emmaunuelle R, Uzal A, Medill SA, McLoughlin PD. Density-dependent, central-place foraging in a grazing herbivore: competition and tradeoffs in time allocation near water. Oikos 2015;124:1142−50.

Sadava D, Heller HC, Orians GH, Purves WK, Hillis DM, editors. Life the Science of Biology 8th Edition. Sunderland, MA: Sinauer Associates Inc.; 2008. p. 1149−51.

Scasta JD, Beck JL, Angwin CJ. Meta-analysis of diet composition and potential conflict of wild horses with livestock and wild ungulates on western rangelands of North America. Rangel. Ecol. Manag. 2016;69:310−8.

Siciliano PD, Gill JC, Bowman MA. Effect of sward height on pasture nonstructural carbohydrate concentrations and blood glucose/insulin profiles in grazing horses. J. Equine Vet. Sci. 2017;57:29−34.

Singer JW, Bobsin N, Bamka WJ, Kluchinski D. Horse pasture management. J. Equine Vet. Sci. 1999;19:540−5. 585-586, 588-592.

Smotherman WP. Odor aversion learning by the rat fetus. Physiol. Behav. 1982;29:769−71.

Sollenberger LE, Newman YC. Grazing management. In: Forages, the Science of Grassland Agriculture. sixth ed.vol. II. Oxford, England: Blackwell Publishing; 2007. p. 651–9.

Teixeira VI, Dubeux Jr JCB, de Mello ACL, Lira Jr MA, Saraiva FM, dos Santos MVF, Lira MA. Herbage mass, herbage rejection and chemical composition of signalgrass under different stocking rates and distances from dung pads. Corp Sci. 2012;52:422–30.

Tyler SJ. The behavior and social organization of the New Forest ponies. Anim. Behav. Monogr. 1972;5:85–196.

Undersander D. Grass varieties for horses. J. Equine Vet. Sci. 2013;33:315–20.

Ungar ED, Noy-Meir I. Herbage intake in relation to availability and sward structure: grazing processes and optimal foraging. J. Appl. Ecol. 1988;25(3):1045–62.

Vallentine JF. Grazing intensity. In: Vallentine JF, editor. Grazing Management. second ed. San Diego: Academic Press; 2001. p. 412–8.

van den Berg M, Lee C, Brown WY, Hinch GN. Does energy intake influence diet selection of novel forages by horses? Livestock Science 2016;186:6–15.

Van Langevelde F, Drescher M, Heitkönig IMA, Prins HHT. Instantaneous intake rate of herbivores as function of forage quality and mass: effects on facilitative and competitive interactions. Ecol. Model. 2008;213:273–84.

Villaba JJ, Provenza FD, Catanese F, Distel RA. Understanding and manipulating diet choice in grazing animals. Anim. Prod. Sci. 2015;55:261–71.

Voisin A. Grass Productivity. Washington, DC: Philosophical Library, France and 1988, Island Press; 1959.

Wang J, Provenza FD. Dynamics of preference by sheep offered foods varying in flavors, nutrients and a toxin. J. Chem. Ecol. 1997;23(2):275–88.

Watts K. What feral horses eat. In: 4th Intl. Conf. on Laminitis and Diseases of the Foot. Nov, 2007, Palm Beach, FL; 2007.

Wiedmeier RD, Provenza FD, Burritt EA. Exposure to ammoniated wheat straw as suckling calves improves performance of mature beef cows wintered on ammoniated wheat straw. J. Anim. Sci 2002;80:2340–8.

Chapter 9

Managing Equine Grazing for Pasture Productivity

Laura B. Kenny[1], Amy Burk[2] and Carey A. Williams[3]

[1]*Extension Division of the College of Agricultural Sciences, Penn State University, University Park, PA, United States; [2]University of Maryland, Department of Animal and Avian Sciences, College Park, MD, United States; [3]Rutgers, the State University of New Jersey, Department of Animal Science, School of Environmental and Biological Sciences, New Brunswick, NJ, United States*

INTRODUCTION

Farm Example 1: Linda is a typical horse owner; she looks out the barn door to her two pastures. The land took a beating last year, and nearly half of the acreage is bare soil, while the other half is either grazed to the ground or weeds. With 10 horses on 8 acres (3.2 ha) of pasture, she wonders what she can do to better manage the pastures so they provide nutrition for her horses and look green and lush (Fig. 9.1).

Farm Example 2: Emily is another property owner, whose pastures look wild and overgrown (Fig. 9.2A). Her original plan to open a horse business fell through, and now her 20-acre (8.1 ha) property only houses her two personal horses. She wanted to rotate her horses through a few existing paddocks, but the forage and weeds grew too fast, and now she worries about letting her horses graze the unknown weeds (Fig. 9.2B).

Both Linda and Emily need to develop a pasture management plan to better utilize the acreage available to them. In this chapter, we will discuss grazing strategies that could help both landowners maximize the use of their pastures.

GRAZING BEHAVIOR

Much of the how and why we manage grazing horses goes back to our understanding of their grazing behavior. A horse will spend approximately 15 h a day grazing, covering up to 7 miles (11. 3 km) as they nibble and consume forages as they move (Mayes and Duncan, 1986). However, horses are not the most efficient grazers and will not uniformly eat all of the available grass in a pasture. Instead, they have preferences for what they like and do not like, and they selectively graze pasture species resulting in a "lawn and rough" pattern with short overgrazed and long undergrazed areas (Odberg and Francis-Smith, 1976). In addition, they do not prefer to graze areas in which they have urinated or defecated, leaving more

FIGURE 9.1 Linda's pasture is overgrazed with many bare spots throughout the pasture. *Photo courtesy of C.A. Williams.*

(A)　　　　　　　　　　　　　　　　　　　**(B)**

FIGURE 9.2　Emily's pasture is underutilized (A), and some areas have many unknown weeds (B). *Photos courtesy of C.A. Williams.*

ungrazed and often weedy land. They also are heavy, athletic animals that can trample and kill vegetation in areas where they congregate such as around hay feeders, waterers, run-in sheds, and gates (Plumb et al., 1984; Bott et al., 2013). If grazing horses are not managed properly, a once productive pasture can quickly become a weedy mess that is devoid of desirable forages (for more on grazing behavior of horses, see Chapter 8).

STOCKING RATE AND DENSITY

When managing horses on pastures, it is important to know how many animals the pasture can handle. This number depends heavily on management of the pastures. The more intensively the horses and pasture are managed, the more horses the land can support. The first step is to calculate the farm **stocking rate**, or the number of animals per unit of land. Important: When determining the units of land area, only include available pasture/grazing land. Do not use total farm size! If a farm has 20 acres (8.09 ha) of pasture for 10 horses, the stocking rate is 20/10 = 2 acres/horse (0.8 ha/horse).

Stocking density is a similar measurement that refers to the animals on a specific area of pasture or grazing land within the farm, at one moment in time. If the 20 acres (8.1 ha) of pasture from the previous example are divided into five paddocks of 2 acres (0.8 ha) each for rotational grazing (the group of horses graze only one paddock at a time), the stocking density of the paddock grazed is 2 ac/10 horses = 0.2 acres/horse (0.08 ha/horse).

These stocking rate and density formulas can also be calculated in the inverse way. Thus, the stocking rate can be expressed as 10 horses/20 acres = 0.5 horses/acre (0.8 horses/ha). Similarly, the stocking density of the paddock can be expressed as 10 horses/2 acres = 5 horses/acre (0.1 horses/ha).

If continuous grazing is practiced, in which all the 10 horses grazed all the 20 acres (8.1 ha) for the whole season, the stocking rate equals the stocking density (Henning et al., 2000).

Another important concept when calculating stocking rate is the **animal unit (AU)**. The AU allows farmers and conservationists to standardize stocking rates across animals of different types and sizes. One AU equals 1000 pounds (453.6 kg) of animal weight. Since the average horse weighs about 1000 pounds, one horse is typically regarded as 1 AU. However, when ponies or draft horses are present, their weights will need to be converted into AU for calculating stocking rate (Westendorf et al., 2016). In some cases, the term "animal unit" may refer to a standard amount of forage dry matter expected to be consumed by an animal in a certain amount of time (Hinnant, 1994).

Let us calculate the stocking rates from our two farm examples.

Linda

10 horses on 8 acres (3.2 ha)
10/8 = 1.3 horses/acre (3.1 horses/ha) or 8/10 = **0.8 acres/horse (0.3 ha/horse)**

Emily

Two horses on 20 acres (8.09 ha)
2/20 = 0.1 horses/acre (0.25 horses/ha) or 20/2 = **10 acres/horse (4 ha/horse)**

Now let us use the AU concept to standardize their stocking rates.

Linda

Two ponies at 600 pounds (272.2 kg) each = 2 × 600/1000 = 1.2 AU
Five horses at 1000 pounds (453.6 kg) each = 5 × 1000/1000 = 5 AU
Three draft horses at 1800 pounds (816.5 kg) each = 3 × 1800/1000 = 5.4 AU
Total farm = **11.6 AU**
11.6 AU/8 acres (3.2 ha) = **1.45 AU/acre (3.6 AU/ha)**

Even though Linda only has 10 horses, she has 11.6 AU. Now, her stocking rate is 11.6 AU per 8 acres or 1.45 AU/acre.

Emily

Two horses at 1000 pounds (453.6 kg) each = 2 × 1000/1000 = **2 AU**
2 AU/20 acres (8.1 ha) = **0.1 AU/acre (0.3 AU/ha)**

Emily's horses are the weight of 1 AU each, so her stocking rate does not need to be adjusted by AU. Her stocking rate does not change.

A question that comes up frequently is, "How many horses per acre are appropriate?" This refers to the **carrying capacity** of the land, or how many horses the forage can nutritionally support over a given period of time without damaging the forage stand. There is no easy answer to this question. It will depend on how the horses and the land are managed, time of year, environmental conditions, etc. A minimum of 1–2 acres (0.4–0.8 ha) per horse is often recommended (Singer et al., 2002), but a farm manager may need more land to grow a productive stand of forage. Additionally, farm operators should always consult with their local planning and zoning office to learn about local regulations. A more realistic guideline is to allow 2–4 acres (0.4–1.6 ha) per horse in a temperate climate to ensure the horse has enough to eat (Wegner and Halbach, 2000). This guideline can be difficult to follow in densely populated areas, and more intensive grazing management is required to maintain forage stands on farms with higher stocking rates. As a general rule, the stocking rate should allow for a pasture to maintain 70% or higher vegetative cover, which protects the soil from eroding (Costin, 1980; Sanjari et al., 2009) and may allow horses to gain nutrition benefits if the cover is composed of well-maintained desirable forage species.

Many farms do not keep horses on pasture 24 h a day, and this complicates the matter of carrying capacity. The longer horses are turned out, the more space and pasture they need to avoid damaging the pasture plants by overgrazing. When pastures are grazed for only a fraction of a day, higher stocking densities can be acceptable as long as 70% vegetative cover can be maintained. Under adverse environmental conditions, such as prolonged hot, dry periods, horses may need to be temporarily removed from the pastures to avoid irreparable damage to the forage stand. Overgrazing, however, is unrelated to the number of animals present in a pasture but is highly related to the time period during which plants are exposed to animals. If animals remain in any one area too long and graze regrowth, they overgraze plants. If a rest period is too short and plants are regrazed before they build up reserve energy, they are overgrazed.

OVER- AND UNDERSTOCKING

When horses are kept at a stocking rate that is not ideal, the result is either over- or understocking. **Overstocking** is defined as pasturing more animals than the land can sustain, without damaging the grasses beyond recovery (Heady, 1970). With minimal pasture management and unlimited turnout, Linda's farm is overstocked (Fig. 9.1) with 11.6 AU on 8 acres (3.2 ha). Poorly managed, overstocked pastures result in the low vegetative cover of 50% seen on Linda's farm. Of the vegetation present, half is overgrazed and half is weeds. Overgrazing allows horses to graze forage so intensively that the

FIGURE 9.3 This pasture is understocked and requires frequent mowing. *Photo courtesy of C.A. Williams.*

plants' growth centers are damaged and reserve energy stores are depleted, and the plants cannot recover. Overgrazing also encourages weed growth because it eliminates desirable grass plants, allowing opportunistic weeds to germinate and fill in the bare ground left behind. When horses overgraze pastures because forage availability is low, they may begin to eat weeds, which could cause health problems if the weeds are toxic. Other signs that horses are being allowed to overgraze a pasture include compacted soil, mud or standing water, and soil erosion (Hubbard et al., 2004). Farm managers who want to turn their overgrazed pasture into a more productive pasture should first consider reducing their animal numbers, then utilizing a grazing system that allows them to better manage their horses and the land (Warren and Sweet, 2003). The alternative is acquiring more acres to manage for the existing number of horses.

Understocking is the opposite problem: having too few animals on a pasture to keep up with the amount of forage produced (Heady, 1970). Emily's farm is understocked, with only two horses (2 AUs) on 20 acres (8.1 ha). From an environmental standpoint, it is more desirable than overstocking because the robust pasture and dense root base minimizes erosion, takes up water and nutrients, and stabilizes soil structure. A thick stand of forages can compete with weeds and prevent weed seeds from germinating (Fig. 9.3). However, mature and overgrown pastures are low in nutritive value and may contain harmful plants if weeds are allowed to establish. Without regular monitoring/or mowing of pastures, weeds will thrive and reproduce.

The biggest problem with understocked pastures is excessive forage and weed growth. Therefore, understocked pastures benefit from regular monthly or bimonthly mowing at heights of 4 inches (10 cm) or higher to keep grass in the vegetative stage and to reduce weeds before they reproduce by spreading their seeds (Undersander et al., 2002). Managers can even make hay from ungrazed fields if they have access to the equipment or another willing farmer.

GRAZING SYSTEMS

A **grazing system** is a part of a planned effort by farm managers to design and lay out lands for animal grazing purposes. The type of grazing system chosen is based on whether the horse farm manager wants the pasture to provide nutrition, exercise, and/or optimal health of the horses. Additionally, the acreage available, level of management required, and type of horse business all are important considerations. For instance, a boarding and lesson facility may choose a different type of grazing system than a farm that grazes broodmares and foals. Likewise, a small (5 acres [2 ha] or fewer) farm may choose to use a different type of grazing system than a farm with more available acreage.

Continuous Grazing

The simplest type of grazing system for horses is a **continuous grazing** system. A continuous grazing system allows horses to graze unrestricted on pastures for long periods of time without periods of rest for the pasture plants (Heady, 1970). In this system, horses have access to the same pasture regularly, regardless of the duration of turnout. Note that we are focusing on the pasture and not the horse. It could be that one group of horses is allowed 12 h of access to the pasture per day, or that three groups of horses go out on the pasture for 4-h blocks during the day. No matter the duration of turnout, the forage in a continuously grazed system never gets rested long enough to regrow (Bott et al., 2013). In the horse industry, it is common to manage horses in a continuously grazed system year-round, even during periods of inclement weather and the winter months.

The main benefit of continuous grazing is that it is simple and relatively inexpensive to design and set up because it only requires a perimeter fence, shelter, water source, and hay feeder. However, there are plenty of drawbacks to continuous grazing. First, continuous grazing allows horses to compact soil and overgraze pasture grasses, which leads to lower forage yields and vegetative cover, more soil erosion, and greater weed encroachment. Also, if manure deposited in the system is not removed regularly, horses will create "bathroom areas" (roughs) that will result in the uneven distribution of nutrients. In addition, in high-density, continuously grazed pastures, horses will encounter feces more frequently, which increases their chances of becoming infected with parasite larvae from manure. One of the biggest drawbacks to continuous grazing is that reduced forage yields may result in a need for additional hay and possibly grain, which increases the farm's expenses. Continuously grazed pastures still require management, including frequent mowing, dragging, or harrowing of manure, and application of soil amendments. The final drawback relates to management. When a continuously grazed system needs to be maintained or renovated, where do the horses go? Some farms have the ability to place horses in stalls for short periods of time, but others do not. Overall, this system is more ideal for the busy farm manager whose main priority is to use the land as an exercise lot with a secondary goal of providing nutrition (Henning et al., 2000; Undersander et al., 2002).

One management practice that can be used strategically in grazing systems to improve the health of pasture grasses is the use of a **stress lot** (Fig. 9.4). A stress lot is a fenced enclosure usually adjacent to or part of a grazing system that is used to house horses when grazing conditions are not optimal. Horses should be housed in stress lots when there is heavy rain, snow, or drought or when pastures are being mowed, renovated, or fertilized. Although the land in the stress lots takes a beating from hoof traffic, it helps the larger pastures stay protected from the wear and tear of horse hooves during inclement weather. Stress lots should provide at least 600 square feet (55.7 m^2) of space for each horse, have a well-drained footing, a shelter that allows for 120 square feet (11.1 m^2) of space per horse, a water source, and a hay feeder (FASS, 2010; HOW, 2007). Because of the heavy foot traffic by the horses, vegetation is not expected to survive in a stress lot for very long. Therefore, consideration should be given to the type of footing to install if dirt and mud will become a problem. Stress lots usually are constructed on a stone base with a surface constructed of compacted, fine crushed stone. Footing materials may be used to cover the compacted stone. Options for stress lot footing materials include sand, sawdust, and wood chips, although the latter holds in moisture and often turns into mud quickly. Horses should not be fed on a sandy surface, as ingesting sand can result in sand colic unless they are fed off the ground. Stress lots should be located on a high, well-drained area and should have a minimum 50-foot (15.2 m) grass buffer outside and downhill of the fence line to filter runoff (water containing manure and soil) coming from the lot. For all of these reasons, stress lots are also called sacrifice areas, exercise lots, loafing lots, animal confinement areas, and dry lots. The addition and proper use of a stress lot in a continuously grazed system would improve grass health and productively as well as flexibility when it comes to managing the pastures.

FIGURE 9.4 This stress lot has stone footing and a shelter, automatic waterer, hay feeders, and laneway with access to pasture. *Photo courtesy of A.O. Burk.*

Rotational Grazing

Another type of grazing system for horses is called **rotational grazing**. This type of system involves rotating a herd of horses through smaller enclosed pastures when forage growth and weather are optimal and then enclosing them in a stress lot during inclement weather or periods of poor grass growth (Henning et al., 2000; Undersander et al., 2002). A key feature of rotational grazing is that horses are only allowed to graze on a previously rested pasture that has an average forage height of 6 inches (15.2 cm) or more, and then they must be removed from that pasture once the average forage height reaches 3 inches (7.6 cm; exact forage height depends on grass species present in the pasture) to allow it to rest again (Fig. 9.5). A typical grazing period for a rotational pasture is 5—7 days, followed by a rest period of 21 days. During periods of slow forage growth (such as the hot summer months if grazing a cool season grass pasture), these grazing periods may become shorter and resting periods longer (i.e., 4—5 days grazing with 30—40 days resting). However, in a very dry environment, the plants in a paddock may only have the capacity to be grazed once in a grazing season and not have enough regrowth until the following season. In a rotational system, there will be a period of time when horses should be confined to a stress lot and fed another forage like hay. Conversely, during periods of rapid forage growth, managers might be tempted to graze longer than 5—7 days; however, longer grazing periods allow horses to regraze new growth: the exact situation that rotational grazing is designed to avoid. Allowing horses to regraze new grass growth stresses the plant, depleting its energy reserves, and effectively increases the time needed to recover and regrow. Repeated grazing of new growth (common in continuously grazed pastures) can kill the plant and is a main cause of overgrazing. In this case, two solutions would be to add more horses to the rotation or subdivide the pastures into smaller fields that will be grazed down faster. The layout of the rotational grazing system will depend on the farm's goals, land available, and existing structures. Designing a rotational grazing system can be as simple as adding some gates and temporary electric fencing to an existing pasture (Henning et al., 2000; Undersander et al., 2002).

To use the design in Fig. 9.5, the manager would close all gates except the one to grazing unit 1 (assuming forage is at least 6 inches high [15.2 cm]). Once the horses have grazed the forage in that unit down to an average of 3—4 inches (7—10 cm) in height, the manager opens the gate to grazing unit or paddock 2 and closes the gate to unit 1. This pattern continues until all grazing units have been grazed. If grazing unit 1 has fully recovered to a height of 6 inches (15.2 cm), the manager can begin the rotation again (Fig. 9.6). If it has not, the manager should confine the horses to the stress lot and provide supplemental forage until one of the grazing units has recovered. Caution needs to be exercised, however, when returning horses back to a field of lush grass after being fed hay for an extended period of time (for more on this, see Chapter 16 on pasture related diseases and disorders). An added benefit to this system is that managers can perform pasture maintenance on an empty field as soon as horses are removed from a grazing unit (paddock): mowing, dragging, fertilizing, applying herbicide, etc. A small farm might have a single rotational grazing system, while a larger farm may use several separate rotational grazing systems for different turnout groups. Depending on growing conditions and weather patterns, horses could be grazing in a rotational system up to 80% of the grazing season (Burk et al., 2011), but the use of a stress lot is necessary for those times when conditions are not conducive for grazing.

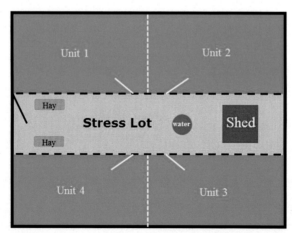

FIGURE 9.5 An example of a rotational grazing system. A rotational grazing system should contain multiple fenced pastures and a stress lot with a shelter, water source, and hay feeder(s). Some systems also use a laneway to help move horses between the pastures and the stress lot. In this example, there are four grazing units with gates leading into a central stress lot. Horses should have access to the stress lot at all times.

FIGURE 9.6 This photo shows horses recently moved into a new pasture unit in a rotational grazing system. *Photo courtesy of C.A. Williams.*

This system seems relatively simple, but there are many planning factors to consider such as number and size of grazing units, appropriate grazing and resting schedules, forage availability, and pasture maintenance.

Number of Grazing Units

The number of grazing units depends on many factors including the size of the available land, the goals of the manager, and the resources available. A general rule of thumb is to have no less than four grazing units per system. However, horse farms with certain limitations can, at the very least, rotate horses between two grazing units to provide some rest for the pasture plants. To calculate the number of grazing units, a simple equation can be used. Take the days of rest a pasture will need to recover under average growing conditions divided by the days of grazing and add 1 (Ball et al., 2015). The number of days of rest needed will vary based on season, weather conditions, and amount of forage growth removed during grazing.

Example: A horse farm manager estimates that he will graze his horses for 7 days per grazing unit and then during times of good growth, rest the pasture for 21 days. In this example, he will need a minimum of four grazing units ((21 days rest/7 days grazing) + 1 = 4).

Size of Grazing Units

When designing a rotational grazing system, there are benefits and drawbacks to the size chosen for a grazing unit. A larger grazing unit will take longer to graze and need less frequent rotation, but it will be grazed more unevenly and forage will be wasted. If horses remain on the same area long enough for grazed plants to begin regrowth or for horses to graze plants a second time, then the grazing unit is too large, and the system is not working properly. A smaller grazing unit may be grazed down in a few days and require more frequent rotations, but the forage will be more uniformly grazed, and the system will be more efficient. The size will also depend on the expected number of horses in the system and the amount of forage available. If horses start to create a lawn/rough pattern, then an additional horse can be added to increase grazing pressure; or if they need to be rotated faster than the manager can handle, a horse can be removed from the system. Grazing units do not need to be the same size.

To determine the size of each grazing unit, the following equation can be used as a guide (Ball et al., 2015). First, multiply four factors together:

1. Average weight of horses in pounds
2. Amount of dry matter (DM) consumed each day as a percent of their body weight (BW)
3. Number of horses grazing
4. Number of days grazing on pasture

Then divide that total by two additional factors that must be multiplied:

1. Pounds of DM available for grazing (pounds DM/inch/acre x forage height at beginning of grazing bout)
2. Percent of the DM utilized by grazing

The amount of DM consumed by horses is usually about 2% BW, but it can be as high as 3% for ponies and other voracious eaters. The number of days grazing pasture in a rotational system is usually somewhere between 1 day for high stocking rates to 7 days for lower stocking rates. DM available for grazing depends on the grass species and forage quality. In Chapter 7, you learned how to estimate forage yields. You can use these same concepts to estimate the acreage needed for a rotational grazing system. A typical yield for low quality/low producing pasture is 150 pounds (68 kg) of DM per inch of forage height per acre of grass available for grazing; and up to 300 pounds DM/inch/acre (336 kg) of grass available for grazing more productive forages. Horses will not consume all of the DM available in the pasture; the amount utilized by horses will vary. However, since it is ideal to leave half of the forage ungrazed when removing horses from a pasture (see Fig. 9.3), we will use 50% as the forage utilization rate.

Example: Ten 1100-lb (500 kg) horses are consuming 2% of their BW for 7 days in an orchardgrass pasture. The grass is currently 6 inches (15.24 cm) and is dense, providing an estimated 250 pounds of forage per inch per acre (110.3 kg/cm/ha) with an estimated 50% utilization rate.

$$\frac{1100 \text{ lbs } \times \ 2\% \ \times 10 \text{ horses } \times \ 7 \text{ days}}{250 \text{ lb/in/acre } \times \ 6 \text{ in } \times \ 0.50} = 2.05 \text{ acres}$$

or

$$\frac{500 \text{ kg } \times \ 2\% \ \times \ 10 \text{ horses } \times \ 7 \text{ days}}{110.3 \text{ lb/in/acre } \times 15.2 \text{ in } \times \ 0.50} = 0.84 \text{ hectares}$$

In this example, a pasture size of 2 acres (0.8 ha) is needed.

Total Acreage Required

The total acreage required is a function of many things including the number of horses, acreage available, grazing unit size, etc. A simple way to make sure you have enough total grazing acres for the rotational system would be to multiply the number of grazing units by the number of acres needed.

Example: Using our examples from before, we multiply four paddocks by 2.05 acres (0.8 ha) each to find that we need a total of 8.2 acres (3.3 ha).

When to Graze and Remove Horses/Length of Grazing Periods

In Chapter 3, you learned that grass in the vegetative state is most nutritious for horses. Therefore, you should begin grazing when the forage is dense and tall but before it starts showing seed heads. In agronomic terms, the best time to graze or mow is during the boot stage, when the seed head is enclosed within the sheath. The grass growth curve shown in Fig. 9.7 shows that grasses generally grow fastest when they are between 3 and 10 inches (7.6–25.4 cm) in height (Crider, 1955).

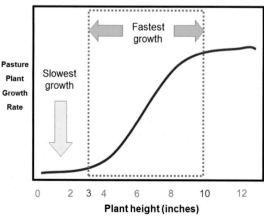

FIGURE 9.7 A graphic depiction of the growth rate of grasses depending on the height of the plant.

The 6-inch (15.2 cm) grazing height recommendation is at the midpoint of the growth curve and gives grasses time to increase their growth rate. Exact heights will vary by grass species (see Chapter 2). The ideal time to graze is in the middle of this fast growth period. Grazing too early is harmful because the grass plant will not have adequate leaf area to photosynthesize the energy needed to continue growth. It will need to tap into energy reserves in the roots and crown to regrow, and doing this repeatedly depletes the reserves, and the plant will die. Grazing too late will likely provide diminished nutritive value as taller plants are more mature. Additionally, grass plants in the stem elongation and reproductive stages no longer produce new leaves (see Chapter 1 for the stages of grass growth). The main growing point of the tiller, previously located at the base of the plant where it could not be damaged by reasonable grazing, rises up the stem as it elongates and can be removed by grazing. The plant must then use its energy reserves to generate a new tiller and reenter vegetative growth (Briske, 1991). It is least stressful to graze grasses in the vegetative stage, as long as adequate leaf area remains after grazing.

How much leaf area should remain after grazing? This question can be answered using the "**take half, leave half**" rule. For example, if the pasture is 8 inches (20.3 cm) tall when you begin grazing, remove horses when stand height reaches 4 inches (10.2 cm). Removing more than 50% of the aboveground biomass reduces the plant's ability to photosynthesize and generate energy for regrowth (Crider, 1955). See Fig. 9.8 for the effect of different levels of aboveground biomass removal on root growth. A deep, thick root system is essential for plants to store energy reserves and access soil water during dry conditions. Allowing horses to graze more than 50% of the leaf volume is overgrazing. There are at least two other good reasons for leaving half or more of the forage height in one grazing period. Maintaining this leafy vegetative cover helps to prevent soil erosion during heavy rainfall and floods. On very hot, sunny days, forage leaves provide shade for the soil and its important organisms. The shorter the forage, the higher the soil temperature can rise, to the detriment of soil organisms and plant roots.

A manager can monitor the height of the grass on a weekly, if not daily, basis by walking the pasture to estimate the average forage height. To keep it simple, managers can walk in a random zig zag pattern, stopping every 10 steps or so to take a height reading (Kenny, 2016). Yard sticks from home improvement stores work well for this purpose. After 10 readings, sum up the measurements and divide by 10 to get an average height. Grazing sticks are also available from many

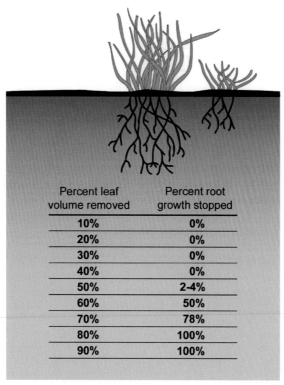

Percent leaf volume removed	Percent root growth stopped
10%	0%
20%	0%
30%	0%
40%	0%
50%	2-4%
60%	50%
70%	78%
80%	100%
90%	100%

FIGURE 9.8 The effect of leaf volume removal on root growth. This provides the reasoning behind the "take half, leave half" rule that was first investigated by Crider (1955). *With permission from NRCS, Bozeman, MT.*

agricultural extension offices. These have ruler markings plus a section that helps the user to assess forage stand density and a chart to estimate forage yield from the height and density.

The length of each grazing period will also depend on the growth rate of the forage. As you learned in Chapter 1, growth rate varies throughout the year, depending on the climate and the type of grass. For example, in the Northeast, cool season forage growth is rapid in the spring, slow in the hot summer months, and moderate in the fall (Briske, 1991). Managers must monitor the growth rate of their forage and be prepared to confine horses to the stress lot, providing supplemental forage, when recovery is slow.

If you already have a rotational system set up and you want to know how long you can graze each section based on available forage, there is a way to be more precise in calculating the length of each grazing bout, although it is still an estimate. For example, say you have a single system that consists of 5 acres of grazing land divided into five units, and you have five average-sized horses at maintenance using that system. The pasture is in good shape and it is a fast-growth season. Your forage yield is estimated to be 250 lbs/in/acre with 6 inches of growth at the time of grazing, and your horses are estimated to consume 2% of their body weight.

- 5 horses × 1000 lbs × 2% BW = 100 lbs (45.4 kg) forage needed per day
- 250 lbs/in/acre × 6 in × 5 acres × 0.5 (take half leave half)/5 grazing units = 750 lbs (340.2 kg) of forage available in each grazing unit
- 750 lbs forage/100 lbs needed daily = **7.5 days**

At this moment in the season, each grazing unit can provide enough forage to feed the five horses for 7.5 days and leave half of the forage. Grazing periods of fewer than 7.5 days provide fewer opportunities for horses to take a second bite of plants. Shorter grazing periods (e.g. Four days) will result in fewer weakened plants and more highly palatable plants. Remember that horses will never graze a pasture exactly the way we expect them to, so these calculations are still only estimates. Horses graze in the lawn and rough pattern, leaving some areas totally untouched (Fig. 9.9). Rotating quickly through smaller grazing units is one method to minimize this behavior because it forces them to eat more of the available forage. Additionally, some of the available forage estimate will not be consumed by horses because of their spot grazing behavior and preference for some grasses over others. It is a good idea to scout pastures and determine what percentage of the vegetative cover consists of weeds, as that can help to adjust the forage utilization estimate. Weeds can have nutritional value; in fact, other livestock graze many plants that horse owners would consider weeds; however, horses tend to prefer grasses and only consume weeds when grass is limited (Olson-Rutz et al., 1996). While the length of grazing bouts can be estimated based on available forage, it is imperative for managers to monitor grazing units frequently to decide when to move horses. Adherence to a strict schedule will result in under- or overgrazing. Using forage height and/or yield to trigger moves of animals among paddocks will promote ideal lengths of grazing time in each paddock.

There are some times when managers should avoid grazing altogether. When horses are turned out in wet weather, their hooves easily slice through the soil, compacting it and damaging plant root systems. Avoid grazing when plants are stressed, such as drought or flood conditions, and whenever grasses have gone dormant, for example, winter in the northern United States (Bott et al., 2013). The horses will continue to trample the dormant plants, damaging crowns and root systems, which will make regrowth slower the following season.

There are many benefits of rotational grazing. First, forage production is increased, resulting in more DM available for the horses. In some rotational grazing systems, the stocking rate can be increased because there is more forage available for grazing. Additionally, less forage and concentrate have to be purchased as a supplement for the horses. Other benefits include higher vegetative cover, fewer weeds, less soil and nutrient runoff, and more even distribution of manure. The

FIGURE 9.9 Lawn and rough pattern grazing in a continuously grazed pasture. *Photo courtesy of L.B. Kenny.*

challenges to rotational grazing include an increased effort in designing and laying out the system, increased capital expenditures for additional gates and fencing, and an increased effort initially in learning how and when to move horses based on weather and forage growth (Henning et al., 2000; Undersander et al., 2002).

GRAZING SEASON

The **grazing season** is the length of time during the year when forage is growing and thus can be used as a nutritional source for grazing animals. The length of the grazing season varies with the climate and growth patterns of the grasses that are grown in that area. In the northern two-thirds of the United States, cool season grasses are primarily used with a grazing season extending from early spring (i.e., April or May) to late summer or fall (i.e., November). In the southern United States, warm season grasses predominate, where temperatures are inhospitable for cool season grasses. In the southern United States, horses are fed hay during the colder winter months when warm season grasses are dormant. A strip across the center of the United States is considered a transition zone that can utilize both warm and cool season grasses for grazing (See Chapter 6 for a map showing the transition zone.).

Management practices that help to extend the grazing season include stockpiling forages for use in the winter, irrigating pastures during the hot and dry periods, and utilization of warm season grasses during the warm summer months to optimize forage production when cool season grasses have reduced growth.

WINTER PASTURE MANAGEMENT

During the winter months in northern climates, when the grasses are dormant and exposed to cold temperatures and perhaps freezing rain and snow, farm managers should refrain from grazing horses and instead place them in stress lots and offer hay. If stress lots are not an option, managers may want to designate one pasture as a "winter pasture" while preserving and resting the other pastures. It is important that the manager understand that the winter pasture will be overgrazed and will need to be renovated and rested the following year.

Another strategy that can be used during the winter months is grazing horses on stockpiled forage. **Stockpiling** is the practice of resting forage in a pasture, starting in late summer, and allowing it to grow so that there is sufficient yield for horses to consume later in the winter (Henning et al., 2000). In the eastern half of North America, the grass commonly used for stockpiling is tall fescue. In cooler, drier parts of the west, Russian wildrye, a grass that cures well without being cut, can be used. Stockpiling may need to occur up to 3—4 months in advance of the winter to provide sufficient forage yield to meet the horse's daily intake requirements. Stockpiling tall fescue can prolong the grazing season between 30 and 60 days. When using stockpiled pastures, it is important to remove horses once the forage is grazed down to avoid excessive trampling and overgrazing. As previously mentioned, it is harmful to keep horses on pasture all winter. Though tall fescue is one of the best types of grasses to stockpile, broodmares should not be grazed on endophyte-infected varieties that produce toxins that affect their reproduction. In the case of the grazing broodmare, endophyte-free or novel endophyte (nontoxic) varieties of tall fescue should be used.

SUMMER PASTURE MANAGEMENT

In the transition zone of the United States, it is possible to use both cool season grasses and warm season grasses to extend the grazing season. Warm season grasses should be established in the late spring to early summer (Johnson and Thompson, 1961) and irrigated well throughout the summer. The grazing season of warm season grasses in the transition zone is generally early summer (June) to early fall (September). This is the time when cool season grasses are dormant due to the higher temperatures. In addition to using seeded varieties of warm season grasses, some varieties establish better if sprigs are used. To produce sprigs, stems containing the stolons or rhizomes from a mature Bermudagrass stand are harvested carefully, and then they are quickly planted into a moist seedbed and packed down to ensure good soil to sprig contact. Although sprigging can result in better establishment, it can also be more expensive and labor intensive than seeding (Munshaw et al., 1998). Finally, warm season grasses, like Bermudagrass, tend to be lower in starches and sugars or "nonstructural carbohydrates" (Kagen et al., 2011), which is a preferred diet for feeding to horses with metabolic conditions usually related to obesity.

Another practice that can be used to extend the grazing season is **irrigation** of pastures. Irrigation up to 1—2 inches (2.5—5 cm) per week helps to encourage grass growth when plants would otherwise be dormant (Fig. 9.10). Placement of catch cans under sprinkler systems is useful to make sure water is being applied uniformly and at the rates that are needed. A variety of irrigation systems exists including wheel lines, hand-moved lines, solid-set underground lines, and portable

FIGURE 9.10 A farm using irrigation to water their pastures during a dry time of year. *Photo courtesy of A.O. Burk.*

systems that can be moved using a tractor or all-terrain vehicle. Some portable sprinkler systems are powered by battery or solar energy. For continuous grazing systems, pastures are often irrigated at dusk or overnight to reduce the risk of evaporation and loss of water into the air. However, keeping grass wet all night (by irrigating in the evening) can favor fungal growth, so early morning is the best time to irrigate without losing water to evaporation (Korb, 2008). Keeping horses on the wet pasture can be detrimental to forage productivity since the saturated soils become soft, allowing horse hooves to damage plant root systems. Therefore, horses in continuously grazed pastures should be stalled or kept in stress lots until the soil dries enough to support the horses' weight after irrigation. For rotational grazing systems, pastures are irrigated after horses graze a pasture and have been moved to another pasture for grazing. The downside of irrigation is that it does take time and effort in laying out the system and costs money to purchase the equipment. Additionally, moving portable systems from site to site takes time.

MORE INTENSIVE MANAGEMENT STRATEGIES

Management-intensive practices are more widely used in grazing livestock operations; however, some could be used on horse farms. One intensive grazing practice is called **strip grazing** (Henning et al., 2000). Strip grazing involves taking a large pasture and giving grazing animals access to a smaller portion, or strip, of that pasture for short bouts until they graze it down to about 3 inches (7.6 cm). Usually, a temporary electric fence is used to control the amount of space to which horses have access. Depending on the amount of forage available and the number of animals, it may take the herd several hours to a day to consume the allotted forage. Once the animals graze down the smaller section of pasture, the fence is again moved to open up a fresh section of the pasture. Strip grazing forces the animals to be less picky when grazing, resulting in more uniform and efficient grazing. The major drawback of strip grazing is the time it takes to frequently move the fence to open up new sections of fresh pasture. Additionally, it can be tricky to design the system so that horses have access to shelter and water while having access to each strip. Incorporating the use of strip grazing with horses grazing stockpiled forages has the advantage of reducing the amount of waste due to selective grazing and trampling by horses.

Mob grazing involves grazing a large number of animals (i.e., several hundred head of cattle) very quickly through smaller pastures each day or twice daily (Henning et al., 2000). Care is given not to allow animals to overgraze, and a sufficient rest period for regrowth is the key to success. This practice is used mainly in beef operations and is not likely useful in horse operations due to the natural agonistic behaviors exhibited by horses to maintain dominance, especially in a large herd confined in small spaces.

ADDITIONAL MANAGEMENT

To keep pastures productive, they need regular maintenance. When rotationally grazing, each grazing unit should be mowed and dragged immediately after horses are removed. Even with small grazing units, horses will still leave some areas uneaten, and mowing ensures that all plants are the same height and return to the same stage of growth when the recovery period begins. Mowing is also one of the best ways to prevent weeds from going to seed or from shading out and killing grasses underneath them. Dragging or harrowing pastures helps to spread out manure piles, making the manure dry faster and distributing manure nutrients more evenly around the field. If horses are allowed to create "bathroom areas," "latrines," or "roughs" in pastures, they will not regraze those areas, and they become a weedy mess (Fig. 9.9). When spreading out manure piles, the risk of internal parasite larvae ingestion temporarily increases, so horses must be on an effective deworming protocol. Harrowing fresh manure on hot sunny days allows ultra-violet rays to kill many parasite larvae. The rest period for each rotational unit also provides an ideal time for applying soil amendments such as lime and fertilizer or herbicides to control weeds (Henning et al., 2000; Undersander et al., 2002).

CONTINUOUS VERSUS ROTATIONAL GRAZING

Researchers have been comparing rotational and continuous grazing systems for many decades. In traditional livestock, effects on animal performance have been mixed, but the rotational systems had higher forage production overall, especially in humid regions of the country (Holechek et al., 1999). Work in horses has shown no difference in horse condition between the two grazing systems, but rotational grazing was found to have higher forage production, lower proportions of weeds, and higher nutritional quality (digestible energy, water soluble carbohydrates, and sugar) (Webb et al., 2009, 2011; Virostek et al., 2015; Daniel et al., 2015; Kenny, 2016). In some situations, the rotational systems may also reduce the need for supplemental feed as hay or grain due to the higher forage production and nutrient content. Given the fact that horses are not raised with the same goals as other livestock (i.e., rapid growth and maximal weight/production), the lack of difference in horse condition is ideal. No manager wants to see their horses becoming obese on pasture, and the manager will likely find ways to restrict an overweight horse's intake on high-quality pasture. Options include fitting horses with grazing muzzles to slow feed intake and limiting the number of hours of pasture access per day.

GRAZING PLANS FOR LINDA AND EMILY

Let us apply these concepts to Linda's and Emily's farms, developing a thorough grazing plan for each.

Linda
11.6 AUs on 8 acres (3.2 ha)

This overstocked farm will need some serious management changes to have productive pastures. First, the pastures need to be reestablished because they have less than 25% desirable grass cover. That means they are not producing much nutrition for her horses. Linda must take soil tests and amend the soil per the results. A complete renovation (see Chapter 6 for more on renovation) would involve killing all existing vegetation and reseeding with appropriate pasture species for her climate. Linda chooses to renovate one-half of her property at a time, so she does not lose use of all the land at once, as she will need to keep horses off the newly renovated pastures until they are well established (at least one grazing season). During this time, her horses will be confined to the stress lots and fed supplemental feed with limited access to the other half of her property.

Once her pastures are productive, Linda should divide them into equally sized rotational grazing units or paddocks to accommodate her horses that cannot be turned out together. She has four groups of horses that get along, and she wants them turned out for a half day. Thus, she will set up two grazing systems and switch the horse groups twice a day. Each 4-acre (1.6 ha) system will be divided into four grazing units and a small stress lot. By splitting her horses' grazing time in half, she now has only five horses on 8 acres (3.2 ha) of pasture at any given time, and her farm is no longer effectively overstocked. This is a case where stocking density of the pastures is more meaningful than stocking rate of the entire farm. Linda knows that even with the rotational system and improved pasture quality, her horses will graze down the forage rapidly. She supplements her horses' rations with hay when they are not turned out to reduce the amount of forage they consume on pasture. She also makes sure to stock up on hay for summer and winter months and other times that are not ideal for grazing. She keeps her horses in the stress lots during the winter and on rainy days.

Emily
2 AUs on 20 acres (8.1 ha)

Emily's plan is to create a rotational grazing system for her two horses and cut hay off the rest of the grassland. This way, if she does acquire more horses, she can section off an additional rotational grazing system. First, she must manage her existing pastures. She soil tests and mows them down. After scouting for weeds, she selects an herbicide and applies it at the appropriate time for the weeds' life cycles. The longer Emily practices good rotational grazing management, the fewer weeds should be growing. Agricultural extension agencies can help to control persistent weeds.

Next, she needs to determine how much land to set aside for the grazing system. Her two 1000-pound horses at maintenance will be outside 24/7, and she would like to keep the horses on a 7-day rotation. She estimates that her pasture will yield an average of 200 lb/in/acre on average throughout the grazing season, and she plans on grazing when they reach 6 inches in height. For the number of grazing units, she estimates a 21-day rest period, divides by 7 for length of grazing bouts, and adds 1 to arrive at four grazing units needed. She calculates the size of her grazing units as follows:

$$\frac{1000 \text{ lbs} \times 2\% \times 2 \text{ horses} \times 7 \text{ days}}{200 \text{ lb/in/acre} \times 6 \text{ in} \times 0.50} = 0.5 \text{ acres } (0.2 \text{ ha})$$

Emily sets aside 2 acres (0.8 ha) of pasture to create four grazing units of a half-acre (0.2 ha) each, plus some extra land for a stress lot. In addition, she fences off two one-acre (0.4 ha) fields to stockpile forage for the winter months and alternate using as a winter pasture. The remaining 16 acres (6.5 ha) are used for haymaking, which will provide supplemental feed for her horses when pasture is unavailable, and she will be able to sell some. In times of slow plant growth, Emily could use temporary electric fence on some of the hay land to provide enough extra pasture. The flexibility to add more grazing areas or to subdivide paddocks allows pasture managers to adapt to changing conditions and needs.

CONCLUSION

Managing horse grazing is important for pasture productivity, environmental stewardship, and horse health. Farm managers must consider the number and size of their animals, the amount of land available for grazing, and the level of management they are willing to put into getting the most from their pasture system. Some overstocked farms will never be able to grow high-quality forage due to the stocking density, but most other farms can manage grazing to provide at least a portion of horse nutrition from pasture. Rotational grazing is a method that allows managers to make the most efficient use of their pasture area while feeding their horses and maintaining high ground cover, which in some states is required by law. It is important to be flexible when implementing a rotational grazing system and remember that forage growth and horse grazing patterns are not perfectly predictable.

REVIEW QUESTIONS

1. Calculate the stocking rate (in animal units) of a farm with 35 acres of pasture and nine 750-pound ponies, 16 1100-pound horses, and five 1500-pound horses.
2. Why might you want to express stocking rate in units of acre/AU instead of AU/acre?
3. Describe one way to manage overstocked farms and one way to manage understocked farms.
4. Which is more harmful to the environment: overstocking or understocking? Why?
5. Under what circumstances should horses be confined to the stress lot?
6. What is the rationale for the take-half, leave-half rule?
7. Why is it important to estimate forage yield several times throughout the grazing season?
8. What is one key management tool that can be used in the winter and one that can be used in the summer to prolong grazing and keep grasses more productive?
9. How many grazing units are required and how many acres are needed for eight 1100-lb horses that will be on a 7-day graze, 28-day rest schedule, if the pasture produces an average of 200 lb/in/acre and will be grazed at 7 inches? Assume a forage utilization rate of 50% and 2% BW consumption.
10. Five 1000-lb horses (at maintenance) are grazing in a 4-acre rotational system with five grazing units. The estimated pasture yield is 150 lb/in/acre and grass is 5 inches tall when grazed. How long can they stay in each grazing unit, assuming they consume 2% BW and a forage utilization of 50%? Should the manager of this farm change anything? What options are there?
11. Refer to Linda's Grazing Plan. Her horses must go out in specific groups so that they do not fight. The groups are (1) two draft horses; (2) one draft horse and one pony; (3) three horses; and (4) one draft horse and two horses. Their weights are given under the Stocking Rate section. Groups 1 and 2 share a rotational system for 12 h each, and Groups 3 and 4 share the other system. Their daily rations will consist of 50% hay and 50% pasture. Given a forage yield of 200 lb/in/acre and using the assumptions from Question 10, calculate how many days these groups can stay in one grazing unit if they are grazed at 6 inches. (Hint: calculate the amount of DM consumed in each group.)

REFERENCES

Ball DM, Hoveland CS, Lacefield GD. Southern Forages: Modern Concepts for Forage Crop Management. fifth ed. Peachtree Corners, GA: International Plant Nutrition Institute; 2015. p. 312.

Bott RC, Greene EA, Koch K, Martinson KL, Siciliano PD, Williams C, Trottier NL, Burk A, Swinker A. Production and environmental implications of equine grazing. J. Equine Vet. Sci. 2013;33:1031−43.

Briske DD. Developmental morphology and physiology of grasses. In: Heitschmidt RK, Stuth JW, editors. Grazing Management: An Ecological Perspective. Portland OR: Timber Press; 1991. p. 85−108.

Burk AO, Fiorellino NM, Shellem TA, Dwyer ME, Vough LR, Dengler E. Field observations from the university of Maryland's equine rotational grazing demonstration site: a two year perspective. J. Equine Vet. Sci. 2011;31:302−3 (Abstract).

Costin AB. Runoff and soil and nutrient losses from an improved pasture at Ginninderra, southern tablelands, new South Wales. Aust. J. Agric. Res. 1980;31:533−46.

Crider FJ. Technical Bulletin No. 1102. Washington D.C: United States Department of Agriculture, USDA-SCS Government Printing Office; 1955.

Daniel AD, McIntosh BJ, Plunk JD, Webb M, McIntosh D, Parks AG. Effects of rotational grazing on water-soluble carbohydrate and energy content of horse pastures. J. Equine Vet. Sci. 2015;35:385−6 (Abstract).

Heady HF. Grazing systems: terms and definitions. J. Range Manag. 1970;23:59−61.

Henning J, Lacefield G, Rasnake M, Burris J, Johns R, Turner L. Rotational Grazing. University of Kentucky, Cooperative Extension Service; 2000. ID-143.

Hinnant RT. What is an animal-unit? A time to conform. Rangelands 1994;16:33−5.

Holechek JL, Gomez H, Molinar F, Galt D. Grazing studies: what we've learned. Rangelands 1999;21:12−6.

Horse Outreach Workgroup (HOW). Sacrifice Lots. Maryland Department of Agriculture; 2007. p. 1−4.

Hubbard RK, Newton GL, Hill GM. Water quality and the grazing animal. J. Anim. Sci. 2004;82(E. Suppl.):E255−63.

Johnson CM, Thompson WR. Fall and winter seeding of lawns. Mississippi Farm Res. 1961;24:4.

Kagan EA, Kirch BH, Thatcher CD, Teutsch CD, Elvinger F, Shepard DM, Pleasant S. Seasonal and diurnal changes in starch content and sugar profiles of Bermudagrass in the piedmont region of the United States. J. Equine Vet. Sci. 2011;31(9):521−9.

Kenny L. The Effects of Rotational and Continuous Grazing on Horses, Pasture Condition, and Soil Properties (M.S. thesis). New Brunswick, NJ: Rutgers, the State University of New Jersey; 2016.

Korb G. Lawn Watering. University of Wisconsin Extension; 2008. http://clean-water.uwex.edu/pubs/pdf/lawnwat.pdf.

Mayes E, Duncan P. Temporal patterns of feeding behavior in free ranging horses. Behaviour 1986;96:105−29.

Munshaw GC, Williams DW, Powell Jr AJ, Dougherty CT. Growth and development of seeded versus vegetative Bermudagrass varieties. In: Annual Meetings Abstracts. Madison, Wisconsin: ASA, CSSA, and SSSA; 1998. p. 136.

Odberg FO, Francis-Smith K. A study on eliminative and grazing behaviour-The use of the field by captive horses. Equine Vet. J. 1976;8:147−9.

Olson-Rutz KM, Marlow CB, Hansen K, Gagnon LC, Rossi RJ. Packhorse grazing behavior and immediate impact on a timberline meadow. J. Range Manag. 1996;49:546−50.

Plumb GE, Krysl LJ, Hubbert ME, Smith MA, Waggoner JW. Horses and cattle grazing on the Wyoming red desert, III. J. Range Manag. 1984;37:130−2.

Sanjari G, Yu B, Ghadiri H, Ciesiolka CAA, Rose CW. Effects of time-controlled grazing on runoff and sediment loss. Aust. J. Agric. Res. 2009;47:796−808.

Singer JW, Bamka WJ, Kluchinski D, Govindasamy R. Using the recommended stocking density to predict equine pasture management. J. Equine Vet. Sci. 2002;22:73−6.

The Federation of Animal Science Societies (FASS). Guide for the Care and Use of Agricultural Animals in Research and Teaching. third ed. 2010.

Undersander D, Albert B, Cosgrove D, Johnson D, Peterson P. Pastures for Profit: A Guide to Rotational Grazing. Madison, WI: Cooperative Extension Publishing, University of Wisconsin-Extension; 2002.

Virostek AM, McIntosh B, Daniel A, Webb M, Plunk JD. The effects of rotational grazing on forage biomass yield and botanical composition of horse pastures. J. Equine Vet. Sci. 2015;35:386 (Abstract).

Warren LK, Sweet C. Manure and Pasture Management for Horse Owners. Edmonton, Alberta, Canada: Alberta Agriculture, Food and Rural Development; 2003. http://www1.agric.gov.ab.ca/$department/deptdocs.nsf/all/agdex9377/$FILE/460_27-1.pdf.

Webb GW, Duey C, Webb S. Continuous vs. rotational grazing of cool season pastures by adult horses. J. Equine Vet. Sci. 2009;29:388−9.

Webb G, Webb S, Duey C, Minton K. Continuous vs. rotational grazing of cool season pastures during the summer months. J. Equine Vet. Sci. 2011;31:285.

Wegner TD, Halbach TR. Manure and Pasture Management for Recreational Horse Owners. Paul, MN: University of Minnesota Extension Service, St; 2000.

Westendorf ML, Williams CA, Mickel RC. Recommended Guidelines for Home Animal Agriculture in Residential Areas. Rutgers Cooperative Extension Fact Sheet #E353. New Jersey Agricultural Experiment Station, Rutgers, The State University of New Jersey; 2016. https://njaes.rutgers.edu/pubs/publication.asp?pid=E353.

FURTHER READING

Foulk DL, Mickel RC, Chamberlain EA, Margentino M, Westendorf M. Agricultural Management Practices for Commercial Equine Operations. Rutgers Cooperative Extension Fact Sheet #E296, New Jersey Agricultural Experiment Station, Rutgers, The State University of New Jersey; 2004.

Chapter 10

Mixed Species Grazing

Paul Sharpe
University of Guelph, Guelph, ON, Canada (retired)

WHAT IS MIXED SPECIES GRAZING?

Mixed or multispecies grazing (MSG) is a system of forage use that includes two or more kinds of ungulate herbivores, domestic or wild, at the same time or at different seasons of the year. Its importance is that it provides an opportunity to achieve uniform plant utilization of a nonuniform plant population, under moderate stocking rates (Merrill et al., 1957; Vallentine, 2001c). MSG is normal for wild species but usually at lower stocking rates than for domestic animals on farms and ranches.

According to the late editor of *The Stockman Grass Farmer* magazine, Allan Nation, monocultures of animals are no more part of nature than monocultures of plants (Nation, 1995). Cattle, sheep, and goats are very complementary to each other in forage selections, having only a moderate overlap in their preferred forages. A grazier in humid climates can add one ewe for every cow on pasture and never miss the grass. A good, eight-wire, sheep- and goat-proof perimeter fence allows options in species mixtures (Nation, 1995). Woven wire horse fence for a pasture perimeter should keep all domestic herbivores in.

Financially, the best time to add a new species to your operation is when the new species you add is at the bottom of its price cycle (Nation, 1995). Sheep tend to be better investments than cattle but require more labor per 1000 kg of animal.

A valuable lesson on multispecies communities in nature is under our feet, in the soil. Plants growing in and extracting water from the soil take their H, O, C, and N from the air. Water absorbed through the roots brings dissolved minerals to the plants. When plant parts die, they contribute these chemical elements to the soil in the form of organic compounds. Organic matter helps soil to hold water, which helps with the growth of more plants. So there is cyclic movement of water and nutrients through soil and plants. Through photosynthesis, plants produce sugars, most of which are used to produce cellulose and other fibrous molecules for building plant cells and the structure of whole plants. In the soil are Glomeromycetes fungi, which form symbiotic associations, called "mycorrhizas," with roots of most vascular plants (Russell et al., 2013). This association benefits the fungi by providing organic nutrients such as sugars from the roots, and it benefits the plants by providing soil nutrients such as P and N to the plants, much more efficiently than the plant roots can do alone. The fungi also secrete growth factors that stimulate roots to grow and branch, plus antibiotics that protect the plant from pathogens in the soil (Campbell and Reece, 2005).

Another symbiotic relationship in the soil is between plant roots and nitrogen-fixing bacteria called rhizobia, providing some plant species (mainly legumes) with a source of fixed N that plants can assimilate into organic compounds.

Mixed species grazing thus mimics natural, mutually beneficial systems such as wild herds and diverse plant populations on grasslands and the diverse assortment of organisms in the soil that interact with plants.

WHAT BENEFITS CAN MIXED SPECIES GRAZING PROVIDE?

MSG provides opportunities to improve biologic and economic efficiency of rangelands utilization (Glimp, 1988). It was widely used in limited areas of the United States for 75 years (1913—88). Multispecies grazing may enhance plant use and animal productivity and may improve economic efficiency. Glimp (1988) reported benefits from adding sheep to cattle in Texas, and if farms had brush, adding goats further enhanced range improvements. More efficient use of pastures, better weed control, control of brush, more parasite control, and more income/acre were all attributed to mixed species grazing by author, blogger, and columnist, Heather Smith Thomas (2010).

Horse Pasture Management. https://doi.org/10.1016/B978-0-12-812919-7.00010-X

Combining sheep (or goats) with cattle (or horses) greatly reduces predator problems. Female donkeys, llamas, and ostriches also make good guardians for sheep (Nation, 1995) (Fig. 10.1). Horses and cattle are known to sometimes repel coyotes, although research in this area is lacking. A principle of managing guardian animals for sheep is to introduce the guardian animals at a young age to the animals needing protection and take advantage of the maternal instincts of female guardian animals.

Improved Forage Utilization Efficiency

Allan Nation (1995) quoted Dr. Lowell Wilson of Penn State University as saying that advantages to co-grazing sheep and beef cattle include lowering market risk, decreasing weeds, increasing total pasture use, and producing an increase in gross production per head. Adding beef cows and calves to sheep almost invariably increased sheep performance in ewe and lamb weight gain while only marginally decreasing suckling calf weight gain. The most successful pasture species mix Wilson found for beef/sheep grazing in Pennsylvania was perennial ryegrass and alfalfa. After 10 years of management-intensive grazing, they still had excellent stands of both forages (Nation, 1995).

Energy costs of grazed grass are about half those required for forage conserved as hay or silage (Mayne et al., 2000). A rotational grazing system, managed intensively, provides opportunities to increase the carrying capacity or stocking rate (SR) of grazing land. Over a select range of SRs, as SR is increased, grazing pressure (number of animals/mass of herbage) increases, herbage allowance (weight of herbage dry matter (DM) per animal) decreases, and level of competition increases. When competition increases, herbage intake per animal decreases and individual animal performance decreases, but efficiency of herbage utilization increases. Conversely, when more herbage is offered (from a reduced SR), efficiency of forage utilization decreases, but intake and performance per animal increase (Mayne et al., 2000).

In practical terms, if you have plenty of forage for the number of horses you manage, adding a few goats, sheep, or cows will increase the SR and the competition, which will also increase herbage utilization efficiency. The slight decrease in intake observed with rising SR in mono-species studies will be relatively insignificant, but the competition may keep horses grazing longer through the day. On horse farms, individual performance (growth rate) is less important than it is on meat livestock farms, so there is little or no disadvantage to keeping horses busy eating through much of the day.

Forward creep grazing (a form of leader-follower grazing), using creep panels to allow young, small animals into an area not yet grazed by adults, accomplishes something similar to mixed species grazing. The young animals, which have high nutrient requirements for growth, can access the most palatable and digestible forage without competition from adults. This technique is used in sheep and beef operations to increase growth rate of lambs and calves (Mayne et al., 2000). Young dairy heifers had higher growth rates and reduced incidence of disease from parasitic worms on a leader-follower system (Leaver, 1970). While in the area excluded from adults, the young are less likely to ingest internal parasite larvae. This can be used on horse farms where several mares have foaled around the same time.

Vallentine (2001a) explained MSG as a way to enhance energy efficiency of a soil-forage-herbivore complex in three stages. In the first stage, there is an increase in conversion of radiant solar energy through photosynthesis into forage plants. Then the increased solar energy in forages is consumed by grazing animals, which are managed optimally to minimize

FIGURE 10.1 Sheep grazing on a horse pasture at a thoroughbred stud near Newmarket, England. The differences in forage preferences between horses and sheep result in more uniform and more efficient pasture use than when only horses graze the pastures.

forage waste. Finally, the animals convert the energy they obtained from plants into animal products usable by humans. This will include kinetic energy released in equine activities.

As a professor of range science, Vallentine was influenced by Alan Savory, whose early grazing experiences were in Rhodesia with herds of wild animals and cattle. Savory (1988) encouraged graziers to consider conversion of "solar" dollars to new wealth and return some of the solar dollars to a chain of energy transfers. This includes returning some organic matter to the soil through animals trampling plants to help with water retention. It also involves ensuring that desirable forage plants have their energy stores largely filled before they are harvested and before they go into a dormant period of winter or drought (Butterfield et al., 2006). Vallentine (2001a) stated principles of grazing management that are the same regardless of the kind of grazing land, explaining that there are optimum levels of SR, seasons of land use, timing of grazing, mixture of herbivore species, and grazing distribution. These optimums vary over time and place. Forage managers have opportunities to manipulate two main resources: (1) forage resources, to better fit the specific kind or class of grazing animals, and (2) grazing animals, to better fit the forage resources. The manager can integrate and apply both approaches. As someone who worked in the arid range environments in Utah, Vallentine indicated that the planning time frame for grazing land must be as long as the projected life of each forage stand, which can extend as short as a few weeks for annual pasture, a century for perennial pasture, or an unlimited time for native rangelands.

Vallentine (2001b) made another important point about energy efficiency, which is a good reason to consider mixed species grazing. Grazing animals use more energy to acquire feed than animals that have feed brought to them, and this extra maintenance energy requirement may be in the order of 125%—170%, depending on environmental conditions. Thus grazing managers can make better use of the energy stored in an assortment of plants by using a variety of animals to graze plants more completely and compensate for the extra maintenance energy requirement.

A list of advantages and limitations of mixed species grazing, based on Vallentine (2001c), is summarized in Table 10.1.

On large ranges, different species may separate based upon topography, but increasing stocking densities can reduce this separation (Ruyle and Bowns, 1985). Competition among individuals of the same species is greater than interspecific competition, even on homogeneous pasture (Huston, 1975; Walker, 1994). Advantages of mixed grazing increase as terrain and vegetation diversity increase. On pastures of mostly orchardgrass grazed by steers and sheep, there was no difference in total gain/acre versus grazing either species alone (Reynolds et al., 1971). In New Mexico, where wild horses grazed with mule deer, pronghorn antelope, and elk, the more kinds of grazing animals on a site, the more likely it was that more plant species would be utilized (Stephenson et al., 1985).

TABLE 10.1 Advantages and Limitations of Mixed Species Grazing

Advantages	Limitations
Complementarity due to differences in forage plant and terrain preferences	Increased facility costs, such as for fencing, watering, and handling
Maintaining a desired balance among forage species	Reduced scale of enterprise resulting in reduced technological efficiency
Providing stability in grazing land ecosystems	Conflicts in labor needs
Providing diversity of income and more uniform cash flow	Need for increased management skills and knowledge
Aiding in control of internal parasites	Possibly greater predator attraction if sheep or goats are added
Developing mutually beneficial interrelationships between animal species	Marketing animals made more complex
Utilization of forage affected by feces of other animal species	Antisocial behavior between animal species in some situations
Elevating some plant species from weed to forage status	Different susceptibility to climate or weather of different species
Increasing grazing capacity	Required proper stocking ratios between animal species
Reducing predator problems when larger animals are added	Trend toward larger and more specialized agricultural enterprises

From Glimp, H.A., 1988. Multi-species grazing and marketing. Rangelands 10 (6), 276—278; Baker, F.H., 1985. Multispecies grazing: the state of the science. Rangelands 7 (6), 266—269; Taylor Jr., C.A., 1986. Multispecies grazing — vegetation manipulation. Tex. Agric. Expt. Sta. Prog. Rep. 4426, 2p.; Walker, J.W., 1994. Multispecies grazing: the ecological advantage. Sheep Res. J. 10 (Spec. Issue), 52—64.

Most mixed species grazing research has been focused on sheep and cattle. In many cases cattle productivity does not benefit as much as sheep productivity, but this can depend on relative SRs and system of grazing management (Bowns, 1989; Olson et al., 1999). When cattle and sheep were grazed on a mix of Kentucky bluegrass and white clover, both plant species persisted, but cattle alone reduced the bluegrass and sheep alone reduced the white clover (Abaye et al., 1994). In the experience of Walker (1994), a manager mixing sheep and cattle could increase SRs by 10% on improved pasture or 25% on native range. By stocking pastures with sheep, cattle, and goats on diverse shrub-forb-grass ranges, they could increase SRs up to 70% more than with a single livestock species. Presumably, domestic horses on pasture would benefit from the company of cattle, sheep, or goats, depending on the predominant types of plants present. More research on the use of cattle, small ruminants, and even deer or elk grazing with horses would be needed to develop recommendations for relative SRs.

Mixed Species Grazing Helps to Control Parasites

The two most important parasites of horses are the Ascaris roundworm *Parascaris equorum* and the group of small intestinal worms within the subfamily Cyathostominae of the family Strongylidae (Peregrine et al., 2014). The Merck Veterinary Manual online indicates about 40 species of small strongyles (cyathostomins), of which 10 are most prevalent (Klei, 2017b) and about six other gastrointestinal parasites, plus liver flukes, that affect horses (Klei, 2017a,b; Lane, 2017). Only Strongyloides (except *S. westeri*) plus Trichostrongylus and the liver fluke (*Fasciola hepatica*) are also parasites of goats and sheep, according to lists in Merck Veterinary Manual (Fox, 2017a,b and Nielsen, 2017 personal communication), so this backs up the assertion that parasite species tend to be (animal) species-specific. Nielsen stated that the drugs commonly used against internal parasites, ivermectin and moxidectin, are no longer effective against ascarids on most farms, and there is evidence that ascarids are also developing resistance against the other two drug classes (benzimidazoles and pyrantel formulations) (Smith Thomas, 2014). Small strongyles were reported to be resistant to ivermectin as early as 1993 (Herd et al., 1993).

Alternate grazing of horses and sheep as a control measure for gastrointestinal helminthiasis was studied in the Netherlands in three grazing experiments during 1981—83 (Eysker et al., 1986). Shetland ponies alone were compared with Shetland ponies that sometimes grazed pastures that had been grazed by sheep from April to July. Groups of ponies moved to sheep pasture acquired considerably lower burdens of nematodes of the subfamilies Cyastominae and Strongylinae, but considerably higher burdens of *Trichostrongylus axei* (a stomach parasite affecting horses and ruminants) than the groups that were not moved (Eysker et al., 1986).

Herd et al. (1981) reported the importance of cyathostomes (small strongyles) in the United States and showed that they often accounted for 100% of the worm egg output of grazing horses. Being that encysted cyathostomes were not susceptible to anthelmintics, there is drug resistance and with it a shorter egg reappearance period. Dr. Herd's research group designed a pasture hygiene approach for horses under intensive grazing conditions where there were constant reinfection problems, based on twice weekly removal of excessive feces, plus a 100% increase in grazing area (cutting stocking intensity in half). Herd (1986) stressed the need for great care in the following:

1. the choice of anthelmintics
2. the timing of treatments
3. the monitoring of fecal egg counts to check the degree of pasture contamination and development of drug resistance.

Herd (1993) recommended aiming primarily at cyathostomes, trying to overcome the problems of short parasite egg reappearance periods, encystment, and ecotoxicity. His two main approaches were the strategic seasonal use of anthelmintics and a nonchemical approach of pasture sweeping or vacuuming. Cost of pasture sweeping or vacuuming could be offset by improved worm and colic control and increased grazing area available due to elimination of ungrazed roughs around feces (Herd et al., 1981; Herd, 1993).

Smith Thomas (2014) interviewed veterinary parasitologists for a magazine article in The Horse.com and reported recent recommendations on using pasture management to supplement treatment with anthelmintics. Some of that article is reproduced here to show a logical progression of ideas toward the recommendation of mixed species grazing. Martin Nielsen, DVM, of University of Kentucky Gluck Equine Research Center warned horse owners that strongyle larvae hatch and migrate onto forage plants. Dr. Andrew Peregrine of the University of Guelph recommended progressive pasture management (Peregrine et al., 2014) and in the interview with Smith Thomas, he stated that "On any horse farm, the majority of these (strongyle) worms (cyathostomes) that plague mature horses … are out on the pasture and not in the horses. How you manage the pasture is more important than deworming when trying to keep parasites down to reasonable levels." Sources in the article by Smith Thomas (2014) recommended using rotational grazing to stop strongyle transmission because the strongyle larvae only access the lower part of forage plants.

Martin Nielsen indicated that parasite numbers are higher where stocking densities are high and horses are overgrazing, which is revealed by the appearance of roughs (areas of tall grass where horses defecate) and lawns (areas of shorter grass where they prefer to graze). He also suggested that overgrazing can be avoided by rotational grazing. Peregrine added that pasture needs more than 3 to 4 weeks of rest and hot, dry weather for strongyle parasites to die off. Thomas Craig, DVM, at Texas A&M University, explained that alternating haying and grazing a field helps to break the parasite cycle. Nielsen indicated a role for mixed species grazing, "Alternating or co-grazing horses and ruminants in a pasture can also help reduce worm populations. These species don't share parasites, except for a couple of very rare ones." "(Sheep), they graze everywhere and don't have specific defecation areas. They also graze down the roughs left by horses. The taller grass in those areas helps protect parasites from heat and drying; it shades them from direct sunlight and helps them hold moisture. Grazing these down with cattle or sheep leaves worms more vulnerable to drying." Smith Thomas (2014) explained that cattle graze the tall grass that horses avoid near manure piles and that the worm larvae on that forage cannot complete their life cycle within the cow, so no eggs will pass. Nielsen and Peregrine indicated that removing feces from pastures once or twice a week will significantly reduce pasture infectivity. They also recommend composting manure because the heat can kill parasite eggs.

Thomas Lane, DVM, (2017) of the University of Florida, writing in the Merck Veterinary Manual, recommended a process to slow the development of drug-resistant parasites. The process includes customizing deworming protocols for the farm and individual horses. Horses most susceptible to parasites (i.e., the high egg shredders) are identified, the overall number of anthelmintic treatments can be reduced, and the efficacy of better-timed specific classes of dewormers is monitored using the fecal egg count reduction test. Lane (2017) listed 12 "nonchemical methods of parasite control" and one of them is "Cross-graze pastures with other species. Cattle, sheep, and goats serve as biological vacuums for equine parasites." Nielsen (2015) discussed challenges for parasite control, pointing out that a therapy protocol against one species of parasite may result in the rise in population of another parasite, as occurred in Denmark (Eysker et al., 1986) and that legislation of a treatment protocol may result in unintended consequences. Thus end users of guidelines need to find a balance between treating too little and too much. Here is a niche for better pasture management, to reduce the dependence upon drugs and to ensure their appropriate use in combating parasites. Transmission of strongyle parasites seems more likely in pasture environments than stall environments, partly because more infective larvae are available on pastures, especially if pastures are shared by multiple mare-foal pairs (Reinemeyer and Nielsen, 2017). Similarly, introducing weanlings to the same pasture that contains the prior year's juvenile horses (which typically have high fecal egg counts) can cause considerable transmission of strongyles to the weanlings. However, if weanlings are placed on a pasture that had mature horses that were low contaminators or high contaminators that were treated selectively, then levels of transmission should be modest. In their review, Reinemeyer and Nielsen (2017) also stated that veterinarians need to have a working knowledge of seasonal strongyle transmission in their practice area, plus the effects of stocking density, pasture rotation, harrowing, and other management options.

Reducing Ecologic Risks of Avermectins

In addition to the concerns mentioned in the previous section about parasites of horses developing resistance to anthelmintic parasiticides, the very nature of these drugs makes them harmful to nontarget organisms living in the horse farm environment and downstream of it. Herd (1995) listed characteristics that make a popular avermectin, ivermectin, potentially ecotoxic. It is highly potent against a variety of nematodes and arthropods (including dung beetles) at low concentrations and is mainly excreted in feces, to which it binds strongly and persists with a half-life of 111−260 days at 68°F (20°C) (Nessel et al., 1983). Herd (1995) reviewed several studies that included biologic assays of ivermectin persistence in dung. Results varied due to differences in formulation, route of administration of ivermectin, and species of animal. A standard dose of ivermectin for cattle, sheep, and horses is 0.2 mg/kg of live weight. Manure of cattle given this dose had detectable ivermectin residues that were toxic to house fly larvae for up to 2 months. The same dosage administered to horses and sheep only inhibited development of fly larvae in excreted feces for 3 or 4 days after treatment. These authors raised concerns about possible ecotoxic effects of ivermectin because insects aid many natural processes in pastures, such as dung dispersal, nutrient recycling, soil aeration, humus content, water percolation, and nematode control. These actions, in turn, have beneficial effects on plant growth. Due to the diversity of insects and other arthropods, the many different soil environments and the widely varying levels of drugs that could move from dung to soil, the variation of effects on growth, fertility, metamorphosis, and other physiologic activities of these soil organisms is vast. While some studies show no direct effects of ivermectin on earthworms, Herd (1995) suggested that ivermectin-sensitive fly larvae are important in aerating fecal pats, which makes the fecal pats attractive to earthworms, which are well-known dispersers of dung. Toxicity of ivermectin to the aquatic species *Daphnia magna* was reported as extremely high (Garric et al., 2007), with adverse effects on growth and reproduction at concentrations below the analytical limit of detection.

Reports of adverse effects of avermectins on dung fauna from 1987 through 2003 were reviewed by Woodward (2005), who concurred that sustained bolus formulations of anthelmintics may offer greater risks than other modes of administration and that dung fauna may be at less risk if treatment of animals with anthelmintics does not coincide with the breeding season of dung fauna. Claims that ivermectin dissipates rapidly in water and binds to sediment were tested by Sanderson et al. (2007). The biodegradation half-life (DT_{50}) of ivermectin in pond water was found to be 3 to 5 days, while pond sediment concentrations of ivermectin increased to a stable level with no assessable DT_{50}. Acute and long term (>229 days) toxic effects were discovered in aquatic sediment-dwelling organisms at or below predicted environmental concentrations of ivermectin.

Avermectins are almost totally excreted in feces, and once on the ground, they are toxic to dung-breeding arthropods of benefit to the ecosystem. Herd et al. (1993) showed that ivermectin treatment of horses in June significantly slowed the rates of dung dispersal over 14 weeks. The area of pasture fouled by horse dung and thus unavailable for horse grazing was initially twice as big for horse dung containing ivermectin than dung containing oxibendazole or no anthelmintic. By 25 weeks after ivermectin treatment, pasture loss for ivermectin treatment was three times greater than for oxibendazole or no anthelmintic. Thus, a toxic effect of ivermectin on arthropods such as dung beetles was evident, and this caused a decline in pasture available for grazing.

Lumaret et al. (1993) investigated effects of a single cattle dose of ivermectin (200 µg/kg body wt.), given to steers, on the fate of that dewormer in feces and on flies (Diptera) and dung beetles (*Euoniticellus fulvus*). The drug was eliminated from steers within 12 days, with a peak in feces at day five. Dung beetles were increasingly attracted to the treated dung from days six to 17, the last few days of this occurring when ivermectin could no longer be detected in the dung. Dung produced 1 and 10 days after treatment inhibited Diptera (fly) larvae in dung, and on day one, dung beetle development was slightly inhibited.

To reduce the negative ecotoxicity effects of ivermectin, Herd (1993) recommended using only ecologically safe anthelmintics in spring, keeping ivermectin-treated horses off pasture for 3 days posttreatment, alternative grazing of horses with cattle or sheep (Eysker et al., 1986), and a selective chemotherapy program to target only those parasites that are evident. To apply only necessary treatments, Herd highly recommended use of quantitative fecal egg count kits.

Having discovered many reports of higher than necessary use of avermectins in livestock in many countries and knowing that moderate exposure to parasites stimulates immunity against them, Herd (1995) suggested that a careful balance between overuse and underuse would protect the environment, reduce parasite burdens, and encourage development of immunity.

Using a combination of these suggestions for anthelminthic use and mixed species grazing, the ecotoxic effects of anthelmintic drugs can be minimized.

Improved Weed Control, Especially with Training

In his book, *Management Intensive Grazing*, Jim Gerrish (2004) explained how adding a second grazing animal species reduced the need for mowing, clipping, and trimming weeds along fence lines. On the Gerrish farm, they started out with just sheep on pastures with a lot of ironweed. After several years of grazing by sheep, little ironweed was seen. After 9 years, the sheep were sold and replaced by cattle. Within 2 years the pasture was covered by ironweed again, and 4 years of mowing were required to bring the ironweed under control.

Smith Thomas (2010) wrote an article for *Canadian Cattlemen* magazine on advantages of MSG in which she said that using multiple species of livestock helps spread grazing pressures across a wider variety of plants. Some of the worst weeds today on cattle ranges (spotted knapweed, leafy spurge, yellow star thistle, etc.) are readily eaten by sheep and goats. Smith Thomas's sources indicated that goats readily eat pigweed, ragweed, poison ivy, dock, sedge, black locust, autumn olive, mulberry, wild roses, blackberry, honeysuckle, and others. Plants toxic to cattle that can be safely eaten by goats include hemlock, poison oak, yellow star thistle, and mustard. Plants toxic to cattle that will not harm sheep include leafy spurge, tall larkspur, tansy ragwort, and pine needles. If these three herbivores can be combined for weed control, there may be a role for one or more of them to help with weed control in horse pastures. Smith Thomas (2017) also reported on a ranch in Oregon that has survived as a family business for five generations, and its owners credit its current success to their use of cattle, sheep, and goats. The owners run 7 to 10 ewes in the same space that one cow would graze. The lamb income from 10 ewes would be $1500 off the same pasture that produced a calf that earned $800. However, they find it is more labor-intensive to raise sheep, and predators are a bigger challenge with sheep than cattle. Breaking the parasite cycle by rotating sheep and cattle groups through pastures is one of the strengths in their system. They have a separate herd of goats that they use for weed and brush control, mainly targeting blackberries and Scotch broom. The goat herd is sometimes hired out to other landowners, communities, and even the US Army Corps of Engineers to control invasive weeds in environmentally sensitive areas. Another claim by these ranchers is that they are "environmentalists and great stewards of the land, and … we support 80% of all wildlife because these animals prefer managed ground over rough scrub." Goats are also very susceptible to predator attacks, and their bleat seems to be attractive to coyotes and cougars, so livestock protection dogs are required.

A coeditor of the online grazing newsletter, OnPasture.com, Kathy Voth (2004), developed a method and has produced information on teaching livestock to eat weeds. This is a valuable resource to encourage horse owners to use other livestock species to effectively control specific problem weeds. Kathy Voth learned from Fred Provenza and colleagues at Utah State University about how animals learn and how they choose what to eat. From studying animal behavior work by Pavlov and Skinner, she developed steps to turn livestock (mainly cattle) into weed eaters. Eventually, she refined a process so that anyone can teach animals to eat weeds in about 8 h spread over 10 days. Beef cows, dairy cows, sheep, and bison have been trained to eat many different weeds across the United States and Canada. The first group of animal trainees teaches their offspring and herd mates. The learning process gives animals the idea that all kinds of things can be food, and they start to eat different weeds on their own. Voth (2004) developed a book, *Cows Eat Weeds*, and a DVD, "Teaching Cows to Eat Weeds," so anyone who wanted to can follow the simple steps to turn problem weeds into forage. At her website, you can learn more about her method, ask questions, exchange ideas, request her to speak to a group, and order her book and DVD. Readers can keep up to date on new developments in this and other pasture management ideas through OnPasture.com. In a recent article, Voth (2014) indicated the scope for greater forage use by training more cows to eat weeds. Economic analysis showed that if one weed species, leafy spurge, was eliminated from an area of 17 states where it is problematic, there would be enough other forage to raise 100,000 more cows per year, and there would be no costs for herbicides or their application.

Do I advocate teaching horses to eat weeds? No, definitely not, since horses suffer poisoning from far more plants and at lower doses than ruminant animals do. However, you can teach cows, sheep, or goats to eat the weeds that are in your horse pastures and introduce those ruminant animals to your horse pastures either before the horses or with them. Continue to work on ensuring that the pastures have healthy, vigorous populations of nontoxic forages that they find highly palatable.

See Voth (2015) for partial lists of weeds that ruminants can eat and weeds that livestock should not eat because they are toxic.

WHY/HOW DOES MULTISPECIES GRAZING WORK?

Plants in a pasture are constantly involved in competition for resources. Those plants that draw on a pool of resources most rapidly or most efficiently when the resources are in short supply will outcompete other plants. Such plants also need to grow rapidly to extend into the surrounding area, to spread their leaves for the purposes of absorbing ultraviolet light and taking in carbon dioxide. Below the soil surface, more competitive plants extend their roots deeper and wider than other plants. The ability to grow in a wider range of temperatures is also an asset in this plant growth tournament. The more competitive a plant is, the more offspring it can leave.

An important influence on the growth, competitiveness, and reproduction of plants is grazing animals, which graze selectively and in patches. This heterogeneous effect causes a pasture to be more complex or diverse than a hay field, in terms of both the plant species mixture and the maturity of plants. Further animal influences on pasture diversity are patches of manure and urine, plus the actions of walking, running, lying, pawing, and all the other contacts with plants and soil. Through these mechanisms, MSG will encourage a broader variety of plant species, plus diversity in maturity within a species, which help to reduce seasonal slumps in forage production.

Feed selection preferences of different animal species are well documented in American research (Glimp, 1988). Generally, grazing cattle eat fewer plant species and a higher percentage of grasses than sheep or goats. Sheep eat a higher percentage of leaf material than cattle and more forbs (broad-leaved, nonleguminous plants) than cattle and goats. Goats select a wider variety of plant species and select more browse (woody plants) than cattle or sheep. Cattle and sheep have both been reported to prefer eating about 70%—74% legumes and 26%—30% grasses (Rutter, 2006, 2010; Torres-Rodrigues et al., 1997). Nonlactating heifers and sheep consumed fewer legumes (64% and 62% of their diets, respectively) than lactating females (75% and 76%, respectively) (Parsons et al., 1994; Rutter et al., 2004a,b), so the physiologic state of animals affects their plant preferences.

In a review of 20 studies by Animut and Goetsch (2008), there were three reports that goats ate more legumes and two reports that sheep ate more legumes than goats (usually clover). Three studies indicated that goats ate more forbs, including leafy spurge and thistle, and one study indicated equal consumption of forbs. Consumption of grass was greater by sheep in 10 studies, but two studies reported that goats ate more grass. The finding that goats ate more browse (including saltbush and broom) than sheep was reported in nine studies. One study showed that sheep ate less browse than goats did in a woodland environment in the wet season, but during the dry period, browse became equally important to sheep and goats. Thus it appears that several factors contribute to relative intake of various plant species by different species of livestock, and these are discussed further subsequently.

Species of plants that different animal species graze in common are referred to as dietary overlap. Sheep, cattle, and goats are reported to share 35—53% overlap. When wild animals shared the same environments as domestic livestock, overlap was reported from 8% (cattle and antelope) to 55% (cattle and elk) (Glimp, 1988).

Mule deer and elk commonly share range in the western United States with cattle. Since the mule deer in Colorado eat mainly sagebrush and cattle eat mainly needle grass, with very little overlap in forage species eaten, the range can be used more efficiently by both animal species than by either animal species alone (Hansen et al., 1977; Holecheck et al., 2010). On many range environments where cattle and sheep graze together, their dietary overlap may be 30% to 60%, or in other environments, over 80% (Holechek et al., 1986). With very high overlap, the grazing by two species is said to be nonadditive (Holechek et al., 2010). The amount of dietary overlap must be taken into account to establish meaningful SRs and carrying capacity. The degree of overlap can change due to the speed of maturing of different forage species or due to such weather events as drought and severe frost. Goats are usually additive when grazed with cattle because they mainly browse on woody and bushy material, while cattle graze mostly grass.

Local geography and dominant vegetation types can influence which animal species can best complement horses in co-grazing. Tall grass prairie and stemmy bunchgrasses are better for cattle than for sheep or goats. Palatable brush is better for goats than sheep, which readily climb to higher elevations and avoid wetlands. Cattle prefer lower elevations and will graze wetlands. The more diverse the forages, the more likely is success with MSG.

Dietary overlap is not constant and can vary across locations and over time. As grazing pressure increases, dietary overlap increases. With seasonal variations in forage availability, palatability, and amount of green versus dead material, animal preferences change, and thus overlap can change.

The sum of influences of each animal species varies due to their anatomy and physiology. The variation in influences within each animal species is due to differences in what each individual animal learned about grazing from its mother, its herd mates, and its own experiences. See Chapter 8 for more information on grazing behavior. Thus the greater the diversity of animals grazing a pasture, the greater the diversity of influences on plants will be. In any population of organisms, the greater the diversity, the higher is the probability that some will be able to adapt to a change in environmental conditions. The environment of any pasture is a series of microenvironments, with differences in soil type, particle size, slope, drainage, pH, and concentrations of water, organic matter, and chemical elements. So diversity of plant species and diversity within plant species are advantages in terms of keeping a healthy, productive sward on a pasture.

Co-grazing one sheep or goat per cow is often recommended to improve utilization efficiency of pasture. Bill Murphy (1998) suggested that pasture deterioration is preventable by having cattle graze either with horses or following horses to eat the forage that horses will not graze and that this may reduce the need to mow, harrow, and fertilize horse pastures.

Cattle seize a mouthful of grass with their tongues and twist it until it breaks, so they suit tall, dense grass. Horses grip grass with their teeth and cut it very short in mouthfuls, as do sheep (although the sheep use a dental pad rather than upper incisors). Voisin (1959b) and Sinclair (1821) described a precursor to modern strip grazing, which involved tethering of livestock on pastures in Scotland and Ireland. Tethered cows were moved only one foot at a time, to prevent trampling of forage. The cows would be followed by tethered sheep, and sometimes, horses were tethered after the sheep (Sinclair, 1821). The tethering technique prevented waste, for following animals were not deterred from eating grass contacted by the dung of another species of animal.

According to *Management Intensive Grazing* consultant, Jim Gerrish (2004), "Variance that occurs in plant communities offers potential for grazing by a number of different livestock and wildlife species." He explained that specialization is valuable and easy where environmental conditions are controlled, in confinement. Outdoors, the value of diversification is more apparent. The variance in plant communities offers potential for grazing by different livestock and wildlife species. A diverse population of animals is likely to graze a diverse plant population more efficiently and more homogeneously than a single grazing species.

If two species of animal have the same grazing preferences and habits, there is no reason for adding the second species. Mixed species grazing can be beneficial when combining species having different diet preferences and foraging habits so that they complement each other (as well as compete). Age and landscape preferences should also be considered. Sheep farmers have learned that young animals are quicker to bond with different species than older animals. Combining the desires of sheep or goats to climb hills with the preference of cattle to stay on lower, more level ground reduces competition and provides opportunities to express natural behaviors.

Since animals have preferences for certain plant species, over time, the preferred species will be grazed more and be weakened more than less-preferred species. Variation in plant species preferences can be altered by shifting the time of the grazing season that a paddock is first grazed or last grazed. Plant variability can also be altered by changing the length of grazing and resting periods or by allowing different animal species to graze. Where animals are kept on a pasture for a whole grazing season (continuous grazing), the pressure on the most palatable species from competing plant species can be lessened by introducing another species of grazing animal that prefers that competing plant species.

Gerrish (2004) provided approximate ranges of forage choices for domestic livestock and wild herbivores, indicating that cattle consume 80%−90% grass, with forbs and browse in equal parts (10%−20%). Sheep consume 50%−60% grass and 40%−50% forbs and browse with a ratio of two forbs to one browse. Goats prefer 60%−70% woody species with a 30%−40% split evenly between grass and forbs. So goats are an ideal choice to manage woody species. Deer and pronghorn antelope are mainly browsers. Elk are primarily grass grazers. Bison have an even stronger preference for grass than cattle.

Feral goats on an island off California showed the following differences in seasonal forage preferences: in winter, 90% browse, 4% forbs, 6% grass, when browse availability was high; and in summer: 8% browse, 18% forbs, 74% grass, when growth of grasses and forbs was rapid (Coblentz, 1977). Several findings help to explain the variable results of forage use in mixed species grazing situations (Rook, 2000). Sheep, goats, and cattle showed a partial (but variable) preference for clover versus grass, consuming 60%−80% clover (Newman et al., 1992; Penning et al., 1995). Cattle adjust their patch residence time on the basis of the intake rate achieved, and this will be influenced by the quality and palatability of the various forages present in each patch (Distel et al., 1995).

Illius et al. (1992) reported that sheep offered grass-white clover swards of varying proportion and height seemed to choose forage patches first on the basis of sward height. Then the sheep switched between patches, as if using information gathered while grazing. Sheep are more selective than cattle and can graze lower into a sward, perhaps due to their narrower muzzles (Grant et al., 1985). In Oklahoma on pastures of mainly Bermudagrass, Johnsongrass, and ragweed, sheep preferred more grasses and fewer forbs (especially ragweed) than goats (Animut et al., 2005). When available, goats select a greater quantity and dietary proportion of browse and shrubs than sheep. There is a greater dietary overlap between cattle and sheep versus cattle and goats. Goats have a shallower depth of biting and they prehend (grab) from the top of the sward or horizon down, with biting and head movements horizontally or from side to side. The percentage of browse species on a pasture or range can be reduced by adding goats to grazing sheep (Animut and Goetsch, 2008). Large-ruminants have the ability to retain coarse forage for a longer time than horses of a similar size or smaller ruminants, so they are able to utilize a more fibrous diet (Demment and Van Soest, 1985).

In summary, if a pasture contains a diversity of plants in terms of species and maturity and a single species of animal grazes that pasture, the preferred plant species will be grazed soon and perhaps again, while less-desired plants will be left to reproduce. If another one or two animal species join the first species on the pasture, their diverse plant species preferences will cause the pasture to be grazed more uniformly. The overlap of plant species preferences by different animal species contributes to competition and grazing pressure, which should stimulate the pasture manager to move the animals to another paddock before regrazing of preferred species of plants begins.

Utilization of Forage Affected by Feces of Other Animal Species

Smith Thomas (2014) quoted Thomas Craig, DVM, of Texas A&M University, indicating that, "free roaming horses rarely eat grass next to manure. When feed supply is short, they will eat grass next to manure piles and pick up a heavy load (of parasites). Generally, in large pastures they get exposed to just a few worms — enough to stimulate some immunity. If the worm load is low, they can handle it." Smith Thomas expanded on Craig's statement, saying, "We intervene in this natural process by confining horses to grazing the same small pasture repeatedly. If you are managing a lot of horses on a small pasture you must be especially careful about managing stocking density and grazing rotations."

A cow will eat grass grown next to horse dung. A horse will not eat forage growing through or next to horse dung pats but eats grass growing through cow dung pats. Thus there is an advantage of associating young horses and cattle, especially in the last group of a rotation. Cows apparently accept grass grown near sheep's dung but refuse to eat grass where a sheep has lain (Voisin, 1959a). Grass cut from a cow dung pat and placed on a clean spot in the pasture will be readily eaten by a cow. Cows will not eat grass grown in a place where there is cow dung during daylight or dark hours. Thus, it seems that the smell of dung from their own species, rather than its appearance, is what repels grazing livestock (Voisin, 1959a).

Other Influences on Workability of Multispecies Grazing

There are situations in which MSG may not be beneficial. High-intensity, low-frequency or short-duration grazing systems may not enhance MSG strategies because animals are on paddocks for such short times and at such high stocking intensities that there is much competition and all forage types are grazed in a short time.

Mixed species grazing works because animal preferences for various plants vary according to many factors as described before. In Chapter 8 on grazing behavior, there are also discussions on animal preferences for plants being influenced by forage height, maturity and yield, nearness of manure, proportions in the sward, concentrations of DM, fiber, nonstructural carbohydrates and protein, season, time of day, and what the animals recently ate.

There may be economic consequences and benefits from adding an additional species. Changes to fencing, handling, and watering systems may be needed. Predator activity may cause use of sheep or goats to be too costly. Better weed control may occur, reducing the need for mowing, clipping, and trimming weeds along fence lines.

There are economic challenges to maintaining any farm. Horse managers look for income opportunities from boarding, giving riding lessons, hauling horses, and perhaps selling hay. If they enjoy working with other animals, it may be worthwhile using one or more other species to help manage pastures. The more plant-diverse and animal-diverse you can make your pastures, the more stable and healthier they will become (Nation, 1995).

WHAT ARE POTENTIAL DISADVANTAGES TO MIXED SPECIES GRAZING ON A HORSE FARM?

Horses and cattle may injure smaller livestock at lambing, watering, supplemental feeding, or handling. Individual aggressive cattle or horses may need to be removed from sheep and goats. Yearling cattle are more likely to disturb the young of other species than are cows with calves. Creep gates may be useful to allow small young animals to get away from aggressive larger animals. Sheep in large bands being herded may disturb other grazing animals (Vallentine, 2001c). Mature male donkeys, purchased to guard sheep, are known to sometimes show their disrespect by biting and picking up sheep by the wool. Recommendations for use of donkeys as guardian animals for sheep or goats include raising female donkeys from a very young age with the sheep or goats, so they bond well. Thus you can expect that horses and other livestock introduced to each other at young ages and continuing to live together will have better relationships than adults suddenly mixed together. If guardian dogs are considered, they must be raised with the species they are intended to protect from a very young age.

In rangeland situations, when feed resources are plentiful, mixed species of grazing animals feed together, but when resources are scarce, the following dominance hierarchy may cause agonistic encounters: bison, horses, cattle, sheep, elk, mule deer, bighorn sheep, pronghorns, and white-tailed deer (Mosley, 1999). Large wild herbivores tend to leave locations when herds of domestic livestock arrive, but there are many observations of moose, elk, or deer spontaneously appearing with grazing cattle. Wild horses in range country may be aggressive to cattle and pronghorn antelope around watering places, especially if water is in short supply or horse numbers are large (Miller, 1983; Plumb et al., 1984; Sowell et al., 1983).

Severe competition for pasture or range forage between horses and other livestock is rare and is most likely with cattle, since horses and cattle share similar grass preferences. Similarities of diets, overlaps of usual grazing sites, increased grazing pressures, and lack of alternative feeds can increase the likelihood of such competition. A low level of forage species overlap with wild herbivores probably prevents horses from competing much with them. In Alberta, where cattle and feral horses shared space and a diet overlap of 66%, horses tended to use feeding sites first and cattle to use them later (Salter and Hudson, 1980). Competition between horses and wild herbivores was minimal.

Smith Thomas (2010) pointed out that a disadvantage of trying MSG is a need for more facilities, labor, and management. Much more research has been done on mixed species grazing that excludes horses than studies that include horses. Thus, much less is known about how to manage mixed systems on horse farms. One significant difference between meat and milk farms compared to horse farms is that on many horse farms, client horse owners want frequent access to their individual horses for riding, grooming, and training. Extracting the horse you want from a herd of horses, sheep, cattle, and goats in a pasture paddock of several acres is more complicated than extracting it from two or three other horses in a small paddock. Horse owners need to ensure that their horses are well trained to coming to them readily, sometimes at a considerable distance, if this grazing arrangement is to be workable. Since most young horses are not familiar with other species of livestock, there is also a need to accustom the young horses to any species of animal that they will graze with later. Fig. 10.2 shows a single horse grazing some distance from a herd of Black Angus cattle.

FIGURE 10.2   A horse grazing near cattle in Kentucky as a reminder that while horses and cattle share a number of preferred forages, there are enough differences in their preferred forages that they complement each other to graze a pasture more uniformly than either species would alone.

HOW IS INFORMATION ABOUT NONEQUINE SPECIES USEFUL IN MANAGING HORSES?

Recent parts of the evolution of wild horses would have occurred in grassland environments shared with other ungulate herbivores, so these animals share some needs and behaviors and thus are able to graze together in present day domestic environments, as long as their needs are met. As mentioned in the chapter on grazing behavior, many of the things learned in the study of one species can be useful in understanding and managing a species with similarities. Far more research time and money has been spent in the last century on domestic species that are raised for meat, milk, and fiber than on horses. We can study and make informed judgements on which parts of the body of mixed species grazing research are applicable to horse management, while planning and fund raising for future research that is more specific to horses. Most farmers are experimenters and innovators to some degree, and they also learn from a wide variety of sources. Cases of livestock species having significant negative effects on each other are rare, so it is logical that some horse managers try analyzing the forage mix available on their properties, and if they think there is an opportunity to improve the efficiency of forage use, introduce one new animal species at a time.

ECONOMIC CONSIDERATIONS OF MIXED SPECIES GRAZING

To benefit economically from mixed species grazing, use regular monitoring of forage yield and factors that affect forage quality (e.g., weed growth). Compare actual grass availability to targets. Alter management tasks on an on-going basis to provide optimal amounts of optimal quality forage and keep accurate records of forage yield and quality before and after management changes. Before adding a new species, perform budget calculations on amounts of forage dry matter (FDM) and acres that are needed and produced just for horses, then add in other animals. If, for example, you are grazing horses and goats together, add in one goat per horse and recalculate. For each 1100-lb (500 kg) horse, add one goat at about 140 lbs (63.5 kg). Since the goat weighs about one-eighth of the horse weight, the increase in animal units per acre will not be great. For details on grazing animal intakes and equivalence, see Vallentine (2001d). What the pasture manager should expect is that a mixed group of horses and goats will have the forage eaten down to a height that triggers moving them to the next paddock slightly sooner than the all horse group. As mentioned earlier, the greatest effect of adding goats will be the eating of shrubs and other woody vegetation in preference to grasses.

Potential Positive Economic Responses

Gllmp (1988) analyzed a series of thoughts on the economics of MSG that are worth considering before bringing another species onto a horse farm. The positive responses to MSG that he reported appear to be more in improved animal performance (growth rate, milk production) and more uniform plant species use than from potential increases in SR.

While horse managers may initially consider that Glimp's ideas are not applicable to a horse farm, they can recognize that ranchers and farmers may also need a considerable effort to adjust to these ideas. Perhaps further analysis is needed by all three groups. According to Glimp, ranchers and farmers are in the business of "appropriate and proper land use," rather than in the cattle or sheep business. Once you have made a significant investment in land, the first principle of managing that land economically is, "What can the land resource base produce on a sustainable basis?" For managers of existing horse farms, the first answer will be about horses. Since horse farm managers already control considerable resources of land, facilities, and labor, why not consider a way to improve the efficiency of their use? Adding one or two more animal species can result in more income. Think about how this can be managed to enhance productivity and appearance of the land and how products can be most efficiently harvested. Then work on how to best market these products. A change to more uniform-looking pastures can increase the value of the land.

A major reason for considering multiple species grazing on a horse farm is in line with the primary objectives of pasture management on livestock farms. That is to supply high-quality forage throughout the grazing season and ensure its efficient utilization (Mayne et al., 2000). The energy costs of producing grazed grass are about half of those required for field-cured hay or silage made from grass, maize, or alfalfa (Mayne et al., 2000; Wilkins, 1980); thus, annual feed costs can be reduced with mixed grazing. For livestock producers, the most obvious potential economic benefits of MSG will be more animal products (and more volume) to market through improved forage use. A principle of agricultural economics says that increasing product diversity improves risk management. A more diverse range of products leads to cash flow spread over more times of the year. Some horse farm managers already look for opportunities to increase income by trucking horses for other horse owners or selling hay. They can bring in cattle, sheep, goats, or even wild herbivores, according to proportions of grass, legumes, other forbs, and browse available on the land. If there is a desire to turn some forested or bush land into pasture, goats, cattle, and horses in combination can reduce the proportion of woody plants considerably, without the expense of using heavy machinery.

Ownership of other species of animals is not necessary. Owners of other species may be looking to rent or lease grazing land. Agricultural extension advisors can help two parties work out grazing agreements that are acceptable to both. As the productivity of land improves, there may be surplus hay to sell.

Potential Negative Economic Consequences

As in any business plan, a manager must determine whether introducing another animal species to a horse farm is biologically feasible and worth the extra costs. For example, determine whether there are technical constraints to adding another species, such as presence of plants poisonous to one species but not another, risk from attacks by predators, and suitability of climate and landforms. Learn where there are adequate market outlets for the potential new products and what the cost of transport to and use of those markets will be. The costs of more facilities and equipment will be higher, and there may be a cost of a broader knowledge base in the labor force. It will be wise to own as little machinery as possible. Look for opportunities to keep costs low by sharing or borrowing machinery or equipment or hiring contract machinery operators, rather than buying things that you will use rarely.

MANAGING MIXED SPECIES GRAZING

How will you use another livestock species on a horse farm? First, recall the goals of MSG and what it can do for your farm. It can improve the efficiency of forage use and land use and promote desired changes in the proportions of various plant species. MSG can help in the control of weeds and parasites.

Examine the plant populations in all pastures and decide what changes you would like to cause. If there are more forbs than your horses eat and you want to reduce their populations or make good use of them, consider adding sheep. If woody plants, whether trees or shrubs are increasing, consider goats, deer, or antelope. If the pastures only contain grasses and you want to keep it that way, the scope for improved efficiency through mixed species grazing is low, so it will not be worthwhile introducing other animals. If soil organic matter level is low, consider sowing an annual crop such as rye, oats, or millet and borrow some cattle to graze and trample it at a high stocking density.

If you know how many acres of grazeable land you have available and divide this by the number of acres needed for a typical or average animal, you arrive at the number of such animals that can be carried or supported for a grazing season. Whether you consider horses, sheep, cattle, or goats, a reasonable estimate of average daily FDM requirement is 2.5% of body weight. Grazing animals should not eat all of the plant material where they graze, so it is reasonable to allow for the animals eating half of the FDM yield and leaving half. The residual FDM includes undamaged leaves and stems that can photosynthesize to support regrowth and trampled material that contacts the soil and becomes part of its valuable organic matter. Calculations involving SRs for horses are covered in Chapter 9.

If there are wild herbivores frequenting the same pastures as the horses, their estimated body weights and physiologic states can also be used to calculate how much feed they need each day. *The Holistic Management Handbook* (Butterfield et al., 2006) provides an interesting method for pacing out the amount of rangeland that a typical grazing animal needs in 1 day and extrapolating that to the amount of rangeland needed by all the animals.

Multispecies herds of livestock can be managed well and simultaneously benefit the pastures, since each species has its own order of preference of grazeable plant species. Each species of animal can be grazed as a segregated herd or all can be grazed together. If the pasture manager is doing a very good job and the herds are put into pasture at its ideal stage of quality and yield, both the domestic and any wild herbivores should benefit.

Implementing mixed species grazing requires observation, measurement, and analysis of changing resources. Records should be kept of pasture species proportions, condition, and height of forages at the start and end of grazing periods. Reviewing pasture records helps managers in making predictions and management decisions.

Estimating yield (see Chapter 7) of forage at the start and end of each grazing period will enable managers to determine whether adding additional animal species is beneficial or not. While availability and yield of different forage species changes through seasons and according to weather, any changes in animal numbers, diversity, and stocking density will affect each forage species uniquely. It is not possible to predict exactly how the interaction of several foraging species of herbivores will affect total forage demand (Walker, 1994), but pasture managers can become better at these predictions with experience.

On public rangelands in the United States and perhaps other countries, movement of wild herbivores through leased grazing land is unpredictable and bound to have significant impacts on plant biomass. Public pressure to maintain significant populations of wild animals on range areas of the western United States, for viewing by tourists and for hunting, can interfere with careful plans for grazing management. Van Dyne et al. (1984) reviewed problems in planning useful and defensible forage allocation procedures in such situations. Thus, adding just one other grazing or browsing species to horses on land with only a few plant species will provide plenty to think about and more tasks than many horse managers may be willing to tackle.

Extra management tasks suggested by Smith Thomas (2010) for MSG include adding an extra electric wire to fences, then ensuring water troughs are low enough for sheep or goats or adding stepping blocks around them. If one species of animal bullies another, rotate one species ahead of another. If you do not own sheep or goats but see an opportunity to benefit from them, rent or lease pasture to someone who owns them. Pasture rental rates for the United States are published for each state and county by the USDA National Agricultural Statistics Service (N.A.S.S., 2016). The online newsletter Onpasture.com frequently publishes a summary of these. If sheep graze with cattle, do not use a mineral mix that contains copper (since copper is far more toxic to sheep than to other species). Do not use any mineral supplements designed for ruminants that could contain additives called ionophores (lasalocid, monensin, or rumensin) if horses can gain access to these supplements, as they are toxic to horses.

Mixed grazing, in which all species are together in the same paddocks, promotes increased animal performance and output per acre (ac) or hectare (Nolan and Connolly, 1989), reduced gastrointestinal parasite burden, diversified animal production, and manipulation of botanical composition (Lambert and Guerin, 1989). With mixed sheep and cattle, animal output/ac (ha) from ryegrass pasture can be increased about 10% (Nolan and Connolly, 1989). However, not all research showed such an advantage (Mayne et al., 2000).

Sequential grazing, in which one species follows another (e.g., when grass heights and pasture quality are too low for cows, sheep graze), allows specific management needs for each species to be applied individually. Examples of management needs include breeding, birthing, vaccinating, pregnancy diagnosis, health checking, and parasite control. In some situations, sheep prefer white clover more than cows or horses do, so a higher clover percentage develops during and after grazing by cattle or horses. (Once taller grass is grazed off, clover is exposed to more sunlight, so its growth rate increases). If market lambs follow horses or cattle in a pasture rotation, this higher clover content can increase their gains by about 30 g/day each (Mayne et al., 2000).

If sheep or beef cows are to be used to improve the pasture for horses, they should be managed both economically and humanely. The Center for Farm Financial Management at the University of Minnesota produced annual labor estimates in 2010 for livestock herds under 50 head, including 11.7 h per head for beef cattle and 5 h per head for sheep (University of Minnesota, 2010). These low costs require disease to be kept at an absolute minimum. This requires a closed-flock policy, so new ruminant animals are purchased very rarely and always from the same source (of high health status). Other costs also need to be kept low. All of the grazing animals should be well trained at an early age to electric fences (Nation, 1995) and to handling systems. An economic analysis of a horse boarding facility in Maryland (Kays and Drohan, 2003) indicated a labor need of 21−39 h of work per horse per month, so 252−468 h per horse per year. Thus the labor requirement for management of meat livestock is very different from the labor requirement for horses, and this may cause some conflict in management discussions.

Income from these sheep, goats, or beef cattle can come from the sale of young, freshly weaned animals or from grass-finished older animals. Fine wool or goat fiber (mohair or cashmere) markets can also be considered if the animals have the appropriate genotypes for premium quality fiber and the horse farm manager has easy, economical access to the expertise for shearing and marketing the fiber. If the pastures are high in legumes or other forbs but the manager does not want to be bothered with wool, then hair sheep, which have no wool, can be used. Llamas and alpacas can also be grazed with horses, although grazing research information on them is scarce.

Pigs and poultry can also be raised on pasture and their meat and egg products can bring premium prices at farmers' markets and through on-farm sales. Fence wires need to be at appropriate heights above ground for the type and size of animals. Pigs can help to break parasite cycles, eat and even kill woody plants, eat acorns or other nuts, plus scavenge orchard fruit and sweet potatoes. Table scraps and byproducts from vegetable production can help pigs to thrive on pasture. Pigs can complement cows, especially if the pasture has a high legume percentage, but for reasonable weight gain or pregnancy or lactation, some grain will need to be fed to pigs. Pigs will eat cow manure and obtain vitamins and minerals from it (Nation, 1995).

Pastured poultry production has been increasing in popularity in North America since the 1980s. Chickens, ducks, and turkeys can be used to reduce the numbers of flies and internal parasites. Good sources of information on pasture raising of pigs and poultry can be obtained through the online newsletter OnPasture.com and the magazine *The Stockman Grass Farmer* (SGF), plus books by the SGF editor, Joel Salatin. His books have helped many farmers diversify their operations and improve efficiency of overall farm production by adding poultry or pigs to a rotational grazing system. One Joel Salatin innovation is the "eggmobile," a 12 × 20 foot (3.6 × 6.1 m) trailer with a mesh floor serving as a henhouse for 100 laying hens (Salatin, 1993). Usually the eggmobiles follow cattle in a pasture rotation (Nation, 1995) and could just as easily follow horses.

MIXED SPECIES GRAZING IN ACTION

The following list of tasks and actions can aid in managing mixed species grazing on a horse farm.

- Learn to identify forages, weeds, and toxic plants.
- Determine whether there is enough diversity in plants to provide opportunities for grazing and/or browsing with animals in addition to horses.
- Plan to add one extra animal species at a time, ensuring that the primary plant species on the land match the grazing preferences of the new animal species.
  - If mostly grasses, use cattle, sheep, or bison.
  - If many broadleaf plants (forbs, including legumes), use sheep or goats.
  - If many woody plants, use goats, sheep, deer, or pronghorns.
  - If elevation is high, use sheep or goats.
  - If elevation is low, use cattle, sheep, or goats.
- Measure plant heights and calculate forage yield. See Chapter 7 and Appendix 2.
- Calculate feed requirements of all expected animals, allowing at least 2.5% of body weight as grazeable feed per animal per day. For lactating or rapidly growing animals, allow up to 3.5% of body weight per animal per day.
- Calculate stocking rates in terms of acres or hectares needed per animal for the whole grazing season. See Chapter 9. Consider these stocking rates as carrying capacity of the land at the current forage yields and harvest efficiency.
- Make a yearly feed budget (including grazed and preserved forages) to ensure enough feed for all animals. Make adjustments either on a regular basis or whenever changes occur in animal needs or in the predicted supply.
- Use a rotational grazing system with paddocks small enough to allow most of the plants to be grazed to half their "start grazing" height within 1 to 5 days. Prevent animals from taking a second bite of plants during the same grazing period. If a permanent fence is lacking, use temporary electric fences. Plan for enough paddocks so that plants have enough rest and regrowth time to reach ideal grazing heights before being grazed again. This may be as short as 14 days in spring and longer than 80 days in a dry climate or drought condition. In the driest climates, some areas should only be grazed once per year.
- If some plant species are not grazed or only slightly grazed, take advantage of this opportunity to do a follow-up grazing with another species of animal right away.
- After a paddock is grazed by two or more species, if the vegetation is uneven or if there are many tall stems or seed heads present, mow the remaining vegetation to a height of about 4 inches (10 cm).
- If manure is not removed from paddocks routinely, harrow the manure after clipping the grazed forage, preferably on a hot, sunny day to spread organic matter and nutrients and to expose parasite larvae and eggs to drying and ultraviolet light.

- Watch for development of latrines or a lawn and rough pattern. This indicates that horses have been on a paddock too long and have overgrazed the lawn areas. Prevent this pattern by shortening the grazing periods and increasing the stock density (number of animals/ac at one time). You may need to make paddocks smaller with temporary electric fence.
- If you only have a few animals of each species, it may be best to accustom them to each other in confinement before the grazing season, and then graze them as one herd. If you have enough horses for one natural-sized band (about 5 to 15), consider a leader-follower system in which the animals with the needs for the highest nutrient concentrations graze an area first, followed by a separate herd of animals with lower nutrient needs. Siciliano et al. (2017) suggested a leader-follower system to help lower the nonstructural carbohydrate concentration of forages grazed by horses.
- If weeds are a significant problem, train cattle, sheep, or goats to eat the weeds, using Kathy Voth's method, and see her articles at http://onpasture.com (Voth, 2004, 2014, 2015).
- Consult a veterinarian on control of parasites. Be prepared to collect fecal samples and either submit them to a lab or learn to analyze them through a fecal egg test. Follow the veterinarian's instructions, including dosage and timing, for use of anthelmintic drugs.
- Nonchemical methods that assist parasite control include these:
  - Alternate use of paddocks between grazing one year and haying the next.
  - Remove feces weekly (if paddocks are small and close to a manure storage facility). On larger paddocks that have been mowed following grazing, use a harrow or drag to break up and spread manure on a hot, sunny day.
  - Continue to monitor parasite egg numbers in manure.
  - Avoid having the youngest animals following slightly older animals or yearlings.
- Observe the condition of animals and plants frequently and whenever there are disruptions or significant environmental changes, including seasons, harvests, and major weather events. Learn how to body condition score all of the animals that you manage. Take pictures of pastures and animals and file them in a record system with dates and locations to help with analysis of the grazing season and planning.
- Pasture management records should provide information that helps you recall what happened in response to your management. Record predominant species in pastures at the start, middle, and end of grazing seasons, weather changes, dates of animal movements, calculated forage yields, and dates and number of days that each paddock was grazed and rested. Keep an inventory of animals grazed and enter changes as soon as they happen. Record which paddock was used first at the start of grazing and rotate this year to year.
- Install fences with wires or rails at heights appropriate for each type of livestock. Install drinking water sources where animals do not need to expend too much time or energy to get to them. Permanent waterers are good value in the long term. Hoses and small portable troughs allow water to be available in every paddock. Make adjustments to waterer height so that animals of all sizes can access them. Train all new animals to use waterers and respect fences.
- Prevent bullying by introducing animals to each other before they go onto large pastures, initially across a fence, then in uncrowded situations with places for subordinate animals to hide, evade, or escape dominant animals.
- If you recognize a need for an animal species and are not able or willing to buy them initially, try to rent animals or rent grazing land to owners of ruminants, with yourself as the manager.
- Be careful with distribution of any feed supplements. Medications designed for ruminants to improve feed efficiency and control coccidia, called ionophores (monensin, rumensin, lasalocid, etc.), are highly toxic to horses. Ensure that no mineral mixes designed for cattle, goats or poultry can be eaten by sheep because the copper levels in these are toxic to sheep.
- Take advantage of plant properties. For example, if a mixed stand of grasses and clovers is grazed by cattle or horses, removal of the grasses provides the shorter clovers with more sunlight, and their growth increases rapidly. This provides an opportunity in a leader-follower system for young sheep to grow rapidly and for lactating sheep or goats to produce milk by eating the clovers that they find very palatable.
- Manage both economically and humanely. Develop realistic expectations for extra labor required for additional animal species. Ensure that staff are knowledgeable, skilled, and enjoying their work. Perform economic analysis of proposals to introduce new species, using tools such as partial budgets and help from extension agents.
- Protect small ruminants and young horses from predators with a blend of effective predator management techniques.
- Explore opportunities for income from new animal species, such as breeding stock sales, wool, cashmere, mohair, and milk. Ensure that such markets are attainable and have potential for growth. Consider pastured pigs, poultry, llamas, and alpacas.
- Communicate your plans and efforts to neighbors. Ask for and respect their advice in areas where they have experience and concerns.
- Keep active in lifelong learning. Meet with USDA and university agricultural extension staff in the United States and consult provincial agricultural extension staff in Canada.

WHAT RESEARCH STILL NEEDS TO BE DONE?

In a review of equine grazing research methodologies, Martinson et al. (2017) listed four research methods that can be useful in mixed species grazing enquiries: evaluating grazing systems, forage intake assessment, grazing behavior analysis, and tracking health parameters. Since there are so many variables affecting relative yields of forages subjected to different grazing pressures and environmental conditions, it is not practical to do experiments to generate exact numbers of sheep, goats, and cattle that should be placed with horse herds of a given size. However, for each major climatic region of the United States, it is useful to know what happens in pastures containing the most common few grasses and legumes for that region, when grazed by horses and when small numbers of sheep, goats, and cattle are added. Plant-related factors to measure can include total daily intake of forages and proportions by plant species. Animal-related factors to measure include grazing behaviors (hours spent grazing and browsing, species preferences, sward height preferences), health parameters such as changes in body weight, body condition score, backfat thickness, and concentrations of glucose and insulin. Variables should include different stocking rates of horses along with small numbers of sheep, cattle, goats, or deer per horse. Standardized research methodologies for such experiments were proposed by Martinson et al. (2017).

It is clear from many of the reports summarized earlier that heavy dependence on frequent anthelmintic treatment of horses has contributed to widespread resistance of parasites to avermectin and other drugs. Hygienic procedures on horse farms can be valuable in preventing worm burdens from rising to dangerous levels. Much more needs to be learned about the quantitative efficacies of reducing the risk of parasite transmission via various management procedures in pastures, paddocks, and stalls (Nielsen, 2016; Reinemeyer and Nielsen, 2017). Fecal egg counts can be performed during studies of relative stocking rates of mixed animal species. This can provide information that will help in the design of experiments to test mixed versus leader-follower grazing and to test effects on horse parasite burdens of one or more grazing rotations by sheep, goats, or cattle before horses rotate back to a paddock.

REVIEW QUESTIONS

1. Name three species of domestic livestock and three species of wild herbivores that should be considered or accounted for in a mixed species grazing scenario on horse farms in North America.
2. What benefits can mixed species grazing provide?
3. Explain how mixed species grazing can cause less loss of pasture land near dung piles.
4. How can mixed species grazing help to reduce parasite infestations? Explain the process and make suggestions for how this can be managed.
5. List some small nontarget organisms that can be harmed by residues of anthelmintics in horse manure.
6. Describe recommended parasite control practices that integrate well with multispecies grazing and help to reduce ecotoxicity from anthelmintic residues.
7. Describe how multispecies grazing aids in the control of weeds in pastures.
8. Discuss why and how multispecies grazing helps to improve grazing management.
9. Describe how to manage multispecies grazing.
10. Explain some potential disadvantages to mixed species grazing on a horse farm.
11. How is information about nonequine species useful when managing horses?
12. List economic considerations of and potential economic consequences of mixed species grazing.
13. Recommend how to manage economic issues of adding additional grazing species.

REFERENCES

Abaye AO, Allen BG, Fontenot JP. Influence of grazing cattle and sheep together and separately on animal performance and forage quality. J. Anim. Sci. 1994;72(4):1013–22.

Animut G, Goetsch AL. Co-grazing of sheep and goats: benefits and constraints. Small Rumin. Res. 2008;77:127–45.

Animut G, Goetsch AL, Aiken GE, Puchala R, Detweiler G, Krehbiel CR, Merkel RC, Sahlu T, Dawson LJ, Johnson ZB, Gipson TA. Performance and forage selectivity of sheep and goats co-grazing grass/forb pastures at three stocking rates. Small Rumin. Res. 2005;59:203–15.

Baker FH. Multispecies grazing: the state of the science. Rangelands 1985;7(6):266–9.

Bowns JE. Common use: better for cattle, sheep and rangelands. Utah Sci. 1989;50(2):117–23.

Butterfield J, Bingham S, Savory A. Holistic Management Handbook: Healthy Land, Healthy Profits. Washington, D.C.: Island Press; 2006. p. 74–90.

Campbell NA, Reece JB. Biology. seventh ed. San Francisco: Pearson Education Inc; 2005.

Coblentz BE. Some range relationships of feral goats on Santa Catalina Island, California. J. Range Manag. 1977;30:415–9.

Demment MW, Van Soest PJ. A nutritional explanation for body size patterns of ruminant and non-ruminant herbivores. Am. Nat. 1985;125:641–72.

Distel RA, Laca EA, Griggs TC, Demment MW. Patch selection by cattle: maximization of intake rate in horizontally heterogeneous pastures. Appl. Anim. Behav. Sci. 1995;45:11−21.

Eysker M, Jansen J, Mirck MH. Control of strongylosis in horses by alternate grazing of horses and sheep and some other aspects of the epidemiology of strongylidae infections. Vet. Parasitol. 1986;19:103−15.

Fox MT. Gastrointestinal Parasites of Cattle. Merck Veterinary Manual; 2017a. http://www.merckvetmanual.com/digestive-system/gastrointestinal-parasites-of-ruminants/gastrointestinal-parasites-of-cattle.

Fox MT. Gastrointestinal Parasites of Sheep and Goats. Merck Veterinary Manual; 2017b. http://www.merckvetmanual.com/digestive-system/gastrointestinal-parasites-of-ruminants/gastrointestinal-parasites-of-sheep-and-goats.

Garric J, Vollat B, Duis K, Péry A, Junker T, Ramil M, Fink G, Ternes TA. Effects of the parasiticide ivermectin on the cladoceran *Daphnia magna* and the green alga *Pseudokirchneriella subcapitata*. Chemosphere 2007;69:903−10.

Gerrish J. Management-intensive Grazing − the Grassroots of Grass Farming. Ridgeland, Mississippi: Green Park Press; 2004. p. 142−6.

Glimp HA. Multi-species grazing and marketing. Rangelands 1988;10(6):276−8.

Grant SA, Suckling DE, Smith HK, Torvell L, Forbes TDA, Hodgson J. Comparative studies of diet selection by sheep and cattle: the hill grasslands. J. Ecol. 1985;73:987−1004.

Hansen R, Clark RC, Lawthorn W. Food of wild horses, deer and cattle in the Douglas Mountain area, Colorado. J. Range Manag. 1977;30:116−9.

Herd RP, Miller TB, Gabel AA. A field evaluation of seven equine anthelmintics including pro-benzimidazole, benzimidazole, and non-benzimidazole drugs. J. Am. Vet. Med. Assoc. 1981;179:686−91.

Herd RP, Stinner BR, Purrington FF. Dung dispersal and grazing area following treatment of horses with a single dose of ivermectin. Vet. Parasitol. 1993;48:229−40.

Herd RP. Epidemiology and control of equine parasites in northern temperate regions. Vet. Clin. N. Am. 1986;2:337−55.

Herd RP. Control strategies for ruminant and equine parasites to counter resistance, encystment and ecotoxicity in the USA. Vet. Parasitol. 1993;48:327−36.

Herd R. Endectocidal drugs: ecological risks and counter-measures. Int. J. Parasitol. 1995;25(8):875−85.

Holecheck JL, Jeffers J, Stephenson T, Kuykendall CB, Butler-Lance SA. Cattle and sheep diets on low elevation winter range in northcentral New Mexico. Proc. West. Sec. Am. Soc. Anim. Sci. 1986;37:243−8.

Holechek JL, Pieper RD, Herbel CH. Range Management Principles and Practices. sixth ed. Upper Saddle River, New Jersey: Prentice Hall; 2010. p. 165−7.

Huston JE. A loud shout for sheep! Rangeman's J. 1975;2(5):136−7.

Illius AW, Clark DA, Hodgson J. Discrimination and patch choice by sheep grazing grass-clover swards. J. Anim. Ecol. 1992;61:183−94.

Kays JS, Drohan JR. Horse Boarding Enterprise. RES-10 Rural Enterprise Series. Western Maryland Research & Education Center, Maryland Cooperative Extension and Agricultural Experiment Station; 2003. https://extension.umd.edu/sites/extension.umd.edu/files/_docs/programs/woodland-steward/RES_10HorseBoarding.pdf.

Klei TR. Large Strongyles in Horses. Merck Veterinary Manual; 2017a. http://www.merckvetmanual.com/digestive-system/gastrointestinal-parasites-of-horses/large-strongyles-in-horses.

Klei TR. Small Strongyles in Horses. Merck Veterinary Manual; 2017b. http://www.merckbetmanual.com/digestive-system/gastrointestinal-parasites-of-horses/small-strongyles-in-horses.

Lambert MG, Guerin H. Competitive and complementary effects with different species of herbivore in their utilization of pastures. In: Proceedings of the XVI International Grassland Congress, Nice; 1989. p. 1785−9.

Lane TJ. Parasite Control in Horses. Merck Veterinary Manual; 2017. http://www.merckvetmanual.com/management-and-nutrition/health-management-interaction-horses/parasite-control-in-horses.

Leaver JD. A comparison of grazing systems for dairy herd replacements. J. Agr. Sci. 1970;75:265−72.

Lumaret JP, Galante E, Lumbreras C, Mena J, Bertrand M, Bernal JL, Cooper JF, Kadiri N, Crowe D. Field effects of ivermectin residues on dung beetles. J. Appl. Ecol. 1993;30:428−36.

Martinson KL, Siciliano PD, Sheaffer CC, McIntosh BJ, Swinker AM, Williams CA. A review of equine grazing research methodologies. J. Equine Vet. Sci. 2017;51:92−104.

Mayne CS, Wright IA, Fisher GEJ. Grassland management under grazing and animal response. 2000. In: Grass its Production and Utilization. third ed. Oxford, England: Blackwell Science; 2000. p. 247−91.

Merrill LB, Thomas GW, Hardy WT, Keng EB, Rechenthin CA, Langford DC, Booker TA, Johnson N, Teer JG, Pederson RJ, Tatum JE. Livestock and deer ratios for Texas range lands. Tex. Agric. Expt. Sta. Misc. Pub. 1957;221:9.

Miller R. Habitat use of feral horses and cattle in Wyoming's red desert. J. Range Manag. 1983;36(2):195−9.

Mosley JC. Influence of social dominance on habitat selection by free-ranging ungulates. Grazing Behavior of Livestock and Wildlife. Idaho Forest, Wildlife and Range Expt. Sta. Bull, vol. 70; 1999. p. 109−18.

Murphy. Greener Pastures on Your Side of the Fence, Better Farming with Voisin Management Intensive Grazing. fourth ed. Colchester, Vermont: Arriba Publishing; 1998. p. 40−1.

Nation A. Quality Pasture How to Create it, Manage it and Profit from it. Ridgeland, Mississippi: Green Park Press; 1995. p. 249−64.

National Agricultural Statistics Service. Agricultural Land, Land Values and Cash Rents. 2016. No. 2016-9. October 2016, https://www.nass.usda.gov/Publications/Highlights/2016_LandValues_CashRents/2016LandValuesCashRents_Highlights.pdf.

Nessel RJ, Jacob TA, Robertson RT. The human and environmental safety aspects of ivermectin. Proceedings MSD Agvet Symposium on Recent Developments in the Control of Animal Parasites. In: Association with 22nd World Veterinary Congress, Perth, Australia; 1983. p. 98—108.

Newman JA, Parsons AJ, Harvey A. Not all sheep prefer clover: diet selection revisited. J. Agr. Sci. Camb. 1992;119:275—83.

Nielsen MK. Universal challenges for parasite control: a perspective from equine parasitology. Trends Parasitol. 2015;31(7):282—4.

Nielsen MK. Evidence-based considerations for control of *Parascaris spp.* Infections in horses. Equine Vet. Educ. 2016;28:224—31.

Nielsen MK. Personal communication. 2017.

Nolan T, Connolly J. Mixed versus mono grazing of steers and sheep. Anim. Prod. 1989;48:519—33.

Olson KC, Wiedmeier RD, Bowns JE, Hurst RL. Livestock response to multi-species and deferred-rotation grazing on forested rangeland. J. Range Manag. 1999;52(5):462—70.

Parsons AJ, Newman JA, Penning PD, Harvey A, Orr RJ. Diet preference of sheep: effects of recent diet, physiological state and species abundance. J. Anim. Ecol. 1994;63(2):465—78.

Penning PD, Newman JA, Parsons AJ, Harvey A, Orr RJ. The preference of adult sheep and goats grazing ryegrass and white clover. Anim. Res. 1995;44(suppl):113.

Peregrine AS, Molento MB, Kaplan RM, Nielsen MK. Anthelmintic resistance in important parasites of horses: Does it really matter? Veterinary Parasitology 2014;201:1—8.

Plumb GE, Krysl LJ, Hubbert ME, Smith MA, Waggoner JW. Horses and cattle grazing on the Wyoming red desert, III. J. Range Manag. 1984;37(2):130—2.

Reinemeyer CR, Nielsen MK. Control of helminth parasites in juvenile horses. Equine Vet. Educ. 2017;29:225—32.

Reynolds PJ, Bond J, Carlson GE, Jackson Jr C, Hart RH, Lindahl IL. Co-grazing of sheep and cattle on an orchardgrass sward. Agron. J. 1971;63(4):533—6.

Rook AJ. Principles of foraging and grazing behaviour. In: Hopkins A, editor. Grass its Production & Utilization. third ed. Oxford: Blackwell Science; 2000. p. 229—46.

Russell PJ, Hertz PE, McMillan B, Fenton B, Addy H, Maxwell D, Haffie T, Milsom B. Biology: Exploring the Diversity of Life. Second Canadian Edition. Toronto: Nelson Education Ltd.; 2013.

Rutter SM, Orr RJ, Yarrow NH, Champion RA. Dietary preference of dairy heifers grazing ryegrass and white clover, with and without an anti-bloat treatment. Appl. Anim. Behav. Sci. 2004a;85:1—10.

Rutter SM, Orr RJ, Yarrow NH, Champion RA. Dietary preference of dairy cows grazing ryegrass and white clover. J. Dairy Sci. 2004b;87:1317—24.

Rutter SM. Diet preference for grass and legumes in free-ranging domestic sheep and cattle: current theory and future application. Appl. Anim. Behav. Sci. 2006;97:17—35.

Rutter SM. Grazing preferences in sheep and cattle: implications for production, the environment and animal welfare. Can. J. Anim. Sci. 2010;90(3):285—93.

Ruyle GB, Bowns JE. Forage use by cattle and sheep grazing separately and together on summer range in southwestern Utah. J. Range Manag. 1985;38(4):299—302.

Salatin J. Pastured Poultry Profits. White River Junction: Chelsea Green Publishing; 1993. VT 05001.

Salter RE, Hudson RJ. Range relationships of feral horses with wild ungulates and cattle in western Alberta. J. Range Manag. 1980;33(4):266—71.

Sanderson H, Laird B, Pope L, Brain R, Wilson C, Johnson D, Bryning G, Peregrine AS, Boxall A, Solomon K. Assessment of the environmental fate and effects of ivermectin in aquatic mesocosms. Aquat. Toxicol. 2007;85:229—40.

Savory A. Holistic Resource Management. Washington, D.C: Island Press; 1988.

Siciliano PD, Gill JC, Bowman MA. Effect of sward height on pasture nonstructural carbohydrate concentrations and blood glucose/insulin profiles in grazing horses. J. Equine Vet. Sci 2017;57:29—34.

Sinclair J. Code of Agriculture. third ed. 1821 (London).

Smith Thomas H. Advantages of Multi-species Grazing. Canadian Cattlemen. 2010. Oct. 18, 2010, https://www.canadiancattlemen.ca/2010/10/18/advantages-of-multispecies-grazing/(.

Smith Thomas H. Pasture Management for Parasite Control. 2014. The Horse: Your Guide to Equine Health Care magazine/TheHorse.com. March 5, 2014, http:www.thehorse.com/articles/33480/pasture-management-for-parasite-control.

Smith Thomas H. Multi-species grazing benefits Oregon grazier. Stockman Grass Farmer 2017;17(4):15—6.

Sowell BF, Krysl LJ, Hubbert ME, Plumb GE, Jewett TK, Smith MA, Applegate SL, Waggoner JW. Wyoming wild horse and cattle grazing research. Rangelands 1983;5(6):259—62.

Stephenson TE, Holechek JL, Kuykendall CB. Diets of four wild ungulates on winter range in northcentral New Mexico. Southwest. Nat. 1985;30(3):437—41.

Taylor Jr CA. Multispecies grazing — vegetation manipulation. Tex. Agric. Expt. Sta. Prog. Rep. 1986;4426. 2p.

Torres-Rodriguez A, Cosgrove GP, Hodson J, Anderson CB. Cattle diet preference and species selection as influenced by availability. Proc. N. Z. Soc. Anim. Prod. 1997;57:197—8.

University of Minnesota. Labor Hour Estimates. Center for Farm Financial Management; 2010. http://www.CFFM.umn.edu/FINPACK.

Vallentine JF. Introduction to grazing. In: Grazing Management. second ed. San Diego: Academic Press; 2001a. p. 1—27.

Vallentine JF. Grazing herbivore nutrition. In: Grazing Management. second ed. San Diego: Academic Press; 2001b. p. 37—8.

Vallentine JF. Kind and mix of grazing animals. In: Grazing Management. second ed. San Diego: Academic Press; 2001c. p. 324—33.

Vallentine JF. Grazing animal intake and equivalence. In: Grazing Management. second ed. San Diego: Academic Press; 2001d. p. 370—3.

Van Dyne GM, Burch W, Fairfax SK, Huey W. Forage allocation on arid and semi-arid public grazing lands: summary and recommendations. In: Natl. Res. Council/Natl. Acad. Sci. "Developing Strategies for Rangeland Management." Boulder, CO: Westview Press; 1984. p. 1–25.

Voisin A. The cow is a gourmet. In: Grass Productivity. Island Press Edition 1988. Washington, D.C: Island Press; 1959a. p. 99–116.

Voisin A. Tether grazing. In: Productivity G, editor. Island Press Edition 1988. Washington, D.C: Island Press; 1959b. p. 209–12.

Voth K. Cows Eat Weeds. 2004. Self-published, http://www.livestockforlandscapes.com/cowseatweedsbook.htm.

Voth K. What Should I Do about My Pasture Weeds? on Pasture. 2014. http://onpasture.com/2014/03/24/what-should-i-do-about-my-pasture-weeds/.

Voth K. 43% Better – the Economics of Getting Your Livestock to Eat Weeds. On Pasture. 2015. http://onpasture.com/2015/04/13/-43-better-the-economics-of–getting-your-livestock-to-eat-weeds.

Walker JW. Multispecies grazing: the ecological advantage. Sheep Res. J. 1994;10(Spec. Issue):52–64.

Wilkins RJ. Grass utilization for animal production. In: soil-Grassland-Animal Relationships. In: Proc. 13th General Meeting of the European Grassland Federation, Banska Bystrica, June 1980, vol. 1; 1990. p. 22–34.

Woodward KN. Veterinary pharmacovigilance. Part 3. Adverse effects of veterinary medicinal products in animals and on the environment. J. Vet. Pharmacol. Ther. 2005;28:171–84.

Chapter 11

Production and Management of Hay and Haylage

Jimmy Henning[1] and Laurie Lawrence[2]

[1]*Department of Plant and Soil Sciences, College of Agriculture, Food and Environment, University of Kentucky, Science Center North, Lexington, KY, United States;* [2]*Department of Animal and Food Sciences, College of Agriculture, Food and Environment, University of Kentucky, Garrigus Building, Lexington, KY, United States*

INTRODUCTION

Hay is one of the most versatile of feeds for horses because it can be stored for long periods with little loss of nutrients, made from many different species, produced and fed in large or small quantities. Additionally, hay can supply most or all of the nutrients for many horses, and its long fiber has a positive effect on the gastrointestinal health of horses.

Hay is used when pasture or other forage is not available. Depending upon management a horse may receive all of its yearly forage as hay or as a combination of hay and pasture. In general, daily hay allowances will range from 1% to 2% of body weight daily, depending on the nutrient needs of the horse and the use of complementary feeds. The population of horses in North America is very large, with Canada, the United States, and Mexico totaling more than 13 million head (Table 11.1). Using an estimate of 1.5−2.0 tons of hay/horse year, these animals will consume approximately 25−35 million tons of hay per year. There is great opportunity to produce, sell, and buy superior hay for horses.

WHAT IS QUALITY HAY?

The ultimate definition of the quality of hay is how well it supports the nutritional needs and health of the horse. The nutritional value of hay is determined by many factors including species, stage of maturity, soil fertility, and growing conditions. Quality hay should be free from dirt and other contaminants. Hay should also be free from dust and mold, as well as toxic plants/weeds and any materials that could cause physical injury.

FACTORS AFFECTING THE NUTRIENT VALUE OF HAY

Plant Species

Many different forage species can be used to produce quality horse hay, and the preference and availability is often regional. In the southern United States, many horse owners commonly feed Bermudagrass for hay compared to orchardgrass or timothy in more northern regions.

In general, legumes are higher in nutritional quality than grasses harvested at comparable stages of maturity. Because they are more nutrient dense, legumes can often be fed with a lower level of supplementation (concentrate) than grasses. Legumes like alfalfa typically have less total cell wall, corresponding to neutral detergent fiber (NDF) values on forage analysis reports. Legumes have more of their dry matter in the more digestible cell contents. These characteristics contribute to a higher digestibility and higher energy content. Legumes have the additional advantage of fixing their own nitrogen (N), while grasses cannot; this characteristic contributes to the higher crude protein concentrations observed in legumes (Table 11.2). Legumes will have higher levels of calcium relative to grasses. Pure legume hays will have wide Ca:P ratios (4:1 or 5:1); grasses will have ratios closer to 1:1.

Alfalfa is the dominant legume for horse hay and is grown in virtually all of North America. Alfalfa is high in protein and energy and is readily consumed by horses. This perennial legume is high yielding and provides multiple cuttings

Horse Pasture Management. **https://doi.org/10.1016/B978-0-12-812919-7.00011-1**

TABLE 11.1 Horse Numbers in North America

Country	Number (Millions)
Canada[a]	0.3
Mexico[b]	6.4
United States[c]	7.2
Total	13.9

[a]2016. Statistics Canada. Other livestock and poultry in Canada. http://www.statcan.gc.ca/pub/96-325-x/2017001/article/54874-eng.htm.
[b]2016. FAOSTAT database. Food and Agriculture Organization. http://www.fao.org/faostat/en/#data/QA.
[c]2017. 2017 Economic impact study of the US horse industry. American Horse Council. http://www.horsecouncil.org/product/2017-economic-impact-study-u-s-horse-industry/.

TABLE 11.2 Typical Nutrient Content of Hays Fed to Horses (as Fed Basis)

Hay Variety	Digestible Energy (Mcal/lb)	Total Digestible Nutrients (%)	Crude Protein (%)	Calcium (%)	Phosphorus (%)
Alfalfa	0.8 to 1.1	48 to 55	15 to 20	0.9 to 1.5	0.2 to 0.35
Red Clover	0.8 to 1.1	46 to 52	13 to 16	0.8 to 1.5	0.2 to 0.35
Orchardgrass	0.7 to 1.0	42 to 50	7 to 11	0.3 to 0.5	0.2 to 0.35
Timothy	0.7 to 1.0	42 to 50	7 to 11	0.3 to 0.5	0.2 to 0.35
Bermudagrass	0.7 to 1.0	42 to 50	6 to 11	0.25 to 0.4	0.015 to 0.03
Tall Fescue	0.6 to 0.9	40 to 48	5 to 9	0.3 to 0.5	0.02 to 0.035

Based on National Research Council, 1989. Nutrient Requirements of Horses. National Academy Press, Washington.

throughout the growing season, spreading the risk of rain damage. Alfalfa can be grown with cool season grasses like orchardgrass or timothy. These grasses will aid in curing and the binary mix is often preferred by horse owners over pure alfalfa.

Many varieties of grasses are acceptable for horse hay, including Bermudagrass, orchardgrass, timothy, bromegrass, and others. Forage grasses are classified as either cool season or warm season depending on their photosynthetic chemistry and temperature optima. Cool season grasses such as the small grains, orchardgrass, and timothy will generally have higher crude protein and lower total fiber values than warm season species at the same stage of maturity, but the amount of protein in grasses can be readily influenced by nitrogen fertilization. Both warm and cool season grasses can be used successfully for horse hay, with higher quality material being produced when the plants are harvested in early maturity.

Stage of Maturity

The biggest variable affecting hay quality and nutrient content within a type of hay is the stage of maturity at harvest. Very early maturity hay often has a soft texture, is very leafy, and has a high nutrient density and palatability. Plants harvested in early maturity are cut soon after the seedheads emerge (grasses) or before the plant begins to bloom (legumes). As the plant matures, the amount of fibrous stem material increases. Plants harvested in late maturity will have seed heads, coarse, thick stems, and less leaf than plants harvested in early maturity. The more mature the plant is at the time of harvest, the lower the nutrient concentrations, digestibility, and palatability.

The challenge in hay production is harvesting at a stage that produces economical yields without causing forage quality to fall below acceptable levels. Hay yields (tons/acre) rise with advancing maturity, but desirable quality measures decline (Fig. 11.1). In general, the optimum stage for forage harvest is just as the crop changes from vegetative to reproductive (Table 11.3). In most cases, the stage that allows acceptable yields and quality corresponds to first flower (pre- or early bloom) in legumes like alfalfa and boot to early head in forage grasses.

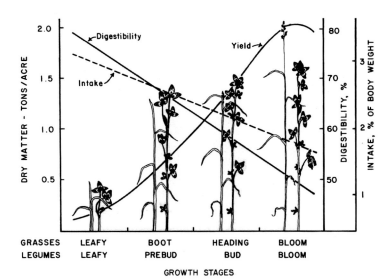

FIGURE 11.1 Effect of forage growth stage on yield, intake, and digestibility. *Used with permission Blaser, R.E., Hammes Jr., R.C., Fontenot, J.P., Bryant, H.T., Polan, C.E., Wolf, D.D., McGlaugherty, F.S., Kline, R.G., Moore, J.S., 1986. Forage-Animal Management Systems. Virginia Agricultural Experiment Station, Virginia Polytechnic Institute and State University, Blacksburg, Virginia, Bulletin 86–87.*

TABLE 11.3 Recommended Stages to Harvest Various Forage Crops for Hay or Haylage

Plant Species	Time of Harvest
Alfalfa	Late bud to first flower for first cutting, first flower to 1/10 bloom for second and later cuttings
Bluegrass, orchardgrass, tall fescue, or timothy	Boot[a] to early head stage for first cut, subsequent cuts at 4- to 6-week intervals as growth permits
Red clover or crimson clover	First flower to 1/10 bloom
Oats, barley, or wheat	Boot to early head stage
Rye and triticale	Boot stage or before
Soybeans	Mid- to full bloom and before bottom leaves begin to fall
Annual lespedeza	Early bloom and before bottom leaves begin to fall
Ladino clover or white clover	Cut at correct stage for companion plant
Bermudagrass	Cut when height is 15–18 inches
Caucasian bluestem	Boot to early head stage
Big bluestem, Indiangrass, and switchgrass	Early head stage

[a]*Boot is the stage of growth of a grass just prior to seedhead emergence. This stage can be identified by the presence of an enlarged or swollen area near the top of the main stem.*
Adapted from AGR 62 Quality hay production.

Visual and Physical Characteristics of High-Quality Hay

Visual characteristics of hay provide some indication of quality. Green color is a good but not perfect indicator of quality hay (Fig. 11.2). A bright, green color is an indication that the hay spent a minimum of time in the field after cutting and experienced little dew and sun bleaching. However, hay bleached in the windrow (a line of cut hay) or from exposure to sun while in the barn can still be nutritious even though the bales lack a bright green appearance. Conversely, if bales have a caramel or brown color in the center, it is an indication of excessive heating due to high moisture content at baling.

FIGURE 11.2   Color is an indicator of quality, but hay may be high quality but lack an external bright green color due to sun bleaching. This alfalfa hay was bleached from sun exposure at the edge of a barn. *Photograph by Jimmy Henning, University of Kentucky College of Agriculture, Food and Environment.*

Quality horse hay will have a fresh, clean smell. A musty or earthy odor is due to mold and/or dirt and is an indication that the hay was baled too wet or was contaminated with dirt during curing and baling. Finally, the most nutritious horse hay will be leafy, have minimal leaf shatter, and have a soft texture.

High-quality hay for horses will be free from known toxins. Toxicity problems for horses associated with forages include the following.

Ergot Alkaloids Associated With the Endophyte of Tall Fescue

Unimproved, established stands of tall fescue are likely to contain an endophytic fungus that produces ergot alkaloids that can cause prolonged gestations, agalactia, and foaling difficulty in mares (Ball et al., 2015). Hay made from infected tall fescue will contain lower concentrations of alkaloids than the standing crop, but it should still be avoided as the main forage for pregnant mares. Tall fescue is most commonly found in the transition zone in the United States and in the Northwestern United States, although it may be found in other locations as well. Tall fescue is coarser in texture than Kentucky bluegrass or Bermudagrass, tends to grow in clumps, and has pronounced ridges on leaves (Figs. 11.3 and 11.4). To avoid problems with tall fescue, pasture and hay fields should be inspected for its presence. Any tall fescue present can be tested to determine whether it is endophyte infected. Some ryegrasses may contain an endophyte, but its effect on horses is unlike that of infected tall fescue.

Glycosides

Sorghum species (johnsongrass, sudangrass, sorghum-sudangrass hybrids) contain glycosides that can cause muscle weakness, urinary tract failure, neural degeneration and, in extreme cases, death (Burrows and Tyrl, 2013).

Cantharadin

Blister beetles contain a toxin, cantharadin, which causes irritation to the digestive tract lining, and ingestion can be fatal to horses (Townsend, 2011). Blister beetles are floral feeders on weeds present in hay fields and are not a pest of alfalfa or

FIGURE 11.3   Tall fescue is a bunch grass with broader leaves than Kentucky bluegrass or Bermudagrass. *Photograph by Jimmy Henning, University of Kentucky College of Agriculture, Food and Environment.*

FIGURE 11.4   Leaves of tall fescue have very visible ridges or veins. *Photograph by Jimmy Henning, University of Kentucky College of Agriculture, Food and Environment.*

other forages. Blister beetles will leave the windrow after cutting if not crushed by the mower conditioner or wheels of the harvesting equipment. Good management (weed control, insect scouting, and control) is needed to ensure hay is free of blister beetles (Fig. 11.5).

Mold and Dust

Dusty and moldy hay should not be fed to horses. Dust or mold can cause respiratory inflammation. Plant material in hay bales provides nutrients for natural bacteria and fungi (including molds and yeasts) on their surfaces. When the moisture content is high enough, bacteria and fungi will grow. Mold growth along with soil on the plant surfaces and shattering of brittle forage contributes to dusty hay.

Mold and dust can aggravate a condition known as recurrent airway obstruction (RAO), which is a chronic lung disease. RAO is also a component of equine asthma and has been called heaves. RAO causes horses to have decreased respiratory function and thus exercise performance. Producing mold- and dust-free hay should be a primary consideration for the horse hay market.

Management to minimize blister beetle problems in alfalfa
• Cut alfalfa at 10% or less bloom • Manage weeds in and around fields, especially pastures. • Consider cropping practices adjacent to alfalfa. Blister beetles can be abundant in soybeans but are usually absent in corn. • Do not grow solanaceous crops near alfalfa – both black and stripped blister beetles can be abundant in tomatoes and potatoes • Sample field margins before cutting – blister beetles usually come from field margins and do not tend to move too far beyond the edges of alfalfa fields • Monitor pastures for grasshoppers. Several blister beetle species develop in grasshopper egg pods. A high grasshopper population can produce high blister beetle numbers the following year.

FIGURE 11.5 Management to minimize blister beetle problems in alfalfa. *Personal communication, Dr. Lee Townsend, University of Kentucky.*

Slaframine

Horses consuming red clover hay will sometimes salivate excessively, or slobber. This condition is caused by slaframine produced by growth of the blackpatch fungus (*Rhizoctonia leguminicola*) on the clover plants. Although unsightly, slobbers is not life threatening and will subside 2 to 3 days after the feed source is removed. Other legume hosts for the blackpatch fungus include white clover (*Trifolium repens*), alsike clover (*T. hybridum*), and alfalfa (Wright, 2004).

Nitrates

Warm season grasses that are subjected to growth stress following nitrogen fertilization may accumulate nitrates. Nitrates are less toxic to horses than to ruminants. However, high-nitrate hays should still be avoided because safe limits have not been clearly established for all categories of horses.

Physically Injurious Plants and Materials

Some plants have the potential to cause physical damage to horses. Hay that contains thistles, awns, or thorns may injure the lips, tongue, or gums. Nonplant materials such as baling wire or twine can also present a hazard to horses. Bale twine/netting which is consumed can create an obstruction in the digestive tract, leading to colic and potentially death (Fig. 11.6).

Chemical Composition and Forage Quality

The best way to accurately assess the nutritional quality of forages for horses is to submit a representative sample of a lot of hay to a laboratory that participates in a quality control program, such as the lab certification program managed by the National Forage Testing Association (NFTA, www.foragetesting.com).

FIGURE 11.6 Hay should be free of foreign matter such as this bale netting, which can cause digestive track blockage, colic, and even death. *Photograph by Laurie Lawrence, University of Kentucky College of Agriculture, Food and Environment.*

A lot of hay is defined as a group of hay bales from a single field that were harvested, cured, and stored in similar conditions. To get a representative sample, use a coring probe to collect samples from 20 bales at random from a lot of hay (Undersander et al., 2016) (Figs. 11.7–11.9). After sampling, place the entire sample into a sealed plastic bag and send to a laboratory with proven proficiency. Be sure to specify that this will be used for horses.

Laboratories may process forage samples using wet chemistry or NIRS (near infrared reflectance spectrophotometry) techniques. The NIRS techniques use infrared light to quickly determine the nutritive value of hay without destroying the sample. Since NIRS is calibrated with data from traditional chemistry techniques, both methods give accurate and similar results for major fractions like crude protein and fiber. NIRS is generally less precise for mineral analysis.

Commercial forage laboratories usually offer a range of services/analyses. Some analyses apply primarily for ruminant nutrition, so it is important to request the analyses that will be most useful in assessing the value of forage for horses.

On most forage analyses reports, the first component listed is moisture expressed as a percent. The moisture value is the percentage of water in the sample. Depending upon the type/size of the bale, hay will usually contain between 15% and 20% moisture at the time of baling (see the later sections on haymaking for more specific information). After baling, moisture will evaporate from the bales. At the time of feeding, most hays will have a moisture content of 8%–12%. The amount of moisture is important because very dry hays may be dusty and more susceptible to leaf shatter. Very moist hay has greater potential to mold.

Dry matter content of hay is calculated as 100% minus percent moisture. At baling, most hays will contain 80%–85% dry matter, and at feeding they will contain 88%–92% dry matter. All of the nutrients are found in the dry matter. Consequently, if you purchase a ton of hay "out of the field" at 85% dry matter, you will be purchasing 1700 lbs of dry

FIGURE 11.7 For accurate hay tests, take 20 cores per lot of hay using a coring probe such as pictured here. *Photograph by Jimmy Henning, University of Kentucky College of Agriculture, Food and Environment.*

FIGURE 11.8 Various models of hay probes are available. Hay probes with collection canisters such as pictured can save time over devices that must be emptied after each core. Heavy-duty, battery-powered drills are very helpful, especially when coring many lots of hay. *Photograph by Jimmy Henning, University of Kentucky College of Agriculture, Food and Environment.*

FIGURE 11.9 After the required number of cores are taken, empty the sample into a plastic bag, seal, and send the whole sample to a forage laboratory with proven proficiency. Be sure to specify the type of animal to be fed. *Photograph by Jimmy Henning, University of Kentucky College of Agriculture, Food and Environment.*

matter and 300 lbs of moisture (water). But, if you purchase a ton of hay that has been stored for 2 months and has a dry matter content of 92%, you will be purchasing 1840 lbs of dry matter and 160 lbs of moisture. Obviously, if the two hays are sold for the same price per ton, you are going to get more nutrients/dollar from the ton of stored hay.

Most forage reports will provide the concentrations of each component on an "as fed" or "as is" basis. The as fed basis gives the concentrations of the chemical components in the hay just as it is fed. Chemical components are also usually expressed on a "dry matter" basis (DM), which is sometimes referred to as 100% dry basis. This basis adjusts for the moisture in the hay and is often used to compare feeds that are different in amount of moisture.

After moisture and dry matter, the next items that appear on most forage reports will be crude protein (CP), acid detergent fiber (ADF), and NDF. CP is determined by measuring nitrogen (N) then multiplying by 6.25. CP is an important quality variable, but ADF and NDF are equally important. ADF is composed of cellulose, lignin, and ash. Lignin has very low digestibility, and cellulose is usually less than 50% digestible. Thus, plants that are high in ADF have a low digestibility. NDF contains ADF and hemicellulose. Hemicellulose is more digestible than cellulose but still below 55%. The cellulose, hemicelluloses, and lignin are part of the cell wall structure of plants, which increases with advancing maturity. When ADF and NDF concentrations in a hay sample are high, it suggests that the plant was harvested in later maturity.

The calorie content of forages is expressed in a variety of ways. The calorie content of forages used for cattle is often expressed in term of net energy, but digestible energy (DE) is used more commonly for horses. In the United States, DE content is expressed in megacalories (Mcal); in Europe, megajoules are used (1 Mcal = 4.184 MJ). DE is not measured

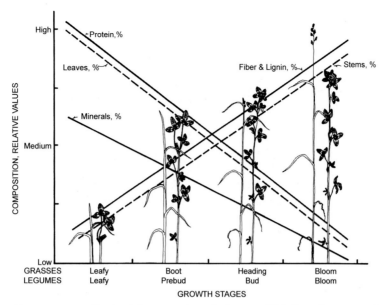

FIGURE 11.10 As grasses and legumes mature, fiber levels increase and protein decreases. High fiber value, measured as ADF and NDF, corresponds to low DE content. *Used with permission Blaser, R.E., Hammes Jr., R.C., Fontenot, J.P., Bryant, H.T., Polan, C.E., Wolf, D.D., McGlaugherty, F.S., Kline, R.G., Moore, J.S., 1986. Forage-Animal Management Systems. Virginia Agricultural Experiment Station, Virginia Polytechnic Institute and State University, Blacksburg, Virginia, Bulletin 86–87.*

directly. Rather, it is estimated from the other chemical components in the feed. There are several equations used to estimate DE in horse hays (see Chapter 3). Most of the equations use a combination of NDF, ADF, and CP as major determinants of the DE content in forages. As the concentrations of NDF and ADF increase, and the concentration of CP decreases, the DE content also decreases. Later maturity forages that are high in fiber (measured as ADF and NDF) and low in CP (Fig. 11.10) will also be lower in DE.

Cellulose and hemicelluloses are referred to as structural carbohydrates, but plants also contain nonstructural carbohydrates. For example, starch is a nonstructural carbohydrate. In general, hays do not contain much starch (<3%). However, forages can also contain mono- and disaccharides, which can be estimated as ethanol-soluble carbohydrates. Some plants, particularly cool season grasses, produce the soluble carbohydrate fructan. The combination of fructan, mono-, and disaccharides can be estimated as water-soluble carbohydrates.

Most forage laboratories can also analyze for mineral content in feeds, most commonly calcium, phosphorus, magnesium, potassium, sodium, copper, zinc, manganese, and iron. Unfortunately, the analytical methods for selenium, iodine, and chlorine are complex and are not available at most laboratories.

Forage analyses may also report total digestible nutrients (TDN) and relative feed value (RFV). These are calculated values that have better application to ruminants than to horses. Neither TDN nor RFV are relevant to specific nutrient needs of horses, but they can be used to make general comparisons among forages. For example, a forage with a TDN of 65 would be expected to be more nutritious than a forage with a TDN of 50. RFV is actually an index that combines estimates of digestibility and intake of a forage based on composition. The index is scaled against full bloom alfalfa hay, which has an index of 100. An RFV of 90 would indicate a forage with lower overall feeding value than full bloom alfalfa, and an RFV over 100 would mean an overall feeding value greater than full bloom alfalfa. Use RFV to compare similar hays (e.g., a grass against other grasses).

PRODUCING QUALITY HAY

The harvest process for hay includes the cutting, swath/windrow manipulation, and mechanical compression into the bale package. Achieving quality hay for horses will depend on the proper implementation of all of the steps from cutting to storage (Fig. 11.11).

The goal of the harvest operation is to mechanically compress the forage into a package that (1) is stable in storage, (2) preserves the maximum amount of nutrients practicable, (3) is convenient for handling and storage, and (4) is appealing to the customer (in the case of hay for sale).

FIGURE 11.11 Proper harvesting, raking, baling, and storage are required to produce quality hay for horses. *Photograph by Clayton Geralds.*

Harvesting

Hay crops should be harvested at the stage that allows for the best compromise between yield and quality, while preserving the persistence of the crop (Table 11.3). The frequency of rainfall events in humid regions will interfere with achieving an optimum harvest schedule for dry hay.

Optimum harvest heights for forage crops vary greatly. Low cutting heights will harm forage grasses more than legumes because they regrow from carbohydrates stored in the stem bases rather than underground crowns like alfalfa and red clover (Blaser et al., 1986). Higher cutting heights will result in quicker regrowth of upright grasses like orchardgrass and longer-lived stands, especially in the mid- to upper south (Fig. 11.12). Operating the cutting implement as low as it will go potentially sprays more soil onto the cut forage than higher cutting heights. Dirt contamination can introduce soil-borne pathogenic bacteria such as *Clostridium botulinum*, which can grow in the preserved forage.

Mowing equipment may have oscillating sickle bar or rotating blade mechanisms. Sickle bar mowers use reciprocating cutter blades that shear the forage crop. They are cheaper and require less horsepower to operate than the newer rotating blade mowers. Sickle mowers tend to have a cleaner cut, especially in light (low-yielding) crops and pull less dust and dirt into the swath. Mowers with rotating blades, often called haybines, have become the prevalent type of mowing equipment because of their higher cutting capacity, lower maintenance requirements, and ability to harvest heavy or lodged crops (Fig. 11.13).

Mowing machines may be outfitted with conditioning mechanisms that either crush (with rollers, Fig. 11.14) or scrape (with impellers) the waxy stems of hay crops (Digman et al., 2011). Conditioning harvesters are more expensive and

FIGURE 11.12 Higher mowing heights encourage quicker regrowth and extend stand life in orchardgrass, especially in the mid- to upper south. *Photograph by Clayton Geralds.*

FIGURE 11.13 Mowers with rotating discs and conditioning rollers are known as haybines. *Photograph used with permission from New Holland, North America.*

FIGURE 11.14 Mowers with rotating discs have become the most common form of hay harvest implement. This model is equipped with intermeshing rubber rollers for crimping and crushing stems to facilitate more uniform drying of the crop. *Photograph by Jimmy Henning, University of Kentucky College of Agriculture, Food and Environment.*

require greater horsepower to operate. However, conditioning is needed to accelerate the drying of the stems of forage crops like alfalfa, especially in humid climates. Conditioned legume forage will dry more uniformly throughout and will have less leaf shatter at baling than unconditioned crops dried to the same moisture content. For example, unconditioned alfalfa hay with an overall moisture content of 20% will have much drier leaves than stems, increasing the likelihood of leaf shatter during baling.

A swath can be defined as the undisturbed forage deposited on the field after being cut by the mowing implement. In general, wide swaths are more desirable because more sunlight is intercepted by the cut forage, leading to shorter drying times (Digman et al., 2011). Swath width behind mower conditioners with less than full-width rollers or impellers will be limited to the width of the conditioning mechanism. Mowing equipment may be equipped to manipulate the cut forage into variable swath widths (Fig. 11.15).

Make wide swaths early in the season (when temperatures are relatively cool) or with heavy crops to maximize sunlight interception by the forage. Midsummer cuttings (when temperatures are relatively hot) can be directly swathed into windrows that can reduce bleaching and help raking operations be done more efficiently.

Swath and Windrow Manipulation

Most harvest operations will require manipulation of the swath prior to packaging. In general, less physical manipulation of the cut forage is desirable, to preserve leaves and to save time. The goal of the physical manipulation of the swath is to

FIGURE 11.15 Mowers may come equipped with moveable panels allowing formation of swaths of various widths. The panels shown are angled inward to form a narrow swath. *Photograph by Jimmy Henning, University of Kentucky College of Agriculture, Food and Environment.*

facilitate uniform drying of the whole swath and to prepare a windrow conducive to efficient, uniform packaging. Windrow is the term applied to the linear formation of hay created by raking for the final drying prior to baling.

Ideally, drying conditions allow swaths to be raked directly into large, fluffy windrows where the final drying takes place before baling. The main types of rakes are the wheel, rotary, or rotating bar types (Figs. 11.16–11.18). Adjust rake height and action so that it does not move loose dirt into the windrow. This dirt is a major source of dust in hay and can introduce undesirable bacteria. The best windrows are uniform in size and narrow enough to be handled efficiently by the pickup reels of baling equipment. Some rakes can invert or turn over a windrow that is dry on top and wet on the bottom.

Under adverse conditions such as untimely rains, heavy dews, or with heavy, matted swaths, tedding may be helpful. Tedding is the mechanical respreading and fluffing of moist cut forage to facilitate uniform drying of the swathed forage (Fig. 11.19). Tedders are similar in design to rotary rakes except they are designed to spread out the crop instead of forming a windrow. Tedding will result in some leaf loss and should be done when the crop is damp. Tedding will nearly always result in a loss of forage quality and yield, especially with forage legumes due to loss of leaves. Tedding should only be as physically aggressive as is needed to spread out and fluff the crop.

Field curing of hay will bleach the green color from the tops of swaths and windrows. This color change does not lower the energy or protein content of hay, only the vitamin A value from the pigments that are bleached. Reducing curing time and swathing direct into windrows will preserve more green color in hay. Remember that green color is a subjective indicator of quality, not a direct measure. Overmature alfalfa hay that has been raked and handled too dry can be very green, but be of little feed value to horses, due to leaf loss.

FIGURE 11.16 Wheel rakes can be side- or center-delivery (shown). Wheel rake advantages include high-speed operation, low maintenance, and low horse power requirements. The ground-driven design of wheel rakes can introduce more dirt and foreign material than other types and can struggle with wet, high-moisture hay. *Photograph used with permission from New Holland, North America.*

FIGURE 11.17 Rotary rakes handle dry and wet material equally well and minimize foreign material and dirt in the windrow since the rotating tines do not touch the ground. They are more costly and complex than wheel or rotating bar rakes and require hydraulics and power takeoff (PTO) to turn the driveshaft. *Photograph used with permission from New Holland, North America.*

FIGURE 11.18 Rotating or roller bar rakes have simple, low-cost designs and are powered by the ground wheels, producing a clean crop. They have limited capacity and sometimes produce a ropey, twisted windrow. *Photograph by Matt Barton, University of Kentucky College of Agriculture, Food and Environment.*

FIGURE 11.19 Tedding respreads and fluffs the swath to facilitate faster drying. *Photograph courtesy of Clayton Geralds.*

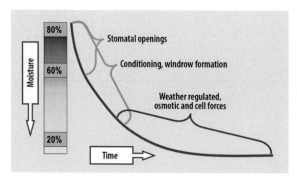

FIGURE 11.20 Initially, stomatal openings and conditioning lead to rapid drying in cut forage. Later, drying rates slow as osmotic forces hold water in plant cells and weather limits hinder evaporation from the swath or windrow. *Digman, M., Undersander, D., Shinners, K., Saxe, C., 2011. Best Practices to Hasten Drying of Grasses and Legumes. University of Wisconsin — Extension. Publication A3927. Used with permission courtesy University of Wisconsin Extension.*

Baling

Baling is the mechanical compression of forage from a windrow into packages for storage and feeding. Baler types vary from the small/medium size square baler to large round and large square machines. Choice of baler will depend on needed capacity (tons/hour), available labor, type of storage, and the needs of the end user.

The best bales come from uniform windrows. In addition, windrows should be as large as equipment will allow, facilitating rapid harvesting. Height of the windrow can be an issue where it must pass directly under the tractor before entering the baler pickup reel. Height is not an issue for side pickup balers since the windrow does not have to pass under the tractor.

Windrow width can be especially important for round bales. The quantity of forage entering round bales should be uniform across the pickup reel so that the top and sides of the bale form a right angle. Some balers automatically redistribute the hay in the bale chamber to keep the packages uniform. Older models may require manual compensation for narrow windrows by weaving the baler left and right to ensure uniform bale density across its width.

Fresh forage is about 75%—80% moisture content on average (Digman et al., 2011). Moisture evaporates more rapidly from leaves than stems because they have numerous openings called stomates and a large surface to volume ratio (Fig. 11.20). Stems have fewer stomates, lower surface to volume ratios, and often have a thick, waxy layer that impedes water loss. Conditioning forage crops by crushing or abrasion will speed the drying of stems relative to leaves. The drying process will often take 48—72 h or longer, depending on weather and density of the crop. When it is not possible for the hay to dry to the ideal baling moisture, preservatives can be applied.

Hay Preservatives

The most proven type of hay preservatives in use in hay production are the organic acid-based products. These products will most often contain propionic acid in combination with other organic acids and buffering agents. Preservatives of this type have been proven effective in baling at moisture contents up to 25% without detrimental heating and mold growth. These materials are applied at baling and must be uniformly distributed throughout the crop to be effective (Figs. 11.21 and 11.22). While higher moisture material can be effectively treated by these materials, bales become excessively heavy, and the cost is often prohibitive. These organic acids are normal products of digestion in the hindgut of the horse and are safe for feeding. Research has shown that horses may prefer nonpreserved hay; however, they readily consume preserved hay after a period of acclimation (Battle et al., 1988). Preservatives are most often applied to higher value hay crops like alfalfa. Baling at slightly elevated moisture levels will reduce leaf shatter and may reduce the curing time, so rain damage is avoided. Therefore, well-preserved hay will have the advantage of retaining more leaves.

Bale Handling

Many options exist for handling bales between the baler and the storage location. Hay operators mechanize the handling of bales as much as possible to reduce labor costs and to move hay to inside storage as fast as possible.

Bale accumulators can be pull-behind models that leave a grouping of bales in the field designed to be picked up by grapples (Fig. 11.23). Self-propelled bale accumulators pick up bales in the field and form them into stackable groupings of

FIGURE 11.21 Hay preservatives are best applied at baling, and application rates are often controlled automatically from bale moisture sensors in the bale chamber. *Photograph by Jimmy Henning, University of Kentucky College of Agriculture, Food and Environment.*

FIGURE 11.22 Application nozzles for hay preservatives must be positioned to fully cover the windrow with product. *Photograph by Jimmy Henning, University of Kentucky College of Agriculture, Food and Environment.*

FIGURE 11.23 Grapples are designed to grip, lift, and load groups of bales onto trailers for transportation to storage. *Photograph by Jimmy Henning, University of Kentucky College of Agriculture, Food and Environment.*

FIGURE 11.24 Self-propelled bale accumulators pick up bales in the field and form them into stacks for transport to storage. *Photograph courtesy of Clayton Geralds.*

bales that can be unloaded directly into barns (Fig. 11.24). It is desirable to store different fields or lots of hay so they can be sampled and accessed independently.

Bale Conditioning and Storage

Bale conditioning is leaving bales in the hayfield for a few days to allow the packages to dissipate the heat that comes naturally after baling. During this period, heat from plant respiration and from microbial growth will drive off moisture, which helps hay be more stable in storage. This process is commonly referred to as "going through the sweat." Ideally bales should have a few days of warm, sunny weather for conditioning before they are placed in covered storage. The greater the moisture at baling, the longer the conditioning period required.

If weather conditions prevent bale conditioning in the field, move hay under covered storage, but leave room for ventilation to allow dissipation of normal bale heat and excess moisture.

Some heating is normal in hay even when baled at appropriate moisture contents. Temperatures up to 130°F (54°C) have little effect on hay quality (Collins and Owens, 2003). In bales above 25% moisture at baling, it may take 2 weeks for temperatures to return to near ambient. At above optimal moisture contents, heating within the bale or haystack can have severe consequences.

Elevated bale temperatures in moist hay can reach levels that damage nutritive value. The heat produced in bales results from the metabolism of carbohydrates to carbon dioxide and water by plant enzymes and microorganisms on the surface of the forage. Loss of these carbohydrates lowers the energy content of hay.

The high heat and moist conditions in wet hay will result in mold growth and dust. Excessive heating will also cause caramelization (browning) in the hay bale. Caramelization is an indication that carbohydrates and protein have begun to react to form an indigestible fraction called the Maillard product. Heat damage is assessed by measuring the nitrogen content of the ADF fraction, sometimes referred to as ADF-N in forage analysis reports.

In addition to damaging hay quality, excessive heating can result in bale combustion and fire within the hayloft or haystack. This heating can occur over a protracted period, which means that combustion could occur many days after hay has been placed in a barn or stack. If temperatures within a stack reach 150–160°F (65–70°C), hay is dangerously close to combusting (Fetzer et al., 2012). At this point, unstacking the hay and allowing it to cool can still prevent fires. At stack temperatures of 175°F, wet hay down, move it away from buildings and other dry hay, and notify fire services. Above 175°F, areas of burning hay in the stack are likely. At temperatures above 175°F, try to have fire services on site before moving hay.

BALE PACKAGES FOR HORSES

Acceptable packages for horse hay include small square, medium square, large square, and large round bales. Traditional small square bales have cross-sectional areas of 14 by 18 inches (36 by 46 cm) and are generally 36–42 inches (91–107 cm) in length (Fig. 11.25). These bales will weigh from 40 to 60 lbs (18–27 kg) depending on the density and

FIGURE 11.25 *Small square* bales weighing 40—60 lbs have been the traditional package for horse hay. More recently, *medium-sized square* bales weighing 100 to 150 lbs have become common for horse hay, especially sources from arid hay-producing regions. *Photograph courtesy of Clayton Geralds.*

moisture at baling. More recently, medium-sized square bales (16 by 18 or 18 by 22 inches/41 by 46 or 46 by 56 cm) have become common for horse hay. Medium bales are wider and longer than the traditional bales and usually weigh between 100 and 150 lbs (45—68 kg). Small and medium square bales are still the preferred package for feeding small numbers of horses individually because they are lighter and flake off into small portions, which are easy to allocate and feed. Small and medium square bales can also be baled at slightly higher moisture than larger bales without detrimental heating and mold development (Table 11.4). However, these bales require more labor to handle and are more expensive to transport. Higher transport costs arise because these lower density bales will not allow a truck to scale the payloads possible with larger rectangular packages. Baling speeds (tons/hour) are lower for small and medium square bales (Hanna, 2016).

Large rectangular bales are increasing in popularity largely to address the limitations of the traditional small rectangular packages (Fig. 11.26). These large bales cross-sectional dimensions include 3 by 3, 3 by 4, and 4 by 4 ft (0.91 by 0.91, 0.91 by 1.2, and 1.2 m by 1.2 m) (height by width) (Table 11.4). Lengths are adjustable and are typically 8 ft. (2.4 m). Length can be matched to either the trailer or barn constraints. These packages have the advantage of being easier to handle mechanically, thus requiring less labor to stack in barns or load out for transport. Balers for these packages have greater capacity (tons/hour) and produce a higher density bale compared to small square or large round bales (Hanna, 2016; Ball, 2016) (Table 11.4). The higher density and rectangular shape enable more cost-effective transportation than for round or small square bales, since trucks can load to maximum legal payloads.

TABLE 11.4 Dimensions, Weights, Densities, and Safe Baling Moistures for Various Bale Shapes (English Units, for Metric Units, See Appendix 8.)

Bale Shape	Height (ft)	Width (ft)	Length (ft)	Volume (ft³)	Typical Weight (lb)	Density (lb/ft³)	Safe Baling Moisture, %
Rectangular	1.2	1.5	3.2	5.5	60	11	18–20[a]
Rectangular	2.7	3	7	56	900	16	12–16[a]
Rectangular	4	4	8	112	1800	16	12–16[a]
Round	4	—	4	50	500	10	15–18[a]
Round	4	—	5	63	850	13	15–18[a]
Round	5	—	4	79	1000	13	15–18[a]
Round	5	—	5	98	1300	13	15
Round	6	—	5	141	1900	13	15

[a]*The lower moisture range is preferred in areas of low humidity; the higher moisture percentage for other areas.*
Adapted from Ball, D.M., Hoveland, C.S., Lacefield, G.D., 2016. Characteristics of hay bales. In: Sulewski, G. (Ed.), Forage Crop Pocket Guide, thirteenth ed. International Plant Nutrition Institute, Peachtree Corners, Georgia, p. 56.

FIGURE 11.26 *Large rectangular* balers have higher baling capacities (tons per hour) and produce bales that are more economically transported and handled than *small square* bales. Their large size and weight limits their usage on small horse operations. *Photograph used with permission from New Holland, North America.*

Disadvantages for large rectangular packages mainly relate to their weight and size. The heavy weights of these bales mandate mechanical handling. Farms with limited storage cannot accept truckloads of these bales, or they cannot place them in the traditional storage space in the lofts of barns. Flakes of these bales are much larger and may be more hay than needed and more physically cumbersome to feed. Some balers are able to combine several smaller bales into one large rectangular package, facilitating the subdividing for feeding (Fig. 11.27).

These smaller bales are still too large (e.g., 2 ft. by 4 ft. by 4 ft.) to be handled manually. The tighter compression of these bales can make them more difficult to take apart for feeding. Haymaking conditions in the humid regions of North America are not conducive to drying hay to the 12%−16% moisture required for these bales. In these areas, treating the hay at baling with hay preservatives is generally required to prevent heating and detrimental mold growth.

Large round bales are commonly used for cattle across North America but can also be used for feeding horses (Fig. 11.28). Large round bales have densities intermediate to small and large rectangular bales and can be made at moisture contents of 18% or less (Table 11.4). Advantages to these packages include baling speed (tons/hour) and lower equipment costs relative to large rectangular systems (Hanna, 2016).

Although their shape naturally sheds water and they can be wrapped with a plastic netting product to provide some additional rain protection (Collins et al., 1997), round bales should be stored inside if they are intended for horses. Round bales that have been stored under cover are suitable for feeding groups of horses if they are placed in covered feeders (Fig. 11.29). Disadvantages include difficulty of stacking and that greater barn space is needed to store equivalent quantities of round versus rectangular packages. Like large rectangular bales, large round bales have limited suitability to small horse operations because they are difficult to handle and divide into individual portions.

FIGURE 11.27 Some *large rectangular* balers can combine several smaller bales into larger packages, facilitating economical transportation, handling, and feeding. The smaller bales are still too large to be handled manually. *Photograph by Matt Barton, University of Kentucky College of Agriculture, Food and Environment.*

FIGURE 11.28 *Large round* bales are most commonly used for cattle but can be useful for horses as well. *Photograph used with permission from New Holland, North America.*

FIGURE 11.29 Round bales that have been stored under cover are suitable for feeding groups of horses if placed in covered feeders. *Photograph by Laurie Lawrence, University of Kentucky College of Agriculture, Food and Environment.*

STORING HAY

Once high-quality hay has been baled, proper storage is needed. A good storage facility should have easy access for delivery trucks and any feeding vehicles (Fig. 11.30). Whenever possible, hay should be stored in a barn that is separate from horses and equipment. In the case of a fire, a separate storage facility is easier to isolate than a loft that is above stalls or a hay shed that is attached to another building. In addition, hay in a separate facility will be less likely to become contaminated by fuel or exhaust from equipment or with particulates that become airborne during stall cleaning. A good storage facility must keep hay dry and should minimize sun exposure to reduce bleaching. A four-sided storage barn provides the best protection, but a three-sided shed can also work as long as the hay is not stacked close to the open end where rain or snow can blow in. Some hay bleaching will be expected in a three-sided shed. In very dry climates, a structure with a roof but no walls may also provide adequate protection for the hay, but again, bleaching is likely to occur. Bleaching is primarily an issue of appearance and thus affects the value of the hay for the seller more than the nutritional value for the horse. The least desirable way to store hay is outside because precipitation will ruin the outer layers of the bale or stack. A plastic tarpaulin can be used to cover hay, but the potential for leakage is high.

FIGURE 11.30 A good hay barn will have easy access for loading and unloading and should be separate from horses and equipment. Barns may be three sided, as shown, as long as rain or snow cannot blow onto stored hay. *Photograph by Jimmy Henning, University of Kentucky College of Agriculture, Food and Environment.*

Rain and snow are the most common sources of moisture damage to baled hay, but ground moisture can also be a problem. Hay should be elevated off of the ground on pallets, and when possible, a vapor barrier should be inserted between the pallets and the ground.

CHOOSING HAY FOR HORSES

Several factors should be considered when deciding what type of hay to feed. Cleanliness is most important, but nutrient value and the type of horse being fed should also be considered. Price, consistent availability, and package size may also influence hay choices on horse operations.

Cleanliness First

The hygienic quality of hay should be the first consideration when choosing hay for horses. Any type of hay can contain dust or mold if it is not harvested and stored properly. Dust and mold have the potential to irritate the respiratory system and thus should be avoided regardless of nutrient value. Other hygienic qualities of importance include absence of insects and animal parts and freedom from toxic plants or injurious items such as bale string, bale netting, thorns, awns, and thistles.

Matching Hay Type to Horse Type

Horses vary in size and nutrient needs, and thus their hay needs vary as well. A hay that is excellent for a mature idle pony may not be the best hay for a lactating mare. The "best" hay for any horse depends on the nutrient needs of that horse. Horse owners should understand the nutrient needs of their horses, and hay producers should understand the types of horses that their clients own.

In general, legume or legume grass mixes that are harvested in early maturity are best for horses with high nutrient requirements. The more nutrients supplied by the hay, the lower the amount of concentrate that is needed. Examples of horses with high nutrient requirements are lactating mares, growing horses, and elite performance horses. Horses that have low nutrient needs can be fed nutrient-dense, high-quality hay, but daily consumption may need to be restricted to prevent horses from gaining weight (Fig. 11.31).

Unfortunately, restricting forage may have some negative effects on the digestive tract and on the behavior of the horse. Thus, grass hays cut in mid to late maturity that are higher in fiber and lower in digestible nutrients may be most desirable for horses with low requirements. Horses with low requirements include nonpregnant mares, retirees, and horses used for light recreational riding. Mid- to late maturity hays are useful for these horses because the horses can eat more hay to satisfy their appetites without getting too fat. Higher hay intakes encourage chewing, which produces saliva that benefits the gastrointestinal tract. More time chewing on hay may mean less time for horses to chew on other objects such as trees or fences. Horses with moderate requirements (pregnant mares, horses in moderate work, breeding stallions) can be fed almost any type of hay effectively. When hay with high nutrient density is fed, the amount of concentrate needed will be reduced. When hay with low nutrient density is fed, the amount of concentrate needed will be increased (Fig. 11.32).

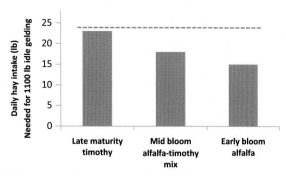

FIGURE 11.31 The nutrient density of different hays affects the amount that is needed to meet requirements. This graph shows the pounds of three different hays that would be needed to meet the calorie need of a mature idle gelding weighing 1100 lbs. The *dotted line* shows the expected maximum amount of hay that the gelding would be able to consume each day. When the mature hay timothy is fed, the amount of hay needed is much higher than when the early maturity alfalfa is fed. The higher amount of hay will take longer for the horse to consume. By comparison, the smaller amount of alfalfa could leave the horse's appetite unsatisfied. If a higher amount of alfalfa is fed, the horse will gain weight.

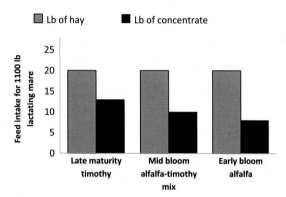

FIGURE 11.32 The nutrient density of hay affects the amount of concentrate needed to meet the needs of horses. In this example, a lactating mare is fed 20 lbs of three different types of hay. Because the hays vary in type and stage of maturity, they provide different amounts of calories. The mature timothy will provide the lowest and the early bloom alfalfa provides the highest amount of calories. Consequently, to meet the calorie needs of a lactating mare, the horse owner will have to feed more concentrate (shown in the second bar of each pair) with the timothy hay and less concentrate with the alfalfa hay. If the two hays were similar in price (or even if the alfalfa is slightly more expensive) the lower concentrate intake will result in a lower total feeding cost when the higher quality hay is fed.

IS HAY A NUTRITIONALLY BALANCED DIET?

Hay provides many essential nutrients, but it is not nutritionally complete for all types of horses. Sodium and chloride concentrations in forages are often low, and a source of free-choice salt (NaCl) is often recommended for animals on all-forage diets. As mentioned previously, legumes are typically higher in calcium and have a wider calcium to phosphorus ratio. Phosphorus concentrations in hay vary with the amount of available phosphorus in the soil, which can result in some wide regional differences. Calcium and phosphorus are important minerals for all horses but especially for broodmares, growing horses, and performance horses. When legumes are fed, the calcium requirements for most horses will be met, but because phosphorus concentrations can vary, supplementation may be necessary, especially for lactating mares and growing horses. Selenium content of forages varies greatly by region, and much of North America has selenium deficient soils. Horses fed hay from selenium-deficient or selenium-marginal areas may need selenium supplementation. Other trace minerals of interest in horse nutrition are copper and zinc. These minerals are often low in common forages. Depending upon the type of horse being fed, mineral supplementation is accomplished through the concentrate feed (when additional calories and protein are also needed) or a daily mineral supplement. Trace mineralized salt blocks and mineral blocks are available, but consumption may vary from day to day, and they may not provide all of the necessary minerals, such as selenium, which is not included in most trace mineral blocks.

Feeding Hay to Individually Housed Horses

Horses are kept in a variety of housing situations. Individual housing is often used for performance horses, stallions, and in urban environments where outside paddock space is limited. Individual housing allows horse owners to monitor appetite and to control the amount of hay that is offered to a horse each day. There are a number of hay feeding options for individually housed horses. The first option is free-choice feeding where the horse is given more hay than it can consume in a day, so hay is always available. This system may be a time-saver for the horse owner if all of the hay is given to the horse at one time, rather than in two or three portions a day. In addition, this feeding system may be the closest mimic to the natural environment where horses would spend several hours a day grazing pasture. Chewing stimulates saliva production, which acts as a buffer against stomach acid and helps protect the stomach from ulcers. In addition, free-choice hay feeding is often used when it is desirable for horses to achieve maximum voluntary dry matter intake. For example, elite equine athletes may have difficultly consuming enough calories to maintain body weight, so maximizing calorie intake from high-quality hay is important. However, if the horse has low nutrient requirements and hay quality is high, then the horse may over consume nutrients and gain weight. Obesity in horses has been linked to several health issues and should be avoided. Therefore, if horses with low requirements have unlimited access to hay, the hay should be low in nutrient density such as a late maturity grass hay. However, free-choice feeding of late maturity hay can have some disadvantages. Horses are selective grazers. They can effectively sort feed particles with their upper lip and are capable of discriminating between leafy material and stemmy material. They may also move hay around to sort the leaves from the stems. Consequently, when they are allowed ad libitum access to mature hay, it is likely that they will consume the most desirable parts and waste the less desirable parts (Fig. 11.33).

One of the main advantages of individually housing horses is the ability to individualize their feeding program. An individualized program can allow the use of more than one type of hay as well as portion control. For example, in some feeding systems a horse may be given free-choice access to grass hay and then a small amount of alfalfa hay once a day. For horses with low requirements, or when weight loss is desired, the amount of hay offered each day can be restricted to the amount that just meets the horse's needs. Particularly when late maturity hay is offered, portion control may minimize waste, as the horse will have to eat all of the offered feed to satisfy its appetite. Feeding individual portions several times a day also allows managers to easily evaluate whether a horse is exhibiting its normal appetite.

Regardless of whether hay is offered free choice or in monitored portions, it can be fed on the floor/ground or from some type of rack, net, or bin (Figs. 11.34 and 11.35). There is no one single feeding method that is the best for all individually fed horses. Feeding on the ground/floor places the horse's head in the most natural position, but it can contribute to the greatest amount of waste because the horse can trample the hay and leaves may be lost in bedding or on the ground. Hay bins/troughs can contain the hay, but the horse's nose will be more exposed to dust and mold. There may also be potential for the horse to put a leg inside the hay bin/trough depending on design and location. Hay racks and nets also contain the hay but are usually suspended up high enough that the horse's legs cannot become entangled in them. However, the high positioning makes it possible for hay particles and dust to fall into the horse's eyes and nose. The best hay feeding system for individually housed horses takes into account the nutrient needs of the horse, health, and safety considerations for the horse and the convenience and economics of the feeding system for the manager/owner.

FIGURE 11.33 Feeding hay with no feeder can result in losses exceeding 50%. Horses can sort hay particles around with their upper lip and are capable of discriminating leaf material and leaving the stemmy material behind. *Photograph by Laurie Lawrence, University of Kentucky College of Agriculture, Food and Environment.*

FIGURE 11.34 Allocating portions of high-quality alfalfa grass hay on the ground. *Photograph by Jimmy Henning, University of Kentucky College of Agriculture, Food and Environment.*

FIGURE 11.35 Utilizing a hay rack with trough for feeding. *Photograph by Jimmy Henning, University of Kentucky College of Agriculture, Food and Environment.*

Horses Fed in Groups

Compared to individual feeding, group feeding can save labor and time. It also reduces investments in facility construction and maintenance. The most effective group feeding systems will group horses by nutrient needs. Grouping by nutrient needs allows managers to feed their most nutrient dense hay to the horses with the highest needs, and their least nutrient dense hay to horses with the lowest nutrient needs. When horses with high and low needs are grouped together, it is likely that either some horses will get too fat or other horses may lose weight. When it is not possible to separate horses with different nutrient needs, it may be necessary to feed hay with a low nutrient density to prevent low-nutrient-need horses from becoming too fat, while providing additional concentrate to the horses with higher nutrient needs.

In nature, horses live in groups (herds or bands) and develop social hierarchies. These hierarchies become apparent when there is competition for resources, such as water or food. Consequently, when horses are fed hay in groups, it is important to account for how social hierarchies will affect the ability of all horses to obtain adequate feed. In general, the dominant horses in the group will have more access to feed than the submissive horses. In addition, competition for feed can lead to injuries from biting or kicking. Therefore, the first goal in group feeding is to provide enough feeding space that all horses can have access to hay at the same time while minimizing competitive behavior. Depending on the number of horses in a group, it may be desirable to have several feeding stations. A "feeding station" could be a hay rack, a hay trough, or site where hay is placed on the ground. Feeding stations should be spaced far enough apart that a dominant horse can only monopolize a single station and in locations that prevent a horse from being trapped in a corner by a more dominant horse (Fig. 11.36).

As with individual feeding, hay can be fed to groups of horses on the ground or from some type of feeder. If hay will be placed on the ground or in open hay troughs/bunks the best practice is to feed it once or twice a day in amounts that will be consumed by the horses before the next feeding. Hay that accumulates on the ground or in an open trough will likely accumulate mold. If hay is being fed on the ground in a pasture, hay can be spread apart and placed on new sites each day to prevent muddy spots from developing.

The labor-saving advantage of group feeding is most apparent when hay handling is minimized. That is, hay is put out for horses once or twice a week instead of once or twice a day. This system is best applied using covered hay feeders. The cover keeps the hay dry, and the hay feeder minimizes waste. Martinson et al. (2016) found that covered hay feeders wasted much less hay (5%–11%) than no feeder at all (57%). There are many different covered feeder designs that are designed and acceptable for horses (Figs. 11.37–11.39). Selection of a specific design will depend on the number of horses that are being fed, the type/size of bale that is being used, and the amount of equipment available to the horse operator for moving the hay. It is important to note that there are also hay feeders available that are designed for cattle, but some may pose safety hazards for horses.

FIGURE 11.36 When allocating hay for multiple horses, space the stations far enough apart to prevent dominant horses from preventing others to have access. *Photograph by Jimmy Henning, University of Kentucky College of Agriculture, Food and Environment.*

FIGURE 11.37 Covered Cradle brand hay feeder. *Photograph courtesy of Krishona Martinson, University of Minnesota.*

FIGURE 11.38 Hayhut brand covered feeder. *Photograph courtesy of Krishona Martinson, University of Minnesota.*

FIGURE 11.39 Waste Less brand covered hay feeder. *Photograph courtesy of Krishona Martinson, University of Minnesota.*

BALING FORAGE CROPS FOR SILAGE

While hay is the most common form of stored forage for horses, hay crop silage may also be used in situations where hay curing is difficult, such as the high-humidity, rainfed areas of North America. It is possible to make high-quality silage or haylage using long (unchopped) forage crops baled and tightly wrapped with plastic to exclude oxygen (Fig. 11.40).

Due to a cool, wet climate in northern Europe, much silage and baleage is made for horses and fed to them there. In a Swedish study of preference of horses for grass conserved as hay, haylage, or silage, silage had the highest rate of consumption, followed by haylage, with hay being least preferred. The longest eating time in minutes/day was observed for silage, and the shortest was for hay. When horses could choose which to eat, silage was the first choice 85% of the time (Muller and Uden, 2007). Similar findings were summarized in Frape (2010a).

FIGURE 11.40 Wrapping high-moisture hay bales in stretch plastic can produce a high-quality feed for horses, especially in the rainfed, humid areas of North America. Shown is a storage yard for baleage wrapped with an in-line wrapper in Northern Kentucky. *Photograph by Jimmy Henning, University of Kentucky College of Agriculture, Food and Environment.*

TABLE 11.5 Protein Concentration, Digestibility and Bale Weight Differences Before and After Storage[a]

	Protein Concentration		Digestibility		Bale Weight		
	Prestorage	Poststorage	Prestorage	Poststorage	Prestorage	Poststorage	
	Percent Dry Matter				Dry Matter (lb)		DM Loss %
Baleage 46% DM	23.7	22.6	63.0	63.8	548	554	Negligible
Baleage 51% DM	23.1	22.3	62.0	65.0	537	541	Negligible
Baleage 57% DM	22.1	21.0	65.1	64.4	587	583	Negligible
Hay	18.2	17.5	67.2	51.9	609	495	18.7

Storage period: May to December. Hay is stored outside on the ground.
[a]*Mike Collins, 1995. University of Kentucky. Unpublished research.*

Round bale silage (or baleage) is the product of cutting forage crops with conventional hay harvest equipment, allowing the forage to wilt to between 40% and 60% dry matter, baling the forage into tight bales, and quickly wrapping the bales in plastic so that oxygen is excluded. The wrap keeps out air, allowing anaerobic microorganisms to ferment carbohydrates to lactic acid, thus lowering the pH, which inhibits the growth of other detrimental microorganisms. The ensiling process uses some dry matter for energy, but this loss is small compared to dry matter losses that result from raking, baling, tedding, and, particularly, storing round bales outside as hay (Table 11.5).

Advantages and Disadvantages of Silage Baling

Advantages of making round bale silage compared to conventional silage or hay include these:

- Plastic cost per bale is low.
- Capital investment required is lower than fixed silage structures.
- Higher quality feed is produced.
- Harvest and storage losses are lower.
- Weather damage is less than hay stored outside.
- Individually wrapped silage bales are more portable.
- Small amounts of forage can be ensiled.
- Baled silage feeding does not require specialized machinery.

Disadvantages of baled silage include the following:

- Long (unchopped) forage crops are harder to ensile (less readily fermentable carbohydrates) than chopped forage.
- Some balers cannot handle wilted (40%—50% dry matter) forage.
- Bales can be very heavy, leading to larger tractor requirements.
- Plastic wrap material can tear or puncture, leading to spoilage.
- Disposal of used plastic is necessary.

Good silage, haylage, or baleage has a clean, pleasant odor due partly to acetic and propionic acids produced during the fermentation process. The pH is usually 4.5 to 5.0 but may be higher in baleage with high legume content. There should be no visible mold, and the color should be green or slightly brownish. Ensiled feeds that are dark brown or smell caramelized or charred have overheated during a poorly managed fermentation. Ensiled feed dry matter should not exceed half of the dry matter fed to horses (Lewis, 2005).

A vacuum-packed, high-moisture forage called "Horsehage" was developed in England to provide high-quality, dust-free forage for an event horse with a cough. It is an expensive specialty product that is produced from selected crops of ryegrass, a high-fiber grass mix, timothy, and alfalfa. All varieties are low in sugar and starch and high in fiber and contain no additives. A similar product called "Tri-Forage Horsehae" is made in Ontario, Canada.

FORAGE REQUIREMENTS

All of the major forages suitable for horses can be harvested effectively as baleage. For good baleage, it is very important to cut at the proper stage of maturity, so the forage contains adequate levels of fermentable carbohydrates for good ensiling (Table 11.3). In general, harvesting forage crops in the transition stage between vegetative (leafy, immature) and reproductive, or flowering, stage will produce the best compromise between yield and quality (Lacefield et al., 1996).

Harvest losses (usually from leaf shatter and loss) are greatest for very dry forage but are low for herbage handled immediately after cutting (Collins and Owens, 2003). However, silage baled too wet is subject to excessive storage losses due to seepage and deterioration. Storage losses arise from microbial activity in moist forage and therefore are generally minimized by harvesting at low moisture levels. Minimum combined field and storage losses are achieved by harvesting forage in the middle of the moisture range, between 40% and 70% moisture. The reasons for field losses in forage harvesting are respiration, leaching, and some leaf loss.

The dry matter levels recommended for baled silage are generally between 40% and 60%, covering the range between wilted silage and haylage. The ideal dry matter content appears to be 40% to 50% because fermentation is adequate and heat damage is minimized. In producing bales for bagged or wrapped silage, it is important to remember that forage in the 50% dry matter range will weigh about twice what the same size bale of hay would weigh. Bale size is frequently reduced to restrict bale weight to 0.75—1.0 ton. Heavier packages may be difficult to transport. In England, 55%—66% DM is considered reasonably safe (Frape, 2010b). Forage crops will typically reach proper moisture contents for baleage in 24 h (Digman et al., 2011). Wilting time can be as brief as 4 to 8 h, depending on the ambient temperature, moisture in the soil, yield of the crop, and the width of the swath. Frape (2010b) suggested 18—24 h of wilting before wrapping in England.

MACHINERY REQUIREMENTS FOR BALEAGE

Equipment needs are similar to those required for making hay. The mower does not need to have conditioning rollers, but adjustable baffles may be useful because they can be adjusted to concentrate the cut forage into a narrow swath. Narrow swaths can help slow crop drying and keep the moisture content in baleage levels longer. Narrow swathing may eliminate the need for raking, saving leaves and time. If the mower leaves a wide swath, it should be raked to ensure adequate pickup into the baler. Rotary or rotating bar rakes may be preferred over wheel rakes since they are better able to handle wet, heavy crops. Heavy duty balers are best because they can handle the heavy bales formed when making baleage.

Since the forage is wet and heavy, bale diameters generally range from 42 to 48 inches to avoid overloading either the baler or the transport equipment. Bales should be formed as tightly as practical. Fixed-chamber hay balers lack the flexibility of variable chamber balers to vary bale diameter as a means of reducing bale weight in wetter crops.

The ground speed of the baler should be lower than speeds used in making field-cured hay. Downshifting one gear should help to guarantee a tighter, denser bale. A typical silage bale (4 ft. in diameter by 5 ft. in length) should weigh 1300 to 1550 lbs (590—703 kg) and contain 600 to 650 lbs (272—295 kg) of dry matter, but it may weigh as much as a ton (907 kg).

Many manufacturers produce balers designed specifically for making baleage (Fig. 11.41). Some recent models of both fixed and variable chamber balers include knife mechanisms to chop the forage, allowing increased density. University of

FIGURE 11.41 An example of a baler designed specifically for silage. *Photograph by Jimmy Henning, University of Kentucky College of Agriculture, Food and Environment.*

FIGURE 11.42 Handling the heavier bales of baleage will require heavier, higher horsepower tractors compared to dry hay. *Photograph by Jimmy Henning, University of Kentucky College of Agriculture, Food and Environment.*

Kentucky research revealed that using a "chopping" fixed-chamber baler increased silage bale weights by about 300 lbs (136 kg) at the same bale diameter (Henning et al., 1998).

Bales of silage can be moved by traditional bale spears prior to wrapping. Tractors should be sized to effectively handle the weight of baled silage (Fig. 11.42). Mounting both front and rear bale spears on a tractor saves time in handling bales, and the bales provide a counterbalance to each other.

Bale-Wrapping and Bagging Equipment

There are many ways to seal freshly baled forage, including individual bags, tubing machines, and individual or group bale-wrapping machines. All operate on the principle of quickly sealing out oxygen from the bale and keeping it airtight until the baleage is fed. Use of plastic manufactured to withstand the damage from ultraviolet radiation in sunlight is strongly recommended. Some plastic manufacturers recommend using untreated sisal twine or plastic twine. In some cases, the oil from treated sisal twine breaks down the ultraviolet radiation inhibitor in stretch-wrap plastic.

Individual Bags

Using individual bags has two advantages: extra equipment is not required, and the bags can be reused to reduce the cost. In practice, however, few bags can be salvaged for use in the next growing season. Disadvantages include the difficulty of getting all of the air out of the bags and maintaining a good seal on the open end of the bag. Making baleage in individual bags is less reliable than with wrapping equipment. Rodent damage also appears to be more prevalent with individual bags compared to wrapped bales.

Long Tubes

Round bales can be loaded mechanically into long plastic tubes that are mechanically stretched during loading but contact around the bales as they are pushed through the implement. The number of bales per tube is flexible (plastic can be cut and sealed). Disadvantages include the need for a uniform ground area for tube placement (if large) and sizing bales to the tube. Also, a hole in a long tube exposes a large amount of silage to potential spoilage. Finally, large bales stored in tubes are less portable than individually wrapped bales.

Individually Wrapped Bales

Individual bales can be wrapped mechanically with multiple layers of stretch-wrap plastic. Each layer of stretch-wrap plastic adheres to the previous one, forming an airtight seal. Wrapping machines vary widely in cost depending on such features as whether they produce a completely wrapped bale and whether they include a self-loading arm.

The cheapest wrappers require a second person (or getting off the tractor) and manually moving the roll of stretch plastic while the bale is rotated on a spear, much like twine is applied to round bales of hay. The plastic is lapped over the ends of the bale about 12 inches.

Single or multiple bales can be sealed by manually stretching plastic across the exposed ends. Jamming multiple bales together (flat end to flat end) allows the plastic from one bale to stick to the next, forming a tube. A uniform, level soil surface is necessary for good bale-to-bale contact and the maintenance of a good seal.

More expensive wrappers completely cover each bale by elevating the bale onto a rotating and revolving platform (Fig. 11.43). Some have hydraulic lifts to elevate the bales onto the platform. Others require a second tractor with lifting capabilities to put the bale on the wrapper.

Consider moving freshly baled forage to the storage area for wrapping. This allows the wrapping process to be done on more level, uniform ground. Bales can "walk off" the wrapping platform if the machine is not level. Minimizing movement of wrapped bales will reduce tearing of the plastic. Wrapped bales can be speared for movement if these holes are resealed afterward. Hydraulic tongs that grasp a bale without puncturing the plastic are available for front-end loaders of tractors.

In-Line Bale Wrappers

In-line bale wrappers have become the prevalent type of baleage equipment because of their wrapping capacity and efficiency of plastic use (Fig. 11.44). These machines feed high-moisture bales through a rotating cylinder that stretches plastic continuously around the row of bales. This process produces a long row of ensiled bales. In-line machines can wrap more tons per hour and require less plastic per bale than individual wrappers.

FIGURE 11.43 An example of an individual wrapper with a rotating platform. *Photograph courtesy of Dennis Hancock, University of Georgia.*

FIGURE 11.44 In-line bale wrappers are gaining in popularity because of the speed of bale coverage and efficiency of plastic use. *Photograph by Jimmy Henning, University of Kentucky College of Agriculture, Food and Environment.*

Other Considerations

Damage to plastic during handling or storage allows oxygen to enter the bale, causing spoilage. Any holes made during bale transport and placement into storage should be repaired immediately by taping. Holes allow oxygen to enter and lead to problems with silage quality due to aerobic deterioration. To minimize storage losses due to spoilage, bagged silage bales should be fed to livestock during the winter following their production.

Do not feed silage that has significantly deteriorated or has a bad odor. Silage that improperly ferments from being too wet can lead to botulism poisoning. To prevent this, do not make silage at moisture contents above 70%. Exposure to oxygen can also lead to deteriorated silage and animal toxicity. Unrepaired holes or having too few layers of stretch-wrap plastic can lead to oxygen infiltration of the bale. Ideally the fields from which ensiled feed are made should not have been grazed or had any manure spread on them in the year before the forage is harvested, to reduce the possibility of pathogenic fecal bacteria contaminating the feed.

Because *Clostridium, Salmonella, and Listeria* pathogens can all be transported with soil particles from the ground into nutrient-rich forage and kept in conditions that encourage their growth during silage fermentation, all reasonable precautions should be taken to prevent soil and manure from contaminating forage to be ensiled. Maintaining an airtight seal on plastic around silage bales ensures an anaerobic fermentation resulting in a low enough pH that these bacteria do not proliferate (Frape, 2010b).

One other rare occurrence that should be avoided is harvesting forage for ensiling from land over which clay-pigeon shooting with lead shot has occurred. Horses fed silage from such fields have been killed because of the high concentrations of lead (Frape, 2010b).

The ability to make baleage allows the harvest and storage of the fall cut of alfalfa or other forages that come in some years during October and November (Fig. 11.45). In most years, this forage goes unused unless these fields can be grazed since curing conditions are too poor to get the forage dry enough to bale as hay. Ensiling conditions are not ideal during this time (low temperatures and low numbers of ensiling bacteria), and fall baleage should be fed first during the winter. Silage inoculants have been shown to improve the ensiling characteristics of fall forage crops.

Time Between Baling and Bagging or Wrapping

The interval between baling and wrapping or bagging is critical to the success of the ensiling process and should be as short as possible. Prior to wrapping, high-moisture forage is subject to very high respiration rates and to the growth of undesirable microorganisms. Respiration reduces forage quality by consuming readily digestible carbohydrates. Significant increases in bale temperature are also associated with delay between baling and bagging of silage bales. As little as an 8-h delay between baling and bagging resulted in greater temperatures during storage compared with those bales bagged immediately after baling (Henning et al., 1998).

Consider identifying different types of baleage and different cuttings by marking with spray paint. Different colors could represent the various crops, while the number of marks (dots or X's, for example) could indicate the cutting (one dot for first cutting, two dots for second cutting, etc.).

FIGURE 11.45 The ability to make baleage allows the utilization of this fall growth of alfalfa in Northern Kentucky. *Photograph by Jimmy Henning, University of Kentucky College of Agriculture, Food and Environment.*

REVIEW QUESTIONS

1. Describe the relationship between forage yield versus forage quality, digestibility, and intake as forages mature.
2. Describe the chemical and visual/sensory characteristics of high-quality hay for horses.
3. Describe major antiquality/toxic factors in forages and how to avoid them.
4. Describe the two main types of hay cutting implements and their advantages and disadvantages.
5. Describe the two main types of mechanical hay conditioners and explain the goals of conditioning hay.
6. What are the three types of hay rakes and their advantages and disadvantages?
7. What are the major types of bale packages and their advantages and disadvantages?
8. What are the characteristics desired in a good hay storage structure?
9. What are the considerations for choosing a good hay for horses?
10. Describe the pros and cons of the following hay feeding methods for individuals: (a) on the ground; (b) hay bins or troughs; and (c) hay racks or nets.
11. What is the best type of hay feeder for groups of horses and why?
12. What are the advantages and disadvantages of producing haylage?
13. Describe the conditions that favor growth of *Clostridium botulinum* in silage.

REFERENCES

Ball DM, Lacefield GD, Schmidt SP, Hoveland CS, Young III WC. Understanding the Tall Fescue Endophyte. Salem, OR: Oregon Tall Fescue Commission; 2015.

Ball DM, Hoveland CS, Lacefield GD. Characteristics of hay bales. In: Sulewski G, editor. Forage Crop Pocket Guide. thirteenth ed. Peachtree Corners, Georgia: International Plant Nutrition Institute; 2016. p. 56.

Battle GH, Jackson SG, Baker JP. Acceptability and digestibility of preservative-treated hay by horses. Nutr. Rep. Int. 1988;37:83—9.

Blaser RE, Hammes Jr RC, Fontenot JP, Bryant HT, Polan CE, Wolf DD, McGlaugherty FS, Kline RG, Moore JS. Forage-Animal Management Systems. Virginia Agricultural Experiment Station. Blacksburg, Virginia, Bulletin: Virginia Polytechnic Institute and State University; 1986. p. 86—7.

Burrows GE, Tyrl RJ. Toxic Plants of North America. John Wiley and Sons, Inc.; 2013. p. 955—7.

Collins M, Owens VN. Preservation of forage as hay and silage. In: Barnes RF, Nelson CJ, Collins M, Moore KJ, editors. Forages — An Introduction to Grassland Agriculture. Ames, IA: Blackwell Publishing; 2003. p. 443—71.

Collins M, Ditsch D, Henning JC, Turner LW, Isaacs S, Lacefield GD. Round Bale Hay Storage in Kentucky. Cooperative Extension Service. University of Kentucky. Publication AGR-171; 1997.

Digman M, Undersander D, Shinners K, Saxe C. Best Practices to Hasten Drying of Grasses and Legumes. University of Wisconsin — Extension. Publication A3927; 2011.

Fetzer LM, Grafft LJ, Hill DE, Murphy DM, Skjolaas C, Yoder AY. Preventing Fires in Baled Hay and Straw. Farm and Ranch eXtension in Safety and Health (FReSH) Community of Practice; 2012. Retrieved from: http://www.extension.org/pages/66577/preventing-fires-in-baled-hay-and-straw.

Frape D. Ingredients of horse feeds. In: Equine Nutrition and Feeding. fourth ed. Ames, Iowa: Wiley/Blackwell Publishing; 2010a. p. 90—135.

Frape D. Grassland and pasture management. In: Equine Nutrition and Feeding. fourth ed. Ames, Iowa: Wiley/Blackwell Publishing; 2010b. p. 265—304.

Hanna M. Estimating the Field Capacity of Farm Machines. Iowa State University Extension and Outreach; 2016. PM 696/Ag Decision Maker File A3-24, https://www.extension.iastate.edu/agdm/crops/pdf/a3-24.pdf.

Henning JC, Collins M, Ditsch D, Lacefield GD. Baling Forage Crops for Silage. Cooperative Extension Service. University of Kentucky. Publication AGR-171; 1998.

Lacefield GD, Henning JC, Collins M, Swetnam L. Quality Hay Production. Cooperative Extension Service. University of Kentucky College of Agriculture. Publication AGR-62; 1996.

Lewis L. Harvested feeds for horses. In: Feeding and Care of the Horse. second ed. Ames, Iowa: Blackwell Publishing; 2005. p. 64–102.

Martinson K, Wilson J, Clear K, Lazarus W, Thomas W, Hathaway M. Selecting a Round-Bale Feeder for Use during Horse Feeding. University of Minnesota Extension; 2016. http://www.extension.umn.edu/agriculture/horse/nutrition/selecting-a-round-bale-feeder/.

Muller CE, Uden P. Preference of horses for grass conserved as hay, haylage or silage. Anim. Feed Sci. Technol. 2007;132:66–78.

National Research Council. Nutrient Requirements of Horses. Washington: National Academy Press; 1989.

Townsend LH. Blister Beetles in Alfalfa. Cooperative Extension Service. University of Kentucky College of Agriculture, Food and Environment. Publication ENTFACT-102; 2011.

Undersander D, Martin N, Howard T, Shaver R, Linn J. Sampling Hay and Silage for Analysis. University of Wisconsin – Extension. Publication A2309.; 2016.

Wright R. Slobbers or Slaframine Poisoning in Horses. Ontario Ministry of Agriculture, Food and Rural Affairs; 2004. http://www.omafra.gov.on.ca/english/livestock/horses/facts/info_slobbers.htm.

FURTHER READING

Keene TC. How to Sample Hay for Analysis – You Tube. Cooperative Extension Service. University of Kentucky College of Agriculture, Food and Environment; 2011. https://www.youtube.com/watch?v=U1fFswpR1kI.

National Research Council. Nutrient Requirements of Horses. Washington: National Academy Press; 2007.

Chapter 12

Climate, Weather, and Plant Hardiness

Paul Sharpe[1] and Edward B. Rayburn[2]

[1]University of Guelph, Guelph, ON, Canada (retired); [2]West Virginia University Extension Service, Morgantown, WV, United States

INTRODUCTION

A pasture that is well suited and managed for horses should be composed of forages that are adapted to the region and soils where they are grown and tolerant of local weather and climate.

Weather is the condition of the atmosphere at one place over a short time period, and climate is a description of the weather at one place over a long time period. Weather is reported in absolute measurements, while climate is expressed as averages or ranges. Measured elements of weather and climate include temperature, humidity, sunshine, wind speed, wind direction, cloud cover, barometric pressure, precipitation (rain, hail, sleet, freezing rain, snow, ice pellets), cloud cover, and storms. Plants can be affected by low and high temperatures, intensity and hours of sunshine, wind, and precipitation. This discussion will begin with solar radiation, which determines the intensity and hours of sunshine, temperatures, and evaporation of water. This leads to humidity and cloud cover, thus to uneven heating of air, causing wind and differences in atmospheric (barometric) pressure. All of these things lead to precipitation, which provides the water needed for the development and growth of plants.

SOLAR RADIATION

Potential solar radiation above the atmosphere on a horizontal surface is a function of latitude and day of the year. At ground level, radiation is reduced by clouds and dust. Daily solar radiation peaks on June 21 (summer solstice) and reaches its lowest level on December 21 (winter solstice), in the northern hemisphere. In the southern hemisphere, these dates are reversed. Because the earth's north–south axis for rotation is tilted, relative to its path around the sun, the angle of the sun's rays changes predictably during each annual trip around the sun. Table 12.1 shows the sun's average energy in watts per square meter where it strikes two different latitudes, approximating those of Florida and New York through Oregon. The months chosen for the table represent the beginning of spring, summer, autumn, and winter.

TABLE 12.1 Differences in Solar Energy on the Earth's Surface at Two Latitudes, Due to Changes in the Angle of Incidence of the Sun's Rays, in Months Representing the Beginning of Spring, Summer, Autumn, and Winter[a,b]

Months	Florida (25–30°N latitude)	New York and Oregon (42–45°N latitude)
March	1183	942
June	1348	1259
September	1183	942
December	747	322

[a]Monthly average of daily solar energy in Watts per square meter.
[b]Adapted from Whittaker, T., 2017. Explore the Effect of the Angle of Incidence on Sun's Energy. Profhorn.meteor.wisc.edu/wxwise/radiation/sunangle.html.

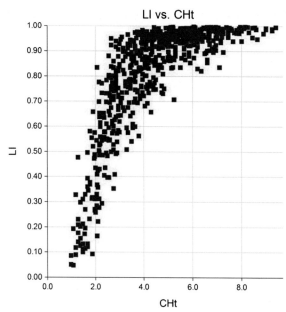

FIGURE 12.1 Fraction of light intercepted (LI) by grass/legume forages versus canopy compressed height (CHt). CHt/0.6 = ruler height (Adapted from Rayburn E, Shockey W, Smith B, Seymour D, Basden T. Light interception by pasture canopies as affected by height and botanical composition. Crop Forage Turfgrass Manag. 2016;2. https://doi.org/10.2134/cftm2016.0013.).

How well a pasture sward intercepts light from the sun is a function of leaf width, angle, and spacing. Legumes intercept more light at low leaf areas than grasses due to their horizontal leaves. Grasses are more efficient in light utilization at high leaf areas since more light will reach into the canopy.

Light interception by a pasture canopy is also a function of forage height, so taller plant canopies intercept more light up to a threshold. For example, as the compressed height of canopy surface rises from 1.7 to 10 or more inches (4.3−24.4 cm), the fraction of light intercepted rises in a curvilinear fashion up to 7.5 inches (19 cm) of height, then levels off at its maximum potential (Fig. 12.1) (Rayburn et al., 2016).

AIR TEMPERATURE

Cool season (mostly C3) grasses and legumes begin growth at temperatures greater than 40°F (4.4°C), have optimum growth near 65−75°F (18.3−23.9°C), then reduced growth to zero as temperature increases to 90°F (32.2°C). In three turns of the Calvin cycle of C3 plants, carbon dioxide (CO_2) is normally fixed (reduced) to 3-phosphoglycerate (3PG) by the enzyme "rubisco," and two of these 3-phosphoglycerates join to become one glucose. Since rubisco will bind oxygen to the same site that binds CO_2, and the relative concentrations of oxygen and CO_2 in the atmosphere are 21%:0.04%, sometimes oxygen binds to rubisco, and this leads to production of phosphoglycolate (a 2-carbon acid), which is respired to carbon dioxide and is lost to plant function. This "photorespiration" step represents a loss of carbon and thus of 15%−40% of the light energy taken into C3 plants. As air temperature rises from 41 to 95°F (5−35°C), the solubility of CO_2 decreases faster than solubility of oxygen, so more oxygenation and photorespiration occurs, causing the increasing loss of carbon and inefficient use of light energy.

Warm season (mostly C4) grasses start growth as the temperature rises above 50°F (10°C) and reach a maximum as temperatures exceed 85°F (29.4°C). C4 plants have the Calvin cycle apparatus to use CO_2 and make glucose, plus they have a C4 pathway that takes in CO_2, uses ATP for energy, and delivers a high concentration of CO_2 to the Calvin cycle. In hot, dry climates, there is usually plenty of sunlight, so plenty of ATP is available. The enzyme bringing CO_2 into the C4 cycle has a high affinity for CO_2 and no affinity for oxygen, so C4 plants are more efficient at fixing CO_2. C4 plants do not need to keep their stomata open as long as C3 plants to fix the same number of CO_2 molecules, so there is less time for water loss. Thus, C4 plants are better suited to arid conditions.

Fig. 12.2 shows relative growth rates of C3 and C4 plants at increasing temperatures. Air temperature can indicate rates of plant growth rate. If mean air temperature in a warm climate is >90°F (32.2°C), no cool season forage growth can be expected. If such hot weather continues, plans can be made to provide alternative feed. In northern regions, where the mean July temperature is 68°F (20°C), warm season grasses will have less growth than where the temperature is >80°F (26.6°C).

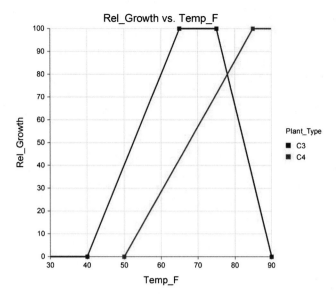

FIGURE 12.2 Effect of mean air temperature on the relative growth rate of cool season (C3) and warm season (C4) forage plants (Rayburn et al., 1998).

Pasture managers following forage management recommendations should ensure that the information is appropriate for their climatic region.

Peak growth rate of cool season grasses and legumes occurs between 65 and 75°F (18.3–23.9°C), whereas peak growth rate of warm season grasses occurs from 92 to above 100°F (33.3–37.8°C). Rises and falls of air and soil temperature, which have effects on seed germination and the rate of metabolism of plants, lag behind solar radiation variations.

If native and tame cool season grasses are allowed to develop to a four-leaf stage before the first grazing in a season, the growing point is elevated and available for grazing. The jointing or long shoot grasses (timothy, smooth bromegrass, and reed canarygrass) also exhibit this trait in aftermath growth. The rate of development and date of reaching a particular stage of growth is dependent on the amount of thermal energy accumulated (Frank, 1996). Before this stage of development, grazing may result in stems devoid of leaves, a low dry matter production, and weak, thin grass stands (Frank, 1996).

Forage plant structure and function, identification, and nutritional value are the topics of Chapters 1–3, respectively. Most of the common forage species used in North America are discussed to some extent there. Pasture managers working in the American South will want to study the book *Southern Forages, Fourth Edition* by Ball et al. (2007). Anywhere in the northern half of the continental United States where summers are hot and/or dry, pasture managers may be able to make use of warm season forages to compensate for a summer slump in production of their cool season forages, so the *Southern Forages* text may be useful there also. Most of the forages used in the northeastern quarter of the United States and the eastern half of Canada are introduced (nonnative or tame). In the hotter, drier regions, native forage species play a greater role than introduced species. We should remember that C3 forages evolved to fit ecologic niches with relatively short growing seasons, freezing weather in winter, and summers that are not excessively hot, whereas C4 forages evolved to fit environments with longer growing seasons, mild winter weather, and summers with long periods of very hot weather. So it should come as no surprise to discover that in Florida, 70% of all native plants are C4 species, and in Manitoba (north of North Dakota and Minnesota), all native plants are C3 species (Russell et al., 2013).

In Iowa, the warm season grasses, switchgrass, Indiangrass, and big bluestem are suited to summer pasture use and are winter hardy. They can be seeded alone or as mixtures, although cool season grasses tend to out-compete these warm season grasses for resources in the spring (Barnhart, 1994). Wherever both warm season and cool season perennial grasses will grow, it is advantageous to grow both and capitalize on their advantages. According to Hoveland (1996), cool season grasses have a greater temperature range for photosynthesis and warm season grasses are more efficient in water use, thanks to deeper roots and the C4 cycle. Warm season grasses also produce more dry matter per unit of nitrogen fertilizer. This apparent advantage is reflected in forage quality since the nitrogen taken up by C4 plants is diluted by the increased amount of dry matter, so protein concentration of warm season grasses is lower than cool season grasses (at the same stage of development), and concentrations of digestible energy and protein decline faster with maturity in warm than in cool season grasses.

Cool season forages and grasslands support horses and production livestock throughout the northeastern United States. Farther south, warm season species are used, in addition to cool season species, in an area called the "transition zone" between the cool-temperate and warm-subtropical regions. Tall fescue is grown on about 10 million acres in this transition zone (sometimes with red clover), and warm season grasses such as Bermudagrass can be used in separate pastures during the hottest part of summer. These two grasses are also grown together in pastures, and their diversity of maturation times may require management adjustments. The annual, crabgrass, is leafy, with higher nutritional quality than some warm season perennial grasses, and it is often grown in mixtures with tall fescue. In the southeast, livestock are fed Bermudagrass and Bahiagrass much of the year, plus cool season annual legumes and grasses in autumn and winter (Sanderson et al., 2012.).

Many generalizations can be made about properties of different forage species and about differences among cultivars (varieties). Many varieties are developed with a particular region or temperature profile in mind. Due to interactions between genotypes and the environments in which they are grown, it is advisable to find variety test results from a research station near you and contact local seed companies to learn which of the locally superior varieties they sell. Some criteria for selecting forage species and varieties include these:

1. Decide whether the most frequent use will be pasture or hay.
2. Ask for information on tolerance to cold temperatures, freezing, hot temperatures, level of precipitation, and frequency of drought that match your environment, including your soil type and drainage.
3. Look for varieties with high resistance to known pests and diseases of plants that are common in your area.
4. Try to match days from starting growth to maturation (heading) among all the forage species within a mix.
5. Pay attention to data on nutritional quality, grazing tolerance, and palatability to horses if such information is available for species and varieties of interest (Martinson and Sheaffer, 2013; Olson et al., 2017).
6. If some of the grazing horses will be lactating, ensure that the forage for them will have high concentrations of digestible energy, crude protein, and calcium.
7. Replace older varieties of endophyte-infected tall fescue with newer, endophyte-free varieties or with varieties containing novel nontoxic endophytes such as Jesup or Lacefield MaxQ II. The endophyte *Neotyphodium coenophialum* contained in older tall fescue cultivars, such as Kentucky 31, causes reproductive problems when fed to mares in late pregnancy, agalactia during lactation, and sometimes deaths of mares or foals (Ball et al., 2007). During a transition between tall fescue varieties, prevent late-pregnant mares from grazing or eating hay from older tall fescue varieties.

GROWING DEGREE DAYS

Breeders of grains and forages use different labels when describing for which areas of the continent a new variety of crop is designed. Forage breeders use "growing degree days" (GDD)$_{alfalfa}$. It is known that the average growing season temperature needs to exceed 41°F (5°C) for alfalfa to grow and develop, so the formula for GDD$_{alfalfa}$ is this:

$$GDD_{alfalfa} = \left(\frac{T_{max} + T_{min}}{2}\right) - 41$$

T_{max} = Daily high temp (°F) and T_{min} = Daily low temp (°F).

To use Celsius temperatures, substitute 5°C for 41°F. GDD values of zero or negative numbers are considered to be zero and indicate no growth or development that day. The sum of all GDD that accumulate through a growing season is the total GDD for that region in that year. Different varieties of alfalfa require different numbers of total GDD to reach specific developmental stages and thus are suited to some regions and not to others (Bootsma, 1984; Baron and Bélanger, 2007). Forage breeders work for governments, universities, international research centers, and private seed companies (Crop Science Society of America) and can cooperate with each other to test new varieties in different locations to determine how many GDD each variety needs. When farmers purchase seeds, they can choose their varieties based upon the number of total GDD they are expecting on their farm in an average year. More information is available from agricultural extension agencies and agronomy extension professors (Noland and Wells, 2017).

Alfalfa

In the Midwest of the United States, the first cut of alfalfa in a growing season should correspond to an accumulation of 700−750 GDD, and this corresponds to a concentration of Neutral Detergent Fiber [NDF] of about 35%. Sanderson (1992) examined alfalfa plant development data from Iowa and Texas and discovered that concentrations of [NDF] and acid detergent fiber [ADF] in alfalfa stems from Iowa were predicted accurately from GDD equations, but for Texas data, GDD was inaccurate and imprecise. He suggested that calibration of equations for geographic or environmental conditions may be necessary. In Rosemount, Minnesota, GDD accumulation reaches 700 around May 25.

Researchers in Ohio, Wisconsin, South Dakota, and New York compared three methods of alfalfa quality estimation: (1) scissor cutting and analysis of NDF; (2) predictive equations of alfalfa quality (PEAQ) derived from stem length and maturity, and (3) predicting alfalfa fiber content from GDD (Sulc et al., 1999). They reported greater NDF predictive ability for PEAQ than GDD and suggested an early spring NDF analysis paired with historic GDD accumulations to predict optimal dates for first alfalfa harvest, followed by PEAQ to estimate [NDF] close to the predicted optimal date (Sulc et al., 1999).

An example of using GDD for predicting harvest dates is that dairy farmers often want to harvest alfalfa at [NDF] of $40 \pm 3\%$. In many areas, this occurs following a GDD accumulation of 750. Michigan State University Extension recommends beginning the first cutting of the season at 680 GDD, corresponding to about 38% NDF (Kaatz, 2011; Allen et al., 2016). GDD is not recommended for predicting [NDF] for second and later cuttings (Lee et al. (2010). The PEAQ technique, facilitated by a specially designed PEAQ stick, is also recommended for first and second cuttings of alfalfa that receive adequate moisture and other growing conditions. PEAQ sticks may provide estimations of NDF, ADF (which can be used in energy calculations), or relative feed value (RFV, which uses ADF and digestibility in its determination and is appropriate only for alfalfa) or relative feed quality (RFQ, which is appropriate for grass/legume mixtures and pure grasses) (Jeranyama and Garcia, 2004). Neither GDD nor PEAQ was reliable for predicting NDF of third-cut alfalfa, but a scissor cut and lab analysis of NDF by wet chemistry was reliable (Lee et al., 2010).

Where alfalfa averages 40% NDF at 750 GDD (base 41°F (5°C)), it takes about 220 more GDD to reach 45% NDF. Planning a first alfalfa cut near 700–750 GDD achieves [NDF] near 35% at harvest, which results in [NDF] near 40% after storage as hay (Noland and Wells, 2017).

Alfalfa GDD figures are different from GDD figures designed for grain crops (Allen et al., 2016).

Grasses

GDD had a statistically significant (regression) relationship with Haun growth stage for five prairie grasses in North Dakota. These grasses had four, five, or six leaves by the time of heading (Frank and Hofmann, 1989). Morphologic development of orchardgrass, smooth bromegrass, reed canarygrass, and tall fescue in Iowa were closely related to accumulated GDD ($r^2 = 0.97$) and to a decline in forage quality, as expressed by concentrations of in vitro digestible dry matter (IVDDM) and crude protein (CP) (Buxton and Marten, 1989). Similarly, development of crested wheatgrass and western wheatgrass in North Dakota during initial growth and after one to three grazings was highly related to GDD ($r^2 = 0.86 - 0.99$), so GDD is useful for making decisions about grazing readiness of grasses (Frank, 1996).

Developmental stages of forage and range grasses were quantified in a system called Mean Stage Count (MSC) by Moore et al. (1991). In predicting leaf/stem ratios and nutritive value of big bluestem, Smart et al. (2001) reported that MSC and GDD were adequate predictors of leafiness, NDF, and CP, but their accuracy was less than in other studies. They also reported that the number of days between May 1 and the day of analysis was a better predictor than the day of the year.

Brueland et al. (2003) recorded developmental morphology and nutrient concentrations of smooth bromegrass in Iowa at different initial grazing dates. CP was linearly related with MSC ($r^2 = 0.87$); IVDMD was linearly related to MSC ($r^2 = 0.83$); NDF and MSC were linearly related ($r^2 = 0.92$).

In British Columbia (B.C.), Canada, GDD have also been used to predict readiness of several native and introduced grasses for range grazing (Fraser, 2006). Leaf-stage development in B.C. was more advanced in plants lightly grazed previously, versus plants that were heavily grazed, possibly due to more leaf litter insulating new plant tillers in spring and to greater root reserves following light grazing. Frank (1996) recommended studying young grass plants, recording development stages, and using this information along with GDD to determine when grazing can begin on rangeland and pastureland, rather than using calendar dates.

Growth temperature affects the digestibility and NDF concentration of temperate grass species, as demonstrated when grasses were grown at day/night temperature combinations of 48/41, 55.4/48, and 62.6/55.4°F (9/5, 13/9, and 17/13°C). The rate of decline of digestibility increased as temperature increased. NDF increased at 0.78 g/kg of DM per day for each degree increase in temperature between 9 and 17°C. Perennial ryegrass and meadow fescue were less susceptible to temperature-related changes in NDF than Kentucky bluegrass, timothy, meadow foxtail, red fescue, and tufted hair-grass. Perennial ryegrass and meadow fescue were also highest in digestibility and had slower declines in digestibility than other species (Thorvaldsson et al., 2007).

GDD provide an estimate of the growth stage of a plant, insect, or microorganism. The warmer the weather, the faster GDD accumulate until the maximum threshold temperature is reached. The base temperature used in the GDD equation can vary with the biology of the organism. Insects of the genus *Lygus* are plant feeders that become yield-lowering pests in a variety of crops, for example, hay in Nevada. Some *Lygus* eggs begin to hatch at 160 GDD, and the peak of hatching in Nevada occurs at about 250 GDD, which is the optimal time to implement a control measure (Cherney and Sulc, 1997; Breazeale et al., 1999).

If you have targets for NDF, ADF, and CP for hay or pasture, comparisons over years can help you in deciding when to cut or graze the forages of greatest importance to you. For some regions, GDD are calculated and maintained by institutions. Many web-based weather sites record historic daily high and low temperatures that you can chart. Weather data to help with GDD analysis can be obtained for Midwest states from the Midwest Climate Center tool, CLI-MATE (http://mrcc.isws.illinois.edu/CLIMATE/).

Graphs of average cumulative GDD at each day through the growing season can be used to predict desired harvest dates for nearby farmers who do not record GDD themselves but who may have information on forage quality and yield (Noland and Wells, 2017).

If you want to use temperatures that you record on your farm, the temperature sensor should be in a ventilated, white enclosure 5 feet (1.5 m) above the ground with the north side and the bottom open and some air circulation space between the temperature sensor and the enclosure (Allen et al., 2016). A data logger receiving data from the thermometer will be useful to prevent loss of data.

Diversity Among Varieties

The effect of new forage varieties is to improve growth potential, but the magnitude of this effect is low compared to fertility and harvest management. Within orchardgrass and endophyte-free tall fescue varieties, the largest variety effect on pasture may be the effect of heading date on early season pasture growth and the decline in nutritional quality that accompanies heading (Rayburn, unpublished data). Late maturing orchardgrass varieties are less productive for early season pasture than early heading varieties.

For use in the northeast quarter of the United States, forages with high winter survival rates include Kentucky bluegrass, reed canarygrass, smooth bromegrass, timothy, alfalfa, and birdsfoot trefoil. Medium winter survival is exhibited by orchardgrass, red top, tall fescue, alsike clover, red clover, and white clover. Only perennial ryegrass has low winter survival, among typical cool season forages used in the United States (Rayburn et al., 1998). Based on testing in Minnesota, cool season perennial grasses that rank highly for horse preference, forage persistence, and yield include orchardgrass, meadow fescue, tall fescue (endophyte-free), and Kentucky bluegrass (Martinson and Sheaffer, 2013). Other grasses that have poor persistence and yield include smooth bromegrass, creeping foxtail, and timothy.

One of the most heat tolerant of the C3 grasses is tall fescue. Endophyte-infected varieties of tall fescue tend to tolerate heat and drought better than the early nonendophyte-infected varieties. An adaptation zone for tall fescue extends south to 32°N latitude in several southeastern states and extends into northern Florida, around 30°N latitude. Some varieties of tall fescue survive winters in Ontario, Canada, around 45°N latitude.

ELEVATION AND TOPOGRAPHIC POSITION

Daytime air temperature decreases with height at a rate of 3 to 5°F (1.7−2.7°C) per 1000 feet (305 m) of elevation. During clear nights, cold air flows off hill tops into valleys, causing night temperatures to be 8−15°F (4.4−8.3°C) lower in small valleys than on the upper slopes. As a result, spring frosts occur later and fall frosts occur earlier in small valleys than on adjacent hill tops, shortening the growing season for frost-sensitive crops. Small valleys will not be the best areas to grow temperature-sensitive warm season grasses.

PRECIPITATION

Average annual precipitation in the northeast United States varies from 32 to 52 inches (813−1320 mm). Pacific coast regions of the United States and Canada tend to have high precipitation from moist air moving off the Pacific Ocean due to prevailing westerly winds and rising as it hits higher elevations. As the air cools, it can hold less moisture, so rain or snow falls. In the east, high rainfall is due to air being forced over lakes Erie and Ontario, then over the eastern mountain ranges. Less precipitation occurs on the eastern "rain shadow" sides of local mountains, such as in the Shenandoah Valley in West Virginia and the Finger Lakes region in New York. Spokane, Washington, is in the rain shadow of the Cascade Mountains and has an average annual precipitation of only 16.5 inches (420 mm). In several eastern states, yield of cool season grasses and legumes varies by about 20% across years, due largely to precipitation variation. There is variation in forage growth rate within the year due to the distribution of monthly precipitation. Lake-effect snows often cover the ground in upstate New York by late November, making grazing impractical after that. Fifty miles east, out of the lake-effect area, grazing is possible in most years until the end of December, if forage is available (Baron and Bélanger, 2007).

SEASONAL CHANGES IN FORAGE QUALITY

Other chapters have revealed details of the decline in concentrations of protein and energy as plants mature and increase in yield. C4 grasses have more fibrous cells, more indigestible lignin, and lower overall digestibility. In cool season grasses, short day length in fall stimulates tiller (stems with leaves) formation, and long day length in the spring stimulates reproductive tiller elongation. Moisture stress or cool fall temperatures can reduce forage growth and cause accumulation of nonstructural carbohydrates (NSC) in forage plant tops and stubble (Rayburn et al., 1998). High NSC can be a contributing factor to laminitis in horses that are prone to it, so alternative forages should be made available to these horses in times of drought and cooling temperatures.

REGIONAL CLIMATIC EFFECTS

In the reference text *Forages, The Science of Grassland Agriculture, Sixth Edition*, Baron and Bélanger (2007) explained systems of classifying portions of North America as "ecoregions" and described the "Koppen Classification Groups" and "types" within groups that are roughly equivalent to each ecoregion. The ecoregions are labelled with numbers, and the types are described in words, for example, "tropical wet" and "temperate oceanic." Koppen climatic regions are defined by combinations of average monthly temperatures, numbers of months above or below 0 or 10°C (0 or 50°F), whether warmest months average above 22°C (71.6°F), and whether there is a dry season. Anyone wishing to learn more about the climatic details of their region and how it affects which forage species thrive there will find Baron and Bélanger (2007) useful. The chapter explains that across North America there is an increase in average July and January temperatures from the northwest toward the southeast, and this is altered by increases of altitude and large water bodies. The average annual temperature decreases by 3.5°F with each 1000 feet (6.4°C/1000 m) of elevation, as mentioned previously. Cool season grasses are most prolific north of about 41°N latitude (the southern border of Wyoming and Iowa). The northern boundary for the warm season Bermudagrass is about halfway up the State of California (35°N). Then it dips farther south in Nevada, almost reaches the southernmost part of New Mexico (32°N), and from there, it runs northeast toward Maryland and New Jersey (39°N) (Baron and Bélanger, 2007).

Maps are available to show the distributions of mean high and low air temperatures by month for all of North America (The Weather Channel, 2017). This series of maps from the Weather Channel uses colored bands to show regions of the United States with average low and average high Fahrenheit temperatures in the 0, 10, 20, 30s, etc. In almost all months, there are four or five bands running fairly straight east and west from the Atlantic coast, westward and curving to the north in the central plains states. In the west coast states the bands continue generally north—south. Other maps indicate rates of precipitation and evaporation.

Past weather data for 122 regions of the United States are available from the National Weather Service (https://www.weather.gov/help-past-weather). Detailed average monthly temperature maps for each of 344 climate divisions of the contiguous United States are available from the National Oceanic and Atmospheric Administration (NOAA, 2017). Also available through this site are maps of "Monthly Temperature Outlook" and "Drought Monitor," which can be useful for deciding when to plant forages. The US Drought Monitor map is updated each Thursday to show intensities and locations of drought. Maps of the "Evaporative Stress Index" can also be helpful to estimate development of drought and wet conditions.

Precise regional climate information for the United States is provided by the National Weather Service (NWS) Climate Prediction Center (CPC), which produces weekly maps for each year, showing average temperature, extreme maximum and minimum temperatures, departure of average temperature from normal, and total precipitation (NWS CPC, 2017). Archives of 1-month and 3-month periods between 1999 and 2015 are available to show total precipitation, percent of normal precipitation, average temperature, and temperature departure.

NOAA's National Centers for Environmental Information (NCEI) supply 30-year batches of historical weather information with recent data available in graph and table form (NOAA NCEI, 2017). Searchers within this site can select the state, then location within state, and the month to obtain monthly, daily, annual, seasonal, and hourly normal values for temperature and precipitation. Canadian climate normal values are available for the 30-year periods 1981–2010, 1971–2000, and 1961–90 (www.climate.weather.gc.ca/climate_normals/).

Cold Temperature Effects

Cold temperatures have greater effects than warm temperatures on forage species distribution, especially north of 37°N latitude. Low winter temperatures of the semiarid northern great plains require plants to have high cold tolerance because there is little snow to insulate the plant crowns from freezing (Bailey, 1996).

Winter hardiness includes freezing tolerance and other factors. Freezing injury can include physical damage to plant cells when ice crystals form inside them or adjacent to them. Growth of ice crystals can dehydrate fluid compartments to the extent that dissolved chemicals, including proteins, precipitate out of solution and are no longer able to contribute to cell function (Volenec and Nelson, 2007a).

Thermal energy stored in the soil can provide enough heat for some plants to survive short, mild periods of freezing temperatures. Solutes such as sugars, ions, and amino acids accumulate in cells during autumn to provide an antifreeze effect, depressing the freezing point to about $-4°C$ (Volenec and Nelson, 2007a). Chilling injury can be avoided by plants that are able to acclimate to cool temperatures over several days prior to very cold days. Plant cell membrane fluidity can increase if the concentration of saturated fatty acids decreases. Large differences in cold temperature tolerance exist among species and cultivars, and it is possible to select genetic lines for more cold tolerance. Shortening photoperiod and decreasing temperatures stimulate the cold acclimation process (Baron and Bélanger, 2007).

Forages planted in autumn need sufficient time to develop at least five leaves for grasses and eight leaves for legumes to avoid winter injury. Ensuring a high enough concentration of available potassium (K) and a low enough concentration of nitrogen (N) will promote autumn dormancy. Some plants, especially legumes in northern temperate areas, need 6 weeks without harvesting to accumulate NSC before dormancy begins. If legumes are grazed or cut during this 6-week critical period, they may not be able to accumulate enough solutes to decrease the freezing point. Regional extension agencies can advise on the critical date range during which alfalfa should not be cut or grazed to promote winter survival. The plant material remaining above ground traps snow, providing insulation and protection from desiccating wind (Volenec and Nelson, 2007b).

North to South and East to West Effects in North America

East of 98°W longitude, average annual precipitation increases from about 20 inches (510 mm) near the border between the Dakotas and Minnesota, to around 51 inches (1300 mm) near the Atlantic coast. In this same section of the United States, the ratio of actual evapotranspiration (AE) to potential evapotranspiration (PET) increases sharply (Baron and Bélanger, 2007; Thompson et al., 1999).

Westerly winds from the Pacific Ocean rise over western mountain ranges, dropping moisture on the coastal side, then descending and warming on the eastern slopes, causing evaporation. Thus deserts and semiarid regions dominate a band from southeastern California to western Texas in the south and extending northward to parts of B.C. and Alberta in Canada. The central prairie states and provinces have moderate rainfall. There is a trend for annual precipitation to increase from 23.6 inches (600 mm) near the boundaries of Minnesota, Ontario, and Manitoba, increasing southward to >47 inches (1200 mm) in Alabama, east of 98°W longitude.

Precipitation alone does not determine the available water supply for plants. AE is the quantity of water actually removed from a surface due to the processes of evaporation and transpiration. AE represents how much water is evapotranspired and is limited by the amount of water that is available. PET is a measure of the ability of the atmosphere to remove water from the surface through evaporation and transpiration, assuming no limit to the water supply. PET is driven by energy received from the sun and wind. PET is also the demand or maximum amount of water that would be evapotranspired if enough water were available from precipitation and soil moisture. AE is always less than or equal to PET. In arid climates, PET is always greater than precipitation. The ratio of AE to PET is highest, approaching 1.0, in areas where AE, PET, and rainfall are high (Baron and Bélanger, 2007).

West of 98°W longitude, AE:PET ratio is near 0.5 because plant water demand is high and water supply is low. Natural grasslands in the eastern prairie region are referred to as tallgrass. To the west, where average annual precipitation and AE:PET are lower, is the mixed grass prairie, and farthest west is the short grass prairie where precipitation is lowest. Big bluestem is a native warm season grass that tolerates drought and high temperatures. Its natural ecotypes became adapted to a zone extending from northern Texas, north to southern Manitoba and Saskatchewan, close to 98°W longitude (Thompson et al., 1999).

North of 40°N latitude and west of the big bluestem area, C3 and some C4 grasses adapted to dryland conditions are common. These include crested wheatgrass, Russian wild ryegrass, and alfalfa. Farther south where heat and drought are more severe, warm season C4 grasses including blue grama predominate.

The Transition Zone

A "transition zone" between predominance of cool season and warm season forages is recognized by forage agronomists and turf scientists. The transition zone is the area that is the transition from where cool season forages dominate to the area where warm season forages dominate (Don Ball, personal communication). Chapter 6 contains a map that shows the location of the transition zone. General climate zones of the United States are considered "cool/humid" in the northwest and northeast, "cool/arid" in the northern prairies, "warm/arid" in the southwest, and "warm/humid" in the southeast. The transition zone for lawn grasses is in the eastern half of the country between the Cool/Humid and Warm/Humid zones (Lawn Care Academy, 2017).

Plants using C3 photosynthesis, actively growing at low temperatures, and with good winter hardiness are predominant in the Cool/Humid areas of northeastern United States and eastern Canada. In the Cool/Dry areas of prairie and western North America, C4 grasses with good winter hardiness are most common. The transition zone consists of areas where introduced species of cool season (C3) and warm season (C4) forages are planted and grazed. High temperatures limit summer growth and persistence of most C3 forages in the transition zone. Cold winters limit growth and persistence of C4 forages. Some native warm season and subtropical grasses survive the mild winters of the transition zone, but few are used for managed grazing.

The transition zone has been referred to as "a warm-temperate (humid mesothermal) region of the east central United States where annual rainfall (from about 39.4−55.1 in) (1000−1400 mm) exceeds annual potential evapotranspiration" (West and Waller, 2007). Its climate was further described as having a frost-free period of 180−210 days per year, overall mean temperature of 55−65°F (13−18°C), and between 70 and 100 days per year with temperatures over 86°F (30°C) (West and Waller, 2007). While the shape of the transition zone is roughly shaped like a broad bean pod or a potato in some descriptions, it is said to exist in mid-Atlantic states and extend westward to 96°W longitude in eastern Oklahoma and from 34°N to 38°N latitudes, except near the east coast where the southern boundary approaches 36°N (West and Waller, 2007). Cherney et al. (2007) stated that the eastern edge of Nebraska and Kansas plus the northern edge of Kentucky and Virginia are in the temperate humid zone, whereas the southern edges of Missouri, Illinois, and West Virginia are in the transition zone. Warm season species predominate from May through October in the southern transition zone. Tall fescue is one of the most heat tolerant of cool season grasses and can survive summers in the transition zone, but most of its growth is limited to fall, winter, and spring.

The American Horticultural Society has a heat zone map for the United States that helps growers avoid planting species and varieties where damage and death from excess heat is likely. There are 12 heat zones corresponding to the average number of days per year that a region experiences temperatures over 86°F (30°C) (called "heat days") and thus suffer physiologically (http://ahsgardening.org/gardening-resources/gardening-maps/heat-zone-map). In zone 1, there is less than one heat day, and in zone 12, there are more than 210 heat days per year. Zones 7 and 8 correspond approximately to the transition zone between warm season and cool season forages and account for elevation (Fig. 12.3) (Rayburn, personal communication).

Tall fescue and Bermudagrass can be well-managed in separate pastures within the transition zone, so that tall fescue provides the majority of the spring and autumn grazing and Bermudagrass provides the bulk of the summer grazing. Cool season grasses grow best between 68 and 77°F (20 and 25°C), whereas 86−95°F (30−35°C) is optimal for warm season grasses. Orchardgrass, Kentucky bluegrass, tall fescue, and annual Italian ryegrass are other cool season forages commonly used in the transition zone, along with alfalfa, red clover, and white clover. Other warm season grasses planted or naturalized in the transition zone include Johnsongrass, crabgrass, and foxtail (West and Waller, 2007). Endophyte-free tall fescue is used in the northern part of the transition zone and overlaps with the adaptation zone for smooth bromegrass, which tolerates drought better than tall fescue (Baron and Bélanger, 2007).

Restrictions to Growth

Warm season grass growth is slowed at 59°F (15°C) and below. Cool season grasses grow slowly below 44°F (7°C), and some may still be growing at the freezing point. Each species and variety within species will have its own ideal range of temperatures, moisture, hours of sun, and soil characteristics.

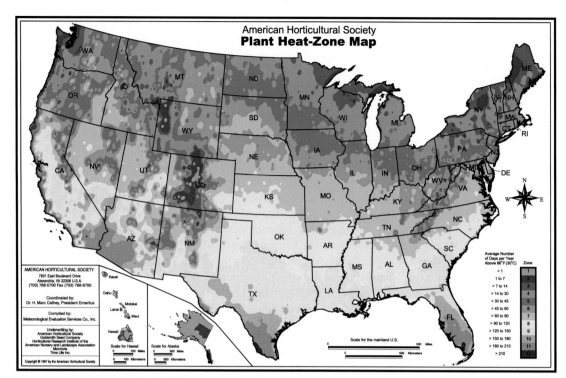

FIGURE 12.3 American horticultural society heat zone map (www.ahs.org).

The length of the growing season decreases from over 300 days in the southern transition zone to about 90 days in parts of western Canada. Plant growth is progressively limited from south to north by the following factors, in order of importance: average dates of spring and fall frosts, mean air temperature below (32°F) (0°C), then by frozen soil. All of these restrictions can take effect north of 45°N latitude. In the southern transition zone, south of 33°N, frozen soil is unlikely, but plant-killing frosts can occur. In the midwestern United States and southern Ontario, cool season grass growth begins between April and June, and this may be as late as May to early July in northern Canada. In cool/arid regions of the Midwest, low summer rainfall and high temperatures can cause cool season grasses to become dormant. Warm season grasses display the peak of their production between June and September. As weather cools off in September through November, cool season species that have almost gone dormant in summer can resume growth (Baron and Bélanger, 2007). In Mediterranean and semidesert climates, the growing season may be limited more by the lack of rainfall in a summer dry season than by a lack of heat and solar radiation in winter.

Fig. 12.4 shows examples of growth rates of forage species in three different North American locations. The top graph illustrates the growth of cool season perennial grasses, alfalfa, and annual grasses or cereals that start growth in that chronological order in western Canada (52°N latitude, west of 95°W longitude). The second graph is typical of a Midwest location (40°N latitude, east of 95°W longitude; approximately at St Joseph, Missouri) and shows cool season grasses starting and peaking in growth earlier than in the top graph. Warm season species peak in early summer, when the growth of cool season species is declining. A warmer climate is represented in the third graph at 33°N latitude, east of 95°W longitude, where warm season grasses have a much longer productive season than in the second location and the cool season species peak in April. Then their growth declines rapidly as the warm season growth is rising. At this latitude, winter annuals can be grown from October with moderate growth until February, when growth rate increases due to increasing day length and temperature, peaking in March while the cool season species are starting to grow. The peak of cool season grass growth begins earlier at decreasing latitudes.

Drought

Aridity is a general, continuous lack of rainfall. According to the Society for Range Management (1989), drought is a period of time with less than 75% of the average annual amount of precipitation. According to that criterion, in the 40-year period 1944−84, drought occurred in 43% of years in the southwest, 13% of the years in the northwest, 21% of years in the

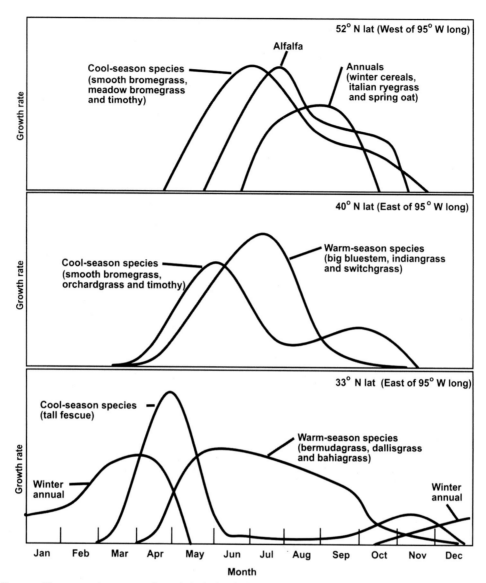

FIGURE 12.4 Patterns of forage growth rates according to latitude, longitude, and management. *Adapted from Fig. 6.11 Baron, V.S., Bélanger, G. 2007. Climate and forage adaptation. Chapter 6. In: Barnes, R.F., Nelson, C.J., Moore, K.J., Collins, M. (Eds.), Forages, the Science of Grassland Agriculture Volume II 6th Edition. Blackwell Publishing, pp. 83–104.*

northern great plains, and 27% of years in the southern great plains. Effects on vegetation can include significant changes in the proportions of different plants. For example, black grama grass that once dominated parts of New Mexico was reduced by 60% and honey mesquite invaded much of that area (Holechek et al., 2011a). Poisoning from toxic plants frequently becomes more frequent during drought due to the scarcity of usual forage plants (Vallentine, 2001). Since much of the western United States, Australia, and parts of Africa can expect drought about 3 years in every 10, it would be wise to implement advance planning in those areas (Holechek et al., 2011b). Forage managers in drought-prone areas should share experiences and accumulate knowledge of weather patterns, price patterns, and forage conditions to facilitate making decisions about reducing livestock numbers, using stored supplemental feed, buying feed, leasing more grazing land, and taking animals off rangeland (Holechek et al., 2011b).

With the large amounts of weather and related data that can be gathered and communicated electronically and wirelessly now, some early warning systems of drought detection should be feasible. Satellite imagery, geographic information systems, rain gauges in several key locations connected to a central agency, and airborne drones equipped with a variety of sensors should be able to provide data to help detect and characterize the severity of drought early in its onset. Measuring pasture growth weekly or biweekly provides a forage inventory that can be used at the farm level to determine if the forage growth rate is adequate for the animal feed demand as a predictor to impending drought.

In arid rangelands of some developing countries, wells have been drilled to provide drinking water for animals in areas not previously grazed. Without considering all the impacts this development of water points could have, the developers have witnessed unfortunate consequences. Livestock and wild animals were attracted to the watering areas and overgrazed the forages because the grazing was not managed. If the water points were in remote areas, maintenance was often insufficient (Holechek et al., 2011c).

Drought is a cause of slowed growth, low forage yields, shorter periods of high-quality forage, reduced vigor, and sometimes death of plants. More exposed soil leads to soil erosion and further drying of soil (Vallentine, 2001). Drought-resistant plants tend to have deeper roots than plants that are highly susceptible to drought or they have subterranean growing points. The physiologic reason for lack of growth is the basic formula for photosynthesis. Chlorophyll contains the enzyme rubisco that converts CO_2 plus H_2O to 3-phosphoglycerate (3PG). Then, two 3PGs are joined to make glucose, which provides energy for other plant functions, including growth. If there is a lack of water, plant functions slow, and if the lack is extreme, the plant will go dormant or die. Shoot growth slows, followed by stomatal closure and decline in the rate of photosynthesis, even in mild drought. In more severe drought, leaf growth stops, and to conserve water, leaves curl to reduce surface area and stomata remain closed. Drought is one situation in which plants reduce metabolism of nitrate, causing it to build to concentrations that can be toxic to animals grazing the plants (Collins and Hannaway, 2007).

Forage species vary in the structure of their roots and in root response to drought. Some species reduce shoot growth after mild drought then increase it after a longer drought. Some legume forages, such as alfalfa, have a long tap root in most varieties, which helps them reach deep water, so they resist drought better than most grasses. Young white clover plants also have a deep tap root, but it can be replaced by shallower roots and stolons in older plants (McGraw and Nelson, 2007). Smooth bromegrass resists drought better than orchardgrass and bluegrass due to its deeper roots and rhizomes (Volenec and Nelson, 2007b).

Plant species evolved to fit environments that varied in amount of soil moisture; thus, pasture managers' choices of which forage species to use must take moisture into consideration. Most of the soil moisture comes from precipitation, and farm managers in some places and times supplement precipitation with irrigation. Irrigation is expensive, so using deep-rooted, native, warm season grasses and grazing management techniques appropriate for them is an economical alternative within the transition zone and farther south. The University of Tennessee has a Center for Native Grasslands Management that does research toward improving deployment of native grasses to ensure optimum ecologic benefits are realized (http://nativegrasses.utk.edu/index.htm).

Drought conditions experienced by plants are not entirely due to a lack of precipitation. High summer temperatures can cause more evaporation than recently acquired rainfall. Significant amounts of rain can fall in a very short time, but due to lack of soil porosity, most of the water runs off. Different species of grasses have different characteristic root depths, so some are less able to reach available water. Some soils have low water-holding capacity. Subsoil acidity can inhibit the growth of roots, limiting their depth, and nematodes are known to "prune" the roots of susceptible plant species, reducing their ability to access water (Ball et al., 2007). Moderately dry conditions can allow vigorous growth of deep-rooted grasses that out-compete shallow-rooted, slow-starting clover, stunting its growth until more soil moisture accumulates.

If warm season grasses will not persist on your farm, even though you have implemented measures to improve soil moisture content, and precipitation is relatively low plus you have considerable slope, then making dams and catch basins to catch runoff water and use it for watering animals and irrigation is worth consideration. Dams of this nature are common in Western Australia and also found in other parts of Australia with enough changes in elevation. Descriptions can be found in Stanton (2005) and Government of Western Australia (2017). Near Swoope, Virginia, on the hills of Polyface Farm, author and editor of *The Stockman Grass Farmer* magazine, Joel Salatin (2017) has built 15 earthen ponds for irrigation and livestock water. Water flows by gravity from these ponds to pastures at lower elevations. Salatin draws water at 16 inches below the surface to avoid debris on the surface and sediment on the bottom. Water pipes surrounded by pond collars penetrate the earthen dams without leakage. Water reaches pastures with a pressure around 80 psi, which is high enough for the irrigation apparatus used and just low enough for the use of inexpensive garden hose valves and fittings to reach livestock waterers. A water catchment pond at the bottom of a sloping horse pasture in Alberta has a windmill for pumping water to a trough, shown in Fig. 12.5. While runoff water will carry some horse manure toward the pond, the thick forage crop will filter out many of the suspended particles.

Since drought may become more frequent during this period of climate change, especially in areas that are already warm and dry, it will be worth learning some techniques of drought management. Increasing soil organic matter has a direct effect on increasing water-holding capacity, and it also helps formation of macropores that increase rainfall infiltration. One of the many recommendations of the National Drought Mitigation Center of the University of Nebraska is to

FIGURE 12.5   Water collects in a catchment pond at the base of a horse pasture in Alberta. *Photo by Paul Sharpe.*

monitor your feed inventory, which requires determining forage yield and carrying capacity throughout the grazing season. Good records of pasture production and use can be very helpful in decision-making. Chapter 7 provides details on forage yield determination. Taking photographs of your forage growth at standardized locations and times of the growing season can help build a history of forage production, which can be used as an indicator of how forage growth appears in most years. Other options can include planting annual forages (if there is enough soil moisture for them to germinate), grazing stubble or crop aftermath, sending animals to areas with more forage, selling your least productive or least valuable animals, and buying feed. Hay prices increase during droughts, so alternative high-fiber feeds can be gradually added to horse diets to supplement but not totally replace the forages you have. Beet pulp and soybean hulls contain fermentable fiber and are used in many commercial horse feeds. Hay cubes can be used as substitutes for hay and pasture until forage production increases. A conclusion of Holechek et al. (2011b) and Vallentine (2001) from reviewing several drought management studies is to use conservative stocking levels before, during, and after drought, with the result that vegetation and enterprise financial integrity will survive. In fact, in many of the studies that Holechek et al. (2011b) reviewed, forage plants on conservatively or lightly stocked ranges survived better than those on areas not grazed. Decisions on whether to confine animals and feed them stored feed should depend on how long the drought has lasted, the price of feed, the price obtainable for animals sold, and other costs related to confinement. Two species of cactus, prickly pear and cholla, have been used as supplementary drought food for cattle, after the spines were burned off.

As rains return following drought, the temptation will be to put animals back onto pastures or ranges as soon as possible. Vallentine (2001) cautioned against doing this too quickly because drought-stressed plants need time to regain their vigor, and many of the plants that develop early may be weeds of low palatability and lower nutritional quality. Reduced stocking rates and extended plant rest periods will probably benefit the regrowth of desired forage plants.

Flood

Excess soil water restricts oxygen supply to roots, and the need for oxygen increases with rising temperatures, so a flood in early spring when plants are dormant may cause little or no loss of productivity, but a flood during the growing season can kill forage plants. Flooded and wet areas can encourage the growth of pathogenic fungi and plants such as sedges and rushes, which are better adapted than forages. A number of toxic plants also grow well in wet soil, such as water hemlock, poison hemlock, and tall buttercup.

Forage plants tolerant of poor drainage include annual ryegrass, reed canarygrass, Johnsongrass, dallisgrass, plus white, berseem, and ball clovers. Grasses with intermediate tolerance to poor drainage are Bahiagrass and tall fescue. The following forages need good drainage: hybrid Bermudagrass, orchardgrass, alfalfa, arrowleaf clover, and crimson clover (Ball et al., 2007).

Livestock should be fenced out of riparian zones except for short periods when the soil is dry and forage is relatively mature. Pastures adjacent to riparian zones that are still subject to some flooding can be considered for installation of drainage ditches or tiles if the flooding shortens the grazing season or interferes with forage yield or quality. Drains can extend the grazing season on fine-textured soils that drain slowly on their own. This will improve soil aeration, which increases microbial activity, resulting in faster organic matter decomposition and N mineralization. Possible disadvantages of installing drainage are, first, that it will decrease quantities of available moisture when plants need it in the driest, hottest periods and, second, that it can speed up transport of sediment and nitrate N to the adjacent water course (Cuttle, 2008).

WINTER HARDINESS

Horticulturists and plant breeders are among the scientists who use "plant hardiness zone" maps, which show average annual extreme minimum temperatures. There are 12 major plant hardiness zones for the United States, each one divided into subzones a and b. The United States Department of Agriculture (USDA) map is based on the lowest average temperatures that can be expected each year in the United States, based on the lowest daily temperatures recorded from 1974 to 1986 (http://planthardiness.ars.usda.gov/PHZMWeb/). Growers of all types of plants can go to the interactive map, insert a zip code, and learn which plant hardiness zone they are in (USDA, 2017). From the whole USA map page, you can also select a state map, which makes it much easier to see zone boundaries. Canada's plant hardiness zone map has nine major zones with a and b subzones (http://planthardiness.gc.ca/). The Canadian index is based on seven climate variables (Natural Resources Canada, 2017a). The differences are explained in more detail by Natural Resources Canada (2017b). Many nursery plants and trees sold in North America carry a label that indicates the zones for which those plants are hardy. For example, a variety within a species may be suitable for USDA plant hardiness zone 3a (minimum temperature from $-40°F$ ($-40°C$) to $-35°F$ ($-37.2°C$) or zone 11a (minimum temperature from $+40°F$ ($+4.4°C$) to $+45°F$ ($+7.2°C$). Looking at the map for your region can show you the boundaries of local hardiness zones and can help to explain why a plant variety used a moderate distance away is not recommended for your region.

Cool season forages have greater winter hardiness than warm season forages. Winter hardiness includes an ability to survive low and freezing temperatures, lack of water, and freeze—thaw cycles. Species and cultivars with these abilities also tolerate fall dormancy and diseases of the root and crown (Baron and Bélanger, 2007). In a winter hardiness study at 42 universities and research stations across Canada, cool season grasses tolerated winter conditions better than legumes, except in the dry prairie region, where the reverse was true (Ouellet, 1976). Rankings in decreasing order for winter hardiness among cool season grasses were Russian wild ryegrass, crested wheatgrass, creeping red fescue, timothy, smooth bromegrass, red top, slender wheatgrass, Kentucky bluegrass, reed canarygrass, tall wheatgrass, intermediate wheatgrass, meadow fescue, orchardgrass, tall fescue, and perennial ryegrass. Hardiness index rankings for legumes in decreasing order were sweet clover, white clover, alsike clover, alfalfa, birdsfoot trefoil, red clover, and ladino clover. Within some forage species like alfalfa and orchard grass, hardiness indices of cultivars differed widely. A winter survival index (as opposed to hardiness) was developed to consider winter hardiness, environmental factors, and general adaptation of species and cultivars to the region. Data were collected by surveying research stations in Canada and northern American states. There was much regional variation, with a tendency for higher winter survival in drier, central continental regions. Cultivars of alfalfa that were developed in a dry climate survived better in that climate than in a more humid one (Ouellet, 1976).

Climatic factors that contribute to winter damage include these:

1. warm, wet conditions in autumn and subsequent growth
2. wide and frequent temperature fluctuations
3. exposure to temperatures above freezing during winter
4. ice sheeting after winter rain (reduced insulation causes heat loss, in addition to the anoxia under a cover of ice)
5. soil heaving due to freeze—thaw cycles damaging crowns of plants, especially legumes
6. lack of snow for insulation (Baron and Bélanger, 2007)

To reduce winter damage of forage plants, allow an accumulation of at least 500 GDD (calculated using a 41°F (5°C) base) to allow plant hardening and ensure that soils have relatively high levels of K but relatively low levels of N before dormancy (Baron and Bélanger, 2007).

Due to the susceptibility of alfalfa to winter-kill in Ontario, Quebec, and the northeastern United States, there are recommended critical periods when it should not be grazed or harvested for hay or silage. Alfalfa plants regrow and accumulate energy reserves in their roots during this time. The critical period is a 6-week period, comprised of 450 GDD, using a base of 41°F (5°C). In Ontario around 47°N latitude the critical period starts on August 15, and in the farthest southern part of Ontario along the western end of Lake Erie, it starts on September 14. Cutting during the critical period can lead to reduced vigor and lower first-cut yields the following year (OMAFRA, 2012). The Alfalfa Management Guide of the American Society of Agronomy suggests no cutting of alfalfa during a critical period in autumn, beginning about 98 days after May 25 (September 6) and ending about October 20 (Undersander et al., 2011).

HIGH TEMPERATURE STRESS

As air temperatures increase, C4 plants have advantages over C3 plants, due to an absence of photorespiration and a higher resistance to water loss through stomata. High temperatures decrease carbohydrate storage and increase respiration rate, resulting in reduced rates of growth and survival of plants (especially C3). In west Texas and Oklahoma, high temperature stress often occurs simultaneously with stress from lack of moisture. For any C4 species, vegetative development is faster in warmer weather, so plants tend to bloom at younger ages and shorter heights than in cooler weather, and this leads to lower yields by the time plants have developed to an ideal harvest stage (Baron and Bélanger, 2007).

ESTABLISHMENT

Establishing forage swards often leads to high germination rates and successful crop development when appropriate forage species and cultivars are sown at appropriate depths, with implements that are properly maintained and adjusted, on soil of appropriate pH, fertility, and moisture level for the forage species, and there is no interference from insects, birds, weeds, wind, flood, drought, or frost (Barker et al., 2012). The Natural Resources Conservation Service (NRCS) has a standard called Conservation Practice 512 Forage and Biomass Planting. In summarizing research related to the NRCS standard, Barker et al. (2012) stated that seeding date is a critical component of success, and the two climatic variables that influence establishment success the most are temperature and soil moisture. To maximize the probability of temperature and soil moisture being ideal at seeding, forage managers should be very familiar with their regional temperature and moisture conditions around seeding times. In northern states such as New York, spring seedings are most common unless following a small grain crop. In the mid-south, fall seeding is generally preferred. Warm season species have higher temperature requirements for germination and seedling growth, so they are planted in late spring. Rainfall in the week after seeding was most closely correlated with time to emergence, according to a summary of several studies, and rainfall in this 1-week period was more important than in the 2 weeks or 1 month after seeding (Barker et al., 2012). This points out how useful irrigation can be in ensuring timely moisture for forage emergence and establishment. Without irrigation, managers can determine from weather records which dates have high probabilities of suitable rainfall and temperatures. Cool season forages should be planted early enough in autumn that they have about 6 weeks of development before a killing frost.

The American Forage and Grassland Council (AFGC) has provided five principles for pasture establishment (Tucker, 2016):

1. Plan. Consider fertility, field history, current plants and soil, current pests, herbicides used recently and needed soon.
2. Prepare. Do research on species and varieties. In stores, read seed bag labels. Check for germination and weed seed percentages. Purchase certified seed. Read Chapter 6 on pasture establishment and renovation. Reread this chapter. Ask questions. Identify current problems and make plans to solve them. Seek help from extension professionals.
3. Plant. Prepare a seedbed properly. Seed at the correct depth and seeding rate on the ideal date relative to past and predicted weather and soil temperature.
4. Prevent. Use integrated pest management against weeds, insects, and diseases. Split applications of fertilizer as recommended for maximum effect. Prevent weakening of young plants by delaying grazing at least until the fourth-leaf stage and ensure that plants are not grazed closer than 6 inches (15 cm) above the soil.
5. Protect. Implement grazing and haying management practices that are recommended, based upon research results in environments and situations similar to yours.

CLIMATE CHANGE

Climate change is predicted to include temperatures that are generally warming, precipitation exhibiting more extremes, and storms becoming more frequent. The Intergovernmental Panel on Climate Change (IPCC) suggested that low emissions of greenhouse gases (GHG) could increase North American temperatures 1.8–5.4°F (1–3°C) by 2100 or 6.2–13.5°F (3.5–7.5°C) if GHG emissions are high (IPCC, 2001a; Baron and Bélanger, 2007). Greenhouse gases include CO_2, methane (CH_4), and nitrous oxide (N_2O or nitrogen dioxide). Rising atmospheric [CO_2] is reported to reduce the flow of water and CO_2 through leaf stomata, improve water use efficiency, stimulate photosynthesis, and increase light-use efficiency by plants (Drake et al., 1997). Average world atmospheric [CO_2] rose from 280 ppm in preindustrial times to 365 ppm in 1998, and it could reach 540–970 by 2100 (IPCC, 2001a). CO_2 has an atmospheric lifetime of 5–200 years (based on different rates of uptake by different removal processes) and has been assigned a global warming potential of 1 (IPCC, 2001b). Table 12.2 shows these same variables for CH_4 and N_2O and reveals their much higher global warming potential than CO_2.

TABLE 12.2 Concentration, Atmospheric Lifetime, and Estimated Global Warming Potential of Atmospheric Methane (CH_4) and Nitrous Oxide (N_2O)

	CH_4	N_2O
Preindustrial concentration	700 ppb	270 ppb
Concentration in 1998	1745 ppb	314 ppb
Atmospheric lifetime	12 yr	114 yr
Global warming potential (IPCC, 2014)	34	310

GHG emissions from agriculture are composed of about two-thirds N_2O and one-third CH_4. Each of these gases has a turnover between the atmosphere and other pools on the earth. The net emission of CO_2 from agriculture can be considered almost negligible if the CO_2 from fossil fuels used to produce inputs and for farm machinery is excluded (Baron and Bélanger, 2007). However, CO_2 still receives much public attention. Pastoral grazing systems have been estimated to contribute 20% of the CH_4 and 16%—33% of the N_2O from all agricultural activities (Clark et al., 2005; de Klein et al., 2008), although Wang et al. (2015) questioned whether earlier estimates of GHG flux took adequate account of the sequestration of carbon that is possible with appropriate pasture management. CO_2 turnover includes incorporating or sequestering carbon in plants through photosynthesis. Following the death of plant parts, some of their carbon is mineralized (attached to nonorganic soil particles) (Lüscher et al., 2005). Most methane that originates from animals is generated by ruminant animal fermentation, and postgastric fermentation by horses can be included. However, the amount of methane contributed to total animal GHG emissions by horses is relatively small. Horses do not have a rumen, which produces considerable methane in cattle, sheep, goats, and other ruminants. Methane production is lower in the colon of horses (23 mmol/mol) than in the rumen of cattle (250 mmol/mol of substrate) (Santos et al., 2011; Demeyer, 1991). Horse manure exposed to air will release some methane and was considered comparable to cattle manure as a feedstock for anaerobic digestion to produce methane for energy generation (Wartell et al., 2012). Daily methane production of Welsh ponies was reported to be about 30 L/pony/day on a total roughage diet (mid- to late-mature grass hay) and it was reduced to 23 L/pony/day when half of the net energy was supplied by a cereal mix (Dansen et al., 2015). William Martin-Rosset, head of equine nutrition research at the French National Institute for Agricultural Research, was quoted as stating that horses only produce 1.5% of the enteric methane in France, compared to 90% by cattle (Lesté-Lasserre, 2013). This is because of lower production per animal and because Food and Agriculture Organization (FAO) statistics indicate only about 453,000 horses in France versus about 10 million head of cattle, as of 2008 (Huyghe, 2012).

Denitrification of nitrate (NO_3^-) in soil produces most of the agricultural N_2O. Increasing concentration of atmospheric CO_2 is expected to stimulate pasture plant growth rate in temperate regions but decrease yields in semiarid and tropical regions (Tubiello et al., 2007). Considerable research was performed at an experimental plot level between 1990 and 2007, but the simple findings may not be useful for complex models designed to predict climate change consequences (Tubiello et al., 2007). There is some evidence that high concentrations of soil N will be needed for the increased [CO_2] to drive higher forage yields (Lüscher et al., 2005). In local areas, the increased frequency and severity of storms and floods following more years of climate change can have negative effects on yield and quality of forages. In Atlantic Canada, where low temperatures limit the growing season, GDD may increase by 400, which is enough to add an additional forage harvest (Bootsma et al., 2001; Baron and Bélanger, 2007). An increase in the legume percentage of a mixture of grasses and legumes was shown to be stimulated by enrichment of CO_2 (Campbell et al., 2000; Baron and Bélanger, 2007).

The fifth assessment report of the IPCC (2014) provided evidence of continued warming of air and oceans, emphasizing that the period from 1983 to 2012 in the northern hemisphere was likely the warmest 30-year period of the last 1400 years. Global anthropogenic CO_2 emissions from forestry and other land use appeared to decline between 2000 and 2010, but the emissions from fossil fuels, cement, and flaring continued to rise. Since 1950, there has been a decrease in cold temperature extremes, an increase in warm temperature extremes, and an increase in the number of high precipitation events in several regions (IPCC, 2014). Continued emission of GHGs is predicted to contribute to further warming through the end of this century.

It appears inevitable that all herbivores will continue to emit CO_2 and CH_4, and it is important to look at the other part of the carbon cycle to determine whether net GHG contributions to the environment can be reduced. Changing the common practice on beef cow operations of continuous grazing to well-managed multipaddock (rotational) grazing showed a potential reduction of GHG emission by 30% through increased grass quality and digestibility plus a larger amount of carbon sequestered (Wang et al., 2015; Teague et al., 2016). Improved grazing management, fertilization, sowing legumes and improved grass species, irrigation, and conversion from cultivation tend to lead to increased soil carbon (Morgan et al., 2016; Conant et al., 2017), partly through ensuring early season moisture to help with plant development.

Pests

Weed and insect problems can potentially become more difficult to manage under higher temperatures and precipitation. For example, midges of the genus *Culicoides* spread bluetongue disease among sheep, goats, deer, and antelope (Tubiello et al., 2007). Low temperatures currently keep most North American bluetongue cases in the United States, but in some years, the Okanogan valley of Washington and British Columbia is warm enough that midges cross the border from Washington and cause bluetongue in B.C. West Nile virus (WNV) can cause West Nile fever in horses and warmer temperatures can cause the spread of the *Culex* mosquitoes which carry the virus. The first North American human case of WNV was discovered in New York City in 1999, and by 2003, it had spread to the Pacific coast and Canada. Several aspects of climate change can contribute to more cases of WNV transmission (Paz, 2015). While the first vaccine against WNV for people just entered human trials in 2015, there are already three commercial WNV vaccines for horses. Since these vaccines are not 100% effective against preventing disease, it is still important to reduce mosquito habitat and prevent mosquito bites. Mountain pine beetle outbreaks have been blamed on climate change, and various ticks that carry pathogens are on the rise in North America (including the black-legged tick that spreads Lyme disease). Effects secondary to temperature and precipitation changes can include shortened plant hardening periods, shorter periods of temperatures less than a certain threshold, plus lower or higher amounts of snow and duration of snow cover for insulating plants against cold temperature damage (Baron and Bélanger, 2007). Variability in these effects will likely be considerable, since they are complicated by combinations of elevation, distance from large water bodies, and other geographic features that cause different ecoregions to exist. Poison ivy has been observed to grow more vigorously when $[CO_2]$ increases. Horse managers may want to either improve their pest management skills or hire pest management experts as their climate continues to change.

Breeding Cultivars for the New Climate

There is a need for forage breeders to develop new varieties of plants to adapt to changing climatic conditions, with respect to drought tolerance, water use efficiency, salinity tolerance, flood tolerance, and different nutrient dynamics (Abberton et al., 2008). Changes in plant function due to elevated $[CO_2]$ may not occur as rapidly as predicted. In more than 60 experiments prior to 1997, involving plants exposed to increased $[CO_2]$, the average rate of photosynthesis increased 58% (Drake et al., 1997; Abberton et al., 2008). However, sward productivity in a later study only rose 15% in response to $[CO_2]$ increased to the same extent (Lüscher et al., 2005; Abberton et al., 2008). Yield responses to increased $[CO_2]$ are largest for legumes, intermediate for broad-leaved, nonleguminous forbs, and least for grasses (Lüscher et al., 2005). Subtle changes may be seen in the relative proportions of C3 and C4 forages that can grow in a region. Higher temperatures favor C4 plant development and yield, but drought from more variable precipitation favors C3 plants over C4 plants (Lüscher et al., 2005).

In a European review of reports on climate change effects on mountain and Mediterranean grasslands, increased $[CO_2]$ did not affect structural carbohydrates and digestibility of forages, but it increased the total NSC and decreased forage [N] in mountains and plains but not Mediterranean areas. Increasing temperature did not affect [N], water-soluble [CHO], [structural CHO], or digestibility in either environment. Forage [N] increased as water availability decreased. During drought, digestibility increased. Authors called for further experiments to determine effects of combined factors and extreme climatic events (Dumont et al., 2015).

Climate change is happening now and will likely contribute to an increased incidence of diseases in distressed animals. Climate change will continue to have effects on livestock, as summarized in Table 12.3, where the changes are indicated by an arrow pointing up to indicate an increase and an arrow pointing both up and down to indicate increased variability of precipitation. The information in Table 12.3 is adapted from a review by Rojas-Downing et al. (2017).

Effects of Climate Change on Horses

Environmental stresses on horses include reduced availability of water and high-quality feed, and temperatures above an animal's zone of thermoneutrality (requiring energy for cooling). High humidity reduces an animal's ability to cool itself. A "temperature humidity index" (THI) was devised to reflect the combined effects of high temperature and humidity on

TABLE 12.3 Impact of Climate Change on Forages and Grazing Horses[a]

↑ [CO$_2$]	↑ [CO$_2$] and ↑ Temperature	↑ Temperature	↑ Temperature and ↕ Precipitation	↕ Precipitation
More forage growth				Longer dry seasons reduce forage growth
Lower forage quality	Seasonal forage development changes	Decreased nutrient availability		Lower forage quality and species diversity
Reduced transpiration	Changing optimal growth rate	Increased C4 growth		Floods affect root structure and leaf growth
Better water use efficiency	Changing water availability			
		Increased water consumption[b]		
		Challenges to health[b]	Proliferation of pathogens and parasites[b,c]	
			More vector-borne diseases[b]	

[a]Adapted from Rojas-Downing et al. (2017). An arrow pointing up indicates an increase, and an arrow pointing both up and down indicates more variability of precipitation.
[b]Effect on horses.
[c]Effects on horses and forage plants.

stressing an animal's ability to cool itself. A Fahrenheit temperature of 80° plus a relative humidity of 60% equals a THI of 140. A THI above 150 reduces a horse's ability to cool itself (Martinson et al., 2017; Ward, 2011), and a THI greater than 180 can be fatal. Horse owners in warm and warming climates can reduce the probability of heat stress by providing plenty of cool water, turnout during coolest times of the day, shade during the day, a cool-down period after workouts, fans in buildings, hay and pasture of higher energy concentration, opportunities to become acclimatized if moved to a warmer region, free-choice salt, electrolytes when exercising or transporting, restricting transport to coolest parts of the day, shade for horses in well-ventilated trailers, and reduced workout length or intensity (Martinson et al., 2017; Ward, 2011). If an adult horse's respiration rate is well above normal (40 vs. 8–12 breaths per minute) or rectal temperature is significantly elevated (over 105°F vs. 99.5–101.5°F), (over 40.6 °C vs. 37.5–38.6 °C) the horse is likely suffering from heat stroke (Martinson et al., 2017).

Effect of Animals on Climate Change

The presence of livestock in the environment has an effect on climate change. Climate change and GHG emissions from agriculture are worldwide problems and issues. The main contributors to these emissions in agriculture are fossil fuel use in machinery, clearing of forests to make croplands and pastures, plus tillage for cropping. Livestock contribute 14.5% of the total annual human-caused GHG emissions worldwide (Gerber et al., 2013; Rojas-Downing et al., 2017). These livestock contributions to GHG include land use change, feed production, animal production, manure, processing, and transport (Rojas-Downing et al., 2017). Of the total amount of N$_2$O emitted into the atmosphere, human activity contributes 29% and livestock contributes 53% of this, thus 15.4% of total anthropogenic emissions of N$_2$O. Of all the CH$_4$ emitted, human activity contributes 44% and livestock contributes 44% of that, so 19.4% of total anthropogenic emissions (Rojas-Downing et al., 2017). Of the three major GHG emissions, enteric fermentation is the single greatest contributor at 39.1% (Rojas-Downing et al., 2017). This is mostly methane expired by ruminant livestock but will include a relatively small contribution from rectal release by horses. There is research occurring to develop agronomic practices, higher quality diets, and feeding systems to reduce this methane production, and some of these have been evaluated and assigned economic value according to their mitigation potential (Gerber et al., 2013; Moran et al., 2011) So far, adding grain and vegetable oil to ruminant diets have had the greatest effects on reducing methane production, but there is a GHG-emitting cost to producing the grains and oilseeds. Someone seeing these figures for the first time may think that owners of horses (which can be considered among livestock for some purposes) should help shoulder the burden for cleaning up this contribution to greenhouse gas emissions. However, it is not the responsibility of every horse owner to clean up the whole world. According to the FAO of the United Nations, the number of equids (horses + asses + mules) in the world in 2016

(59,048,194) was only 2.3% of the total of the six most numerous large domestic animal species (4,944,503,071) (FAO, 2018). According to the 2012 Census in the United States, there were 3.62 million horses and ponies in the United States, and this is only 1.98% of the five most numerous large domestic animal species (cattle, pigs, sheep, goats, horses) (APHIS, 2010). So it is unlikely that horses will make a large collective contribution to GHG emissions. What any horse manager can be expected to do is to look after her own horses and land according to national, state, and local regulations and continue to educate herself on ways to avoid and reduce GHG emissions.

Grazing and Cycling of Greenhouse Gases

In theory, the roots of a forage plants grow in proportion to the shoots, and as a plant is grazed or harvested, the living root mass shrinks proportionately, leaving carbon compounds in the soil, which are part of "soil organic carbon" (SOC). Does this suggest that grazing helps the process of sequestering carbon?

This is not a simple question to answer. There is a cycle of carbon in different forms entering and leaving the soil intermittently, and apparently, the carbon left when roots die has less effect on the carbon cycle than precipitation, soil type, existing SOC, and type of vegetation growing (Voth, 2017). Much SOC cycles back to the atmosphere after a few weeks or months (Gilker and Voth, 2017). "Carbon sequestration" indicates that CO_2 from the air is transferred into long-lived pools and stored securely so it is not immediately reemitted (Lal, 2004). [SOC] increases with plant growth in the spring and decreases in the winter when plants are dormant and CO_2 is respired back to the atmosphere (Gilker and Voth, 2017). The existing [carbon] is very important to the rate of GHG emission and reflects a degree of degradation of soils (Gilker and Voth, 2017). Each soil has a potential amount of carbon that it can sequester during its cycles of acquisition and loss, and this depends on the mineral make-up, bulk density, clay content, and depth (Gilker and Voth, 2017; Xiao, 2015). The amount of carbon that a soil can actually obtain is influenced by the climatic variables, rainfall, temperature, and solar radiation. The actual organic content of a soil fluctuates greatly and can be influenced by agricultural management in terms of plant productivity, rotation strategy, and manipulations such as tillage (Gilker and Voth, 2017). Some soils have lost up to two-thirds of their original SOC since 1750, when land use started to change rapidly and fossil fuel combustion began (Lal, 2004; Voth, 2017). To lower [GHG] through sequestering more carbon in the soil, humans should concentrate on the following carbon absorbers in order of greatest to least potential: degraded soils and desertified ecosystems > cropland > grazing lands > forest lands and permanent crops (Lal, 2004; Voth, 2017). Permanent pasture with a very high [SOC] is analogous to a sponge saturated with water, and degraded soil is analogous to a dry sponge. North American horse farm managers can look for opportunities to promote carbon sequestration in degraded soils and desertified ecosystems, follow recommended management practices on agricultural soils, ensure that grazing lands are covered by plants permanently, restore marginal lands to vegetative cover, and apply composted plant material and manure to pastures (Lal, 2004). Leaving high levels of standing forage on pastures at the beginning of winter and allowing some trampling of forage during the growing season can contribute to soil organic matter content. The level of SOC loss through erosion on a 7% slope under natural rainfall was quantified by Polyakov and Lal (2008). Of the SOC mobilized by overland flow, 44% was washed away, but from sediment deposited on the site, 15% of its SOC was lost by mineralization to CO_2 and CH_4 within 100 days of the rainfall. The rate of mineralization of SOC was eight times greater for coarse aggregates than for fine particles. Thus, organic carbon that was recently deposited in soil or on its surface and became part of course aggregates could be quickly washed away or mineralized to become part of GHG emissions. This illustrates how quickly and continuously SOC can be lost from soil that is unprotected by vegetation.

PROFESSIONAL HELP

Reports of forage variety and management research trials are available from Cornell University, Ohio State University, Pennsylvania State University, and University of Kentucky, plus a few others. There has been a trend of universities dropping this service, so forage growers may need to search for the state university nearest to them that still provides this information. For turfgrasses, there is a National Turfgrass Evaluation Program, partly sponsored by USDA that makes results available at www.ntep.org. Contributors to the reports can provide advice on interpretation.

Among university extension professors and seed salespeople, there are Certified Forage and Grassland Apprentices (CFGA) and Certified Forage and Grassland Professionals (CFGP), with their certification coming from the AFGC. Some of the CFGPs are also Technical Service Providers through the USDA NRCS. The AFGC and NRCS promote education about forages and forage lands through conferences, conference proceedings, competitions, awards, and grazing schools. Among these professionals are individuals who can provide relevant advice on selecting and managing forages for grazing and hay production in various environments and advising on adjustments as climate change progresses. A useful place to look for extension articles is eXtension.org. This is a "knowledge to action service" with contributions from over 50 land grant universities.

REVIEW QUESTIONS

1. Look at Table 12.1 on solar radiation.
    a. What similarities do you see among months?
    b. What is the likely cause of these similarities?
    c. What is the difference between the monthly solar radiation between a northern American state and a southern state?
2. Look at Figure 12.1 on solar radiation. What benefit does a young plant gain as it grows from a seedling to about 4 inches tall?
3. Describe one benefit that C3 plants have over C4 plants in a temperate, humid environment such as Vermont.
4. Describe one benefit that C4 plants have over C3 plants in a warm, humid environment such as South Carolina.
5. Describe an apparent disadvantage of C4 plants that results in lower feed quality.
6. What is a good use of the units called growing degree days (GDD)?
7. Name two cool season grasses tested in the northeastern United States that have medium winter survival and high horse preference.
8. Name one agency in the United States that provides regional weather data and one that provides average monthly weather data.
9. In a warm, arid region, which measure would have a higher value most of the time, actual evapotranspiration (AE) or potential evapotranspiration (PET)? Why?
10. In the "transition zone," which type of forages have their peak production from March to May, warm season or cool season? Which type of forages have their peak production from May through September?
11. Briefly describe at least four drought management or drought prevention measures.
12. Describe at least two practices that can help to improve winter hardiness or avoid winter damage to forages.
13. List the following items that can be measured in the atmosphere, in a column: temperature, $[CO_2]$, $[N_2O]$, $[CH_4]$, precipitation. After each measure, use arrows to indicate whether they are expected to increase ($\uparrow$), decrease ($\downarrow$), or become more variable and extreme ($\updownarrow$) between now and the year 2100.
14. Name four insects or arthropod pests and their effect on people or animals due to climate warming.
15. In a climate change scenario where the average temperature and $[CO_2]$ increased, with no change in precipitation, what are the expected results in a mixture of C3 and C4 plants?
16. List some of the combined effects of climate change on horses.
17. What can horse managers do to minimize the contributions of horses to climate change and GHG emissions?
18. What can grazing managers do to promote carbon sequestration on horse farms?
19. What professional resources are available to grazing managers to help with matching your management practices to your climate? How can you make use of professional resources to aid in minimizing effects of your management on climate change?

REFERENCES

Abberton MT, MacDuff JH, Marshall AH, Humphreys MW. The Genetic Improvement of Forage Grasses and Legumes to Enhance Adaptation of Grasslands to Climate Change. 2008. Prepared for Food and Agriculture Organization, http://www.fao.org/3/a-ai779e.pdf.

Allen M, Leep R, Andresen J. Timing Spring Alfalfa Harvest — the Final Word?. Forage Connection. Michigan State University; 2016. https://forage.msu.edu/extension/timing-spring-alfalfa-harvest-the-final-word/.

APHIS. Overview of U.S. Livestock, Poultry and Aquaculture Production in 2010 and Statistics on Major Commodities. 2010. https://www.aphis.usda.gov/animal_health/nahms/downloads/Demographics2010_rev.pdf.

Bailey RG. Ecosystem Geography. New York: Springer-Verlag; 1996.

Ball DM, Hoveland CS, Lacefield GD. Southern Forages: Modern Concepts for Forage Crop Management. fourth ed. Norcross, Georgia: International Plant Nutrition Institute; 2007. p. 117–25.

Barker DJ, MacAdam JW, Butler TJ, Sulc RM. Forage and biomass planting. In: Conservation Outcomes from Pastureland and Hayland Practices; 2012. https://www.nrcs.usda.gov/Internet/FSE_DOCUMENTS/stelprdb1080494.pdf.

Barnhart SK. Warm-season Grasses for Hay and Pasture. Iowa State University; 1994. University Extension Pm-569, https://store.extension.iastate.edu/.../Warm-Season-Grasses-for-Hay-and-Pasture-PDF.

Baron VS, Bélanger G. Climate and forage adaptation. Chapter 6. In: Barnes RF, Nelson CJ, Moore KJ, Collins M, editors. Forages, the Science of Grassland Agriculture Volume II 6th Edition. Blackwell Publishing; 2007. p. 83–104.

Bootsma A, Gameda S, McKenney DW. Adaptation of Agricultural Production to Climate Change in Atlantic Canada. Final Report for Climate Change Action Fund A214. 2001. http://res2.agr.gc.ca/ecorc/clim/20010613_e.pdf.

Bootsma. Forage crop maturity zonation in the Atlantic region using growing degree-days. Can. J. Plant Sci. 1984;64:329−38.

Breazeale D, Kettle R, Munk G. Using Growing Degree Days for Alfalfa Production. Fact Sheet 99-71. Reno: University of Nevada; 1999 (Cooperative Extension.).

Brueland BA, Harmoney KR, Moore KJ, George JR, Brummer EC. Developmental morphology of smooth bromegrass growth following spring grazing. Madison Crop Sci. 2003;43(5):1789−96.

Buxton DR, Marten GC. Crop quality & utilization. Crop Sci. 1989;29:429−35.

Campbell BD, Stafford Smith DM, GCTE Pastures, Rangelands Network members. A synthesis of recent global change research on pasture and rangeland production: reduced uncertainties and their management implications. Agric. Ecosyst. Environ. 2000;82:39−55.

Cherney JH, Sulc RM. Predicting first cutting alfalfa quality. In: Silage: Field to Feedbunk, North American Conference. February 11−13, Hershey, PA; 1997.

Cherney JH, Baker EV, Kallenbach RL. Forage systems for temperate humid areas. Ch. 18. In: Barnes RF, Nelson CJ, Moore KJ, Collins M, editors. Forages, the Science of Grassland Agriculture Volume II 6th Edition. Blackwell Publishing; 2007. p. 277−90.

Clark H, Pinares PC, deKlein CAM. Methane and nitrous oxide emissions from grazed grasslands. In: McGilloway DA, editor. Grassland: A Global Resource. Plenary and Invited Papers from the XX International Grassland Congress. Dublin, Ireland. 26 June−1 July, 2005. Wageningen, the Netherlands: Wageningen Academic Publishers; 2005. p. 279−93.

Collins M, Hannaway DB. Forage-related animal disorders. Ch18. In: Barnes RF, Nelson CJ, Moore KJ, Collins M, editors. Forages Volume I: An Introduction to Grassland Agriculture. sixth ed. Blackwell Publishing; 2007. p. 415−41.

Conant RT, Cerri CEP, Osborne BB, Paustian K. Grassland management impacts on soil carbon stocks: a new synthesis. Ecol. Appl. 2017;27(2):662−8.

Crop Science Society of America. https://www.crops.org/about-crop-science/crop-breeding.

Cuttle SP. Impacts of pastoral grazing on soil quality. In: McDowell RW, editor. Environmental Impacts of Pasture-based Farming. Wallingford, Oxfordshire, UK: CABI; 2008. p. 33−74.

Dansen O, Pellikaan WF, Hendriks WH, Dijkstra J, Jacobs MPT, Everts H, van Doorn DA. Daily methane production pattern of Welsh ponies fed a roughage diet with or without a cereal mixture. J. Anim. Sci. 2015;93:1916−22.

de Klein CAM, Pinares-Patino C, Waghorn GC. Greenhouse gas emissions. In: McDowell RW, editor. Environmental Impacts of Pasture-based Farming. Oxfordshire, UK: AgResearch, Invermay Agricultural Centre, Mosgiel, New Zealand. CABI. Wallingord; 2008. p. 1−32.

Demeyer DI. Quantitative aspects of microbial metabolism in the rumen and hindgut. In: Jouany JP, editor. Rumen Microbial Metabolism and Ruminant Digestions. Paris: INRA Publications; 1991. p. 217−37.

Drake BG, Gonzàlez-Meler MA, Long SP. More efficient plants: a consequence of rising atmospheric CO_2? Ann. Rev. Plant Physiol. Plant Molec. Biol. 1997;48:609−39.

Dumont B, Andueza D, Niderkorn V, Luscher A, Porqueddu C, Picon-Cochard C. A meta-analysis of climate change effects on forage quality in grasslands: specificities of mountain and Mediterranean areas. Grass Forage Sci. 2015;70(2):239−54.

FAO. Live Animals. 2016. Food and Agriculture Organization of the United Nations; 2018. http://www.fao.org/en/#data/QA.

Frank AB, Hofmann L. Relationship among grazing management, growing degree-days and morphological development for native grasses on the Northern Great Plains. J. Range Manag. 1989;42(3):199−202.

Frank AB. Evaluating grass development for grazing management. Rangelands 1996;18(3):106−9.

Fraser DA. Determining Range Readiness and Growing Degree-days (GDDs). B.C. Min. for. Range, Range Br., Kamploops, B.C. Rangeland Health Brochure 11. 2006. https://www.for.gov.bc.ca/hra/publications/brochures/Rangeland_Health_Brochure11.pdf.

Gerber PJ, Steinfeld H, Henderson B, Mottet A, Opio C, Dijkman J, Falcucci A, Tempio G. Tackling Climate Change through Livestock − a Global Assessment of Emissions and Mitigation Opportunities. Rome: Food and Agriculture Organization of the United Nations (FAO); 2013. http://www.fao.org/3/a-i3437e.pdf.

Gilker R, Voth K. Does Grazing Sequester Carbon? Part 2- Some Background. On Pasture. 2017. http://onpasture.com/2017/10/09/does-grazing-sequester-carbon-part-2-some-background/.

Government of Western Australia. Excavated Tanks (Farm Dams). 2017. https://www.agric.wa.gov.au/water-management/excavated-tanks-farm-dams.

Holechek JL, Pieper RD, Herbel CH. Ch. 3 rangeland physical characteristics. In: Range Management: Principles and Practices. sixth ed. Upper Saddle River, NJ: Pearson Education, Prentice Hall; 2011a. p. 31−48.

Holechek JL, Pieper RD, Herbel CH. Ch. 13 range livestock production. In: Range Management: Principles and Practices. sixth ed. Upper Saddle River, NJ: Pearson Education, Prentice Hall; 2011b. p. 262−81.

Holechek JL, Pieper RD, Herbel CH. Ch. 16 range management in developing countries. In: Range Management: Principles and Practices. sixth ed. Upper Saddle River, NJ: Pearson Education, Prentice Hall; 2011c. p. 337−50.

Hoveland CS. Warm Season Perennial Grasses. Ga. Cattlem.; July 1996. p. 30−1.

Huyghe C. Country Pasture/Forage Resource Profiles, France. 2012. http://www.fao.org/ag/agp/agpc/doc/Counprof/France/france.htm#4.RUMINANT.

IPCC. In: McCarthy JJ, et al., editors. Intergovernmental Panel on Climate Change. Climate Change 2001: Impacts, Adaptation, and Vulnerability: Contribution of Working Group II to the Third Assessment Report of the Intergovernmental Panel on Climate Change. Ottawa, Canada: Intergovernmental Panel on Climate Change; 2001a. p. 735−800.

IPCC. Intergovernmental Panel on Climate Change. Climate Change 2001. Third Assessment Report. Working Group I: The Scientific Basis. C. The Forcing Agents that Cause Climate Change. C.1 Observed Changes in Globally Well-mixed Greenhouse Gas Concentrations and Radiative Forcing. Table 1. 2001. https://www.ipcc.ch/ipccreports/tar/wg1/016.htm.

IPCC. Climate Change 2014 Synthesis Report. Summary for Policymakers. 2014. www.ipcc.ch/pdf/assessment-report/ar5/syr/AR5_SYR_FINAL_SPM. pdf.

Jeranyama P, Garcia AD. Understanding Relative Feed Value (RFV) and Relative Forage Quality (RFQ). Extension Extra, Cooperative Extension Service, College of Agriculture & Biological Sciences. South Dakota State University Article EXEX8149. 2004. http://agbiopubs.sdstate.edu/articles/ExEx8149.pdf.

Kaatz P. Setting the Cutting Schedule for Your Alfalfa. Michigan State University Extension; 2011. http://msue.anr.msu.edu/news/setting_the_cutting_schedule_for_your_alfalfa.

Lal R. Soil carbon sequestration to mitigate climate change. Geoderma 2004;123:1−22.

Climate Zones of the U.S. Lawn Care Academy; 2017. http://www.lawn-care-academy.com/climatezones.html.

Lee K, Allen M, Leep R. Predicting Optimum Time of Alfalfa Harvest. 2010. Extension.org. http://articles.extension.org/pages/25471/predicting-optimum-time-of-alfalfa-harvest.

Lesté-Lasserre C. Does Horses' Waste Help or Hinder the Environment?. 2013. TheHorse.com, http://www.thehorse.com/articles/32259/does-horses-waste-help-or-hinder-the-environment.

Lüscher A, Fuhrer J, Newton PCD. Global atmospheric change and its effects on managed grassland systems. In: McGilloway DA, editor. Grassland: A Global Resource. Wageningen Academic Publishers; 2005. p. 251−64.

Martinson K, Sheaffer C. Cool-season Perennial Grasses for Horse Pastures. University of Minnesota Extension; 2013. https://www.extension.umn.edu/agriculture/horse/pasture/docs/pasture_grasses_final.pdf.

Martinson K, Hathaway M, Ward C, Johnson R. Managing Horses during Hot Weather. Horse Extension. University of Minnesota Extension; 2017. https://www.extension.umn.edu/agriculture/horse/care/managing-horses-during-hot-weather/index.html.

McGraw RL, Nelson CJ. Legumes for northern areas. Ch 8. In: Forages, Volume I: An Introduction to Grassland Agriculture. sixth ed. 2007. p. 171−90.

Moore KJ, Moser LE, Vogel KP, Waller SS, Johnson BE, Pedersen JF. Describing and quantifying growth stages of perennial forage grasses. Agron. J. 1991;83:1073−7.

Moran D, MacLeod M, Wall E, Eory V, McVittie A, Barnes A, Rees R, Topp CFE, Moxey A. Marginal abatement cost curves for UK agricultural greenhouse gas emissions. J. Agric. Econ. 2011;62(1):93−118.

Morgan JA, Parton W, Derner JD, Gilmanov TG, Smith DP. Importance of early season conditions and grazing on carbon dioxide fluxes in Colorado shortgrass steppe. Rangel. Ecol. Manag. 2016;69:342−50.

National Drought Mitigation Center. drought.unl.edu/.

Natural Resources Canada. Climate Variables Used to Develop Canada's Plant Hardiness Zones. 2017. http://planthardiness.gc.ca/?m=15&lang=en.

Natural Resources Canada. Extreme Minimum Temperature Models. Plant Hardiness of Canada. 2017. http://planthardiness.gc.ca/?m=17&lang=en.

NOAA NCEI. Data Tools: 1981−2010 Normals. 2017. https://www.ncdc.noaa.gov/cdo-web/datatools/normals.

NOAA. Data Snapshot Details: Average Monthly Temperature. 2017. https://www.climate.gov/maps-data/data-snapshots/data-source-average-monthly-temperature.

Noland RL, Wells MS. Using Growing Degree Days (GDD) to Plan Early Season Alfalfa Harvests. Forage Production. University of Minnesota Extension; 2017. https://www.extension.umn.edu/agriculture/forages/utilization/use-gdd-to-plan-harvest/.

NWS CPC. Regional Climate Maps: USA. 2017. http://www.cpc.ncep.noaa.gov/products/analysis_monitoring/regional_monitoring/usa.shtml#weekly.

Olson GI, Smith SR, Phillips TD, Lawrence LM. Cool-season Grass Horse Grazing Tolerance Report. University of Kentucky College of Agriculture, Food and Environment; 2017. p. 2016. Agricultural Experiment Station report PR-718, http://www2.ca.uky.edu/agcomm/pubs/PR/PR718/PR718.pdf.

OMAFRA. Fall Cutting of Alfalfa. Ontario Ministry of Agriculture, Food and Rural Affairs; 2012. Order No. 91−9072, http://www.omafra.gov.on.ca/english/crops/field/forages/fallcuttingalfalfa.htm.

Ouellet CE. Winter hardiness and survival of forage crops in Canada. Can. J. Plant Sci. 1976;56:679−89.

Paz S. Climate change impacts on West Nile virus transmission in a global context. Phil. Trans. R. Soc. B 2015;370. 20130561, https://doi.org/10.1098/rstb.2013.0561.

Polyakov VO, Lal R. Soil organic matter and CO_2 emission as affected by water erosion on field runoff plots. Geoderma 2008;143:216−22.

Rayburn EB, Hall MH, Murphy W, Vough L. Pasture production. In: grazing in the northeast: assessing current technologies, research directions and education needs. In: Krueger CR, Pionke HB, editors. Proceedings from the Grazing in the Northeast Workshop. Camp Hill, Pennsylvania: Northeast Regional Agricultural Engineering Service (NRAES), United States Department of Agriculture; March 25−26, 1998. p. 13−50.

Rayburn E, Shockey W, Smith B, Seymour D, Basden T. Light interception by pasture canopies as affected by height and botanical composition. Crop Forage Turfgrass Manag. 2016;2. https://doi.org/10.2134/cftm2016.0013.

Rojas-Downing MM, Nejadhashemi AP, Harrigan T, Woznicki SA. Climate change and livestock: impacts, adaptation, and mitigation. Climate Risk Manag. 2017;16:45−163.

Russell PJ, Hertz PE, McMillan B, Fenton B, Addy H, Maxwell D, Haffie T, Milsom B. Chapter 7 photosynthesis. In: Biology: Exploring the Diversity of Life. Second Canadian Edition. Toronto, Ontario, Canada: Nelson Education; 2013. p. 139−60.

Salatin J. Personal Communication. 2017.

Sanderson MA, Jolley LW, Dobrowolski JP. Pastureland and hayland in the USA: land resources, conservation practices and ecosystem services. In: Nelson CJ, editor. Conservation Outcomes from Pastureland and Hayland Practices: Assessment, Recommendations and Knowledge Gaps. Lawrence Kansas: Allen Press; 2012. p. 25−40.

Sanderson MA. Crop quality and utilization. Crop Sci. 1992;32:245−50.

Santos AS, Rodrigues MAM, Bessa RJB, Ferreira LM, Martin-Rosset W. Understanding the equine cecum-colon ecosystem: current knowledge and future perspectives. Animal 2011;5(1):48−56.

Smart AJ, Schacht WH, Moser LE. Predicting leaf/stem ratio and nutritive value in grazed and nongrazed big bluestem. Agron. J. 2001;93:1243−9.

Society for Range Management. A Glossary of Terms Used in Range Management. third ed. Denver, CO: Society for Range Management; 1989.

Stanton D. Farm dams in western Australia. Western Australia, Perth: Department of Agriculture and Food; 2005. Bulletin 4609, http://researchlibrary.agric.wa.gov.au/cgi/viewcontent.cgi?article=1048&context=bulletins.

Sulc RM, Albrecht KA, Owens VN, Cherney JH. Predicting harvest time for alfalfa. In: Proc. Tri-state Diary Nutrition Conf., Fort Wayne, in. 20−21 April. The Ohio State University; 1999. p. 167−77. https://fyi.uwex.edu/forage/files/2014/01/alfqualest.pdf.

Teague WR, Apfelbaum S, Lal R, Kreuter UP, Rowntree J, Davies CA, Conser R, Rasmussen M, Hatfield J, Wang T, Wang F, Byck P. The role of ruminants in reducing agriculture's carbon footprint in North America. J. Soil and Water Conserv. 2016;71(2):156−64.

The Weather Channel. Average Monthly Temperatures. 2017. https://weather.com/maps/averages/normal-temperature.

Thompson RS, Anderson KH, Bartlein PJ. Climate-vegetation Atlas of North America − Introduction. USGS Professional Paper 1650-A. 1999. 17 pp, http://pubs.usgs.gov/pp/1999/p1650-a/atlas_intro.html.

Thorvaldsson G, Tremblay GF, Kuneliuis HT. The effects of growth temperature on digestibility and fibre concentration of seven temperate grass species. Acta Agric. Scand. Section B Soil Plant Sci. 2007;57(4):322−8.

Tubiello FN, Soussana J-F, Howden MS. Crop and pasture response to climate change. Proc. Natl. Acad. Sci. U.S.A. 2007;104(50):19686−90.

Tucker JJ. The 5 P's to Pasture Establishment. The Forage Leader. American Forage & Grassland Council; 2016. http://www.afgc.org/newsletter/TFL_May2016.pdf.

Undersander D, Cosgrove D, Cullen E, Grau C, Rice ME, Renz M, Sheaffer C, Shewmaker G, Sulc M. Alfalfa Management Guide. American Society of Agronomy, Inc; 2011. https://www.agronomy.org/files/publications/alfalfa-management-guide.pdf.

USDA. Plant Hardiness Zone Map. United States Department of Agriculture, Agricultural Research Service; 2017. http://planthardiness.ars.usda.gov/PHZMWeb/.

Vallentine JF. Ch. 13 grazing intensity. Grazing Management. second ed. San Diego, CA: Academic Press; 2001. p. 435−8.

Volenec JJ, Nelson CJ. Ch. 3 physiology of forage plants. In: Barnes RF, Nelson CJ, Moore KJ, Collins M, editors. Forages Volume II: The Science of Grassland Agriculture. sixth ed. Blackwell Publishing; 2007a. p. 37−52.

Volenec JJ, Nelson CJ. Ch. 5 environmental aspects of forage management. In: Barnes RF, Nelson CJ, Moore KJ, Collins M, editors. Forages Volume I: An Introduction to Grassland Agriculture. sixth ed. Blackwell Publishing; 2007b. p. 99−124.

Voth K. Does Grazing Sequester Carbon? Part 1. On Pasture. 2017. http://onpasture.com/2017/10/02/does-grazing-sequester-carbon-part-1/.

Wang T, Teague WR, Park SC, Bevers S. GHG mitigation potential of different grazing strategies in the United States Southern Great Plains. Sustainability 2015;7:13500−21.

Ward M. Horse Health Care in Heat and Humidity. Holistic Horse; 2011. http://holistichorse.com/health-care/horse-health-care-in-heat-and-humidity/.

Wartell BA, Krumins V, Alt J, Kang K, Schwab BJ, Fennell DE. Methane production from horse manure and stall waste with softwood bedding. Bioresour. Technol. 2012;112:42−50.

West CP, Waller JC. Forage systems for humid transition areas. Ch. 21. In: Barnes RF, Nelson CJ, Moore KJ, Collins M, editors. Forages, Volume II: The Science of Grassland Agriculture 6th Edition. Ames, Iowa: Blackwell Publishing; 2007. p. 313−21.

Xiao C. Soil Organic Carbon Storage (Sequestration) Principles and Management − Potential Role for Recycled Organic Materials in Agricultural Soils of Washington State. Olympia, Washington: Waste 2 Resources Program. Washington State Department of Ecology; 2015. Publication no. 15-07-005, https://fortress.wa.gov/ecy/publications/documents/1507005.pdf.

FURTHER READING

Hall M. 2016 Forage Trials Report. PennState Extension. 2017. https://extension.psu.edu/forage-variety-trials-reports-1.

Noland R, Wells MS. Alfalfa Management Planning with Growing Degree Days. UMN Extension Forage Team; 2016. https://www.extension.umn.edu/agriculture/forages/presentations/docs/gdd-and-alfalfa-harvest.pdf.

Soil Quality Pty Ltd. Soil Quality Website. 2017. www.soilquality.org.au.

UNFCCC (United Nations Framework Convention on Climate Change). Global Warming Potentials. 2014. http://unfccc.int/ghg_data/items/3825.php.

Chapter 13

Matching Plant Species to Your Environment, Weather, and Climate

Edward B. Rayburn

West Virginia University Extension Service, Morgantown, WV, United States

INTRODUCTION

When reseeding a pasture, select a combination of forage species that are complementary and adapted to the site's soil and climate. The pasture community needs to include legumes for fixing nitrogen for the grasses and forbs in the stand and possibly for improving forage quality. The pasture community also needs rhizomatous grasses for developing strong self-repairing sods. Once a new seeding is established, proper grazing and soil fertility management are essential for maintaining the stand and determining which plant species will persist and flourish in the pasture.

At times, livestock producers look at the condition of a pasture and think that the pasture needs to be reseeded. In many cases, this is not the most practical solution to improving the pasture's condition. The mixtures of plants growing in a pasture represents the species best adapted to the current site conditions and management. If a different plant community is desired, then changes in the grazing management or soil fertility are appropriate.

When considering reseeding a pasture the manager should first look at the limiting factors that are impacting the current stand.

- What is the fertility and pH status of the soil?
- Is the current grazing management the cause of poor pasture condition?

Either of these can be changed. Poor soil fertility or pH can be corrected by proper soil testing and application of needed plant nutrients and lime. Grazing management may be more difficult to change since we have to change the mindset of the humans controlling the livestock grazing the pasture. Once grazing management and soil fertility are corrected, pastures frequently regain the plant species desired and improve in condition. Changing grazing management and modifying soil fertility is less expensive and lower risk than reseeding; and these management factors need to be changed prior to any reseeding effort. If the soil needs additional nutrients or liming and this is not accomplished, then a new seeding will fail. If grazing management is not adjusted to the needs of the desired species, then the new seeding will fail. In a large on-farm forage establishment study conducted in the Northeast, the number one reason new seedings failed was improper grazing management after establishment (E. B. Rayburn, unpublished data).

At times, there are good reasons to reseed a pasture. For example, a breeder has too much toxic endophyte tall fescue in a pasture where they maintain pregnant mares. The manager can confine the mares and feed nontoxic hay but prefer to have the mares out on pasture for the exercise, sunlight, and improved nutrition. Another case is when a farm is purchased that was previously in cultivated crops and the new owner wants to convert the land to pasture and hay land.

When choosing plant species to be included in a seeding mix, there are three important factors to consider. These are the plants' adaptation to the site's climate, soil, and management; yield potential; and nutritional quality.

Adaptation to the Site and Management

When choosing forage species for a new stand, it is important to evaluate the species based on their:

- ecologic niche

- growth habit and management adaptation
- growth distribution
- stand longevity

Ecologic niches of forage species differ and are related to climate, soil fertility, soil drainage, grazing management, and pest tolerance (Table 13.1 and Table 13.2). Cool season species grow best when the mean daily air temperature is between 50°F (10 °C) and 70° F (18 °C). Warm season species grow best when the mean air temperature is over 80°F (27 °C) and do poorly when mean air temperature is below 60°F (16 °C). Mean air temperature is the average of the daily low before sunrise and the daily high in midafternoon. This can be measured with a max-min thermometer or looked up on many of the web-based weather services. Within the cool season forages, there are differences in tolerance to hot and cold weather. Tall fescue is the most tolerant cool-season grass at both ends of the heat spectrum in the eastern US. In northern dry environments, without winter snow cover, tall fescue will likely be winter-killed. Alfalfa is a cool-season legume that grows well in hot weather.

Some forage species tolerate poorly drained soils, while other species tolerate excessively drained soils. Both surface drainage and internal drainage are important. Poor surface drainage, which lasts for short periods of time, has less impact on plants than does extended periods of flooding. In the first case, timothy, bluegrass, and fescue do all right. In the latter case, reed canarygrass is more tolerant. Poor internal drainage results in wet lower soils that kill plant roots. Reed canary grass tolerates poor internal drainage. For sandy soils that are excessively drained, deep-rooted plants are required. These include the cool season forages alfalfa and smooth bromegrass and warm season forages such as Bermudagrass.

Growth habit determines the forage species' response to grazing or hay harvest. Orchardgrass, tall fescue, and bluegrass head out in the spring then tiller in later growths, keeping their growing points near the ground. Because the growing point is near the ground, they tolerate close, frequent defoliation (Fig. 13.1). Timothy, smooth bromegrass, and reed canarygrass joint in the aftermath. This means that the growing point in the tiller moves up above the soil surface in each growth cycle and can be injured if grazed or hayed at the wrong time (Fig. 13.2). Alfalfa, red clover, and birdsfoot trefoil have growing points that form on basal crowns (Fig. 13.3). These are good legumes to use in hayfields and rotationally grazed pastures with orchardgrass, tall fescue, timothy, smooth bromegrass, or reed canarygrass. White clovers (tall-growing ladino and short-statured Dutch) and prostrate-type birdsfoot trefoil are good legumes to use with orchardgrass, tall fescue, and bluegrass in continuously stocked pastures. Due to the competition between legumes and grasses, legumes do best under rotational grazing.

Another growth habit is the ability of a plant to move about in the pasture. Rhizomes are plant stems that grow under the soil surface (Fig. 13.4). At nodes on the rhizome, roots will form, and a new plant will come up above the soil surface. Stolons are stems that grow along the soil surface and at intervals the tip will set down roots and form a new plant (Fig. 13.5). A few plants like Bermudagrass have both rhizomes and stolons. These vegetative growth habits enable the plants to revegetate open soil areas in a pasture.

Growth distribution of forage species is a response to climate, weather, and day length. Some forage species grow better in cool weather; others grow better in warm weather (Fig. 13.6). Tall fescue is more tolerant of hot and dry weather than orchardgrass; timothy is least tolerant of these conditions. Tall fescue is most tolerant of freezing weather in the fall and is the best perennial grass for fall and winter grazing. Smooth bromegrass and timothy can go dormant due to short days before freezing weather in the fall. The flush of spring growth in grasses differs among species. Orchardgrass heads out earlier than does timothy. Cool season forages have the bulk of their growth in cool, moist weather. In southern environments, they may have little growth in July and August when the mean air temperature is over 80°F (27 °C). Warm season grasses will have the bulk of their growth occurring in July and August. In cool environments, warm season grasses will have little growth when mean air temperatures are below 60°F (16 °C). There are differences between varieties within a species. Hallmark orchardgrass goes to head sooner than Pennlate and has more vigorous fall growth.

Local climate and forage type determine the growth pattern of forages on the farm. For example, two farms within 50 miles (805 m) of each other can have considerable differences in forage growth (Fig. 13.6). Terra Alta has a mean July temperature of 68°F (20 °C) and a 54-inch (137 cm) annual rainfall. Moorefield has a mean July temperature of 75°F (24 °C) and a 32-inch (81 cm) annual rainfall. These differences are due to Terra Alta being on the highlands of the Alleghany plateau and Moorefield being east of the plateau, in its rain shadow, and at a lower elevation. In the southern Allegheny Plateau, mean July temperature decreases 3.3°F (5.9 °C) and annual rainfall increases 2.3-inches (5.8 cm) for each 1000-ft (305 m) increase in elevation. Parts of the plateau rise to elevations greater than 4000 ft (1200 m). The Allegheny Plateau is a western section of the Appalachian Mountains that includes parts of New York, Pennsylvania, West Virginia, and Ohio. For cool season forages having the same potential growth rate, more forage is produced at the higher elevation due to cooler weather and more rainfall (4.1 vs. 3.3 tons dry matter/acre/year) (9.1 vs. 7.3 mt/ha/yr). At the cool site, growth starts later in the spring and displays less summer slump than occurs at the warmer, drier site. At the cool site, half of the annual production is achieved about July 14 (day of year 195), while at the warm site, half of the annual production is achieved about June 21 (day of year 172). At the warm, dry site the risk of yield

TABLE 13.1 Selected Cool Season (CS) and Warm Season (WS) Forage Species and Their Adaptation to Soil Environments

Forage Species	Tolerance to Heat	Optimum Soil pH	Tolerance to Poor Drainage	Tolerance to Flooding	Tolerance to Excessive Drainage	Tolerance to Drought
Legumes						
Alfalfa	CS	6.5–7.0	P	P	E	E
Birdsfoot trefoil	CS	6.0–6.5	E	P	G	G
Lespedeza, Korean	WS	5.8–6.5	G	P	P	G
Lespedeza, sericea	WS	5.8–6.5	G	P	G	G
Red clover	CS	5.8–6.5	G	P	G	G
White clover	CS	5.8–6.5	E	P	P	P
Grasses						
Bermudagrass	WS	5.8–6.5	P	P	E	E
Bluegrass, Kentucky	CS	5.8–6.5	G	G	P	P
Bromegrass, smooth	CS	5.8–6.5	G	P	G	E
Orchardgrass	CS	5.8–6.5	G	P	P	G
Ryegrass, perennial	CS	5.8–6.5	P	P	P	P
Ryegrass, annual	CS	5.8–6.5	P	P	P	P
Reed canarygrass	CS	5.8–6.5	E	E	G	E
Tall fescue, E+	CS	5.6–6.5	E	E	G	E
Tall fescue, E−	CS	5.8–6.5	E	E	P	G
Timothy	CS	5.8–6.5	G	P	P	P

E = excellent, G = good, P = poor

Tall fescue, E+ = endophyte infected

Tall Fescue, E− = endophyte free

TABLE 13.2 Forage Species Characteristics and Management Adaptation

Forage Species	Growing Point	Aftermath Reproductive Growth	Establishment Vigor	Growth Habit	Management Adaptation			
						Grazing		
					Hay	Rotational	Continuous	Forage Quality
Legumes								
Alfalfa	Elevated	Yes	Medium	Crown	E	E	P	G-E
Birdsfoot trefoil	Elevated	Yes	Low	Crown	E	E	G	G-E
Lespedeza, Korean	Elevated	Yes	Medium	Crown	G	G	G	E
Lespedeza, sericea	Elevated	Yes	Low	Crown	G	G	G	P-G
Red clover	Elevated	Yes	High	Crown	E	E	P	G-E
White clover	Ground	Yes	Medium	Stolons	G	E	E	E
Grasses								
Bermudagrass	Ground	Yes	Medium	Stolons Rhizomes	E	E	E	P-E
Bluegrass, Kentucky	Ground	No	Low	Rhizomes	P	E	E	G
Bromegrass, smooth	Elevated	Joints	Medium	Rhizomes	E	G	P	G-E
Orchardgrass	Ground	No	Medium	Tillers	E	E	G	G-E
Ryegrass, perennial	Ground	No	High	Tillers	G	G	G	G-E
Ryegrass, annual	Ground	No	High	Tillers	P	G	G	G-E
Reed canarygrass	Elevated	Joints	Low	Rhizomes	E	G	P	P-E
Tall fescue, E+	Ground	No	Medium	Tillers Rhizomes	G	G	G	P-G
Tall fescue, E−	Ground	No	Medium	Tillers Rhizomes	E	E	P	G-E
Timothy	Elevated	Joints	Medium	Tillers Corms	E	G	P	G-E

E = excellent, G = good, P = poor

Tall fescue, E+ = endophyte infected
Tall fescue, E− = endophyte free

FIGURE 13.1   Bunch grasses like orchardgrass grow by producing tillers off the base of other tillers. Other than during reproductive growth the growing points stay near ground level.

differing from average is greater. In a moist year, yield can be good, but in a drier than average year, it can be quite poor. The warm Moorefield site is a good site for warm season forages (4.0 tons dry matter/acre/year) (8.9 mt/ha/yr). These forages do well in the warm environment and grow better in dry soils than do cool season forages, with about half of their production occurring about the middle of July. At this site the warm season forages do well and complement the cool season forages, which have a noticeable summer slump in this environment.

Stand longevity is the result of species tolerance to grazing by management, soil fertility and drainage, disease and insects, and the species' ability to grow into openings in the stand (rhizomes and stolons). Some forage species increase plant size by tillering (orchardgrass, timothy) or by growing larger crowns (alfalfa, red clover, birdsfoot trefoil). Other species invade open areas by rhizome growing points moving underground (Kentucky bluegrass, tall fescue, Bermudagrass) or stolon growing points moving above ground (white clover, Bermudagrass). Species that move to fill openings repair damage to the sod that occurs from hoof action and winterkill. Also, species and varieties that have good spring and fall growth (early heading orchardgrass and tall fescue) are more competitive with winter annual and biennial weeds, resulting in fewer weeds in the stand when properly grazed.

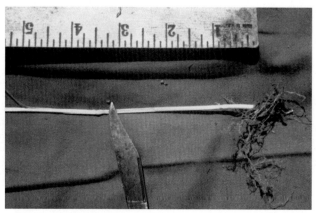

FIGURE 13.2   Grasses that joint in nonreproductive growth include timothy, smooth bromegrass, and reed canarygrass. The small bulge or joint just left of the knife blade is the elevated growing point on this timothy tiller.

FIGURE 13.3 Red clover, alfalfa, and birdsfoot trefoil are legumes that expand as new buds form in the crown. This crown has been cut in two to show the new buds.

The longest-lived grasses in the mid-Atlantic United States when managed properly are orchardgrass, Kentucky bluegrass, endophyte-infected tall fescue, timothy, bromegrass, reed canarygrass, and Bermudagrass, depending on the climate and soil. The best legumes for general use are white clover, red clover, alfalfa, and birdsfoot trefoil, again depending on the climate and soil.

Species diversity in a pasture or hay field is important for maintaining productivity across the landscape and across years differing in weather. In a field having variable drainage, it helps to mix two or more species that complement each other relative to drainage. For example, birdsfoot trefoil or ladino clover mix well with alfalfa in fields with variable drainage. The trefoil or clover will predominate in poorly drained areas of the field, and the alfalfa will thrive in the well-drained areas of the field. Mixing Kentucky bluegrass with orchardgrass provides a good sod-forming species that fills in around the bunch-type orchardgrass. Having sod-forming species in a pasture helps the pasture heal itself when hoof action damages the sod. Legumes in the stand provide nitrogen for the grasses, so no nitrogen fertilizer needs to be purchased. Palatable forbs, which usually come into a stand on their own, such as dandelions and plantain, are good sources of the microminerals copper and zinc, which are often low in grasses and legumes.

(A)   **(B)**

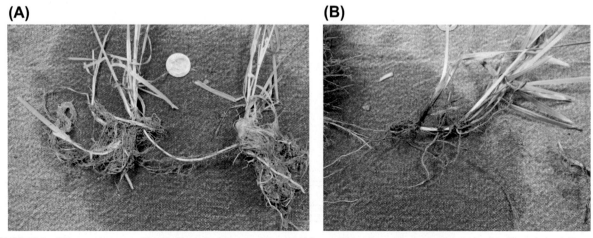

FIGURE 13.4 Rhizomes in grasses like this Kentucky bluegrass (A) and tall fescue (B) produce sods that hold up to hoof traffic and mend themselves when damaged.

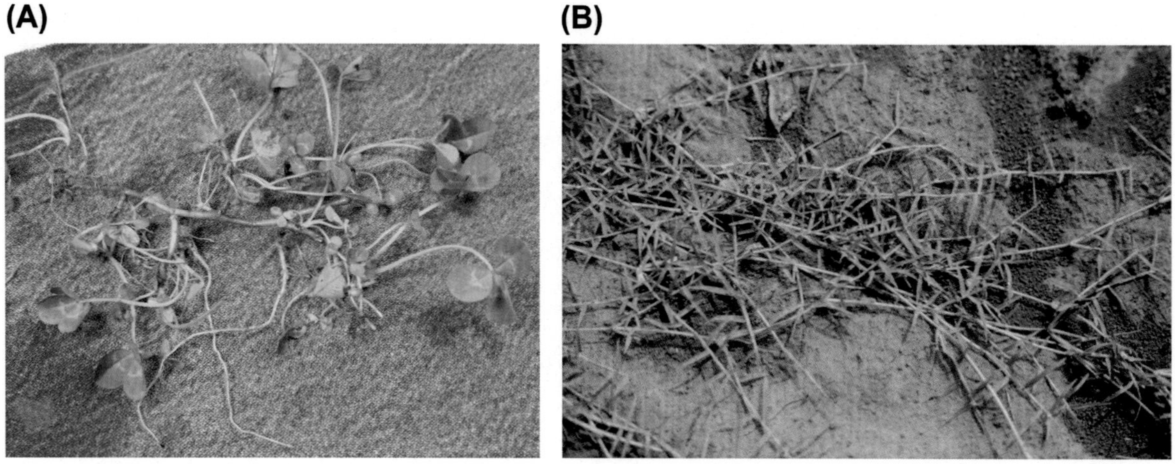

(A)

(B)

FIGURE 13.5 Legumes like white clover (A) and grasses like Bermudagrass (B) spread on the soil surface using stolons.

Species diversity helps increase stand longevity. Seeding only a single forage species results in other plants invading the stand. These may be "weeds," palatable forbs, or native forages. Proper mixes will help reduce weed encroachment. For example, a little white clover (1 lb/acre) (1.12 kg/ha) in an alfalfa-grass mix will fill in openings as the alfalfa thins, reducing weed invasion. A small amount of Kentucky bluegrass serves the same purpose in an orchardgrass clover seeding. When mixing forage species, use ones that are compatible or complementary in growth habit, palatability, growth distribution, and ecologic niche. The manager needs to select forage species that work best in the region's climate and on the soil existing in the field and implement the grazing management needed by the pasture plant community.

Forage Yield

Forage yield is largely determined by the following:

- soil type (potential plant rooting depth and plant-available soil water-holding capacity)
- precipitation (annual total and seasonal distribution)
- soil fertility (optimal-fertility soils are more productive than low-fertility soils)
- forage species (deep-rooted species are more productive than shallow-rooted species)
- nitrogen availability (legume fixed or applied fertilizer)
- grazing and haying management

A farm's climate, soil, and management affect what forage species are most adapted to the site. Soil type, its depth, and drainage can be determined from the county soil survey or web soil survey sites (https://casoilresource.lawr.ucdavis.edu/gmap/). This information should be used when choosing forage species. Soil fertility can be determined by soil testing and applying the needed fertilizers and lime.

For optimal growth, grasses and legumes need optimum soil test levels of phosphorus and potassium in the soil. These forages will also do best when the soil pH is above 6.0 (6.5 for alfalfa). Grasses also require nitrogen (N) from legumes, soil organic matter, manure, or mineral N sources. Legumes provide N to the grasses through N fixation and N-cycling through soil organic matter. In established sods a good grass legume mixture, 25%–30% or more legume, will be about as productive as the grass alone fertilized with 150 lbs of N per acre. When commercial N is used, it should be applied at 50–60 lbs (23–27 kg) N per acre at spring green up and again for each cut of hay or as needed in pastures. Use no more than 180–200 lbs N/a/year (202–224 kg N/ha/yr) on cool season grasses, more on well-managed Bermudagrass. Excess N can result in nitrate leaching into the groundwater, causing health risks to humans and livestock consuming well or spring water coming from the contaminated groundwater. Likewise, if manure is used as the N source, plant-available N should be limited to what the grass is able to use.

Grass and legume species differ in their potential productivity (Table 13.3 and Table 13.4). Local site conditions determine which species will be more productive or longer lived. In cool, dry areas, tall fescue and smooth bromegrass are excellent for pasture and hay production. In cool, moist areas, orchardgrass and timothy are highly productive. On poorly drained sites, reed canarygrass is most adapted. On well-drained soils in hot environments, Bermudagrass is an excellent species. The most productive legume on deep, well-drained soils is alfalfa. On shallow or poorly drained soils,

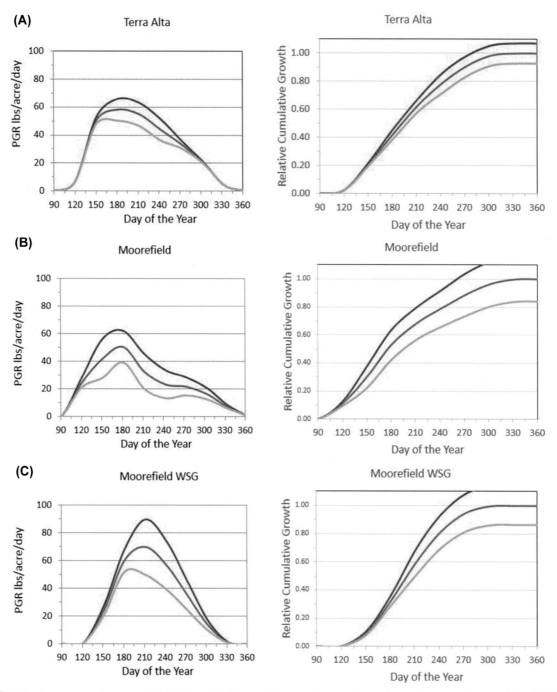

FIGURE 13.6 Cool season pasture growth rate (PGR) and relative cumulative growth across the year in a cool moist environment (Terra Alta) (A) and a warm dry environment (Moorefild) (B) and PGR for a warm season grass in the warm dry environment (C). The average PGR (middle line) is a reflection of the local climate. Annual weather differences cause production to fall within upper and lower lines 4 in 6 years (two-thirds of the time), while production falls above the upper line or below the lower line about 1 in 6 years.

red clover will be as or more productive than alfalfa. White clover is one of the least appreciated legumes for the value it has in pasture or hay fields. In Virginia, Kentucky bluegrass white clover pastures produce as much grazing per year as orchardgrass clover pastures and 83% as much as orchardgrass fertilized with 200 lbs N/acre/year (224 kg/ha/yrs) (Table 13.4).

Grazing and haying management affects total forage yield, forage yield in the first cut, and the proportion of forage available in aftermath growth. Taking first-cut orchardgrass at early head growth stage will reduce annual yield slightly compared to cutting at post bloom, but the nutritional quality will be higher, and there will be more forage regrowth for

TABLE 13.3 Average Yield (lbs of Dry Matter per Acre per Year) of Grass (Fertilized With 180–220 lbs N/a/year) and Legume Species Grown in the Northeast United States

Plant Species	Site Years[a]	Yield	SD[b]	Yield Relative to Orchardgrass[c]
Grasses				
Tall fescue	55	5.05	1.38	1.07
Orchardgrass	68	4.80	1.22	1.00
Reed canarygrass	36	4.78	1.46	0.92
Smooth bromegrass	35	4.39	1.08	0.87
Timothy	54	4.25	1.24	0.87
Perennial ryegrass	25	3.58	1.39	0.73
Legumes				
Alfalfa	122	6.23	1.18	1.26
Red clover	46	4.35	1.47	0.94

Grasses fertilized with about 200 N/acre/year.
[a]Site years: the number of years across locations where data was gathered.
[b]SD: standard deviation, the range where two out of three observations occur.
[c]Yield relative to orchardgrass: yield compared to orchardgrass where the two species were grown at the same site in the same year.

TABLE 13.4 Animal Grazing Response to Grass-Clover and Nitrogen-Fertilized Grass Pastures (200 lbs N/acre/Year) in Virginia Over 10 Years

Mix	700-lb Animal Days Grazing/A	ADG/Head	Gain/A	Relative Yield, Days Grazing/A[b]	Relative Yield, Gain/A[b]
Orchardgrass clover	257	1.28	329	0.83	0.99
Orchardgrass N	311	1.07	333	1.00	1.00
Tall fescue clover	303	1.02	309	0.97	0.93
Tall fescue N	403	0.91	367	1.30	1.10
Bluegrass white clover	258	1.21	312	0.83	0.94

Grazing days given in terms of 700-lb yearling steers. Average daily gain (ADG) given in terms of lbs/head/day. Gain per acre is the product of grazing days and ADG.[a]
[a]Roy E. Blaser et al., Managing Forages for Animal Production. Res Bul. 45. Va. Tech Blacksburg, VA.
[b]Relative yield by ratio to orchardgrass fertilized with N.

aftermath grazing. Grazing aftermath growth on hay meadows is a significant way to manage pasture summer slump by providing more acres for grazing when there is a lower pasture growth rate.

Forage Quality

Forage quality is determined by the following:

- plant growth stage at harvest (as plants mature quality goes down)
- legume content (at a given growth stage, legumes are higher quality than grasses)
- grazing intensity or harvest damage in hay
- plant species (some forage species have antiquality components)

Forage quality is little affected by forage species. The single most important factor determining forage nutritional quality is plant maturity. As a plant matures the forage increases in fiber and decreases in crude protein and digestibility

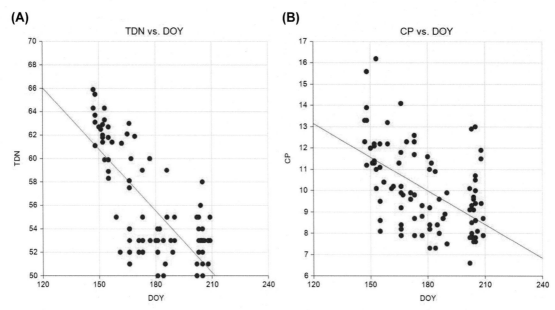

FIGURE 13.7 Effect of harvest date expressed as day of the year (DOY) on total digestible nutrient (TDN) and crude protein (CP) content in dry hay across elevation, latitude, and years in West Virginia (DOY: 151 is May 31, 181 is June 30). TDN and CP are expressed as percentages of forage dry matter.

(Fig. 13.7). At midlatitudes, first-cut hay digestibility decreases 0.33−0.50% points for each day harvest is delayed. Delaying harvest also reduces the amount of aftermath growth, critical in a grazing system to reduce the summer slump in pasture availability.

The second-most important factor in determining forage quality is legume content of the stand. Legumes are of higher nutritional quality than grasses when harvested at the same growth stage. Legumes are lower in cell wall fiber (neutral detergent fiber, NDF), which allows animals to eat more. Animals that eat more forage gain weight faster or produce more milk. However, when they eat more, it takes more forage to feed the same number of animals (Table 13.4). Legumes are higher in nonstructural carbohydrates (sugars and starches that are nearly 100% digestible) and protein since there is less cell wall fiber present. When pasture or hay does not supply the nutrients the livestock require or if the cost of supplemental feed is too high, there will be an economic response to earlier harvest or grazing and the introduction and management of legumes in the stand.

Grazing management impacts forage availability and forage intake by the grazing animal. When pastures are grazed to below a forage mass of 1200 lbs dry matter per acre (1345 kg/ha) (about a 4-inch (10.2 cm) height), livestock forage intake is reduced since they cannot eat as much forage per bite as they can at a higher forage mass. At forage mass over 2400 lbs dry matter per acre (2690 kg/ha) the grazing animal is able to selectively graze. Selective grazing is the ability of animals to select forage of higher quality than the average quality in the pasture. This can be good for a lactating mare or growing yearling. It may be of no value to an easy keeping mare or gelding used occasionally for riding.

Forage quality differences between grass species occur due to date of heading. This is a management concern once the grass most adapted to the site is chosen and needs to be considered in management. One characteristic of late-heading grass species and varieties is that they are less competitive with weeds.

There are negative quality components in some grasses, such as toxic alkaloids in endophyte-infected tall fescue and wild-type reed canarygrass. These problems can be avoided by using endophyte-free or endophyte-enhanced tall fescue and new varieties of palatable low-alkaloid reed canarygrass.

When mixing forage species, use those that are compatible in growth habit and palatability and which complement each other in growth distribution and ecologic niche. When deciding what seeding rate to use for establishing a new forage seeding, consider the percentage seed germination, seedling vigor, and the size of the mature plant. Pure seed and seed germination percentage are listed on the seed bag's tag for inspected seed. Vigor of the seedlings depends on the species. Larger seeded plants usually have more vigorous seedlings than small-seeded plants. Most forage crop seeds are more vigorous the year after harvest than they are if they have aged too long. The size of the mature plant depends on the genetic ability of the species to increase in size and by the amount of open space around the plant. Some forage species increase in size by tillering or by growing larger crowns. Other species are able to invade open areas of the stand by moving growing points underground or aboveground. Increasing the seeding rate to compensate for poor seedbed preparation is usually not justified. If seeds fall on an area of ground that is not suitable for germination and survival, two seeds will do no better than one. On good soil, if two seedlings emerge close to each other, they will compete with each other and may do no better than one seed.

In general, seeding rates of individual species are lower when combined in a mix than when seeding alone, but the total seeding rate of the mix may be higher. When mixing forage species, use those that are compatible in growth habit and complement each other in growth distribution and ecologic niche.

REVIEW QUESTIONS

1. You live in a hot climate (July average temperature 80°F (26.7 °C)) and the soil in your field is excessively drained. What grass is most adapted to your site?
2. You live in a cool environment and the soil in your pasture is poorly drained but not subject to flooding.
   a. What are the grasses most adapted to this site?
   b. What grass may be good enough but not the best?
3. In 2b, which grass is least tolerant to continuous grazing?
4. Which legume is most tolerant to continuous grazing?
5. Which legumes are least tolerant to continuous grazing?
6. Which cool season grass has the highest production potential in the Northeast United States?
7. Which cool season grass has the lowest production potential in the Northeast United States?
8. If I am growing young horses on pasture, which pasture type is best suited: bluegrass, white clover, or orchardgrass fertilized with nitrogen?
9. In what type of environment is a cool season forage—based system most productive?
10. In what type of environment is a warm season forage—based system most productive?

Chapter 14

Managing Manure, Erosion, and Water Quality in and Around Horse Pastures

Laura B. Kenny[1], Michael Westendorf[2] and Carey A. Williams[2]

[1]Extension Division of the College of Agricultural Sciences, Penn State University, University Park, PA, United States; [2]Rutgers, the State University of New Jersey, Department of Animal Science, School of Environmental and Biological Sciences, New Brunswick, NJ, United States

INTRODUCTION

Jamie thought his situation was perfect. Keep the horses on pasture and he will never have to clean stalls! However, a year after moving his nine horses to the farm with a 5-acre (2 ha) pasture, he is running into some problems. The pasture quickly got covered by more manure than he had expected, and the initially lush pasture is now overgrazed with bare spots. The fly population on Jamie's farm has also gotten out of control. A stream runs through his pasture for the horses' drinking water, and the downstream neighbors called because they have noticed that the pond at the end of the stream is turning green. Jamie needs to find a better way to manage his manure to protect the stream, care for his pasture, and make his operation more environmentally friendly.

An average size horse, which weighs 1000 lbs (454 kg), will defecate 4–13 times each day, producing up to 50 lbs (22.7 kg) of manure. This horse will produce over 9 tons (8.2 tonnes) of manure yearly, which can easily be doubled when soiled bedding is also added to it (Westendorf and Krogmann, 2013). In addition to considering horse health, nutrition, and forage plant growth, managers need to minimize the environmental risks associated with grazing horses. Environmental concerns are often ignored, either due to lack of awareness or low priority, but in the long run, they are critically important and affect everyone, not just the current landowner. Decisions made today can affect the productivity of the land and health of waterways generations from now.

Two major environmental concerns due to poor grazing management are soil erosion and contamination of ground and surface water by unmanaged manure. These are sources of **nonpoint source pollution**, contaminants that enter surface water through runoff rather than direct discharge.

EROSION

When water flows off an uncovered or bare field, it often carries soil particles with it in a process known as erosion. Wind can also transport soil particles. These soil particles represent a loss of topsoil, the most productive layer of soil where plant roots grow. The process begins when raindrops strike bare soil, dislodging soil particles from their matrices. Intense rainfall is more likely to cause erosion than light rain, and plant cover reduces erosion by intercepting raindrops and infiltrating rain water into the ground through channels made by plant roots. The threat of erosion could be intensified when the soil is compacted by animals and equipment. There are four types of erosion: sheet, rill, gully, and streambank (USDA-NRCS, 2003). Sheet erosion is the uniform removal of soil in thin layers, and it occurs when soil particles are carried evenly over the soil surface by rainwater that does not infiltrate into the ground. The sediments in this runoff can also fill in cracks in the soil surface, creating an impervious "crust" that further reduces water infiltration. Rill erosion is removal of soil by concentrated water flow, and it occurs when the water forms small channels in the soil as it flows off site. If rill erosion is not controlled, it can develop into gully erosion, where runoff forms deep, wide channels (Fig. 14.1). The last type, streambank erosion, occurs when the flow of water in a stream is strong enough to degrade the bank and detach sediments. It can be caused by increased water flow from runoff or human activity, decreased bank vegetation, and trampling by livestock (USDA-NRCS, 2003).

FIGURE 14.1 A photo of rill erosion that over the years has turned into gully erosion. This not only has environmental implications but can be hazardous to the safety of horses. *Photo courtesy of M. Westendorf.*

In addition to losing valuable topsoil, erosion pollutes waterways (Fig. 14.2). Sediments are transported by storm water runoff into surface water where they cloud up the water. This results in elevated water temperature, reduced sunlight penetration to aquatic plants, and clogged fish gills. If the sediment settles to the stream bed, bottom-feeding life may have a hard time finding food, plus the water body becomes shallower and water level will rise, increasing the risk of flooding (Welsch, 1991). Soil particles carried by rainwater can also transport nutrients such as nitrogen and phosphorus into water bodies, which we will discuss in the next section.

FIGURE 14.2 This fence line erosion is caused by a combination of water runoff and horse traffic. Note the difference in ground level on either side of the fence. *Photo courtesy of C.A. Williams.*

CONTAMINATED WATER LEACHING/RUNOFF

Horse manure contains nutrients and fecal pathogens. As rainwater flows through manure, it picks up these nutrients and pathogens and carries them along its path. This nutrient-laden water coming from an uncovered manure pile is called leachate. If you have ever seen brown, odorous fluid around a manure pile, that is leachate (Fig. 14.3). If the leachate infiltrates into the soil, the nutrients can be taken up by plant roots or incorporated onto soil particles. However, conditions such as low plant cover, concentrated areas of manure, sandy soil, and highly concentrated leachate can cause the fluid to reach and pollute groundwater. Nitrogen is highly mobile in soil and tends to dissolve and leach as nitrate, while phosphorus adsorbs strongly to soil particles (85% of phosphorus is attached to soil and organic particles) and is more likely to be lost by runoff erosion (Westendorf and Krogmann, 2013; Welsch, 1991).

If the nutrient-laden runoff does not infiltrate (due to heavy rainfall and/or low soil infiltration capacity) and there is a sensitive area (surface water, wetland, well) downslope, it continues to travel over the surface and into the water body. The presence of excess nutrients in surface water can result in a process called **eutrophication or algae bloom** (Fig. 14.4). The first step is an algal bloom due to the elevated nutrient levels in the water. This is what Jamie's neighbors were noticing in their stream due to the excess runoff from his property. Phosphorus is usually the limiting nutrient to algae growth in fresh water. The bloom blocks sunlight and shades aquatic plants, which can no longer photosynthesize and eventually die. The dead plant material (including the algae) is then decomposed by bacteria in a process that consumes oxygen. Once water oxygen levels drop far enough, much of the remaining aquatic life cannot survive. In addition, some anaerobic organisms thrive and release hydrogen sulfide and methane, which are toxic to yet more aquatic life. Fish kills and reduced ecosystem diversity are symptoms of severe eutrophication of a water body (Welsch, 1991).

While phosphorus is mostly responsible for eutrophication, nitrogen in surface water and groundwater is also quite harmful. Babies suffer methemoglobinemia (blue baby syndrome) from drinking water with high nitrate levels, as nitrate is converted to nitrite in their stomachs, which binds to hemoglobin, reducing the blood's oxygen-carrying capacity. Cattle also suffer from a similar condition when drinking water or consuming feed high in nitrates (Welsch, 1991). Horses can also be poisoned by nitrates in water, but they require a higher dosage than ruminants before toxicity occurs. As nitrate is quite soluble, it is easily leached to groundwater through infiltration.

Runoff from horse farms can also contain pathogens from horse manure. Luckily, humans and horses do not share many common water-borne pathogens, and transmission of zoonotic diseases from horse manure to humans is rare. Examples of zoonotic pathogens that can be found in horse manure are *Salmonella* spp., strains of *Escherichia coli*, *Listeria monocytogenes*, *Giardia*, *Cryptosporidia*, and *Campylobacter* spp. (Brophy et al., 2008; Jenkins et al., 2015). The Centers for Disease Control and Prevention recommend washing your hands with running water and soap after touching or doing chores with horses, in order to help prevent infections by *Salmonella* and other bacteria.

Many factors influence runoff and erosion, such as soil structure, soil moisture content, infiltration capacity, slope, climate, intensity and duration of rainfall, vegetative cover, etc. In addition, intense livestock grazing can increase erosion by compacting soil and breaking soil aggregates, leading to lower water infiltration and therefore increased runoff. This management can also decrease plant cover and alter plant species composition, both of which have an effect on reducing infiltration and increasing runoff (USDA-NRCS, 2003).

FIGURE 14.3 Leachate around an uncovered manure pile after a rainfall. *Photo courtesy of C.A. Williams.*

FIGURE 14.4 Algae bloom turns a beautiful pond into an unpleasant view. *Photo courtesy of M. Hashemi.*

PARASITE CONCERNS

Pasture management can have direct effects on horse health as well, from toxic plants and metabolic issues to parasites. The life cycles of some gastrointestinal parasites of horses depend on eggs being deposited on pasture and larvae being consumed as horses graze. Currently, a parasite of major concern for adult horses is the small strongyle or cyathostomin. Eggs are deposited onto pastures through manure. Under ideal conditions, the eggs hatch and go through three larval stages; during the infective L3 phase, they can survive some extreme weather conditions. Equine "lawn and rough" grazing behavior (see Chapter 8 for a definition) limits exposure to larvae near manure piles, but water runoff can transport them downhill. The practice of dragging or harrowing pastures also spreads larvae throughout the entire pasture. If temperatures are over 25°C and fecal balls have been broken by dragging a harrow, the infective L3 larvae will only survive for a few weeks; however, most farm owners do not restrict grazing for this long after dragging. Most horses are already infected with small strongyles, but they vary in immunity to infection and rate of egg shedding. Twenty percent of horses shed 80% of small strongyle eggs (AAEP, 2013).

Most horse owners use an anthelmintic (deworming) schedule to control small strongyles, large strongyles, tapeworms, roundworms, etc. Large strongyles (*Strongylus vulgaris*) have been largely eradicated in the United States due to the frequent deworming protocol used by most horse farms, but this same protocol is not as effective on small strongyles and has led to parasite resistance to dewormer products. If a horse is treated with a dewormer and parasites develop resistance to that product, then only resistant eggs will be shed in the pasture, creating a large resistant population. The common practice of rotating deworming products has led to parasite populations that are resistant to several products. One key to battling resistance is to maintain a population of untreated parasites in the pasture called refugia. To accomplish this, the "low-shedder" horses are treated less frequently with dewormers. Eggs from parasites that are sensitive to the dewormers will be deposited in the pasture and genes from the developing parasites can "dilute" the genes of the resistant population. It is also important for horse owners to perform fecal egg count reduction tests (start by contacting your veterinarian) to evaluate the effectiveness of deworming products on their farm, and only use effective products on moderate- and high-shedding horses for small strongyle control. Resistance has also been observed in ascarids (*Parascaris equorum*), which are a primary concern for foals (AAEP, 2013).

How can we manage pastures to reduce parasite infection? There are a few management practices farm owners can use, such as removing manure from pastures, not rotating horses immediately after deworming, rotating pastures into hay or ruminant grazing for at least a year, dragging (see earlier), and maintaining high-quality pastures. However, it is nearly impossible to eradicate small strongyle infection in horses that graze, so more effort should be focused on managing resistance. Most horses on a farm will have strong immunity to these parasites and require only one to two deworming treatments per year. Limiting the use of dewormers will help to maintain pasture refugia. The remaining high shedders can be treated more often as needed based on egg reappearance in fecal egg counts (AAEP, 2013).

OTHER ENVIRONMENTAL CONCERNS

Insects: Flies, mosquitos, and ticks are disease vectors for horses and humans, in addition to being a nuisance. Several types of flies lay their eggs in manure piles, so the manure storage area is usually where they will be found. However, pastures or lots with large, permanent piles of manure can serve the same purpose, which is why Jamie's property has seen an increase in the fly population. Eggs are laid in the top few inches of moist manure, so keeping a manure storage covered from rain can aid in reducing fly populations (Wheeler and Zajaczkowski, 2009). Mosquitos lay eggs in standing water, including permanent puddles in areas of poor drainage, and eggs hatch within 4−14 days. Managing pastures to improve drainage, in addition to removing other sources of standing water, can reduce mosquito populations on the farm (Swinker, 2015). Even water troughs and deep hoof prints can serve as breeding areas for mosquitos. Ticks tend to congregate in wooded areas with tall grass and shrubs, clinging onto passing humans or animals as they brush against the foliage. Ticks in pasture can be controlled by fencing horses out of wooded areas, clearing brush in pasture, keeping a mowed buffer around fields, and controlling fence line vegetation to reduce tick habitat. Additionally, there are insecticide sprays that can be applied to the vegetation in areas bordering pastures (Swinker, 2015). Bot flies lay their eggs on horse hair so that the larvae may be ingested, then burrow into the gums and attach themselves to stomach linings. Eggs must be removed from the horse manually, and deworming products can be used for internal larvae (Swinker, 2015).

Salinity: In addition to nitrogen and phosphorus, horse manure contains salts. These salts can build up in the soil and have negative effects on forage plant growth (Li-Xian et al., 2007), especially in areas of low rainfall. In some arid areas in the Western United States, salt can work its way to the soil surface as soil water evaporates, and adding manure to the soil will intensify the salinity problem. Salt-intolerant plants can suffer from drought-like symptoms and toxicities from high levels of soluble salts. Water diffuses into plant roots by osmosis through semipermeable membranes, and high salt levels in the soil can either make it harder for plants to take up water, or even draw water out of plant roots due to the difference in solute concentration across the membrane. Additionally, plants can take up the dissolved salt components. A plant that has accumulated toxic salt levels will appear to have burned edges on its leaves (Fig. 14.5). When reading a soil or compost test result, electrical conductivity can provide information on soluble salts. Salt tolerance varies among pasture species and even cultivars. Saline soils can sometimes be treated by leaching with low-salt water (Provin and Pitt, 2001).

Weed Seeds: Seeds from hay or pasture, after consumption by horses, can often survive in the digestive tract and emerge unharmed in manure. This allows for the introduction of new, potentially invasive weeds into pastures. Harmon and Keim (1934) found that 10%−12% of weed seeds fed to horses are passed intact, and 9% of the seeds fed were viable and germinated. Viable seeds included velvet weed, bindweed, sweet clover, and peppergrass. Weaver and Adams (1996) germinated 29 different weed species from horse manure piles left on trails in Australia. Little research has been done to explore this process in horses.

Aesthetics: A bare lot, littered with large piles of manure, is less appealing than a lush green pasture. While this may seem inconsequential, horse farm owners need to consider the importance of neighbor relations. Farms near metropolitan and suburban areas may border backyards and public spaces where a concerned homeowner or citizen feels the farm is

FIGURE 14.3 High salinity around manure pile prevents the growth of any vegetation; if vegetation does grow the leaves will appear burned. *Photo courtesy of M. Hashemi.*

being improperly managed and reports it to public authorities. In fact, most states do have comprehensive manure and nutrient management regulations that all farmers should know and follow.

BENEFITS OF MANURE ON PASTURE

If manure on pasture is managed correctly to minimize environmental impacts, it can have great benefits for the farm.

Nutrient Recycling: Horses consume nutrients in feed and pasture, digest about 60% by weight, and any unused nutrients are excreted in manure (Westendorf and Krogmann, 2013). When pasture is grazed, the plants require water, nutrients, and sunlight to regrow. Grazing allows for recycling of nutrients by depositing manure directly back onto pastures and providing nutrients for forage plants. This reduces the need for manure removal and fertilizer purchases. However, it is important to keep the cycle in balance by using appropriate stocking densities and rotational grazing to minimize risks from the concerns described in the previous section.

Improves Soil Quality: Adding manure to the soil provides plant nutrients and increases organic matter in the soil, improving soil physical and biologic characteristics. In addition, adding organic matter to fast-draining soil is helpful because it acts like a sponge and helps the soil hold water in the root zone for longer periods. However, too much organic matter in poorly drained areas such as high-traffic areas can contribute to and/or intensify the mud problems because of this water-holding capacity. Organic matter also adsorbs nutrients, which can later be released for plant uptake (USDA-NRCS, 2008). The carbon in organic matter also provides an energy source for soil organisms, which aerate the soil, break down organic matter, improve soil structure, and mineralize nutrients into plant-available forms (Graham et al., 2009).

Reduces Farm Costs: If enough pasture land is available, keeping horses outside has a number of advantages, including less labor spent on stall cleaning. However, as Jamie found, overstocking his pasture to avoid cleaning stalls has turned into an environmental disaster. Other areas of cost savings with increased grazing include purchased bedding, feed, fertilizer, manure removal, and improved property value.

SPREADING MANURE ON PASTURE

At this point, it should be clear that horse manure can be an excellent soil amendment for pasture. It contains essential macro- and micronutrients for plant growth. It is important to test the soil for current nutrient levels and pH before adding any amendment. Exact nutrient composition varies based on the horse's diet and exercise level, but average fertilizer amounts for manure from a 1000-lb exercised horse are shown in Table 14.1. It is much more difficult to estimate these numbers when the manure has bedding mixed in, as is usually the case with stall waste. Manure and compost can be tested in a laboratory for nutrient content, and this is a good idea when using it as a source of nutrients for a crop. It is important to note that overfeeding nitrogen and phosphorus increases the amount of these nutrients in manure, so care should be taken to balance the horse's diet to its nutritional needs (Westendorf and Krogmann, 2013).

However, we cannot forget about the downsides of adding manure to pasture, such as the spread of pathogens, weed seeds, and parasite eggs, as described earlier. One solution to this problem is to compost manure before spreading. **Composting** is an aerobic process whereby bacteria and fungi break down organic matter and produce humus, a stabilized material that provides the benefits of manure. When composting is done correctly, it kills parasite eggs, pathogens, and weed seeds through the high temperatures achieved by the microbial activity (Wheeler and Zajaczkowski, 2009). Composting requires proper management including the right materials (proper carbon to nitrogen ratio), temperature, moisture, aeration, and oxygen.

TABLE 14.1 Daily and Yearly Amounts of Fertilizer Nutrients in Horse Manure, per Horse and as Excreted

Horse Manure	N	P_2O_5	K_2O
Pounds per day	0.31	0.15	0.23
Pounds per year	113	55	84
Pounds per ton of manure	12	6	9
Percent by weight	0.6%	0.3%	0.5%

Daily manure production was assumed to be 52 lb/day. K_2O, potash; P_2O_5, phosphate. Compare percent by weight to commercial fertilizer N–P–K values and be aware that not all N from manure is immediately available.
Adapted from USDA-NRCS. 2008. Agricultural Waste Management Field Handbook. Washington, D.C., pp. 4–22. http://www.nrcs.usda.gov/wps/portal/nrcs/detailfull/national/water/?&cid=stelprdb1045935.

True finished compost is dry, odorless, and the nutrients are in a more stable form, so they are released into the soil over a longer period of time and with less risk of nutrients leaching out of the soil (Krogmann et al., 2006).

Spreading Guidelines: There are some important guidelines for spreading manure or compost on cropland that are collectively called **nutrient management**. When spreading manure, the application rate must be based on current nutrient levels in the soil and limited to the nutrient uptake rate of the crop from the soil (Westendorf and Arogo-Ojego, 2014). Pasture grasses will only take up limited amounts of nitrogen, phosphorus, and potassium in a grazing season; therefore applying excessive manure may result in potential environmental pollution risk from contaminated runoff and nutrient buildup in the soil. Care must be taken to balance the nutrient content of the manure with the nutrient removal of the crop. Nutrient content can be estimated from average nutrient content (Table 14.1), or samples of the manure can be analyzed by testing labs for improved precision.

It is important to understand that not all nutrients in manure are immediately available for plant use. The amount of nitrogen available is a function of the percentage of nitrogen in the manure, whether or not it is incorporated in the soil, and the rate of organic matter decomposition of the manure. Nitrogen availability will range from 20% of the total nitrogen when manure is spread on the soil surface to 60% when immediately incorporated into the soil. Surface-applied manure will continue to release plant-available nitrogen for about 4 years after application. Approximately 80% of the phosphorus from phosphate (P_2O_5) is available and about 90% of the potassium from potash (K_2O). If removing manure from horse stalls with additional bedding mixed in with the manure, this will have an effect on the nutrient utilization. Sawdust or wood shavings are high-carbon materials that require a great deal of nitrogen to break down. This process can tie up available nitrogen, rendering it unavailable to plants or crops (Westendorf and Arogo-Ojego, 2014). In this situation, pasture plants can become stunted or even die.

Horse manure alone will not perfectly meet a pasture's (or other crop's) nutrient needs; manure contains approximately twice as much N as it does P, yet productive pasture grasses may require three to four times as much N as they do P. The appropriate amount of manure or compost applied to any given field will vary due to a number of factors: the nutrient content and mineralization rate of the amendment, existing nutrient levels in the soil, previous applications of slow-release fertilizer or organic matter, presence of nitrogen-fixing legumes, the risk of phosphorus contamination of surface water, and more. If the risk of phosphorus reaching surface waters is high, then applications must be limited to the amount of phosphorus the crop will take up or avoided altogether. This is often evaluated using the "**Phosphorus Index**," which estimates the risk of water contamination on a field-by-field basis based on phosphorus levels in the soil, proximity to surface water, hydrologic characteristics of the soil, risk of soil erosion, method of manure application, and other factors that vary state by state (USDA-ARS, 2003). It is important to test the soil regularly to monitor soil phosphorus levels. Software exists to help nutrient managers perform all of these calculations and determine the correct rate of manure to spread on each field or paddock on the farm, as well as additional fertilizer necessary to meet the crops' needs.

Once the appropriate manure application rate is calculated, it is necessary to calibrate the spreading equipment to ensure that the rate of application is correct. There are a number of methods to calibrate a solid manure spreader. One method is to calculate the volume of a single load of manure, spread it, and then calculate the area of land that received manure. The volume must then be converted to weight based on the density of the manure, which can be challenging based on the type and amount of bedding present. A more precise calibration is called the "tarp method." The applicator lays down several tarps of a known area in the path of the spreader. The spreader is then driven over the tarps and the manure collected on each one is weighed (Fig. 14.6). The overall manure application rate can be calculated from the weight of manure on each and the known surface footage of the tarp. For more information on the procedure of calibration, see Komar and Westendorf (2009).

FIGURE 14.6 Using the tarp method of manure spreader calibration. *Photo courtesy of M. Westendorf.*

Manure must not be spread too close to any sensitive areas: surface water, wetlands, wells, livestock drinking water, sinkholes, or any other direct outlet to surface water. Many states have regulations on how wide these "setback distances" must be. They should be vegetated, and the width must increase with higher slopes. Care must also be taken in the timing of application. Compost or manure should not be spread immediately before a heavy rain to avoid washing it away and generating runoff. Likewise, it should not be spread any time the ground is frozen or snow-covered. The best time to apply any kind of fertilizer to established pasture is when plants are actively growing and roots can take up the nutrients being added to the soil (Westendorf and Arogo-Ojego, 2014).

MANAGING PASTURES TO ALLEVIATE ENVIRONMENTAL CONCERNS AND PROTECT WATER QUALITY

As mentioned previously, the most effective way to reduce erosion and runoff is to maintain a high vegetative cover on pastures. Research has shown that the cover threshold for controlling erosion is 70% (Costin, 1980; Sanjari et al., 2009). Even the type and habit of the plants present have an impact on runoff. Decomposing vegetation provides organic matter for the soil, which increases water-holding capacity.

Listed next are some additional Best Management Practices (BMPs) that protect water quality. There are two basic concepts in protecting water: keeping "clean" (uncontaminated) water clean and treating or filtering the "dirty" (contaminated) water.

- **Removing manure.** The easiest way to ensure that nutrients from manure do not enter surface water is to remove it frequently. This is most important in unvegetated stress lots where soil is often compacted and water does not infiltrate quickly, so it flows off the lot, carrying contaminants with it. Manure removal in pastures is less common, although some specialized equipment exists for the purpose, such as pasture vacuums. Removing manure also reduces parasite burdens on pastures. However, urine cannot be removed and contains high levels of urea and potassium, so proper siting of pastures and heavy use areas is still important.
- **Constructing heavy use pads.** A heavy use pad is a structure that improves drainage in an area where animals concentrate and vegetation cannot be grown to filter runoff. In general, they are constructed by removing a section of topsoil and adding a layer of coarse rock or gravel and a layer of footing on the surface. This allows water to collect at the bottom of the pad instead of the surface (as in Fig. 14.7), where it can infiltrate or be removed by drainage pipes (as in Fig. 14.8). Heavy use pads are highly effective at improving drainage and controlling mud in stress lots or near gates and feeders where animals congregate (Westendorf and Williams, 2013).
- **Designing proper manure storage.** There are many options for suitable manure storage, and they all have a few characteristics in common. Location is key: manure should not be stored within 100 ft of property lines and surface water (different states and municipalities may have their own regulations on this issue). It should not be stored in low areas prone to flooding or on steep ground. The storage should be easily accessible for stall cleaners and removal. If a tractor and bucket loader are used for removal, it helps if the storage has three walls and a solid base that does not form muddy tire ruts (Fig. 14.9). It should be sized appropriately for the amount of manure produced and period of time between removals. Manure storage should have a nonpermeable base. Whether it is cement or simply packed earth, the idea is to keep fluid from leaching into the ground directly under and around the pile. There should be a plan for leachate; the leachate should be diverted to an area where it can be filtered (Fig. 14.10), or the pile can be kept covered, so rainwater does not flow through the manure (Fig. 14.11; Kelly and Westendorf, 2014). Sometimes horse farm managers will use dumpsters or bins that are filled with manure and then removed from the farm for convenience; these can also be a source of leachate.
- **Using vegetated buffers.** It is inevitable that stormwater will flow over unvegetated areas on a farm at some point. A vegetated buffer, or filter strip, will filter the stormwater of nutrients and sediments as it passes through the strip. This BMP should be used around any unvegetated areas where manure is deposited, including stress lots, barnyards, manure storages, and riding arenas. A vegetative treatment area is an area where animals are not allowed to graze. Manure contaminated runoff will enter this treatment area and nutrients will be taken up by the plants. The area is large enough to prevent manure contamination outside of the area. A rain garden is also a type of buffer or treatment area that

FIGURE 14.7 This is what can happen around feeders and water sources without a heavy use pad in wet weather. This can be unhealthy for horses and unsafe for human handlers/workers. *Photo courtesy of C.A. Williams.*

FIGURE 14.8 Newly constructed heavy use pads under hay feeders in a pasture will keep this area mud-free due to proper footing and drainage. *Photo courtesy of C.A. Williams.*

FIGURE 14.9 This is an example of a manure pit properly designed as a temporary storage of manure for the appropriate number of horses and days. *Photo courtesy of C.A. Williams.*

FIGURE 14.10 The manure storage is slightly sloped with the rear downslope and has drainage holes at the back of the pit (see *arrows*) to allow leachate to drain into a vegetative treatment area for filtration. *Photo courtesy of C.A. Williams.*

can be used in areas where grass might not be the best option for growth (Figs. 14.12 and 14.13). A rain garden is defined as follows:

A landscaped, shallow depression that captures, filters, and infiltrates stormwater runoff. The rain garden removes nonpoint source pollutants from stormwater runoff while recharging groundwater. A rain garden has two main goals. The first goal is to serve as a functional system to capture, filter, and infiltrate stormwater runoff at the source, and the second goal is to be an aesthetically pleasing garden. Rain gardens are an important tool for communities and neighborhoods to create diverse, attractive landscapes while protecting the health of the natural environment.

Obropta et al. (2015).

- **Keep clean water clean.** This refers to uncontaminated stormwater that falls from roofs, and it can be accomplished with several BMPs that prevent this water from flowing into stress lots or other unvegetated areas where it can become contaminated. A 1-in. (2.54 cm) rainfall will drop 0.6 gallons (2.27 L) of water on each square foot 0.0929 sq m of a roof. A 12 × 24-ft (3.66 × 7.32 m) run-in shed would drop 173 gallons (655 L) of rainwater during a 1-in. rainfall! Gutters are the first line of defense in rerouting roof runoff. From there, it can be collected in rain barrels or tanks and used for irrigation or arena watering, or routed via

FIGURE 14.11   This is an example of a large manure storage with a roof to eliminate rainwater flowing through the manure. *Photo courtesy of C.A. Williams.*

FIGURE 14.12   Constructing a rain garden alongside a stress lot. *Photo courtesy of C.A. Williams.*

underground pipes to a suitable outlet such as a long grassed area where the water can infiltrate. French drain systems are another good way to reroute stormwater to a suitable outlet (Fig. 14.14). A perforated pipe is laid underground and backfilled with gravel, so water can easily reach the pipe, plus the perforations allow water to infiltrate as it travels through the pipeline.

- **Construct berms and waterways to divert water flow.** A berm is a raised barrier, and a waterway is a sunken channel. Both can be constructed to divert stormwater away from sensitive or contaminated areas such as the manure storage or stress lot. These can be particularly useful in areas with a steep slope, where vegetative cover alone cannot slow runoff flow enough to prevent it from entering surface water.
- **Fence off open water and wet areas.** Horses are not as likely to wallow and defecate directly into open waters as some other livestock, but there are a number of good reasons to fence them out. First, placing a fence back from water provides a vegetated buffer to filter pasture runoff before it enters the waterway (Fig. 14.15). Second, horse hooves damage stream banks and cause erosion quickly when they drink from or cross a stream. It is not always possible to completely block off a stream when it cuts through a pasture or across a property, but fencing off portions of it and constructing stream crossings and culverts go a long way in protecting water quality and making it easier to move livestock and

FIGURE 14.13   The finished product (rain garden) after a few months of growth. *Photo courtesy of C.A. Williams.*

FIGURE 14.14   Installation of new gutters and a French drain system around a barn and paddock to reroute the clean roof runoff. *Photo courtesy of C.A. Williams.*

FIGURE 14.15   A swale fenced off from horses in the adjacent pasture. This will keep the manure out of the water and maintain an active vegetative buffer. *Photo courtesy of C.A. Williams.*

equipment. A stream crossing stabilizes the streambank and streambed using gravel, geotextile fabric, and other stabilizing materials. This is one BMP that Jamie should consider to limit the access of his horses to the stream on his property.

GRAZING NEAR STREAMS: RIPARIAN BUFFERS

The closer an agricultural field is to a stream, the greater is the chance for nonpoint source pollution of nutrients or sediment. The word **riparian** refers to an area along a stream or riverbank, and a riparian buffer is essential to preserving water quality on farmland. A riparian buffer could be as simple as a grassed filter strip, but maintaining a streamside forest is one of the most effective ways to filter pollutants. These are complex ecosystems that provide a number of benefits to water quality, including filtration of runoff and groundwater flow, shading and temperature control of streams, and fostering rich biologic diversity in stream communities. The USDA defines a riparian forest buffer as follows:

An area of trees and other vegetation located in areas adjoining and up gradient from surface water bodies and designed to intercept surface runoff, wastewater, subsurface flow and deeper groundwater flows from upland sources for the purpose of removing or buffering the effects of associated nutrients, sediment, organic matter, pesticides or other pollutants prior to entry into surface waters and groundwater recharge areas

Welsch (1991).

As runoff enters the riparian forest, water flow is slowed down by plants and litter. This gives sediment a chance to settle out of the water and increase water infiltration into the soil, which can then be held by high amounts of organic matter. The filtering of sediments is particularly useful in reducing phosphorus in surface water, which binds tightly to soil particles. In fact, approximately 80% of phosphorus in runoff is filtered in this way.

In addition to mechanical filtration, riparian buffers provide biologic and chemical filtration of nutrients and pesticides in runoff (Welsch, 1991). Bacteria and fungi convert nitrogen into mineral forms, which can be used by plants and other organisms, or gaseous forms that return to the atmosphere. Pesticides can be degraded by microbial decomposition and a number of other chemical and biologic transformations.

As they accumulate and utilize nutrients in runoff, riparian buffers serve as a nutrient sink. Nutrients are utilized by soil organisms and plants; for example, uptake by trees and other plants provides long-term nutrient storage (Welsch, 1991). Nutrient removal occurs when trees are logged and wildlife feeds on plants. In addition, dissolved organic content and plant parts that fall into streams provide food for aquatic organisms.

Not only do riparian buffers, forests specifically, serve to ameliorate pollutants and prevent surface water contamination, they are also essential to the health of the stream ecosystem. Shade from the forest keeps water from heating too much due to solar exposure. Warmer water increases fishes' need for dissolved oxygen, thereby reducing dissolved oxygen in the stream. Forests also provide habitat structure for aquatic life as branches and other debris fall into the stream, creating areas suited for rearing young and hiding from predators. They are an important food source for the stream ecosystem, and plant roots stabilize stream banks, reducing water sediment levels (Welsch, 1991).

MAINTAINING AND MANAGING RIPARIAN FOREST BUFFERS

In order for the buffer to work properly, livestock should be fenced out of it. Riparian forest buffers are divided into three zones (see Fig. 14.16). Zone 1 is directly adjacent to the stream and composed of undisturbed forest that provides shade and detritus to the stream. It should be 15 ft wide and planted with native trees and shrubs to stabilize the streambank. Livestock should be kept out of Zone 1 unless there is a properly constructed stream crossing (Welsch, 1991).

Zone 2 begins at the edge of Zone 1 and is a minimum of 60 ft wide. It consists of managed forest, where harvesting of trees removes sequestered nutrients and encourages rapid growth of younger trees. Zone 2 is where the majority of the filtering action takes place and consists again of native trees and shrubs. Livestock should also be excluded from Zone 2 except for stream crossings (Welsch, 1991).

Zone 3 is a minimum of 20 ft wide with the purpose of runoff control. It is composed of grasses and forbs and often graded with a shallow slope and water diversions, much like a vegetated buffer strip. It slows and spreads out the concentrated flow of water from farmland while infiltrating water, taking up nutrients, and collecting sediment. The vegetation must be kept dense and vigorously growing, and any clippings must be removed from the ground to remove those nutrients. Zone 3 can be intensively grazed for short periods of time as long as the soil is dry and firm (Welsch, 1991).

FIGURE 14.16 Illustration of forested riparian buffer zones. *Used with permission from Welsch, D.J. 1991. Riparian Forest Buffers: Function and Design for Protection and Enhancement of Water Resources. U.S. Dept. of Agriculture Forest Service, Radnor, PA. Publication # NA-PR-07–91. http://www.na.fs.fed.us/spfo/pubs/n_resource/riparianforests/index.htm.*

On the other side of Zone 3 lies the cropland or pasture (Welsch, 1991). To minimize nutrient and sediment runoff, care should be taken to manage the land and nutrient applications according to the guidelines presented earlier in the chapter.

CONCLUSION

Consider owning a farm downstream from a neighbor who does not manage his or her farm for water quality (like Jamie). You may experience odors and eutrophication, and you would be wary to let your animals drink from the stream. If each landowner on the stream utilizes BMPs designed to protect water quality, the entire stream ecosystem will be healthier, and you will be contributing to cleaner rivers and estuaries. In addition, maintaining dense pastures and buffer strips will benefit your horses and the aesthetic value of your farm, and stormwater control will reduce mud on the farm.

Please see the questionnaire, "Is My Farm Environmentally Friendly?" in Appendix 9. It can be used to determine a quantitative level of environmental management on a farm.

REVIEW QUESTIONS

1. Review the Introduction about Jamie's farm problems.
   a. Why is the pond at the end of the stream turning green?
   b. What other symptoms might be seen in the pond if Jamie does not change his management?
   c. Name four ways that Jamie could improve water quality on his farm.
   d. Jamie's stream cuts his property in half. How can he limit horses' access to it and still use his whole property?
   e. Name two reasons that Jamie should not use the stream as the only water source for his horses.
2. Why is erosion an environmental concern?

3. How can we manage pastures to reduce parasite infection?
4. What are three benefits of using manure on pastures?
5. What are three important guidelines to consider when spreading manure on pastures?
6. List ways to keep roof water clean and out of pastures and paddocks.
7. During a 3-in. rainstorm, how much water will fall from a barn roof that is 30×50 ft?
8. Briefly describe the environmental challenge of grazing animals in riparian zones.
9. List five roles or functions of riparian areas in environmental protection.
10. Describe a set of conditions in which animals can be allowed to graze riparian zones with a minimal probability of causing environmental damage.
11. Contact the owner of a local farm and obtain permission to perform an environmental risk assessment using the "Livestock Farmer Survey: Is My Farm Environmentally Friendly?" and develop a management system that includes recommended improvements for at least one area on the farm where they could be more environmentally friendly.

REFERENCES

AAEP. AAEP Parasite Control Guidelines. 2013. http://www.aaep.org/custdocs/ParasiteControlGuidelinesFinal.pdf.

Brophy F, Bridgeman C, Quinn A, Erb L. Diseases of Humans and Horses (Zoonosis). 2008. http://atfiles.org/files/doc/DiseasesHumansHorses09.doc.

Costin AB. Runoff and soil and nutrient losses from an improved pasture at Ginninderra, Southern Tablelands, New South Wales. Aust. J. Agric. Res. 1980;31:533—46.

Graham E, Grandy S, Thelen M. Manure Effects on Soil Organisms and Soil Quality. Michigan State University Extension; 2009. http://msue.anr.msu.edu/uploads/files/AABI/Manure%20effects%20on%20soil%20organisms.pdf.

Harmon GW, Keim FD. The percentage and viability of weed seeds recovered in the feces of farm animals and their longevity when buried in manure. J. Am. Soc. Agron. 1934;26:762—7.

Jenkins M, Brooks J, Bowman D, Liotta J. Pathogens and Potential Risks Related to Livestock or Poultry Manure. EXtension. 2015. http://articles.extension.org/pages/8967/pathogens-and-potential-risks-related-to-livestock-or-poultry-manure.

Kelly F, Westendorf M. Storing Manure on Small Horse and Livestock Farms. Rutgers Cooperative Extension; 2014. Fact Sheet # FS1192, http://njaes.rutgers.edu/pubs/fs1192.

Komar S, Westendorf M. Calibrating Manure Spreaders Using the Tarp Method. Rutgers Cooperative Extension; 2009. Fact Sheet # FS1103, http://njaes.rutgers.edu/pubs/publication.asp?pid=FS1103.

Krogmann U, Westendorf ML, Rogers BF. Best Management Practices for Horse Manure Composting on Small Farms. Rutgers Cooperative Extension, Bulletin Series, E307; 2006. https://njaes.rutgers.edu/pubs/publication.asp?pid=e307.

Li-Xian Y, Guo-Liang L, Shi-Hua T, Gavin S, Zhao-Huan H. Salinity of animal manure and potential risk of secondary soil salinization through successive manure application. Sci. Total Environ. 2007;383:106—14.

Obropta CC, Bergstrom PE, Boyajian AC, Higgins CS, Salisbury KV, Young WE. Rain Garden Manual of New Jersey. Rutgers Cooperative Extension Water Resources Program; 2015. http://water.rutgers.edu/Rain_Gardens/RGWebsite/RainGardenManualofNJ.html.

Provin T, Pitt JL. Managing Soil Salinity. Texas A&M AgriLife Extension; 2001. Publication # E-60, http://soiltesting.tamu.edu/publications/E-60.pdf.

Sanjari G, Yu B, Ghadiri H, Ciesiolka CAA, Rose CW. Effects of time-controlled grazing on runoff and sediment loss. Aust. J. Agric. Res. 2009;47:796—808.

Swinker A. Insect Pests. Penn State Extension; 2015. http://extension.psu.edu/animals/equine/news/2015/insect-pests/pdf_factsheet.

USDA-ARS. Agricultural Phosphorus and Eutrophication. second ed. PA: University Park; 2003 ARS-149 https://sdda.sd.gov/legacydocs/Ag_Services/Agronomy_Services_Programs/Fertilizer_Soil_Amendment_Program/NRCSAG_Phos_Eutro_2.pdf.

USDA-NRCS. National Range and Pasture Handbook. 2003. Washington, D.C, http://www.nrcs.usda.gov/wps/portal/nrcs/detail/national/landuse/rangepasture/?cid=stelprdb1043084.

USDA-NRCS. Agricultural Waste Management Field Handbook. 2008. p. 4—22. Washington, D.C. http://www.nrcs.usda.gov/wps/portal/nrcs/detailfull/national/water/?&cid=stelprdb1045935.

Weaver V, Adams R. Horses as vectors in the dispersal of weeds into native vegetation. In: Proceedings of the 11th Australian Weeds Conference, vol. 30; 1996. p. 383—97. http://caws.org.au/awc/1996/awc199613831.pdf.

Welsch DJ. Riparian Forest Buffers: Function and Design for Protection and Enhancement of Water Resources. Radnor, PA: U.S. Dept. of Agriculture Forest Service; 1991. p. 07—91. Publication # NA-PR-, http://www.na.fs.fed.us/spfo/pubs/n_resource/riparianforests/index.htm.

Westendorf M, Arogo-Ojego J. Managing Manure on Horse Farms: Spreading and Off-farm Disposal. Rutgers Cooperative Extension; 2014. Fact Sheet # FS1193, http://njaes.rutgers.edu/pubs/fs1193/.

Westendorf M, Krogmann U. Horses and Manure. Rutgers Cooperative Extension; 2013. Fact Sheet # FS036, http://njaes.rutgers.edu/pubs/fs036/horses-and-manure.asp.

Westendorf M, Williams C. Managing Manure on Horse Farms: Exercise or Sacrifice Lots for Horses. Rutgers Cooperative Extension; 2013. Fact Sheet # FS1190, http://njaes.rutgers.edu/pubs/fs1190/Managing-Manure-on-Horse-Farms.asp.

Wheeler E, Zajaczkowski JS. Horse Stable Manure Management. PA: Pennsylvania State University, University Park; 2009. http://extension.psu.edu/animals/equine/horse-facilities/horse-stable-manure-management.

Chapter 15

Fencing and Watering Systems

Paul Sharpe

University of Guelph, Guelph, ON, Canada (retired)

PURPOSES AND DESIRED FEATURES OF FENCES

One day while I was visiting my parents, my mother and I went for a drive. Soon, we saw a horse grazing on the shoulder of a busy, hilly highway. I was able to approach the horse and walk it to the closest farm, then put it behind a fence. A horse running on a public road, especially at night, is a frightening prospect because of the human and equine suffering that can happen and the consequent financial obligations following a collision between a horse and a motor vehicle.

Thus, the first purpose of fences around pastures is to keep horses in the pasture and safe from danger, reducing the probability of injury and death. To do this, a fence around the perimeter of a pasture needs *visibility*, *strength*, and *durability*. *Visibility* is especially important when horses are running. Large surface areas, contrast against background, and high light reflection will all contribute to high visibility. Among interior fences, used to divide a whole pasture into paddocks, semipermanent and temporary fences may use electric fence tapes and ropes that move in the wind and thus help to increase visibility. *Strength* in a fence is obtained from using strong materials and designs that are proven to withstand pressures along and across them and thus keep animals from passing through them. High d*urability* or long life of a fence is a result of using strong materials and proven designs so that a fence resists forces, pressures, impacts, and both physical and chemical degradation of materials, remaining effective for a long time.

In addition to keeping horses in pastures, a desirable fence will keep dangerous things out of the pastures and away from the horses. Wild animals (including coyotes, wolves, cougars, and bears) and domestic dogs are probably the most likely animal threats to horses (except parasites). Calves, lambs, goat kids, and poultry are more common victims of animal predation than foals are. Some fences provide a reasonable degree of resistance to predators, but no fence should be considered "predator-proof." Of course, people can potentially cause a great amount of damage to pastured horses, but few horse owners will consider spending enough money to build a pasture fence that can keep people out.

A search was made of reported attacks by large predators (bears, cougars, coyotes, and wolves) on cattle, sheep, and horses in the Canadian province of Ontario. These data were compared to the numbers of cattle, sheep, and horses in the province. The data in Table 15.1 show the actual number of reported predator attacks in 2015 (Ontario Ministry of Agriculture Food and Rural Affairs, 2015) and the 2016 Canadian Census numbers of each type of livestock in Ontario (Statistics Canada; Canadian Horse Defense Coalition, 2016). Reported large predator attacks per 100,000 animals were much higher for cattle and sheep than for horses, and the only confirmed predator on horses was the coyote. The number of horses killed by predators was less than 8 per 100,000 in 2015.

Cougars may be an occasional predator on feral horses in the western states and provinces. Turner and Morrison (2001) reported extensive predation over 11 years by four to eight cougars on foals (45% of those born) in a feral horse population along the California—Nevada border. However, Blake and Gese (2016) studied six GPS-collared cougars for 2 years in Wyoming and Montana and did not detect any predation of feral horses from a herd of about 170.

National statistics on predation of cattle and sheep in the United States are maintained by the United States Department of Agriculture, Animal and Plant Health Inspection Service, Veterinary Services through their National Animal Health Monitoring System (NAHMS). The list of reports at the NAHMS Equine Studies site (including one called "Equine Mortality in the United States, 2015") does not include reports of predation of horses (NAHMS, 2017a). Cattle and calf death losses to predators in the United States are largely due to coyotes (40.5% of cows, 53.1% of calves, respectively), dogs (11.3% of cows and 6.6% of calves), and wolves (4.9% of cows and 3.4% of calves) (NAMIIS, 2017b). Well-designed and maintained electric fences can reduce predation by coyotes, dogs, and wolves. A combination of

Horse Pasture Management. https://doi.org/10.1016/B978-0-12-812919-7.00015-9

TABLE 15.1 Reported Attacks by Large Predators on Cattle, Sheep, and Horses in Ontario in 2015[a,b]

Livestock Species and Population (x 100,000) [a]	Bears	Cougars	Coyotes	Wolves	Total by Large Predators
Cattle (20.897)	25 (1.2)	1 (0.05)	616 (29.48)	40 (1.91)	682 (32.64)
Sheep (4.76)	6 (1.26)	4 (0.84)	1389 (291.8)	30 (6.3)	1429 (300.2)
Horses (0.645)	0	0	5 (7.75)	0	5 (7.75)

Actual numbers of reported attacks in 2015 by predator species and number per 100,000 domestic animals (from 2016 Census) in parentheses.
[a]Statistics Canada, from 2016 Census. Tables: 003–0083 Cattle Statistics, 003–0094 Sheep Statistics, 004–0224 Horse Statistics.
[b]Ontario Ministry of Agriculture Food and Rural Affairs, 2015. Ontario Wildlife Damage Compensation Program.

"unknown predators" and "other predators" totaled 26.8% of cows and 15.5% of calves reported killed. Regionally specific predators, large birds, bears, wild cats, and foxes composed the remainder of reported causes of cow and calf deaths.

The practical message here is to learn from neighbors and extension personnel which predators are in your community, and if the predominant predator is the coyote, build some coyote resistance into your fences where foals and small horses or ponies are kept. A five-wire to nine-wire high-tensile fence with 6-inch (15.2 cm) lower spacings and 10-inch (25.4 cm) upper spacings, and including alternating ground and live wires, will discourage many coyotes. If you are using horse netting, live electric wires on offset insulators protruding outward, at 6 inches above ground and 6 inches above the top of the netting will discourage coyotes from digging under and jumping over the fence. No fence, however, is 100% predator-proof, especially if it is not maintained well.

Heather Smith Thomas (2017) wrote a useful article on insuring horses that raises a question, "Will insurers reduce the premium on horses if a farm has recently installed new, more effective fencing, especially if it is installed by a professional company?" For horse owners, this question is worth asking of their insurance agents.

Fences help to define property boundaries and in providing a safe place for horses, also provide the opportunity to exercise and graze. Ideal fence designs prevent injuries to horses that contact them, ensuring very low probabilities of horses getting their heads or hoofs caught or being cut on sharp edges (Worley and Heysner, 2015).

While planning and before building any fences, it is worth learning whether a fence is subject to any laws. The following points are summarized from Vitale (2009). While a fence often marks a boundary, it will not be taken at law as absolute proof of a boundary line. In some places, it is considered the duty of landowners to erect a fence where the absence of a fence would constitute a nuisance. In rural environments, it is the duty of animal owners to prevent their animals escaping onto the land of their neighbors. Where two owners of adjoining lots in Alberta, Canada, want to erect a fence for their common advantage, they shall bear the expense of erection, maintenance, and repair in equal shares. Municipal by-laws may specify heights or materials to be used or not to be used and setbacks from roadways.

Horses, when bored, frustrated, or with an ulcer or nutritional deficiency, may chew wooden boards, so alternative materials should be considered. For many farm owners and admirers, well-constructed fences have great aesthetic value or eye appeal. Desirable features of pasture fences include keeping horses in a safe place, keeping dangerous animals out, avoiding injuries to horses, adding aesthetic value to the farm, lasting a long time, and having a reasonable price. When deciding how much to budget for pasture fences, one fencing contractor from Ontario suggests considering the following questions:

1. What investment are you comfortable with?
2. Is it long term or short term?
3. Is the land owned, leased, or rented?
4. How may your plans change?
5. What do you want to look at for the next 10–20 years, and how will it affect the value of the property if you sell?

PLANNING FENCES

Any building project needs a good plan to accomplish the goals of the owner. The contractor from Ontario would also ask you the following:

1. How much security do you need? (Consider the dollar value of the horses, whether stallions need to be kept from breeding mares, horse heights, and activity levels.)

2. How durable do you need the fence to be?

3. How important is safety?

4. How important is appearance?

Three basic types of farm fences include these:

1. permanent (usually nonelectric but could be supplemented by electric)

2. semipermanent (usually electric)

3. temporary (electric)

The property perimeter fence is permanent and should have great strength and ability to keep horses from escaping. An attractive appearance is important for sections of fence near a main road and the farm entrance. Stallions, geldings, mares, and foals may all have different fencing requirements. This chapter will concentrate on fences around pasture paddocks. A well-planned layout of paddocks allows ease of moving horses, removing manure, and transporting field equipment in and out of paddocks (Robertson and Mowrey, 2005; Fabian, 2017). Farm equipment needs lanes at least 16 feet wide. Where snow falls, leave room for snow storage or removal. Poor planning related to machinery can result in damaged fences (Fabian, 2017). Permanent fences are appropriate in the following locations:

1. around the perimeter of the property, especially next to busy roads

2. along laneways where horses, vehicles, and machinery travel frequently

3. bordering major segments of a farm, dividing different functions

4. where physical security is of great importance

5. where eye appeal is very important

A perimeter fence, including gates, should be able to contain any loose horse on the property, including at the end of lanes and at the public entry. Some horse farms use double fencing on the perimeter and some use it between breeding pastures (Fabian, 2017). It could also be used around a quarantine paddock.

Calm horses grazing plentiful forage are unlikely to put much physical pressure on a fence, but an excited, playful, or fearful horse may not see a fence and crash into it. Thus, resistance to collision and materials unlikely to injure horses are important. A fence with elasticity can be useful in that situation. Fences can be made highly visible by using wide surfaces and light colors. To make fences secure enough to contain a running horse, they should be high enough to discourage jumping and be free of sharp projections. Perimeter fences should be 5 feet (1.5 m) high and dividing fences should be at least 4.5 feet (1.38 m) high (Robertson and Mowrey, 2005; Fabian, 2017).

An overall fence plan should include a layout diagram, showing fence lines, gates, water bodies, water lines pathways, traffic routes, and access for field equipment. Topographic constraints, property boundaries, predominant vegetation, and soil types can be useful additions to the diagram. Indicate where fences need to be permanent or temporary, electric or nonelectric, and make notes on the types and materials chosen (Brown, 2015). Other farm features such as barns and riding arenas should be on the diagram to help illustrate where horses, people, and machines will be moving. Indicate whether laneways will be grass or gravel. Rectangles and square corners are easy to construct and provide great strength, but paddocks big enough to allow horses to run should have rounded corners to prevent horses crashing into fences and to prevent dominant horses trapping subordinate horses in square corners. Posts on curved corners can be tilted outward to help resist inward forces of wire under tension (Fabian, 2017).

Once you have a complete fence plan, you can count and calculate the amounts of materials you will need (Andrews, 2015). A magazine article by Burt (2015) quoted well-known cattle grazing consultant Jim Gerrish and Gallagher Fencing territory manager, Kevin Derynck, on seven common errors in electric livestock fencing and how to prevent them (Table 15.2).

LOCATIONS OF FENCES

The first fence to plan and build is the permanent fence around the perimeter of the area that is intended to contain animals. It will likely be one of the strongest and highest. Many of the internal fences can be of lighter construction to divide grazing paddocks and to define features such as corrals and arenas for training and riding. Other internal fences can keep horses away from water, buildings, feed storage areas, sensitive areas, steep slopes, residences, and dangerous areas (Fabian, 2017). Fig. 15.1 shows fences used to keep horses away from sinkholes and to keep them from stripping bark off a tree.

TABLE 15.2 Seven Common Mistakes in Electric Fencing and Their Prevention

Mistakes	Prevention
Corner posts not big enough or resistant enough to leaning	Post diameters: net wire: 8 inch; 5—6 strand high-tensile: −6—7 inch; 1—2 wire: −4—5 inch; 10 foot × 4 inch brace post
Post spacings are too close, which is expensive	Line posts 50—70 feet apart with "stays" or "stringers" sitting on ground and holding wires apart
Energizer is the wrong size	Select a low-impedance energizer with output of 1 J per mile of fence; voltage does not need to exceed 7000
Ground rods are too close together	Three feet of ground rods per joule of energizer output; space the ground rods throughout the whole network of fencing
Built a fence to keep moose out	Be wildlife-friendly; use flexible posts and adjust wire heights for minimal damage to the fence where there are many large wild ungulates coming through
Making gates carry electric current	Run insulated live and ground wires in a trench 1 foot deep under the gate
Relying on steel posts, insulator breakage or accidental live wire contacting the post causes current to leak into the ground	Use flexible plastic, fiberglass, or wood-plastic combination posts

Adapted from Burt, A., 2015. 7 Common Cattle Fencing Mistakes. Beef Magazine. http://www.beefmagazine.com/pasture-range/grazing-programs/0301-common-fencing-mistakes.

FIGURE 15.1   Painted board fences are used to separate paddocks, to protect horses from a sinkhole, and to keep horses from stripping bark off a tree at the University of Kentucky. *Photograph by Paul Sharpe.*

If rotational grazing will be practiced, each management group of horses should have at least as many paddocks as you derive from the following equation:

((Days of rest required)/(Days grazing in one paddock)) + 1 = Minimum number of paddocks required.

The more paddocks available to horses, the more flexibility you have to adjust for changes in forage yield and animal numbers per group. For example, during drought, an appropriate response is to create more paddocks, shorten grazing times, lengthen resting times, and reduce animal numbers. In dry weather, if 40 days of pasture rest are required between grazings and the current stocking rate and carrying capacity allow 4 days grazing per paddock, the minimum number of paddocks required is (40/4) + 1 = 11.

Fences dividing grazing paddocks can define major divisions of pasture with semipermanent fences and smaller divisions with temporary electric fences. It is temporary electric fences that provide the greatest flexibility in rotational grazing management because they can be moved frequently to provide any size of grazing area.

MATERIALS TO CONSIDER

There is no perfect horse fencing material. Each material has advantages and disadvantages relative to others in terms of availability, ease of use, safety, strength, flexibility, visibility, durability, resistance to moisture, heat, and freezing, probability of causing injury, appearance, and cost.

Stones and Trees

Some of the earliest materials to be used for fencing were the most natural: stone and living trees. Stone for fences and walls has been readily available in many places and is practical to use locally. If the pieces are small enough for one person to lift, they can be easy to use, and some people develop great expertise at creating dry stone walls, with no mortar to hold the stones together. Being hard, dense, and strong but inflexible, stone is not a material that is safe for horses to run into. Its visibility may be obvious at times in good light, but in some situations, it may blend into the landscape. Most stone lasts a long time, with some types being quite resistant to heat and freeze-thaw cycles and other types more susceptible to breaking apart. Appearance of a relatively new or old but well-maintained stone fence is usually impressive and should add to the value of a property (Fig. 15.2). Its cost can be limited to the labor to move it around on the property where it is found, but its density means that transportation is expensive. If you become the manager of a farm with stone fences that are not in good condition, it would probably be less expensive to have them repaired by an experienced stonemason than to remove and replace them.

Living trees and shrubs have been used where they were found or transplanted and cultivated into hedges in much of Europe and England. Many have lasted hundreds of years, so they have been highly available and easy to use through much of their life. If not too thorny, they can be safe, strong, and flexible. Visibility may be limited as they blend into the surrounding greenery. Hedges were an iconic part of the landscape in England for centuries, providing a satisfying appearance (Fig. 15.3). However, many hedges have been removed in recent decades. Some of the plants have died. Hedges are great habitats for a variety of wildlife, but unfortunately, badgers living in hedges in England are carriers of *Mycobacterium bovis*, which causes tuberculosis in cattle and rarely in horses (Thoen, 2016; Buick, 2006). Fences constructed of modern materials in the 21st century are generally considered more practical than stone or hedge fences.

Wood

Wood was possibly the most common fencing material for horses in the 20th century. Wood is available anywhere that trees grow and is widely traded as a building material. It is easy to use, as rails, split rails, or boards can be used to create fences with planned amounts of strength and flexibility. Wood is highly visible when new or when painted, and can resist temperature and humidity changes to last hundreds of years in some conditions (Fig. 15.4), but if the wrong wood is used, it soon rots. Wooden board fences, freshly painted, can have a very attractive appearance and add value to a property (Fig. 15.5). The fence in Fig. 15.5 has an extra board on the inside top of the fence and a bevel on the top of the posts to drain away water. An all-wood fence can be expensive to build and requires considerable maintenance, so it is often used for relatively short spans of fence or for areas where people are working with horses daily and can quickly tend to needed

FIGURE 15.2   Stone fence in need of rebuilding. *Photograph by Paul Sharpe.*

FIGURE 15.3   A hedge serving as a fence. *Photograph by Paul Sharpe.*

FIGURE 15.4   An untreated oak wood fence of boards provides physical strength, reasonable visibility in good light, and rustic appearance. *Photograph courtesy of Andy McDonald, Highland Electric Fence Systems.*

FIGURE 15.5   A sturdy, painted wood fence, reinforced with an extra top board. *Photo by Paul Sharpe.*

FIGURE 15.6 Pressure-treated wooden fence posts. *Photo courtesy of Andy MacDonald, Highland Electric Fence Systems.*

maintenance. Horses that are bored or habitual cribbers will cause considerable chewing damage to wood, and damaged wood can result in a physical hazard to the horses. Wood is frequently combined with a variety of other materials.

Moisture at or in the ground causes most wood to decay, although red cedar, white cedar, black locust, the red heartwood of California redwood, and cypress are resistant to decay. Oak is often used for fences where it is locally most available, and white oak posts last far longer than those made from red oak. Protecting the in-ground parts of fence posts by filling the post holes with concrete extends their life and provides more fence strength. In dry climates, the concrete can come up all the way to the surface, but in wet climates (including Georgia and Ontario), the combined effects of repeated wet and dry cycles at ground level and acidity from concrete cause such posts to rot just above the ground. So if concrete is being used to make the underground portion of posts more resistant to movement and to bending, then only fill the hole with concrete up to about 6 inches (15 cm) from the top.

"Pressure-treated posts," which are treated with chemical preservatives under pressure, are very commonly used for fences (Fig. 15.6). Different preservatives and their amounts are appropriate and rated for different uses. Copper-based preservatives cause wood to have a light green or, more recently, a brown color. They are effective because they are toxic to many organisms found in the soil. Some preservatives are rated for aboveground use, some for soil, and some for freshwater contact (Robertson and Mowrey, 2005). Wood sold for residential use is not necessarily suitable for farm fence posts and rails. The wood preservatives "alkaline copper quat" and "copper azole" are relatively corrosive to metals, so fasteners used with such treated lumber should be hot-dipped galvanized or stainless steel (Robertson and Mowrey, 2005). Another wood preservative, chromate copper arsenate, will slowly release arsenic into the environment.

Creosote is a category of toxic and carcinogenic chemicals made by distilling tars and heat-treating both wood by-products and fossil fuels (usually coal). Creosotes are black, sticky, corrosive to rubber, and strong-smelling. They are typically used to preserve seagoing wood structures, bridgework, and railroad ties. Some creosotes are also used as pesticides. Creosote-treated posts will last a long time in the ground, and it should be noted that some of the creosote will end up in groundwater. To be an effective wood preservative, creosote should be applied under pressure and not just brushed on. Creosote-treated wood should not be used in any place that horses can crib, bite, or lick the wood. No treated wood should be in direct contact with feed or drinking water. Guidelines on use from manufacturers and government agencies should be consulted before deciding to purchase treated wood products. While wood is relatively inexpensive to buy, it is one of the most expensive materials to maintain (WindRiver Fence, 2017).

Metals

Metals, usually steel, are denser and stronger than wood. They may require more specialized methods of fastening (welding), so ease of use is more difficult than with wood. Metal fencing materials can be purchased from many locations, but certain types, such as surplus pipe from oil exploration operations, may be cheaper at their source than if they need to be shipped very far. The biggest disadvantage of iron-based metals (steel) in the presence of moisture is their tendency to oxidize (rust). Thus, steel used for fence posts or rails should be galvanized or painted with a rust-proof coating, as seen in the photo of a chain link fence with galvanized posts at the RCMP horse breeding farm (Fig. 15.7). Steel fasteners should be galvanized or stainless. Uncoated steel pipe fencing is one of the most expensive options to buy and will rust, even in arid Arizona. Even when pipe comes coated with powder or plastic, it tends to rust at connections, spoiling the appearance. In locations with considerable frost, posts can be heaved out of the ground, causing unevenness.

FIGURE 15.7   Fence made from steel pipe and chain link. *Photo by Paul Sharpe.*

Most wire fencing is galvanized steel, but some aluminum is also used. Aluminum does oxidize in the presence of moisture, but the oxide forms a stable, thin layer that prevents further oxidation. The oxide does not flake off like iron oxide does. Galvanized steel and aluminum have low toxicity and high strength. Degree of flexibility depends upon thickness. Galvanizing is a coating of zinc. The thicker the zinc coating, the longer the time before rusting starts. Classes of galvanizing are 1 (lightest) through 3 (heaviest) (Worley and Heusner, 2015). Metal post and pipe fences are very strong but inflexible, thus dangerous to horses that run into them. Metal wire fences are less strong but more flexible, and special wire mesh fences have been designed for horses with safety as a top priority. Metal fences, especially galvanized wire components, can reflect sunlight moderately when new (Fig. 15.8), but after some oxidation and because of their small diameter, they have relatively low visibility. Fig. 15.9 shows a wire mesh fence in the foreground that is difficult to see and a black polymer-coated single wire held by an offset insulator on the other side of the fence, which is much more visible. Plastic coating on wire also improves attractiveness and resistance to deterioration. When properly coated to hinder oxidation, metal fence components have very long durability, potentially stretching to centuries, partly due to their high resistance to extreme temperatures and freeze-thaw cycles. On their own, metals do not have highly attractive appearances, but like with visibility, they can be greatly improved with coatings and incorporation into plastics. Per unit of weight, metals are more expensive than wood, but there is some compensation due to their strength and durability. Many horse fences comprised of wooden posts and metal wire are supplemented with plastic components to improve visibility (Fig. 15.10).

FIGURE 15.8   A roll of moderately reflective galvanized wire on a spinning jenny, ready to be installed. *Photo courtesy of Gallagher North America.*

FIGURE 15.9 A combination of mesh wire with low visibility in the foreground and a more visible polymer-coated wire on the other side of the fence. *Photo courtesy of Andy MacDonald, Highland Electric Fencing.*

FIGURE 15.10 Wire fence supplemented with high-tensile polymer rail ("flex rail") on a curved corner. This fence has a combination of a 5-inch (12.7 cm) Flex Rail on top for visibility and below that, alternating live braid and vinyl-coated (neutral) strands. *Photo courtesy of Andy MacDonald, Highland Electric Fencing.*

Plastics

The use of plastic polymers in fencing materials has expanded considerably since the 1980s, and various plastic products are readily available. Polyethylene (PE) and poly vinyl chloride (PVC) are the most commonly used plastics in many household items, water pipes, and fencing components. PE can be either low density (LDPE), due to much branching of the ethylene polymers or high density (HDPE) due to the majority of polymers being oriented in straight lines. LDPE is softer and appropriate for coatings, whereas HDPE is stiffer and stronger and more appropriate where physical stress resistance is important in containers and pipes. PVC is more resistant to oxidation than PE and has high resistance to inorganic chemicals. PVC retains its strength for decades and is more resistant than many other plastics to deformation under constant pressure. PVC is readily embossed with different surface textures (e.g., wood grain) and can be molded to a wide variety of shapes. Flexibility, elasticity, and impact resistance of PVC can be modified by addition of plasticizers, additives, and modifiers (Sevenster, 2016). Compared to other plastics, PVC can be considered to be dense, hard, strong, and cheap (Creative Mechanisms, 2016). Plastic fencing components are easy to use. Posts and rails can be nailed or screwed. Plastics resist cribbing and never need painting but may need washing with mildew-removing agents (Worley and Heusner, 2015). Some plastic components can be fitted into molded slots. Plastic wires and tapes can be joined with hardware or simply tied. Plastics are generally nonconductive to electricity, plus resistant to moisture and many chemicals. Their appearance can be very much like wood and have great eye appeal, without the same maintenance requirement as wood (Fig. 15.11). Considering their relative weight, strength, and durability, plastic components should be very competitive with wood for fencing.

FIGURE 15.11 Close-up of extruded hollow PVC rail fencing. The white braided polymer rope on the ground was later mounted on insulators and electrified. *Photo courtesy of Andy MacDonald, Highland Electric Fencing.*

Fencing contractors have remarked that when PVC rails break, the resulting pieces have very sharp edges, especially if the PVC is recycled (Gourlay, 2011). One brand of HDPE fence claims resilience from 150°F down to −106°F (65.5°C down to −76°C) (WindRiver Fence, 2017). HDPE is more flexible than PVC, but this means it can expand and contract more. Virgin PVC and HDPE are better products than recycled materials. Check the guarantee of recycled components before buying.

For water pipe, HDPE is more suitable for lower pressure installations, and PVC is more suitable for direct burial in the ground (Difference Between, 2017). LDPE is flexible and well-suited to carrying water on the surface or underground to pastures. Some companies market PVC pipe for assembling into fences. It should be remembered that PVC pipes are not designed for the stresses on a fence, but other configurations are. Plastic is used for temporary fence posts, and most plastic posts need an incorporated metal insert to be driven into the ground (Fig. 15.12).

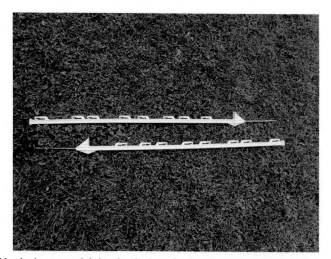

FIGURE 15.12 Flexible plastic posts and their galvanized metal spikes for temporary electric fences. *Photo by Paul Sharpe.*

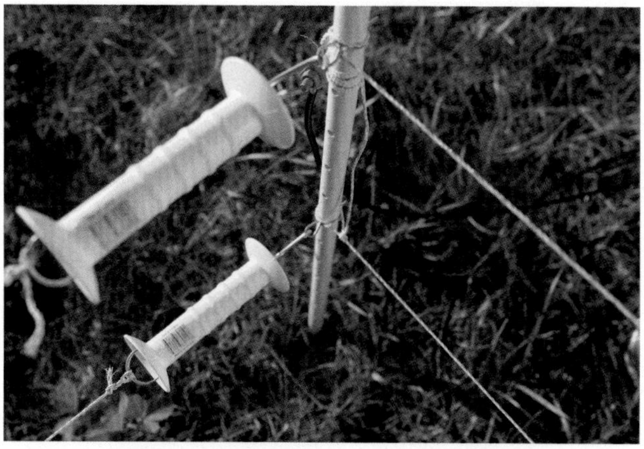

FIGURE 15.13 Fiberglass posts are suitable for semipermanent fences and as line posts in permanent fences. This one is drilled to accommodate metal pins, forming loops. Conductive metal hooks extending from insulated gate handles can be attached to the loops. *Photo courtesy of Gallagher North America.*

Fiberglass

Fiberglass is a nonconductor, more dense than plastic, and provides greater strength, as well as flexibility. Many fiberglass posts can be driven into the ground without the need of a metal spike attached to their bottom end. Some are predrilled so wire clips can be installed at various heights, and some are coated in plastic for protection against ultraviolet light (Fig. 15.13).

Concrete

Concrete fence posts are common in parts of Australia where bush fires are common because they do not burn. Concrete posts and rails are reinforced with interior steel bars. They can have notches or holes to accommodate fence wires. Concrete fence components are very strong, heavy to use, and inflexible, so they are not safe for collisions from horses. The light natural color of concrete provides reasonable visibility. Concrete resists temperature extremes, moisture, chewing pests, damage from animals, and most chemicals. Its appearance is utilitarian, perhaps giving the impression of an attempt to prevail forever, and it is expensive.

Plastic Coatings

Some fencing materials involve a plastic coating on pressure-treated wood. There are also plastic caps that can be placed over square wooden posts or metal t-posts. This improves weather resistance, visibility, and appearance (Fig. 15.14). There is a need to ensure that water does not penetrate the boundary between the wood or metal and the plastic, or decay will begin. Some PVC fence posts and rails are reinforced with steel or aluminum, adding considerably to the price (Gourlay, 2011).

Plastic coatings on galvanized steel wire have been developed into several attractive fencing products offering high degrees of visibility, flexibility, strength, durability, and resistance to extreme temperatures and moisture. A fence wire may be included within a plastic coating to increase flexibility and visibility. Fig. 15.15 shows a double fence bearing a product composed of three high-tensile wires within a wide polymer band, sometimes called "HTP rail" or "flex fence" or "flex rail." It also shows two electric polymer ropes mounted through plastic insulators to pressure-treated wood posts. Special hardware holds the flexible rail to the posts, and a buckle for joining two ends of HTP rail is shown in Fig. 15.16.

FENCE PLANNING, TYPES, DESIGNS, AND DESCRIPTIONS

Until the 1930s, almost all fences for animals were strictly physical barriers that might or might not add eye appeal to a property. Most of these fences were either wooden post and rail or wooden posts and metal wire. Numerous extension studies of fencing costs have been published, including one from Saskatchewan (2015) that lists prices of individual post and wire fence components and total material, labor, and machinery costs for 1 mile (1.6 km) of fence.

FIGURE 15.14 A plastic case to put over top of a wooden post. The combination provides weather resistance, high visibility, attractive appearance, and great strength. *Photo courtesy of Gallagher North America.*

FIGURE 15.15 Three high-tensile steel fence wires are coated with flexible plastic polymer in these bands or rails (HTP rail) for better visibility and eye appeal. Two strands of electrified braided polymer rope are mounted on insulators between the rails. *Photo courtesy of Andy MacDonald, Highland Electric Fencing.*

In the 1930s a psychological barrier was developed, the electric fence, which did not need a great deal of physical strength to contain animals. Thick posts, wide rails, and strong corners are of secondary importance for a psychological barrier. In some situations, a single electric wire will contain large horses or cattle that are well trained to their use. The development of electric fence technology also meant that many fences of reasonable but not extreme strength could be reinforced by a single electric wire to prevent animals leaning or pressing on them.

There is reluctance by some people to expose their horses to electric fence, for fear that the electric shock will harm the horses. Modern electric fence energizers were first developed in the 1930s by Bill Gallagher, using alternating current (AC), giving long pulses and sometimes unpredictable voltages. Another New Zealander, Doughy Phillips, improved the energizer design using capacitors and solid-state circuits. In the 1980s the shock pulse was made much shorter, only a few milliseconds (with a frequency around once per second), and the amperage was reduced, making fences more predictable

FIGURE 15.16 Hardware for joining two ends of flex rail together. *Photo courtesy of Andy MacDonald, Highland Electric Fencing.*

and safer. When electric fencing was compared to wooden fencing around paddocks, horses spent less time in proximity to the fence if it was electric and less time engaged in social contact with neighboring horses (Moors et al., 2010). A similar experiment resulted in this same observation with horses in square paddocks that were just 11.5 feet (3.5 m) or 19.5 feet (6 m) on a side and made of either electric fence or wood. It revealed that when horses were in the smaller paddocks, they exhibited more pawing, stamping, and kicking (Glauser et al., 2015). However, there was no difference between electric and wooden fences in physiologic stress parameters exhibited by the horses, including heart rate, heart rate variability, and salivary cortisol concentration. A practical application of this knowledge may lead a horse farm manager to allow a slightly larger paddock area per horse when installing electric fencing than board fencing.

Fig. 15.17 shows two fences close to their junction on a horse farm in Alberta, both supported by pressure-treated wooden posts, one having 3-inch-wide (7.8 cm) wooden half-round rails and the other with 8-inch-wide (20.4 cm) horizontal boards. Both of these designs are strong, visible physical barriers. The board fence is also strengthened with five wires. The second and fourth wires run through insulators and are electrified or "live," while the first, third, and fifth wires are grounded. A horse pushing its head between any pair of wires will contact one live and one ground wire, causing a significant flow of electricity through the head or neck and a memorable shock.

Horse fences should have their lowest horizontal element (rail or wire) around 8—12 inches (20.4—34 cm) from the ground to avoid trapping a hoof yet discourage reaching under the fence for grass and to prevent foals from rolling under it. The top element should be 54—60 inches (137—152 cm) above ground for most horses. Stallions, very tall horses, and

FIGURE 15.17 Two different wooden fence types. One has wooden rails, and one has wooden boards on the outside and both live and grounded electric wires on the inside. *Photo by Paul Sharpe.*

FIGURE 15.18 Tennessee Walking Horse stallion standing by a five-board fence to the top of its withers. *Photo by Paul Sharpe.*

those that are inclined to jump may need a greater fence height. The top rail or wire should be around the height of the withers of the tallest horse. Figs. 15.18 and 15.19 of a Tennessee Walking Horse stallion on a breeding farm show how five fence rails, with the top at wither height, appear to be necessary to convince this stallion to stay in the pasture. If the constructed height of a fence is too low for certain horses, there are offset insulators for electric wire that can be fastened to the tops of posts (similar to those in Fig. 15.20 but oriented vertically) to increase the effective height.

The placement and number of fence wires should not allow horses to reach through or over a fence to reach grass on the other side. Keep the open spaces to 12 inches (30.5 cm) or less. For electric fences, this distance can be increased to 18 inches since horses will not often be touching the fence. A single strand of electric wire can be run on offset insulators, 4–6 inches (10–15 cm) above or just inside the top rail of nonelectric wire to prevent horses from reaching over a fence (Fabian, 2017) (Fig. 15.20).

To see example diagrams of fence layouts or to study tables that compare various fencing types and materials, see Brown (2015), Fabian (2017), Robertson and Mowrey (2005), or Worley and Heusner (2015). Several manufacturers and installers of fencing components have instructions for installation on their websites. The following fence types are taken from a British Columbia Fencing Factsheet by Brown (2015).

FIGURE 15.19 Tennessee Walking Horse stallion rearing by a five-board fence, showing how the fence height provides a disincentive to jump over the fence. *Photo by Paul Sharpe.*

FIGURE 15.20 A single strand of galvanized 12.5-gauge (ga) electric fence wire on offset insulators protects the fence from horses leaning on it and from reaching beyond it. *Photo courtesy of Gallagher North America.*

Post and Rail Fences

Vertical posts, usually of wood or metal, sunk into the ground at least 3—4 feet (0.91—1.22 m) at corners and 2—3 feet (0.61—0.91 m) at other locations (line posts) should extend above the ground about 5 feet (1.52 m). Typically, line posts are 8 feet long and corner posts can be 9 or 10 feet long (2.74 or 3.05 m). Fig. 15.21 shows a painted fence with just three wooden boards to hold thoroughbred mares and their foals at a stud farm in England. These valuable horses are probably only on the pasture during daylight and when staff are nearby. Holes for posts can be dug by hand (cheap, time-consuming, and tiring) or by an auger, which may be powered by hand, an engine mounted on it, or by the power take off (PTO) shaft of a tractor (Fig. 15.22). Alternatively, posts of small diameter (preferably with sharpened ends) may be pounded down into the ground by a hand-operated post driver (Fig. 15.23). Hydraulic post drivers for large-diameter wooden posts are usually mounted on a trailer and attached to the hydraulic system of a tractor. To supplement efforts with a manual post hole digger, a post hole spoon and a digging bar are helpful. Augers powered by gasoline engines are readily available at equipment rental businesses. Post hole augers are available for mounting on the three-point hitch of a tractor or the boom of an excavator. Hydraulic augers can be mounted on a front-end loader, a skid steer loader, or any machine with a hydraulic drive system. Hydraulic post drivers are usually owned by farmers and fencing contractors.

FIGURE 15.21 A fence with three wooden boards containing a mare and foal. *Photo by Paul Sharpe.*

FIGURE 15.22 A post hole auger. *Photo courtesy of Gallagher North America.*

FIGURE 15.23 A hand-powered post driver, suitable for small diameter posts. *Photo courtesy of Gallagher North America.*

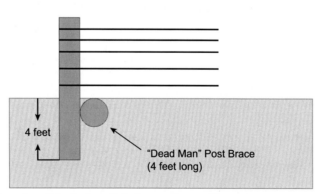

FIGURE 15.24 Position of a "dead man" post lying horizontally in the ground to provide more support to a corner or gate post. *Adapted from Worley, J.W., Heusner, G., 2015. Fences for Horses. University of Georgia Extension. Bulletin 1192. extension.uga.edu/publications/detail.html?number=B1192.*

Some gravel placed in the bottom of post holes before the posts can help drainage and thus reduce rot (Robertson and Mowrey, 2005). Posts needing a great deal of strength can have the hole filled with concrete. Extra underground bracing of posts for corners, ends, and gates can include "dead man" posts, which are short posts placed horizontally, either at the bottom of the vertical post on the side away from the direction of tension or just below ground level on the side toward the direction of fence tension (Fig. 15.24). More common methods of bracing are described subsequently.

Rails can be round, half-round, or rectangular boards (planks), usually made from wood, but metal pipes, plastic pipes, fiberglass, and plastic boards are also useful. When constructed evenly and with symmetry, rail fences are attractive to look

FIGURE 15.25 A relatively new wooden rail fence on thin soil over bedrock, showing a structure that does not require digging post holes or driving posts into the ground. Posts and rails are both white cedar, which is commonly used for these fences in eastern Canada. Many have survived over 100 years. *Photo by Paul Sharpe.*

at, highly visible, and fairly safe (Worley and Heusner, 2015). Horses colliding with post and rail fences are not usually physically harmed unless the collision breaks the rail. Horses are not likely to catch a hoof or their head between horizontal rails. Wooden rail dimensions are often sold as 2×6 inches (5.28×15.24 cm). The wood is rough sawn to these dimensions. After they are planed to make them smooth, boards tend to be of significantly smaller dimensions than these. Thinner planks are more flexible and cheaper but more prone to warp and break. Fig. 15.25 shows a fairly new cedar rail fence of a very old design on thin soil over bedrock in Ontario. The posts are not driven into the ground but are propped in groups of three or four like in a teepee and with a horizontal log added just above ground level. Round wooden rails are wired or nailed to the posts. Historically, some rail fences were constructed so that they could be disassembled without tools and moved. One of these designs includes a zig-zag or snake pattern across the ground with joint-to-joint distances of 16.5 feet, which is one rod (5.03 m). More than a dozen different designs were developed in the 1800s, and some were patented. Most used split rails of cedar, chestnut, or whatever rot-resistant wood was plentiful.

Wood Posts and Steel Strand Wire

Many fences designed for cattle consisted of three strands of barbed wire. This is quick and simple to install but does not necessarily keep animals from reaching between the wires to eat grass on the other side. Many horse injuries are due to contact with barbed wire, so it has no place on a horse farm.

Corner posts should be properly braced to prevent the tension from the fence wire pulling posts off the vertical orientation. Fig. 15.26A shows a typical, modern single-span brace assembly with directions of loads, and Fig. 15.26B highlights the location of hardware, galvanized 3/8-inch pins to hold the horizontal bar in place and the inline strainer to tighten the diagonal wire (Bading and Lenz, 2017). Other instructions and diagrams for horse fence posts and braces were presented by Ferris (2015). He showed how raising the height of the diagonal wire on the end post places it above the level where horses might catch their feet on it. He also eliminated the need for a double loop of diagonal wire and covered the wire with Insul-Tube to reduce injury in case of a horse contacting it. Fig. 15.27 shows a double brace for fences greater than 650 feet (198 m) long (Gay and Heidel, 2009). Notice the double brace wires on each brace. For a double brace in fence lengths between 200 and 650 feet, Gay and Heidel (2009) recommended single brace wires in each brace. They also illustrated how to select correct post sizes, insert staples properly, plus accommodate contours and curves in fence construction. At fence corners, there are two brace assemblies sharing a corner post and at right angles to each other. A top horizontal brace post should be two times as long as the height of the post above the ground. The brace wire is usually two but possibly more wraps of 12.5-gauge galvanized high-tensile wire, twisted to increase tension, and held in place by a strong stick or by specialized galvanized metal tensioner hardware. The corner and brace posts should be notched or have staples driven into them to hold the brace wires in place. Extra brace assemblies should be installed at intervals of 1320 feet (400 m) or less. Horizontal braces can be held in place by notching the vertical posts or preferably by drilling the ends of the horizontal brace and the inner sides of the brace and corner post, then inserting a straight 3/8 inch (9.5 mm) galvanized metal pin. The pin in the corner posts and brace posts should be cut flush with the post surface to prevent injuries.

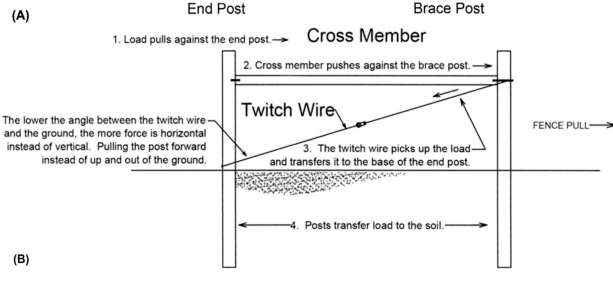

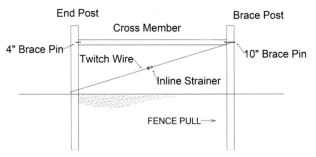

FIGURE 15.26 A single-span fence brace assembly. (A) Wires (not shown) under tension pull the upper part of the end post toward the right. The cross member pushes part of the load from the end post to the brace post. The diagonal twitch wire transfers some of that load back to the end post at ground level, distributing it between horizontal and vertical directions. The portions of the posts in the ground transfer part of the load to the soil. The larger the proportion of post length that is in the ground, the greater the proportion of the load that is borne by the soil. (B) Holes should be drilled completely through the brace post and partly through the end post and into the ends of the cross member to accommodate 3/8 inch diameter galvanized metal brace pins. The twitch wire can be a loop of wire around both posts, held above the 10 inch brace pin on the brace post and to the end post with a fence staple. The inline strainer can be tightened with a wrench. *Images courtesy of Texas and Southwestern Cattle Raisers Association, Bading, K., Lenz, R., 2017. Proper Fence Brace Construction. The Cattleman Magazine. http://tscra.org/what-we-do/the-cattleman-magazine/ranch-diy-proper-fence-brace-construction/.*

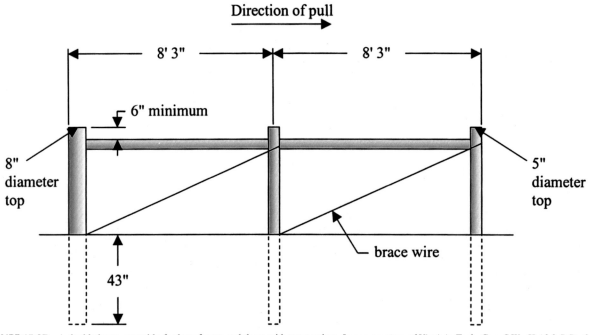

FIGURE 15.27 A double brace assembly for long fences and those with many wires. *Image courtesy of Virginia Tech, Gay, S.W., Heidel, R.D., 2009. Constructing High-tensile Wire Fences. VCE Publications/442/442-132. https://pubs.ext.vt.edu/442/442-132/442/132.html.*

FIGURE 15.28 A mixture of five strands of bare, high-tensile, galvanized fence wire on fiberglass posts and one strand of polymer-coated wire. Note the relatively poor visibility of the bare wire and the higher visibility of coated wire. *Photo courtesy of University of Georgia Extension, Worley, J.W., Heusner, G., 2015. Fences for Horses. University of Georgia Extension. Bulletin 1192. extension.uga.edu/publications/detail.html?number=B1192.*

Smooth galvanized 12.5-gauge wire under tension is a slight improvement on barbed wire, but horses rapidly contacting bare, smooth wire can be badly injured. If used for horses, such fences should have five to eight horizontal wires. Connecting the wires to posts through plastic insulators and electrifying most of the wires, then training animals to keep away from the wire is a more effective step that also prevents horse injuries. However, all of these post and simple horizontal wire combinations lack visibility, so polymer coatings can be added to wire to improve it (Fig. 15.28). An improvement to polymer-coated galvanized wire that conducts electricity is called "Hotcote" and "Shockline." Gaps in the polyethylene coating are filled with a black high-carbon mixture that carries electric current from the wire to anything that touches it (Fig. 15.29). Both coated products have high visibility for a single wire and should be less damaging to horses than bare wire. The product with the ability to shock serves as both a physical and a psychological barrier for relatively low cost.

When attaching high-tensile wire to posts, the attachments should be loose enough to allow the wire to move horizontally as it expands and contracts due to temperature changes. Most insulators allow this expansion. Stapling wire firmly to posts will not allow much expansion. High tension fences usually have cross bracing that includes diagonal wires on corner assemblies. These diagonal wires or wood provide triangular spaces in which feet or heads can be trapped. Fence design should prevent horses from having access to such triangular spaces. Another brace design has fence wire on the outside of posts and the diagonal brace wire covered in Insul-Tube, a nonconducting plastic tube that can surround the brace wire to reduce the chance of a wire injury.

Wire Gauge Sizes

American Wire Gauge was established in 1857 to standardize the descriptions of the thicknesses of cables. This measurement is used specifically for electrically conductive wire (see table below). The measured thickness is of the metal wire only, not of insulation around it.

There is apparently a slightly different system for galvanized wire for fencing, although there is also a claim that this is the American Wire Gauge, based upon the Birmingham Wire Gauge and the Stubs Iron Wire Gauge from early 19th-century England (Page, 2009). Historically, gauge refers to the number of times that a large diameter wire rod from a rolling mill is drawn through a series of increasing smaller dies to reduce its diameter. Thus the first draw is the thickest and is called 1 gauge and a relatively thin wire is 14 gauge at 0.080 inch (0.2032 cm) (Page, 2009). Wire of 8.5 gauge is quite thick at 0.155 inch (0.3937 cm), and 12.5 gauge is moderately thick and very common for high-tension fence wire at 0.099 inch (0.2515 cm). Conductors of electricity and conductors of water share a similarity: as the diameter increases, the resistance to flow decreases.

AWG Wire Gauge Sizes for Solid Conducting Wire*

AWG Gauge	Conductor Diameter (Inches)	Conductor Diameter (mm)
0	0.325	8.25
1	0.289	7.35
2	0.258	6.54
4	0.204	5.19
6	0.162	4.11
8	0.129	3.26
10	0.102	2.59
12	0.081	2.05

*Adapted from Power Stream, Wire Gauge and Current Limits Including Skin Depth and Strength. https://www.powerstream.com/Wire_Size.htm.

FIGURE 15.29 Galvanized wire coated with polyethylene and strips of a black carbon-rich material that conducts electricity, providing shocks to animals that touch it. *Image courtesy of Savannah Petrachenko, System Fence.*

Wood Posts and Steel Woven Wire

Woven wire designed for horses has small openings, about 2 × 4 inches (5.08 × 10.16 cm), which prevent foals and bigger horses from catching a hoof or a head in the wire. Fig. 15.30 shows rectangular woven wire (2 x 4 × 60 inches, "Keepsafe") (5 × 10 × 152 cm) horse fence that prevents hoofs and heads from protruding. When installed on the inside of the fence, mesh will prevent horses from accessing the hazardous triangular spaces. The woven wire should have continuous (one piece) vertical strands that are knotted or twisted at joints to horizontal wires. This mesh fence has a board along the top to prevent horses from damaging the mesh when leaning on it (Fig. 15.31). Welded joints have no flexibility and are prone to breaking. The openings may also be diamond-shaped, which helps to avoid entrapment (Fig. 15.32). To keep horses from leaning on the top of a mesh fence, electric wires can be installed instead of or on top of a board (Fig. 15.33). Many diamond mesh horse fences like the one in this photo have a board on top and small plastic insulators on top of the board to carry an electric wire.

There are also polymer grid products with 2 × 2 inch (5.08 × 5.08 cm) openings (Brown, 2015; Worley and Heusner, 2015).

Nonelectric metal fences, including straight wire and wire mesh, should be properly grounded to protect horses from lightning. Horses leaning on a nongrounded metal fence that is struck by lightning could be injured. If the fence is supported by metal posts (without insulators), the fence is adequately grounded. If other, relatively nonconducting materials like wood or plastic are used for posts, grounding electrodes should be installed into the ground and connected to the metal fence at intervals of no more than 150 feet in dry soil and 300 feet in moist soil (Robertson and Mowrey, 2005). Another way to reduce the probability of horses receiving a lightning shock from a nonelectric metal fence is to break the fence into sections, perhaps of 1000 feet (305 m) or less, by wooden panels, gates, or other insulating barriers (Robertson and Mowrey, 2005).

Wood Posts and Polymer-Coated Wire "Boards" or "Rails"

Steel wire is coated with a plastic polymer that is usually white for high visibility. In addition to a single coated wire product, there are products in various colors appearing as rails that contain two or three parallel 12.5-gauge wires within a polymer coating that is 4–6 inches (10.16–15.24 cm) wide (Figs. 15.10 and 15.15). This provides high visibility, strength, and enough flexibility to prevent many contact injuries. These "high-tensile polymer" rails (sometimes called "HTP Rail"

FIGURE 15.30 Rectangular horse mesh on the pasture side of a fence prevents horse access to the triangular spaces in a brace. This fence was constructed to keep dogs in. *Photo courtesy of Andy MacDonald, Highland Electric Fencing.*

FIGURE 15.31   A structure of woven wire horse fence with parallel vertical wires. *Photo courtesy of John Worley, University of Georgia Extension.*

FIGURE 15.32   A structure of woven wire horse fence with nonparallel vertical wires. *Photo courtesy of John Worley, University of Georgia Extension.*

FIGURE 15.33   A diamond horse mesh fence in Kentucky with wooden boards along the top to prevent bending the mesh when horses lean on the fence. Several similar fences also have an electric wire and insulators on top of the board. *Photo by Paul Sharpe.*

(A)   **(B)**

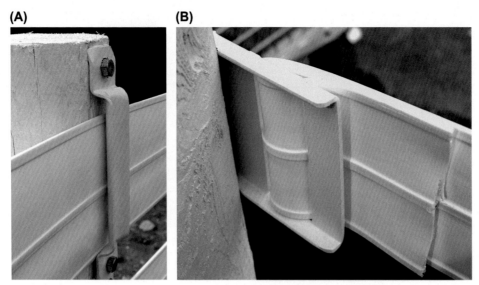

FIGURE 15.34 (A) high-tensile polymer fence (Flex Fence) rail held to the post with a specific flex corner bracket. (B) A flex end buckle holds the end of a flexible rail. Tensioners can be part of this buckle or located farther along the rail. *Photos courtesy of Savannah Petrachenko, System Fencing.*

or "Flex Fence") are less expensive than wooden or rigid polymer rails, are very strong, with high visibility (depending on color). Their appearance, at least when new, is quite attractive, and the cost seems competitive since the components are well-established and their use seems to be increasing. Since they contain wires and expand with heat and contract with cold, they must be tightened periodically with inline fence tensioners or spoolers to maintain proper tension. The polymer/wire flexible rail is held to posts by specialized brackets that allow the strip to slide horizontally (Fig. 15.34A). Ends of flexible rail fence are held by end buckles (Fig. 15.34B). A foal running into the rails of this product will bounce off and keep running (http://www.centaurhtp.com/index.html). A conductive polymer coating (similar to Hotcote and Shockline) can be specified that allows electricity to flow through it and thus to provide both a physical and a psychological barrier (HorseFencing.net, 2016). Some distributors of these products offer installation service, and some supply video instructions for installation.

Wood Post and All Polymer Strand

A pure polymer strand of 8.5 gauge (4 mm), with a breaking strength greater than barbed wire but less than high-tensile 12.5-gauge galvanized steel is available as an attractive, low-maintenance white wire. It is also called "polymer line." It is not capable of carrying electricity, thus it would not dissuade a persistent horse from pushing on it. It is joined by knotting or simple metal splicers and can be stapled tightly to fence posts, run through holes drilled in posts or through insulators. Polymer line does not lose tension between −40 and 122°F and easily releases if a horse becomes entangled in it (https://www.finishlinefence.com/index.html#).

All Plastic Polymer Posts and Rails

All plastic components imitate wood post and rail fences, so they have high visibility and aesthetic value but will cost more than wood. Their maintenance cost is lower than wood, and durability is generally greater than many wood products. In Fig. 15.35 a polymer post and rail fence, which is already attractive, highly visible, strong, and durable, has been supplemented with insulators, so electrified wires can be added and horses will receive reminders to avoid it.

So far PVC and HDPE are the main polymers used. Virgin polymer apparently has greater strength than products with some recycled plastic content. A disadvantage of some polymer products is that the rails can pop out of their posts if a horse leans on them heavily (HorseFencing.net, 2016).

Rubber Belt

Rubber strips or belts from old conveyor belts or tires are soft and yielding but they tend to sag, especially in hot weather, so they require considerable maintenance to keep them tight. The appearance is not as attractive as most other alternatives,

FIGURE 15.35 Posts and rails of all polymer, supplemented with a combination of insulators, designed by a fencing contractor in Ontario. *Photo courtesy of Andy MacDonald, Highland Electric Fencing Systems.*

but since this is a used product, the cost should be low. Horses have nibbled on fibers in the belts, causing colic and impaction (Worley and Heusner, 2015).

Permanent Electric

For permanent electric fencing, corner and line posts are usually of wood, which should be of a long-lasting species or pressure-treated (Kammel and Pedersen, 1989). Corner posts should have tops that are at least 6–7 inches (15.5–18 cm) in diameter for fences with five or six high-tensile wires or that are at least 4–5 inches (10.2–12.7 cm) in diameter for a fence with only one or two wires. Corner posts should be at least as deep in the ground as the height of the top wire. Brace posts (next to corner and gate posts) should be at least 5 inches (12.7 cm) in diameter. For fences under 42 inches (106.7 cm) high the horizontal brace posts should be 8 feet (244 cm) long, and for fences 42 inches (106.7 cm) or taller the horizontal brace should be 10 feet (305 cm) long. Round wooden braces bear loads better than square or rectangular cross-section lumber. Where soil is sandy or frequently wet, posts (especially at corners and gates) should be set deeper and perhaps supported by concrete. Fence wire should be 12.5-gauge, high-tensile, with the highest class of galvanizing you can access (Class 3 rather than 1 or 2). There is a temptation to use 14-gauge wire, which is smaller in diameter, cheaper, and easier to work with than 12.5 gauge. However, the finer wire conducts less current, is less visible, not as strong, kinks more easily, and does not last as long. Line posts can be much smaller in diameter than corner and brace posts and should be about 25 feet (7.62 m) apart on level ground. Short fiberglass or wooden stays (also called stringers) can be placed between line posts to maintain spacing and improve visibility. Steel posts, usually T-shaped in cross-section, can be used as line posts to save money, but it should be noted that they bend more easily than wooden posts and they stay bent. If these are used, there should be a stronger wooden post periodically along the fence, especially at stress points such as changes in elevation. On ground with hills and hollows, a screw-in tie down, also referred to as an earth screw, can be used to hold a small-diameter line post or dropper close to the ground and prevent it rising. Many state and university extension services have bulletins on proper fence construction (Wheeler et al., 2005).

Inline wire stretchers or tensioners and springs should be installed as recommended by manufacturers. Wire tension can be relaxed during winter and tightened in spring. At least one strand should be highly visible material. Alternating live and ground wires on such a fence ensures that animals reaching through the fence receive an effective shock. Ensure that the top wire is live and the bottom wire is grounded to reduce the possibility of live wires being grounded by vegetation (Worley and Heusner, 2015). This is especially effective in dry conditions, when grounding is reduced.

Electric Fence Connections and Current

Ensure that your electric fence does not run close to and parallel with electric power lines. Do not connect an electric fence directly to a main power supply. Always use a modern energizer (controller) that has an approval rating by a certifying agency such as Underwriters Laboratories (UL) or Canadian Standards Association (CSA). Modern electric fence energizers produce very short pulses (about 3/1000 of a second) of high voltage and low amperage, preventing damage to animals and humans. Energizers can get their power from standard 120-V AC outlets or from batteries. Small systems use dry cell batteries. Medium-capacity energizers for use away from AC power use a deep cycle or marine type battery and a solar recharger. Automobile batteries will not last as long as deep cycle batteries. Good quality energizers give the output voltage delivered at varying loads measured in ohms (Ω) and should deliver 1000 V at 100 Ω of resistance. When sizing controllers, multiply the number of miles of fence by the number of wires to get total miles of wire, then add 25% to this to offset power drain (Worley and Heusner, 2015).

Fig. 15.36 shows the path that electric current follows from the positive (+) terminal on the right side of the white energizer attached to the building, through an insulated wire to three live wires on the fence. An animal's nose touching the top live wire connects with the current, which travels along the path of least resistance through the animal to the ground. The current moves through the ground to three ground rods and from them via an insulated ground wire back to the energizer's ground terminal.

To ground the controller and fence, at least three copper or, preferably, galvanized steel ground rods should be driven into the ground at least 10 feet (3 m) apart according to the manufacturer's specifications (Fig. 15.36). These rods must be in contact with moist soil, so their ideal length depends upon the distance of permanent soil moisture from the soil surface. Several publications recommend a diameter of one-half to three-quarters of an inch (12−19 mm) and 6−8 feet (1.8−2.4 m) long. The Gallagher Electric Fencing Manual (2017) recommends a minimum of three earth stakes (ground rods) for energizers up to 15 joules (J), at least five stakes for energizers up to 25 J, and at least seven stakes for energizers up to 35 J. In addition, it suggests 4 m (13.1 ft) between stakes, a minimum length of 2 m (6.6 ft) for stakes, and one wire connecting all the earth (ground) stakes. Energizers with earth monitors require an independent earth stake to monitor the earth system (Gallagher EFM, 2017).

There are several manufacturers of good quality electric fence components, and it is worth taking the time to learn which products are available locally and compare the literature from the different companies. Neighbors and mentors can tell you what has worked well for them. Lightning arrestors and diverters are necessary components to protect the energizer. A good quality current testing meter that shows voltage and preferably the direction of current leakage is a necessary maintenance tool. The wires that you use to connect components of an electric fence should be those designed

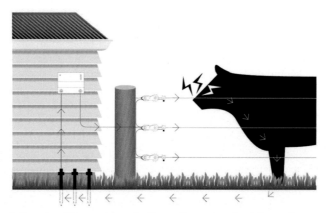

FIGURE 15.36 Electric current travels from the white energizer to live fence wires to the animal, then to the ground, through ground rods and a ground wire to the energizer, completing the circuit. *Illustration courtesy of Gallagher North America.*

for electric fence use and not for household use. House wiring has insulation rated for 400–600 V, and a fence energizer may produce 6000–10,000 V. Using the wrong type of insulated wire for connections may contribute to a problem called "stray voltage," which means that current can be detected, and shocks can be felt in unexpected places, such as "ground" wires, gates, and waterers.

Safety in Electric Fence Construction and Maintenance

Most large distributors of electric fence equipment have documents and videos on proper design and installation of electric fences. Beware of internet videos that are produced by individual users, rather than corporations or institutions that have a vested interest in the accuracy of their information, because some individuals have made electric fence construction videos available to the public that show unsafe practices. Reputable manufacturers ensure that their energizers are certified as safe by UL or the CSA in North America and similar appropriate agencies in other countries.

I once was tempted to wear rubber gloves to handle live electric fence wire. Unless the gloves have been designed and tested to insulate against the maximum voltage of the energizer, this is not safe. It is safe to turn off the current using a conveniently placed cutout switch. Plastic and metal/plastic composite pliers designed for handling live wires are available, but many of these are only rated to protect from a maximum of 1000 V AC or 1500 V DC. A small electric fence energizer may be putting out just 1500 V and will keep your horses in, but a high-capacity energizer may be putting out over 6000 V DC. Rubber gloves designed for use by electricians and linesmen have various ratings and corresponding prices. Class 0 rubber gloves, selling for about $50/pair, are only rated for 1000 V AC; class 1 rubber gloves, costing about $86 will handle 7500 V AC/11,250 V DC, and both of these require an additional pair of leather linesman's gloves over them to protect the rubber. Furthermore, there is no way to protect parts of your body from shock other than your hands. Remember that insulated wire designed for use in the ground under electric fence gates and between electric fence components is rated for 10,000 V rather than the 600 V of protection offered by insulation on wire for use in buildings. Insulation on wire under gates can be further protected by placing the insulated wire inside plastic water line or electrical conduit.

Training Horses to Electric Fencing

Since electric fences are more of a psychological than a physical barrier, the animals to be contained need to be trained to the fence. If untrained horses are transported into a pasture contained by electric fence, they may be travelling at considerable speed when they make their first contact with the fence, and this will increase their anxiety, possibly leading to injured horses and damaged fencing. Training can take place in a small confined area to which the horses are accustomed. Run a small amount of electric wire of the same type that will be in bigger paddocks and connect it to a suitable energizer that is properly grounded. The training area should have good visibility and enough room for horses to move quickly away from the fence without injuring themselves, other animals, or damaging the facilities. Erect a short segment of the type of electric wire and posts that horses will be subjected to in the pasture. The segment should project into an open space where animals walk by. Connect the fence segment to a grounded energizer and turn the energizer on. Some people will put a little bit of feed near the fence to encourage horses to touch it. Two or three days of training should be sufficient for horses to respect the fence in relatively calm situations. If horses initially knock the electric fence down, turn the power off, put the fence back up, and check the fence periodically. Once the fence stays up for an extended period, the horses will have learned to avoid it. If very excited or frightened, a horse may forget about the electric shock it obtained in the past and break through an electric fence, but these occasions are rare. This is why a perimeter fence should be a strong physical barrier.

Maintaining the Current

Once you have selected, read the manual about, and installed an electric fence system, you will have an idea how many volts of electricity you should be able to measure on the live fence wires. When this number is too low, you probably have a leakage of current. It may be due to the wrong type of wire, as described before, to loose connections, or to vegetation touching live wires. Very commonly, the lack of current is due to inadequate grounding. Instructions from the energizer manufacturer should indicate how to test whether the grounding is adequate (Duncan, 2011). Suggestions for ensuring adequate grounding and high fence voltage include these:

1. Use at least 3 feet (1 m) of ground rod per joule of output from the energizer. If you have a 15 J output, you need $15 \times 3 = 45$ feet (14 m) of total ground rod in the ground. Use galvanized steel or copper rods, insulated electric fence

connecting wire (rated for 20,000 V), and ground rod clamps. Rods and clamps should be made of the same metal. Each rod should be 4–6 feet (122–183 cm) long and at least 0.5 inch (1.3 cm) in diameter. Check the connections for tightness and lack of corrosion every spring.

2. Use three separate, connected ground rods, at least 10 feet (3 m) from each other. Select a grounding site at least 50 feet (15 m) from any other utility grounding (power poles, well houses, etc.), from building foundations or buried water-lines. The grounding site should have moist soil, such as a low spot near a pond or creek.

3. In very dry soil, dig holes for each ground rod at least 3 inches (8 cm) in diameter and fill them with a powdered bentonite paste. Then insert the rods in the center. Alternatively, use longer ground rods or do not insulate the bottom fence wire and ground it through the posts and to ground rods.

4. If you have a very long electric fence (greater than about 2600 feet (790 m), with at least one ground wire as part of it, connect that ground wire to a grounding rod in moist soil near the farthest point from the energizer or every 1300 feet (396 m), whichever is less.

5. Avoid using metal T-shaped fence posts with plastic insulators for electric fences. A wire out of place or a cracked insulator can cause current to leak through the metal post to the ground.

6. Walk the fence frequently, looking for grass or any vegetation growing up, down, or sideways to touch live wires. Remove that vegetation. Remove branches or other items that fall on the fence. Do this every time you check on the condition of the pasture and the well-being of the horses. Look for places where wildlife may have dug under or damaged the fence.

7. Check the fence voltage daily and after thunderstorms. Keep the voltage at or above 1500 V (1.5 kV) for horses alone and at 2500 V if they are accompanied by sheep. To reduce the need to go all the way to the fence, install a device that hangs on a live wire and shows a bright light when the fence is carrying a high voltage and a dim light when the voltage is reduced.

8. Many electric fences carry current in all wires (all hot), but there are situations where alternating hot and grounded wires should be used. In dry conditions and over frozen or snow-covered ground, there is poor conduction from an animal to the ground, thus having adjacent wires alternating between hot and grounded causes animals to be shocked when they contact both a hot and a ground wire.

To reduce or eliminate the risk of being shocked while working on a fence, the electric current to that section of fence should be turned off. The switch on the energizer can be turned off, and the live wire can be disconnected from it. For large, complex fences that have two or more sections, cutout switches should be installed on posts leading to sections, so the power can be cut to one section at a time. Ensure that the wiring allows power to reach fence sections farther away from the section to be worked on. Electric cutout switches are also useful at gates and close to the energizer (Fig. 15.37). Some of these switches are two-way (double throw), so they can carry current to one section or another. Another use of double throw switches is to allow switching between a grounding system of rods in the soil and a system of alternate fencing wires. The ground rod system may work best in times of high soil moisture, and the alternate wire grounding system may work better when soil moisture is low.

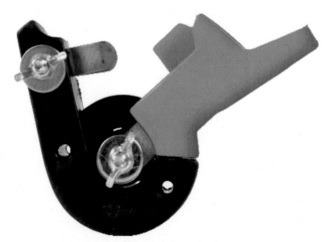

FIGURE 15.37 An electric cutout switch to direct and cut off electric current in a fence. *Photo courtesy of Gallagher North America.*

In flood-prone areas, electric fences are susceptible to power loss and animal escapes when water rises to the fence. Power to bottom wires can be disconnected or electrified flood gates can be installed across seasonal waterways. These include a flood gate controller between the fence and floodgate, a cutout switch, and a fence wire crossing between the tops of both banks of the waterway, with chains that hang down into the waterway (Gallagher EFM, 2017).

Some components of electric fence systems can cause interference with radio signals and land telephone lines. Electric fence lines that must run parallel to telephone lines should be at least 350 ft (100 m) away from them. It is acceptable to have an electric fence cross perpendicular to a phone line. Energizer manufacturers' manuals can provide suggestions for preventing and diagnosing interference (Gallagher EFM, 2017).

Energizers that are capable of obtaining power from AC and DC (battery) sources provide some extra assurance of electric fence function during AC power outages. If you already have a large AC-powered energizer, buying a small DC-powered energizer to power just the sections of fence containing animals can meet your needs during a power outage.

Semipermanent and Temporary Electric Fences

The previous section discussed fences with significantly braced corners, using large wooden posts that you hope to never remove. This is suitable for perimeter fences and for defining major pasture areas. For flexibility in managing your farm, you may want some semipermanent fences that are constructed of lighter materials or fewer large posts. These can differentiate major pasture areas (Figs. 15.38 and 15.13). To provide the ultimate in flexibility for adjusting paddock size within a large pasture, temporary fencing is very useful. In Fig. 15.39, temporary polymer electric tread-in fence posts with only one electric polymer rope have prevented grazing horses crossing from the left to the right side of the fence, effectively resting the forage on the right.

Corner posts for semipermanent fence can be about 4–5 inches (10.2–12.7 cm) in diameter. The greater the number of wires, the longer the fence, and the greater the amount of tension on the wires, the more substantial the corners or ends need to be. The number of products suitable for semipermanent and temporary electric fencing has expanded rapidly, with several designed specifically to increase visibility to horses.

The simplest electric fence wire product sold for semipermanent and temporary confinement of livestock is generally called "polywire" because it is a series of twisted plastic polymer strands with some fine stainless-steel, copper, or aluminum wires blended in. The polymer is usually yellow or white and there may be some black fibers. This is lightweight and easy to use. Pieces can be joined together and fastened to insulators by tying knots in it. It is much weaker than 12.5-gauge high-tensile wire but also much cheaper. Due to the smaller diameter of the wires, it does not conduct as much electricity as the high-tensile wire. This is a good example of the principle of resistance being inversely proportional to diameter. Electric wires of bigger diameter have greater capacity to carry electric current because they have less resistance to it. Similarly, bigger diameter pipes and hoses have less resistance to water flow and greater capacity to deliver water. Resistance to carrying current increases with length, and similarly, resistance to carrying water increases with length.

Polywire is only about one-sixteenth to one-eighth of an inch (1.6–3.2 mm) in diameter, so it does not have much visibility, especially to running horses, thus larger diameter cord, braid, and ropes plus a polymer tape were invented (Fig. 15.40). The polymer may be polyester, which is soft and easy to handle. The thicker it is, the more conducting wires it

FIGURE 15.38 A semipermanent fence using fiberglass line posts, which require no insulators. These solid posts are notched and have orange clips to hold the wires. *Photo courtesy of Gallagher North America.*

FIGURE 15.39 Temporary electric fencing comprised of a polymer rope containing conductive wire on polymer tread-in posts. In the drought year that this photo was taken, overgrazing of the left side is greater than on the right side, which has some regrowth of green grass and white clover, yet horses respected the fence. *Photo by Paul Sharpe.*

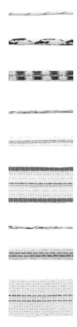

FIGURE 15.40 A comparison of the shapes, sizes, and color patterns of poly wire, thicker poly rope, and poly tape products of different widths, all containing conductive wire. *Photo courtesy of Gallagher North America.*

can carry, and the more protection is supplied to the wires, presumably increasing its useful lifespan. The first polymer tape developed was a plastic woven material that is three-quarters of an inch wide. The extra width provides much greater visibility than thin wire, but it also catches wind and thus moves more easily (which further helps visibility but might be a detriment in a storm). The downside of catching wind is that it puts more strain on fence insulators and posts. Other electric fence tape products that are 2 inches (5.1 cm) wide have been developed specifically for horses. These widest of the woven products have more wires within them, giving them increased capacity to carry electrical current. While this wide tape can be tied, it will look better and be easier to manage if its ends are fastened with specific hardware that keeps its shape. Appropriate insulators are designed for wider tapes.

Temporary fence wire or tape can be measured and cut to fit specific locations, or for more flexibility, it can be held on a metal-reinforced plastic geared reel that dispenses and gathers wire quickly. Reels can be placed on high-tensile electric

FIGURE 15.41 Three reels for electrified poly wire, poly tape, or poly rope, all attached to one standard to help create a three-strand temporary electric fence. *Photo courtesy of Gallagher North America.*

wire, where the reel hook makes the electrical connection. Connecting wires with alligator clips can also be used to power the wire. Groups of three reels can be placed on a metal stand that is placed into the ground and usually needs to be tied to a fencepost or a ground stake (Fig. 15.41).

Hardware fittings are available for connecting polymer tapes and ropes. In Fig. 15.42 a single line of polymer rope has been wrapped around a painted wooden post, with an insulated tube around all parts of the rope that can contact the post.

FIGURE 15.42 An insulating polymer tube surrounds part of an electric polymer rope that is wrapped around a fence post. Two connectors are used on the rope, one to form a loop and one to splice a break in the rope, presumably because the long rope was cut too short. *Photo by Paul Sharpe.*

FIGURE 15.43 Electric netting for controlling poultry, small ruminants, horses, small predators, and bears. *Photo courtesy of Gallagher North America.*

Electric netting (Fig. 15.43) has electrified horizontal poly wires and nonelectrified polymer vertical strands at close intervals of only a few inches, and spaces are smaller near the bottom than the top. Typical heights are 36−48 inches (91−122 cm). Slightly different versions are designed and marketed to keep poultry, sheep, and goats in and to keep racoons, canine predators, and bears out. It does not have a great deal of physical strength or visibility, so it is not often used for horses, but one company does market a version for horses (https://www.premier1supplies.com/p/horse-quikfence-4-48-24-electric-netting?criteria=netting+for+horses).

Line posts for temporary fences are mostly made from fiberglass or plastics, some of them reinforced with metal at the base for ease of inserting them into the ground. Some of these posts can carry multiple wires and/or tapes, and some are designed for only one wire, having a pig-tail shape or split ring at the top. Other plastic insulators can be installed on these posts, but one of their highlights is the speed with which they can be installed. For mature horses that are well-accustomed to their environment and to electric fence, a one-wire temporary fence is a very convenient tool for rotational grazing. The Gallagher company recommends a three-wire temporary electric fence for horses, with wire heights starting at 28 in (700 mm), then at 38 in (950 mm), and 48 in (1200 mm) from the ground.

Manufacturers of electric fencing include, but are not limited to, Gallagher, McGregor, Patriot, Powerflex, Premier, Speedrite, Stafix, and Timeless. Most farm supply and feed stores carry some electric fence products. For a large project, it may be time-efficient to contract a local fence building company that has been recommended for its good work in your community. Many of these companies have good videos on fence construction and maintenance on their websites and on a YouTube channel. At least one manufacturer makes energizers allowing main power or battery backup, remote control to switch on or off from anywhere, and an alarm for a broken or shorted fence that is auto-dialer compatible.

Electric fencing components have become more innovative and diverse since the 1980s. Plants growing under the fence that contact live wires cause significant draining of current into the ground. Thus, time and money are often spent on trimming or spraying that vegetation. Very early energizers had high enough amperage and voltage and long enough pulses to burn off much of this vegetation, but modern energizers are safer and less capable of this due to lower amperage, more appropriate voltage, and shorter pulses of current. A recent development in energizers is the "energy limiter" that allows current pulses to flow normally through upper wires but delays the current from going to the bottom wire, which is the one in greatest contact with vegetation. It prevents the shorted wire from grounding the entire fence. Energy limiters can also be used in flood situations, along areas of public foot access, places where children are playing, and other areas where accidental human contact is probable. Plastic warning signs about the electric current hazard can be hung on sections of a fence near public access (Fig. 15.44).

FIGURE 15.44   Warning sign to be hung on electric fence wire to reduce the probability of people and pets being shocked. *Photo courtesy of Gallagher NA.*

Definitions of Electrical Terms (Testguy.net, 2017)

Alternating Current (AC): An electric current that reverses its direction many times a second at regular intervals. Wiring in houses and barns uses AC.

Ampere: A unit of measure for the intensity of an electric current flowing in a circuit. One ampere is equal to a current flow of 1 Coulomb per second (Kencove.com. 2017a).

Circuit: A closed path in which electrons flow from a source of voltage (current). Circuits can be in series, parallel, or in any combination of the two.

Circuit Breaker: An automatic device for stopping the flow of current in an electric circuit, usually due to excess current flow. To restore service, the circuit breaker must be reset (closed) after it has broken (opened) and after correcting the cause of the overload or failure.

Conductor: Any material where electric current can flow freely. Conductive materials, such as metals, have a relatively low resistance to electric current. Copper and aluminum wire are the most common conductors. Wooden posts are relatively poor conductors, but water within them can conduct electricity. Ceramics, fiberglass, and most plastics are very poor conductors.

Coulomb: (C) (a) The SI unit quantity of charge transferred in 1 s across a conductor in which there is a constant current of 1 A. (b) The equivalent of 1 A-second. Conversely, an electric current of 1 A represents 1 C of unit electric charge flowing past a specific point in 1 s. The unit electric charge is the amount of charge contained in a single electron; 1 C = 1 A × 1 s.

Current: The flow of an electric charge (electrons) through a conductor. An electric current can be compared to the flow of water in a pipe. Current is measured in amperes.

Direct Current (DC): An electric current that flows in only one direction and generally comes from batteries.

Electrolysis: Corrosion between different metals in a wet environment such as between copper and galvanized wires on a fence line. Avoid this by only using galvanized wire (Gallagher EFM, 2017).

Electron: A tiny particle that rotates around the nucleus of an atom. It has a negative charge of electricity.

Fuse: A circuit-interrupting device consisting of a strip of wire that melts and breaks an electric circuit if the current exceeds a safe level. To restore service after a fuse wire melts, the fuse must be replaced, using a similar fuse with the same size and rating, after correcting the cause of the failure.

Generator: A device that converts mechanical energy into electrical energy.

Ground (Earth): The reference point in an electrical circuit from which voltages are measured; a common return path for electric current or a direct physical connection to the earth.

Ground Fault Circuit Interrupters (GFCI): A device intended for the protection of personnel that functions to deenergize a circuit or portion thereof within an established period of time when a current to ground exceeds some predetermined value that is less than that required to operate the overcurrent protective device of the supply circuit.

Induction: Power transfer without contact. For example, the charging of dead or neutral fence wires that run parallel to live ones. The closer the live and neutral wires and the further they travel together, the greater is the amount of inductance (Gallagher EFM, 2017).

Insulator: Any material where electric current does not flow freely. Insulating or nonconducting materials such as glass, rubber, air, and many plastics have a relatively high resistance to electric current. Insulators protect equipment and life from electric shock.

Inverter: An apparatus that converts direct current into AC.

Joule: A measurement of energy or work. In electronics, 1 J is (a) 1 W of power, applied for 1 s or (b) the work required to move a coulomb of electrical charge through an electrical potential difference of 1 V (Kencove.com. 2017b). Joules are the most important measure of the power of an energizer (Gallagher EFM, 2017).

Continued

$1 J = kg \times m^2/s^2 = W \times s = C \times V = N \times m = Pa \times m^3$, where J = joule, m = meter, s = second, W = Watt, C = Coulomb, V = Volt, N = Newton, and Pa = Pascal.

Leakage: Energy loss caused by poor insulation, shorts, and/or vegetation on the wires, resulting in a drop in voltage (Gallaher EFM, 2017).

Load: Anything that consumes electrical energy, such as lights, transformers, heaters, and electric motors.

Ohm: (Ω) A unit of measure of resistance to current. One ohm is equivalent to the resistance in a circuit transmitting a current of 1 A when subjected to a potential difference of 1 V. In the resistance scale a lower value indicates increased shorts to ground. Zero ohms is a direct short to ground, while a very large resistance to ground indicates that the fence is well insulated (Gallagher EFM, 2017).

Ohm's Law: The mathematical equation that explains the relationship between current, voltage, and resistance (V = IR), where V = volts of potential difference between two points, I = the current through the conductor in amperes, and R = resistance in ohms. Also, I = V/R and R = V/I.

Ohmmeter: An instrument for measuring the resistance in ohms of an electrical circuit.

Open Circuit: An occurrence when a circuit is broken, such as by a broken wire or open switch, interrupting the flow of current through the circuit. It is analogous to a closed valve in a water system.

Polarity: A collective term applied to the positive (+) and negative (−) ends of a magnet or electrical mechanism such as a coil or battery.

Power: The rate at which electrical energy is transferred by an electric circuit. Power is measured in watts. Power in watts = potential difference in volts x current (I in amperes).

Protective Relay: A relay device designed to trip a circuit breaker when a fault is detected.

Pulse: A brief electrical current or shock emitted by an energizer. Each pulse is on for about 0.0003 of a second. Pulses are spaced about 1 s apart.

Resistance: The opposition to the passage of an electric current. Electrical resistance can be compared to the friction experienced by water when flowing through a pipe. Resistance is measured in ohms.

Resistor: A device usually made of wire or carbon that presents a resistance to current flow.

Service: The conductors and equipment used to deliver energy from the electrical supply system to the system being served.

Short Circuit: An occurrence in which one part of an electric circuit comes in contact with another part of the same circuit, diverting the flow of current from its desired path.

Solid-State Circuit: Electronic (integrated) circuit that utilizes semiconductor devices such as transistors, diodes, and silicon-controlled rectifiers.

Volt: A unit measure of voltage. One volt is equal to the difference of potential that would drive 1 A of current against 1 Ω of resistance.

Voltage: An electromotive force or "pressure" that causes electrons to flow and can be compared to water pressure that causes water to flow in a pipe. Voltage is measured in volts.

Voltmeter: An instrument for measuring the force in volts of an electrical current. This is the difference of potential (voltage) between different points in an electrical circuit. Voltmeters have a high internal resistance and are connected across (parallel to) the points where the voltage is to be measured.

Watt: A unit of electrical power. One watt is equivalent to 1 J per second, corresponding to the power in an electric circuit in which the potential difference is 1 V and the current 1 A. W = J/s and W of power = V × current (I in amperes).

GATES

Gates allow horses, people, and machinery to move through fences conveniently. Hardware for holding gates closed can be homemade or purchased, lockable or not, operable from horseback or not, and even openable remotely, like a garage door. Every pasture paddock may need cutting, harrowing, or fertilizing and thus should have one or more gates big enough to allow vehicles and machinery to enter and leave. Gates should be hung on posts as large and deep as those at the corners of the fence. Ensure that gate posts are supported by sufficient bracing that the weight of the gate will not cause the gate posts to lean and the gates to drag on the ground (Fig. 15.45). Excessively heavy or long gates can have a wheel installed under the swinging end to help bear some of the weight. If an extremely long gate is required, a mechanism that rolls the gate sideways rather than allowing it to swing will be suitable. Clearance under gates should be adequate to swing a gate after a typical snowfall without having to clear the snow away but the clearance should not be so great that animals you are trying to constrain can pass under the gate. The top of the gate should be as high as the top of the fence to discourage horses from trying to jump over it.

Useful information can be displayed at a gate, such as the paddock numbers on the gate post and the notice that the paddock contains no tall fescue, so it should be safe for mares in late pregnancy, as in Fig. 15.46. This photo also shows a simple gate latch, securely fastened into a horizontal component of the gate and fitting into a hole drilled into a post.

FIGURE 15.45 A horizontal brace for a gate, showing vertical post, horizontal brace post, and diagonal wire to transfer force to the most stable component. This gate is braced on both sides, contains mesh to keep horse hoofs off it, and is at least as high as the adjacent fence. The wire mesh is on the side next to the lane, suggesting that horse contact may be expected from that side. This 2 × 4 × 48-inch (5 × 10 × 122 cm) fence was constructed to keep children safe inside it. Photo courtesy of Andy MacDonald, Highland Electric Fencing.

FIGURE 15.46 A simple gate with a latch on a thoroughbred breeding farm in Kentucky, showing useful information on the gate post (2 A) and adjacent brace board (No Fescue). There is also an electrified bare wire attached to the post to keep horses off the top of the fence. *Photo by Paul Sharpe.*

Horses tend to congregate around gates, perhaps because they are sometimes rewarded with food coming through open gates or the opportunity for exercise once they have passed through them. Increased traffic around gates leads to erosion, low spots, wet spots, and mud. There are effective methods to eliminate such wet spots, which involve improving the drainage, digging out about 10 inches of soil, and replacing it with layers of gravel, separated by geotextile fabric (Greene, 2007). You should avoid locating gates in low areas that accumulate water. Where larger numbers of horses gather for long periods of time, gates need to be particularly strong (Robertson and Mowrey, 2005).

For a single horse and handler to pass through safely, a gate should be at least 4 feet wide. To allow modern agricultural machinery through and to turn, widths of 16−25 feet (4.8−6 m) should be considered (Worley and Heusner, 2015; Fabian, 2017). Gates on laneways should be as long as the width of the laneway, so they can block the laneway and direct animal movements. Gates can be hung so they swing both ways or just in toward the paddock. A second latch should allow the gate to be fastened in the fully open position against the fence (Fabian, 2017). Gaps between gates and their supporting posts should be too narrow for horses to catch a hoof or head in the gap (Worley and Heusner, 2015).

A gate should not be a weak area in the fence. It should be at least as high, strong, and safe as the rest of the fence. Wooden gate kits and metal tube gates are readily available from building supply stores. Plans to build your own gates are available through retail outlets and extension organizations (for example, HingeandLatch.com). Edges of gates should be smooth, and corners should be rounded to prevent injuries. A low-maintenance finish such as powder coating in a bright color or hot-dipped galvanizing will help a gate to last a long time. Gates in permanent fences should be physically strong and well-braced. For appearance sake, gates should have a similar appearance to the rest of the fence. Metal tube gates can have horizontal and vertical struts connecting the perimeter tube and spaces may be filled in with horse mesh to prevent the catching of hoofs and heads.

Gates in semipermanent or temporary fence can consist of a coil spring that carries electric current with a plastic, nonconducting handle (Fig. 15.47). Where a more permanent fence has at least one electrified wire and a gate, an insulated, undergate cable should be installed according to manufacturer's directions, with a cutout switch on one or both posts, so the current can be turned on or off (Fig. 15.48). Insulated wire designed for use in the ground under electric fence gates and between electric fence components is rated for 10,000 V rather than the 600 V of protection offered by insulation on wire for use in buildings. The cable should be buried at least 12 inches (30.5 cm) deep. Some fence builders will slide the undergate cable through plastic water pipe to protect it from downward force when heavy machinery passes over it. The ends of the pipe should be turned downward to keep rain out of it.

Where people need to pass through farm fences frequently, sturdy gates can be built, such as the one in Fig. 15.49, at Avebury Historic Site in England.

Many paddock fence plans call for gates in the corners of paddocks, and this can work where groups of horses are herded along a fence, into, and out of the paddock. In contrast, a study of herd animal behavior provides a reason for putting gates in the middle of paddock fences, rather than in the corners. As herd animals that needed to escape predators, wild horses tended to herd together and acted as a group when pressured or threatened. On the other hand, horses desire some space between themselves and the next horse as they move. The balance between these behaviors, herding and spacing, is influenced by pressure exerted by a predator or herder. More pressure causes more herd behavior, and less pressure allows more spacing. When a herd of horses is herded to a gate in the corner of a paddock, they herd together to get through the gate, and as soon as the first horses are through the gate, they tend to spread out and can do this in only one direction. Horses waiting to get through the gate see the horses on the other side of the fence moving off to the side and tend to follow them, rather than go through the gate. If the gate is in the middle of the fence, there is less tendency for horses to space out and move sideways from the gate once they have passed through (Marble, 2017). Locating gates in the middle of fences also prevents the trapping of horses in a corner near a gate. A tractor and implement travelling from one paddock across a laneway into another paddock should be able to go straight across without turning.

Next to roads, gates should be located where there is good visibility both along the road for drivers leaving the paddock and into the paddock for drivers on the road. Gates from a road into a paddock should be at least 40–60 feet from a corner, to allow a driver to park a vehicle and trailer off the road while opening the gate (Fabian, 2017).

Research is needed on which gate latches to install on permanent fence gates. There are some excellent products on the market, some allowing only one-way swinging and some allowing two-way. Any gate latch should be openable with just one hand but should not be so simple that a horse can learn to open it. The hardware should be tough enough to withstand horses leaning on it and to last many years (Fabian, 2017). An alternative to a gate latch is a ground anchor tube and a vertical rod mounted on the gate end that fits into the tube.

FIGURE 15.47 Two insulated electric gate handles with coiled spring live wires in them are keeping these cattle in place. *Photo courtesy of Gallagher North America.*

(A)

Electric Fence Diagram - Conventional gate

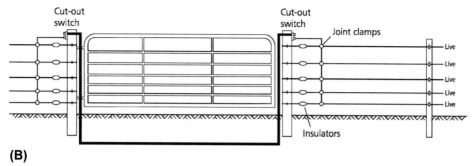

(B)

Electric Fence Diagram - Spring gate, bungy gate or tape

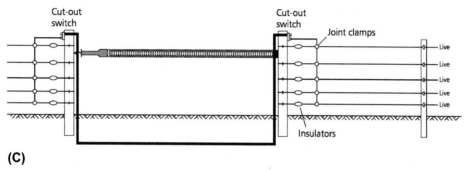

(C)

Electric Fence Diagram - Ground wire return system under gate

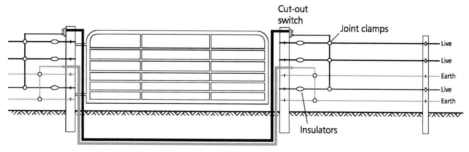

FIGURE 15.48 Placement of insulated undergate cables at gates in electric fences. Notice the heavy black line indicating the specific undergate cable that is coated with extra insulation. (A) At a conventional gate in a conventional multiwire electric fence, notice the connections of all live wires with joint clamps and the connections from there to cutout switches. (B) At a spring, bungy, or tape gate, notice that the power comes into it just at the handle end, so that when it is unhooked, the gate is not powered. There is a risk of a horse's tail getting caught in a spring gate, so some people recommend against using them where horses are kept. (C) Where some of the fence wires are ground wires, there is a need for a separate ground wire underground. The gray line indicates the same quality of cable as the one carrying power. Notice that it is connected just to ground wires, which are not insulated from the posts. *Illustrations courtesy of Speedrite and Tru-Test Group.*

MAINTENANCE OF FENCES

While fencing materials vary in their need for maintenance, any fence that is constructed will probably need some maintenance. Generally, the more you pay for materials and the stronger you build your corners, braces, and gate posts, the less maintenance will be required. Steel components should be hot-dip galvanized to prevent rust. If they are damaged and start to rust, remove the rust and spray on a cold galvanizing coating. Worley and Heusner (2015) suggested the following items to consider in a fence maintenance program:

1. Repair or replace anchor (brace) post assemblies when they show signs of weakness. Usual evidence of this is corner or gate posts leaning toward the middle of the fence and fence wires drooping.

FIGURE 15.49 A strong wooden gate and latch that can be used by people on foot and can be used from horseback in a farm fence at Avebury Historic Site in England. *Photo courtesy of the author's wife, Helen MacGregor.*

2. Paint wooden fences as needed for longer life and appearance.
3. Wash PVC or vinyl fences with chlorine bleach to remove mildew.
4. Refasten loose wires to posts and splice broken wires.
5. Keep fence wires properly stretched, especially high-tensile wires. The start and end of the grazing season and the hottest part of the summer are times to check tension.
6. Keep weeds and brush cleared from fence lines, especially on electric fences. After storms, check for trees and branches that may have fallen on a fence.
7. Check voltage on electric fences regularly, and when voltage is lower than usual, inspect connections, grounding, and the fence lines.
8. Plan and follow a regular inspection routine.

A fence maintenance tool kit can be very useful. It can include your voltmeter, tools used for gripping and cutting wire, a crimping tool and joiners for joining two ends of wire, a wrench or two for tightening any connectors and spare connectors, inline wire strainers, leather gloves, wires, and insulators. Bigger tools that will be needed occasionally are a big hammer or post driver, a fence wire stretcher, and something for cutting tree branches.

Fig. 15.50 shows a number of maintenance issues that need attention: a broken board and a lightweight gate that is lower than the top of the fence and has been bent. The pasture is overgrazed and trampled next to the fences. Trees have been killed, possibly by horses stripping bark from them.

FIGURE 15.50 Fence, gate, and pasture maintenance issues that need attention. *Photo by Paul Sharpe.*

FIGURE 15.51 The gate height is the same as the fence and has not been bent, even though the bare soil shows plenty of horse traffic near the gate. *Photo by Paul Sharpe.*

FIGURE 15.52 This gate is not quite long enough for the width of the gateway, so an unpainted board has been attached to the post to prevent a hoof from being trapped in the space. Lower rails have been bent, presumably by horses stepping on them. This gate can be fastened by hanging the chain into the slot and by wrapping the chain around the gate and clipping it back onto a link. *Photo by Paul Sharpe.*

The gate in Fig. 15.51 is the same height as the fence and has not been bent by horses leaning over it, even though the bare soil shows plenty of horse traffic near the gate. It is possible that horses did not lean on this gate because it is adjacent to a gravel lane with no feed on the lane side, but other gates on nearby paddocks are bowed toward the lane, indicating horses had pushed against them. Fig. 15.52 shows the addition of an unpainted board to fill in a gap between a metal gate and the adjacent post, eliminating a space for catching a hoof. The board also stops the gate from swinging outward.

DRINKING WATER SYSTEMS

Horses obtain water from three sources, their feed, drinking, and from metabolic processes. High-moisture feed such as green grass and silage contribute substantial amounts of water to daily requirements. Fresh grass can contribute 68–132 lbs (31–60 kg) of fluid per day, approaching a mare's daily water requirement (NRC, 2007). A table of estimated water needs of 1100-lb (500 kg) horses shows a daily average total water intake as follows:

1. idle, eating 16.5 lb (7.5 kg) of feed at 68°F (20°C)/6.6 gal (25 L)
2. idle, eating 16.5 lb (7.5 kg) of feed at 86°F (30°C)/12.7 gal (48 L)
3. idle, eating 16.5 lb (7.5 kg) of feed at −4°F (−20°C)/11.1 gal (42 L)

4. lactating, eating 33 lbs (15 kg) of feed at 68°F (20°C)/17.2 gal (48 L)
5. 1 h moderate exercise, eating 24.2 lbs (11 kg) at 95°F (35°C)/21.7 gal (82 L) (NRC, 2007)

These water intakes include metabolic water and water in feed. There can be day-to-day variation of water intake among similar horses and within individual horses of 20%–64%. Adding grain and reducing hay in a diet reduces the water intake. During transport, horses eat and drink less than usual. Limiting water intake reduces feed intake, so a horse refusing feed may indicate an inadequate water supply. Water and salt intakes have a positive linear correlation, but adding salt to a diet does not necessarily affect water intake or urine volume.

Foals will start drinking water at around 1 week of age. Drinking episodes for housed horses have been reported to occur about two to eight times per day and last about 10–60 s per episode, but another study by the same authors on housed pregnant mares recorded drinking episodes 18–39 times per day and lasting 23–26 s (McDonnell et al., 1999). Lactating mares on pasture had drinking bouts averaging 23.4 s, and their foals drank water for 20.4 s. Increasing ambient temperature caused drinking frequency of these mares to increase, for example to every 1.8 h (13.3 × per day) when temperatures were 86–95°F (30–35°C) (Crowell-Davis et al., 1985; NRC, 2007). Horses at cool ambient temperatures drank less near-frozen water than water that was 66°F (19°C), but when ponies were kept at warm temperatures, they drank similar amounts of warm or icy water (National Research Council (NRC), 2007). When adult military horses were monitored in England, their daily water consumption on grass pasture was 3.2 gal (12 L) per 1100-lb (500 kg) horse, and when they were stabled and fed hay, this increased to 8.45 gal (32 L) per 1100-lb (500 kg) horse (Williams et al., 2015). While on pasture the horses were not exercised, but while stabled, they had 60 min of exercise per day.

Drinking Water Quality

Water suitability is primarily based on the concentration of total dissolved solids (TDS), which measures all the ions collectively. Table 15.3 shows acceptable TDS for livestock drinking water (NRC, 1974). Higher TDS levels may be due to salts from seawater or soil. Canadian guidelines (Olkowski, 2009) indicate a TDS of 6500 constituting common minerals and is generally considered the safe limit for horses. Water hardness, due to high concentrations of calcium, magnesium, sodium, and potassium salts, can adversely affect palatability. Horses tolerate fluoride to about 4 ppm. Selenium should be below 0.0005 ppm. Other minerals known to be toxic should be avoided. If people or animals object to the smell of water or animals are reluctant to drink it, an analysis is a good idea. Contact an agricultural extension agent or public health agency for information on water testing and local advice on potential contaminants.

An elevated sodium concentration in water may be due to naturally high soil sodium concentration in alkali lakes and prairie areas. Less conventional excess sodium contamination of livestock water can come from sodium chloride salt used to soften household water or deicing salt used on roadways in winter. Some prairie waters have had sodium concentrations that would supply four times the sodium needed by an idle horse (McLeese et al., 1991; NRC, 2007). For a table of elements and chemicals with their recommended upper safe levels, see Olkowski (2009).

Nitrates and nitrites from fertilizer, stored manure, or human septic systems can potentially contaminate livestock drinking water, but cases are rare. Eutrophication of surface water by several species of cyanobacteria (formerly called "blue-green algae") is becoming more common throughout North America. These organisms can cause a bright green

TABLE 15.3 Guidelines for Total Dissolved Solids (TDS) or Total Soluble Salts (TSS) in Drinking Water for Livestock

TDS or TSS (mg/L)	Guideline
<1000	Safe and should pose no health problems
1000–2999	Generally safe but may cause a mild temporary diarrhea in unaccustomed animals
3000–4999	Water may be refused when first offered to animals or may cause temporary diarrhea; animal performance may be less than optimum because water intake is not optimized
5000–6999	Avoid these waters for pregnant or lactating animals
7000	This water should not be offered; health problems or poor production may occur

NRC (National Research Council), 1974. Nutrients and Toxic Substances in Water for Livestock and Poultry, National Academy Press, Washington, D.C., 93 p.; NRC (National Research Council), 2007. Water and water quality. In: Nutrient Requirements of Horses 6th Revised Edition. The National Academies Press, Washington, D.C., 132.

appearance, plus bad tastes and odors in water, in addition to the hepatotoxins called microcystins, which can poison livestock and humans. Australian and New Zealand trigger values for microcystin-LR in water for horses is 2.3 μg/L or 11,500 cells/mL (ANZECC, 2000; NRC, 2007). Water contaminated with algae should be avoided as a source of drinking water for animals. Cyanobacteria poisoning may cause photosensitization, weakness, bloody diarrhea, tremors, convulsions, and sudden death (Olkowski, 2009). Cyanobacteria are sometimes killed by an application of algaecides, but the toxins they produce can persist in the water.

A threshold for bacteria in livestock water is 100 thermotolerant coliforms/100 mL (ANZECC, 2000), but NRC has not established a bacterial limit for drinking water for horses. Urinary excretion of *Leptospira* bacteria by rodents (more likely on feed than in water) can cause abortions in mares and the deaths of foals (Olkowski, 2009). Veterinary descriptions of managing water quality for cattle are available by Morgan (2011) and Wright (2011).

Horses should be provided with fresh, clean, palatable water at all times. Water moving from a well can warm up considerably if it flows through black pipe exposed to the sun on a hot sunny day. Water pipes underground or covered by forage will result in less warming of the water. If the resulting water enters a large trough, there is opportunity for cooling by evaporation, but if a small drinking device with no reservoir is used, water consumption may be reduced because the water is too hot. Large water troughs of 100 gallons (387 L) or more for a few horses are quite common. While they likely will not restrict water intake because of high temperature, they are difficult to empty and to clean. If water troughs or tubs are small enough, they can easily be emptied, cleaned, and moved among paddocks. Drinking water and salt should be available in all paddocks.

Horses can become dehydrated if they do not have enough good water to drink, if they have been exercising vigorously, if they do not have enough salt available, if no shade is available, if they have diarrhea, and if the temperature is high. These effects will likely be additive. You can test for dehydration by checking whether the pulse rate is elevated and whether a pinched tent of skin along the neck in front of the shoulder takes more than 2 s to retract. In more severe dehydration, a horse may have sunken eyes and a tucked-up appearance to the abdomen (Beckstett, 2017).

Water Sources

Water for horses can be accessed from a variety of sources, including obvious lakes, ponds, streams, and springs. Water seeping out of a hillside, drainage tile, or ditch can be directed through gravel, perhaps in a pipe, then through a geotextile filter fabric to remove fine particles and collected for drinking (OMAFRA, 2000). Surface runoff water can carry contaminants, so if it is directed into a reservoir, contaminants may settle out. Ponds can be dug and enhanced with small dams in areas where water naturally flows. Advice and permission from a local conservation authority or environmental agency may be required to construct a pond. Wild animals often drink at natural water sources (streams, rivers, ponds, and lakes), which farms also used for centuries. Because horses and cattle erode the banks and sometimes cause manure to enter natural water bodies, farm managers are now encouraged to provide drinking water away from natural water bodies.

Well water is the most likely source of water on farms, and it should be tested for quality when major changes to a farm or surrounding farms have taken place. Disturbances to soil such as land-clearing, the start of annual cropping, or spraying chemicals or manure on the land can potentially affect well water quality. A catchment pond at the bottom of a hill can be used to trap runoff water, and windmills or other power sources can be used to run pumps which send this water to drinking troughs (Fig. 15.53).

Moving Water from Source to Pasture

There is no cheaper way to move water than to let gravity do it. Fig. 15.54 shows an example of a gravity-based system that can collect water from a spring or hillside uphill from where you establish a stock tank. A large reservoir can be part of the system if freezing is not a problem. If possible and affordable, draw water from surface sources and drill wells on the highest point of land that is practical, to take advantage of gravity in the distribution of water to pastures. A certain amount of water flow resistance is desirable through water supply pipes, to help prevent sudden pressure changes. Where the grade (slope) is 0.5%−1.0%, the pipe should be at least 1.5 inches (3.81 cm) internal diameter. For grades over 1% a minimum pipe diameter is 1.25 inches. Grades less than 0.2% will not carry water by gravity alone satisfactorily. The grade or slope of the supply pipe should be uniform to prevent air locks from forming. Flexible PE water pipe can be encased in more rigid pipe through sloped areas to keep the slope uniform. PE pipe contracts in cold and expands in warm temperatures, so it should be laid on the pasture on cool days in the early morning. This way, PE pipe is in its most contracted state, and any connections you make will not be pulled apart by further contraction. Junctions to create individual sections of pipe should be fitted with ball valves that can be used to turn off water to those sections and for blowing out the pipes with compressed

FIGURE 15.53   A catchment pond and windmill at a horse farm in Alberta. *Photo by Paul Sharpe.*

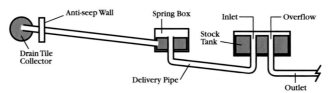

FIGURE 15.54   A gravity-powered watering system. *Illustration used by permission of Publications Ontario.*

air at the end of the grazing season. To facilitate flexible pasture moves and allow water troughs to be moved around, saving forage from trampling damage, quick-connect fittings should be installed in locations under pasture fences for hose connections.

A wide variety of pumps and systems can be used with goals of reliable flow, freedom from freezing, and economy (OMAFRA, 2004). Pumps can be powered by windmills, gasoline or diesel engines, animal power, or electricity. Electrical power can be from an AC source or solar panels. For getting "free" solar power to pump water in remote areas, some innovators have mounted solar panels on carts, wheeled them to a water source (which may be small and seasonal), then used the power to pump water into troughs for livestock (Fig. 15.55).

One example of an animal-powered pump is called a nose pump (Fig. 15.56). Animals push a plunger with their nose, and this causes a rubber diaphragm to move and check valves to prevent backflow. Only one animal can drink at a time, but a nose pump can lift water up to 25 feet (7.6 m).

Hydraulic ram pumps use falling water for power (Fig. 15.57). The weight of the falling water drives a small amount of water above the water source. Water output of 700 to 3000 gallons (2692−11,538 L) per day is possible (OMAFRA, 2000; Hydraulic ram, Wikipedia). Hydraulic ram pumps can be purchased (http://www.theramcompany.com/), and there are several websites with instructions on building your own. Sling pumps are powered by flowing water, so they are suitable for streams with reliable flow, at least 2 cubic ft/s, and a minimum depth of 16 inches. Sling pumps (river pumps) can raise water more than 80 feet (24.4 m). Capacity of sling pumps can reach about 950−3900 gal (4320−17,725 L) per day (https://www.animatedsoftware.com/pumpglos/gl_sling.htm.).

Spring floods may cause problems from high flow rates and debris for hydraulic ram and sling pumps. One manufacturer of both nose and river pumps is Rife Hydraulic Engine Mfg Co (http://www.riferam.com/pumps.html). There is also a frost-free version of the nose pump built for Canadian winters (www.frostfreenosepumps.com/). More information on these systems is available in a booklet put together by the Great Lakes Basin Grazing Network and Michigan State University Extension (Bartlett). Enclosed water tanks on trailers, hauled by motor vehicles or draft animals, is another water transport method that may be suitable while a better system is being financed and built. On rented land or pastures that a land owner does not intend to use frequently, transporting water is probably more economical than installing a pipeline system. Such mobile water tanks can have attached water drinking devices on them (Fig. 15.58), or a hose can be connected to a water trough (Bowser Supply, 2017). The trough can be filled quickly, so the mobile tank can be moved

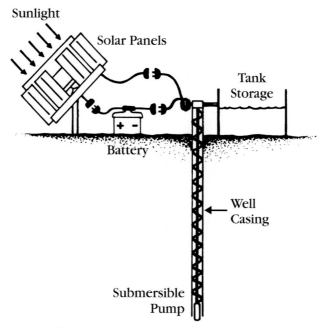

FIGURE 15.55 Schematic of a solar-powered pumping system. *Illustration courtesy of Publications Ontario.*

FIGURE 15.56 A nose pump for livestock to drink one at a time and which requires no other power. *Illustration used by permission of Publications Ontario.*

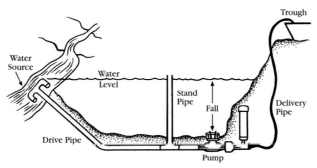

FIGURE 15.57 Hydraulic ram pump. *Illustration courtesy of Publications Ontario.*

FIGURE 15.58 Bowser brand mobile water tank and two station drinker from England, comes in 1000 and 2250 L sizes. *Photo courtesy of Bowser Supply Limited, Rochdale, Lancashire, England.*

away or the tank can stay in place near the trough, delivering water as long as it is needed or until it is nearly empty. Carrying water one or two buckets at a time by human power is still surprisingly common. The advantage is that it provides an opportunity to monitor the amount of water consumed. The disadvantage is the amount of labor required, which can be spent on other aspects of management.

Rules of thumb from Ben Bartlett for intensively grazed pastures: where animals drink individually and the distance from water to the farthest corner of the paddock is < 900 feet (300 m), provide a flow rate that can accommodate all animals in 4–8 h and use a small tank or trough that allows 2%–4% of the herd to drink at once. Where animals are continuously grazed and can walk >900 ft (300 m) from water, provide a water tank holding at least one-fourth of the daily needs of all animals and that accommodates 5%–10% of the herd at once. NRCS (1997) stated that water must be within a quarter mile of the forage-producing site on level to undulating topography to get maximum use of the forage by cattle. On slopes >25%, watering sites should be <600 feet from forage, thus <1200 feet between watering sites. Other barriers to easy walking will cause forage past those distances from water to be grazed lightly if at all (NRCS, 1997). The tank should refill within 1 h. Livestock that are waiting for water become bored, perhaps even irritated, and start to chew and play around with the trough and water supply apparatus, often ending in breakage. Research from Missouri and personal observations from Jim Gerrish show that pastured cattle tend to spend most of their time near water and shade, and thus parts of a paddock that are farthest from water and shade will be grazed the least. Thus, temporary fence and mobile watering devices can encourage animals to graze smaller sections of pasture to obtain better utilization of forage. While there has not been an equivalent amount of research on drinking behavior of horses, the ranges of daily water requirements for pastured beef cattle and horses at temperatures from 40–90°F (4.4–32.2°C) are very similar (NRC, 1996, 2007).

Water pipes (waterlines) can be laid on the surface or buried in the ground. The larger the diameter of pipe that you start with at the source of the water, the less resistance there will be and thus the less friction and loss of pressure and flow rate there will be. Rigid PVC can be used near the source. Where bends, flexibility, and expandability are important, flexible polyethylene water pipe is appropriate, with a minimum pressure rating of 125 psi (862 kpa). Smaller branches to individual paddocks can be supplied with smaller diameter (cheaper) waterlines. Before purchasing, decide where your waterlines will be on the surface and where they will be buried because this determines the grade of plastic chosen for the waterlines. Material that is to be buried does not need to be resistant to ultraviolet rays from the sun and is usually PVC. Waterlines that need to be crossed by machinery and trucks should be buried and protected inside rigid metal pipes. For lying on the ground, a grade of plastic pipe rated for irrigation is suitable because it is stabilized against ultraviolet light and somewhat resistant to freezing.

There is friction between water and the walls of a pipe or hose through which it is flowing. The larger the pipe diameter, the lower is the rate of friction loss. Table 15.4 provides amounts of friction loss per 100 feet (30.5 m) at two water delivery rates. More detailed calculations of friction head loss, for example, using the Hazen–Williams equation, are available through websites (The Engineering Toolbox, 2017).

TABLE 15.4 Friction Loss for 100 Feet of Plastic Pipe (in psi)[a]

Delivery Rate (gal/min)	Pipe Diameter (Multiply × 2.54 for cm)			
	¾ inch	1 inch	1 ¼ inch	1 ½ inch
8	5.8	1.3	0.4	0.2
10	8.4	2.7	0.7	0.3

Pounds/square inch. Divide by 6.89 to convert to kilopascals.
[a]*Bartlett, B. Watering Systems For Grazing Livestock.*

Watering Devices

What sort of vessel should pastured horses drink from? The variety includes the simple (dew on the grass; puddles) to the new and complex (insulated, heated, frost-free, stainless-steel, and plastic manufactured waterers). Feed and farm supply stores carry a variety of the latter. See Information Sources listed later for a list of manufacturers. Helpful hints on selection and installation of pasture waterers can be found in Equisearch (2017). For large numbers of animals, large troughs or open tanks can be used. Open tanks and troughs should have a mechanism to regulate the level of water in them, unless you like standing beside them with a hose while they fill up. There are many different mechanisms available from agricultural and plumbing supply businesses. Some float mechanisms are hidden under a housing that is an integral part of the trough or tank. Since bored animals will sometimes play with and destroy devices attached to the side of a water trough, some design that either hides or blocks access to the water level regulator (often a float attached to a valve) should be included. Since there will be a lot of hoof traffic around them, permanent water tanks are normally installed in the middle of a packed gravel, crushed stone, or concrete pad. Fig. 15.59 shows a permanent, stationary (on demand) waterer on a gravel pad that is typical of many thoroughbred farms near Lexington, Kentucky. Water comes into the device from an underground pipe. Gravel and crushed stone in the surrounding pad will allow water to drain down through them, rather than flow to the edge like it would from concrete. Unless the pad is very big, there is usually some bare soil around the pad, due to hoof action. If the pad is concrete, there is a higher probability that the soil around the pad will be muddy much of the time. Grazing consultant Jim Gerrish has suggested that the pad surface should (1) be about 12 inches (30 cm) above the surrounding ground and (2) have a big enough radius that the largest animals can have all four feet on the pad, back up a couple of steps, then turn around, while still on the pad (Gerrish, 2016). This central part of the pad should be fairly level, and the outer portion should slope down to grade level. So he suggests that the distance from the edge of the watering device to the outer rim of the pad should be about 15 feet (4.6 m). Some people have installed water tanks on bare ground, then piled up crushed rock for a protective pad around them. Gerrish (2016) explained with photos that this can cause livestock to push the rock backward with their feet, causing a circular dam outside the pad. This causes water from spills, overflows, and rain

FIGURE 15.59 A stationary waterer on a gravel pad. *Photo by Paul Sharpe.*

FIGURE 15.60 A round post-shaped stationary waterer on a gravel pad. *Photo by Paul Sharpe.*

to build up and create a ring-shaped mud hole. With a lower than normal height of the tank or trough, animals may be tempted to walk right into it.

For use by smaller numbers of animals, many commercial, permanent, stationary pasture waterers are boxy in shape, and some are columnar or like a round post (Fig. 15.60). These are relatively expensive to buy and install, but they provide many years of reliable use. The simplest and cheapest are neither insulated nor heated. In areas with significant frost, electric heating elements are one solution. Heavily insulated models can work in a relatively frost-free mode. These require water supply lines to be buried below the frost line and may involve ingenious flaps or ball-shaped closures that animals manipulate to reach the water. Some also involve a method for spilled and surplus water to run down into the ground beneath them. Any water lines on the soil surface or buried so shallow that they can freeze should be drained and possibly blown out with compressed air at the end of the grazing season. As water freezes, it expands, potentially causing damage inside pipes, fittings, or valves.

In a rotational grazing system, there should be water in every paddock when horses are present. If the paddocks are relatively large and accommodate about 12 or more horses at a time, permanent waterers may be justified. If many smaller temporary paddocks are used for small numbers of animals and short grazing periods (<5 days), it is probably more practical to use water troughs that are small, light, and portable. A farmer can purchase a trough for every defined grazing paddock or just a few that are emptied when the grazing period is finished for a paddock, then moved to another paddock. Any permanent system of water troughs on pastures should have water lines running under fence lines, either buried or covered by vegetation, with hydrants at intervals corresponding to pairs of paddocks on opposite sides of the internal paddock fences. Many livestock equipment retailers stock water troughs with volumes of 40—400 gallons (about 15—1500 L). The bigger troughs will accommodate a large herd of very thirsty animals. When it is time to clean bigger troughs or to move them to another place so a huge circle of forage is not destroyed around them, they are not easy to deal with. People tend to leave big things alone and not empty and clean them frequently. The more sun that falls on a water trough, the more algae will grow and continue to grow until it is cleaned out. Some farms locate watering devices under a shade structure, to reduce algae growth in the water. A long-handled boot brush, like the ones veterinarians use, is just right for cleaning water troughs. A flexible hose, flowing water, and the ability to tip a trough over or open a drain will facilitate trough cleaning and removal of algae. Some people recommend keeping goldfish in water troughs because the goldfish eat the algae. Pecha.com (2017) suggested that the success of a few goldfish in a 150—300-gallon tank is due to the fish being underfed and underpopulated and that nitrifying bacteria on the tank walls accommodate all of the ammonia produced by just a few fish. Ammonia is volatile and probably does not accumulate to toxic levels in open water tanks. That source suggested that six goldfish in 300 gallons would not be excessive.

There have been many discussions about the pros and cons of putting fish into a livestock watering tank, but peer-reviewed research reports on this practice are (in 2017) as scarce as unicorns. Consider the ecosystems in which wild horses, deer, and other ungulates live. There are many different organisms inhabiting streams and ponds where these ungulates drink. Resources for some organisms may be waste products for others, such as ammonia from fish being used by nitrifying bacteria (Pavlis, 2015). So a water trough or tank can be considered a simple ecosystem. Water flows through it, so it is not stagnant. Some sunlight reaches it, so photosynthetic organisms such as cyanobacteria and algae can flourish.

Many insects will lay eggs in it. Compare the water tank to a more complex ecosystem in a pond, which may be natural or dug for the purpose of capturing runoff water to supply water troughs for livestock. The pond can have some level of various species of bacteria, algae, vascular plants, insects, amphibians, fish, and other living creatures. The health of ponds and streams is often analyzed by aquatic biologists who collect and survey the bottom-dwelling (benthic) organisms. The greater the number and diversity of benthic organisms, the healthier the ecosystem. Thus I do not think that pasture managers need to worry about the manure from a few goldfish in their horse water trough.

If you were planning a pond to enhance your garden, an ecologic mindset would help you match the fish species to the region, pond size, and aquatic plants. The fish species that you select could include fathead minnows from the American northeast, which tolerate low oxygen levels and high nutrient levels, if the pond has little turnover of water. Natural-colored fish tend to be better camouflaged than goldfish and koi, so they are less vulnerable to predation by raccoons, otters, and birds such as eagles, ospreys, and herons. Fatheads eat mosquitoes, insect larvae, and algae. Johnny darter minnows live on the bottom, eating snails and insect larvae and preferring cool, clean water. For southern ponds, pygmy killifish, golden topminnow, and banded pygmy sunfish are all insectivores. For ponds with a surplus of aquatic weeds, sterile grass carp (originally from China) can reduce the weed population. Some states require a permit for their release (Tregaskis, 2017). To prevent exotic fish from getting into natural waterways where they have no other species that controls their numbers, only native fish species should be used in water troughs near drainage ditches and creeks. Four species of Asian carp that were introduced to fish farms in the American south to control snails have escaped and become very numerous in the Mississippi River watershed. They are causing disruption of native aquatic ecosystems. In her article on fish for garden ponds, Tregaskis (2017) suggested that one can establish a resilient ecosystem by mixing and matching appropriate fish and plants in a simple aquatic environment. Thus, a stock tank with a few native fish to control algae and mosquitoes seems like a reasonable idea. Guppies have also been suggested to help goldfish control algae and mosquito larvae in water troughs, but since they are native to South America, they are probably not cold tolerant. On the other hand, they should probably not be used in warm areas because of the risk of them getting into a waterway and reproducing.

A theory I would like to see tested is that small (<50 gallon or 192 L) water troughs could accommodate groups of about four to 12 horses in rotationally grazed pastures. Each adult riding horse will need about 12 gallons (46 L) of water/day in hot weather. Water flow rates should be enough to refill troughs within half an hour (12 gal/30 min equals only 0.4 gal/min). So a 50-gallon trough could accommodate four horses drinking their whole daily water demand, then another four horses, once the trough is refilled. Of course, horses will not wait around for troughs to fill. As mentioned before, they drink in short intervals of perhaps 15 s about 10 times per day to acquire their 12 gallons, thus taking about 1.2 gallons each time. If horses are awake for 16 h, then on average, each horse needs 1.2 gallons about every 1.6 h. Twelve horses will need $12 \times 1.2 = 14.4$ gallons every 1.6 h. To refill 50 gallons in 30 min requires a flow rate of $50/30 = 1.67$ gallons/minute. A spreadsheet for calculating water flow rate, assuming PVC pipe and a pressure drop of 20 psi along a pipe 100 ft in length, shows that pipe with 0.75-inch interior diameter will carry 13 gallons per minute. A 1-inch pipe will deliver 28 gal/min (108 L/min) and a 1.5-inch pipe will deliver 81 gal/min (312 L/min) (Bengston, H., 2011). Thus a 50-gallon (192 L) trough with a 1-inch hose is theoretically more than big enough to accommodate 12 horses. The booklet by Bartlett contains some tables and calculations that are useful in determining your pasture's water supply needs. It is possible to follow recommended calculations and determine what diameter of water pipe you should have over required distances and determine what volume of water trough you should have, given the number of horses drinking from it. If your needs are large or unusual, you should probably consult an agricultural engineer, since the calculations require knowledge of water pressure changes and can become complex. Further recommendations for managing pipelines and troughs and hauling water for livestock can be found at AAFC (2015).

Fig. 15.61 shows a popular stock tank of about 100 US gallons (378 L) that is 53 inches long and 25 inches tall. The 50-gallon (189 L) version is about the same length but only half the height. At 8.34 lbs/US gallon, 50 gallons of water weighs 417 lbs (189 kg), which takes some effort to push over to empty. Drain plug kits can be inserted into these tanks to facilitate emptying. The float valve pictured is from a different manufacturer than the tank.

Homemade water supply systems are very common among beef cattle farmers and ranchers. Often, these systems accommodate large numbers of animals with minimal costs in materials and energy. Steve Kenyon, an Alberta rancher who also writes articles for *The Stockman Grass Farmer* and *Canadian Cattleman*, describes a winter watering system he uses for his horses and chickens at home and for herds of custom-grazed cattle on many different farms. He creates a continuous-flow water system between a water trough and a water source. The water source may be a well or a dug pond called a "dugout" in the Canadian prairies. For convenience, some pasture managers dig a hole close to the pond or dugout and insert a culvert vertically, creating a "wet well" that will contain ground water at the same level as the surface of the pond. For a power source, he uses 110 V AC if it is available and, if not, a small generator (2000−4000 W) or a

FIGURE 15.61 A livestock water trough with a water inlet and a float valve to control the water level, of about 100 US gallon capacity, which for some people may be too large for easy emptying and moving. *Photo courtesy of Gallagher North America.*

solar-powered battery charger and a small pump run by a DC motor. In the case of a dugout, he cuts a 1.5 × 1.5-foot (46 × 46 cm) hole in the ice about 15 feet from shore where there is a reasonable water depth, places two big square bales of straw over the hole and squeezes a 1.5-inch or larger water line between the bales and into the water. The water line has a submersible pump on the end of it, and the electrical cord from the pump must reach out of the water, through the ice, and up to the bales where it is tied and connected to an extension cord. The water line runs onto the shore and up a slight slope to a water trough, which also has big square bales around it for insulation. The trough is on a slight slope, so overflow water can run down to the dugout and into another hole in the ice, close to its edge. The continuous slow flow prevents the system freezing in most winter weather, northwest of Edmonton, Alberta. Since things can go wrong, Steve also runs an electric heat tape along the water line for a short distance where it enters the ice and water. If the line does freeze, applying power to the heat tape for 15−20 min usually thaws it. Steve lives off the power grid, so he watches the drain on his power resources carefully. He has learned that it is far cheaper to pump water in a continuous flow system than to heat the same amount of water. The flow rate can be just a trickle. Steve also plans to use a solar-powered, motion-controlled waterer. It does not require a heat source, regardless of how few animals use it. The submersible pump only operates when the motion detector indicates presence of an animal. The bowl drains dry after the last animal leaves. Water in the casing stays below the frost line. The water source can be from a pond or a pressurized line.

Good descriptions of pasture watering systems for cattle are provided by NRCS (2006). If you can bear in mind the differences in how horses and beef cattle are managed, there is a lot of useful information there. For great pictures and descriptions of both how to and how not to set up permanent stock water troughs, see Gerrish (2016). For a description, design, and performance information on a homemade portable watering system, see Blomfield (2014).

Dr. Clair Thunes (2017) compiled a very useful list of water trough maintenance hints for cold winter weather for readers of TheHorse.com:

1. Locate your trough for sun exposure.
2. Insulate your trough. Styrofoam board and/or foil-covered insulation works well and can be wrapped around the outside of the trough.
3. Place a float in the trough to keep water moving as it bobs about, making it harder to freeze.
4. Bury your trough
5. Heat the trough. There are several options including battery, electric, or propane heaters. But before trying these, you could try putting manure under your trough. Composting manure generates heat, and the thinking is that if you have a several inches thick layer of manure under your trough, as it breaks down, it will help warm the trough.

Here are some additional tips to prevent water systems from freezing. Keep a trough out of the wind. Hay and straw also work as insulation, in addition to construction materials. As for digging a hole and sinking a trough into the ground, it makes use of heat from the ground. Several commercial watering systems require water lines to be buried below frost line, then have an insulated column containing the water line running vertically to the trough.

When heating your trough via compost, to maintain a reasonably high composting temperature, some mixing and aerating of the compost will help.

Electric heaters that are approved by UL or the CSA and used as directed are usually a safe, but expensive way to prevent water from freezing. In some situations a heating device combined with shelter and insulation is suitable.

Keep water flowing, similar to Steve Kenyon's system described earlier. Water that runs through such a system needs to flow to a place that will not recycle contaminants from the trough into the water source, if that source is potable. Ideally, water that overflows or runs through a system would be of a low volume and directed underground or into a pond.

INFORMATION SOURCES

Manufacturers of Electric Fencing

Gallagher https://www.gallagherfence.net/
Patriot https://www.patriotglobal.com/
Powerflex https://www.powerflexfence.com/
Premier https://www.premier1supplies.com/
Speedrite https://www.speedrite.com/
Timeless http://www.plastic-innovation.com/
Tru-Test https://www.tru-testgroup.com/en/electric-fencing-solutions.

Manufacturers of Waterers

Bowser Supply (England) http://www.bowsersupply.co.uk/water-bowsers/water-bowser/
Doug's Waterers (http://horsewaterer.com/)
Drinking Post (https://www.dpwaterer.com/)
Jug waterers http://jugwaterers.com/
Nelson (https://www.nelsonmfg.com/horse-equipment/)
Petersen (http://www.petersenwaterers.com/secure-petersenwaterers/Scripts/default.asp)
Ritchie (https://ritchiefount.com/product-category/horses/)
Kelln Solar (http://www.kellnsolar.com)

REVIEW QUESTIONS

1. Describe situations and locations where each of these three types of fencing is appropriate on horse pastures: (a) permanent, nonelectric; semipermanent, electric; temporary, electric.
2. List three necessary features of a perimeter fence around a horse pasture.
3. List desirable features of perimeter fences, in addition to the necessary features mentioned before.
4. Which predators have been recorded as sometimes killing horses in North America?
5. When planning fences, extra space should be allocated for several things. List four.
6. Describe three common problems with electric fence function, and explain how you can prevent and/or fix each problem.
7. Discuss the types of electric fence wire that may be used to confine horses to a pasture, including the issues of conductivity, longevity, and visibility. Wire types include (a) bare galvanized steel, (b) thin polymer string, (c) thicker polymer rope, (d) polymer tape of varying widths, and (e) steel coated with a conductive polymer.
8. Describe some recommendations for the height, width, and location of gates in horse pastures.
9. Why should a person not put on rubber gloves and handle live electric fence wire?
10. If you were to begin a job managing a horse farm, how would you determine whether the water quality was suitable for horses?
11. Describe several desirable features of a watering system for horses on pasture.

12. Describe the relative advantages and disadvantages of permanent versus temporary portable watering devices for horses on pasture.

13. This question is a self-learning exercise. The next time you are looking at a fence on a farm, regardless of the types of animals there, make a mental list of the weakest points in the fence, where breaks and maintenance needs are most likely to occur. Decide how you would fix them.

REFERENCES

AAFC (Agriculture and Agri-Food Canada). Remote Watering Sites Far from Water Source. 2015. http://www.agr.gc.ca/eng/science-and-innovation/agricultural-practices/water/livestock-watering/water-systems-for-range-livestock/remote-watering-sites-far-from-water-source/?id=1371147693608.

Andrews J. How to build the perfect livestock fence. Farmers Wkly. August 28, 2015;164(4):60−2.

ANZECC (Australian and New Zealand Environment and Conservation Council). Livestock drinking water guidelines. In: Guidelines for Fresh and Marine Water Quality, vol. 3; 2000. p. 1−32.

Bading K, Lenz R. Proper Fence Brace Construction. The Cattleman Magazine. Texas, Southwestern Cattle Raisers Association; 2017. http://tscra.org/what-we-do/the-cattleman-magazine/ranch-diy-proper-fence-brace-construction/.

Bartlett, B. Watering Systems for Grazing Livestock. Great Lakes Basin Grazing Network and Michigan State University Extension. 1−24. https://efotg.sc.egov.usda.gov/references/public/CT/WateringSystemsforGrazingLivestockBook.pdf.

Beckstett A. Horse Hydration FAQs. 2017. theHorse.com. https://www.thehorse.com/149081/horse-hydration-faqs/.

Bengston H. Excel formulas to Calculate Water Flow Rates for Different Pipe Sizes. Bright Hub Engineering; 2011. http://www.brighthubengineering.com/hydraulics-civil-engineering/73748-excel-formulas-to-calculate-water-flow-rates-for-different-pipe-sizes/.

Blake LW, Gese EM. Cougar predation rates and prey composition in the Pryor Mountains of Wyoming and Montana. Northwest Sci. 2016;90(4):394−410.

Blomfield K. Water Systems for Livestock. The Conscious Farmer; 2014. http://www.theconsciousfarmer.com/stock-water-options/.

Brown L. Pasture Fencing for Horses. British Columbia Ministry of Agriculture and Food; 2015. p. 260−3. Factsheet Order No. 307, www2.gov.bc.ca/.../farming.../farm.../307260-3_pasture_fencing_for_horses.pdf.

Buick W. TB in domestic species other than cattle and badgers. Government Vet. J. 2006;16(1):87−91. http://webarchive.nationalarchives.gov.uk/20110318125545/http://www.defra.gov.uk/gvs/publications/gvj/pdf/gvj-vol1601.pdf.

Burt A. 7 Common Cattle Fencing Mistakes. Beef Magazine; 2015. http://www.beefmagazine.com/pasture-range/grazing-programs/0301-common-fencing-mistakes.

Canadian Horse Defense Coalition. Inventory of Horses. 2016. https://canadianhorsedefencecoalition.files.wordpress.com/2016/07/horsesupply_june-2016-english.png.

Crowell-Davis SL, Houpt KA, Carnevale J. Feeding and drinking behavior of mares and foals with free access to pasture and water. J. Anim. Sci. 1985;60:883−9.

Difference Between. Difference between HDPE and PVC. 2017. http://www.difference between.net/object/difference-between-hdpe-and-pvc/.

Duncan G. Stray Voltage and Electric Fencing. PasturePro blog; 2011. http://www.pasturepro.com/blog/2011/06/stray-voltage-and-electric-fencing/.

Equisearch. Automatic Waterers Provide Fresh Water for Your Horse. 2017. https://www.equisearch.com/articles/automatic-waterers-provide-fresh-water-your-horse.

Fabian. Fence Planning for Horses. Penn State Extension; 2017. http://extension.psu.edu/animals/equine/horse-facilities/fence-planning-horses.

Ferris J. Fencing Fundamentals: Posts & Braces. Canadian Horse Journal 2015-08-11; 2015. https://www.horsejournals.com/fencing-fundamentals-posts-braces.

Gallagher Electric Fencing Manual. https://gallagherelectricfencing.com/blogs/news/free-gallagher-power-electric-fence-manual#temp_created_link; 2017.

Gay SW, Heidel RD. Constructing High-tensile Wire Fences. VCE Publications/442/442-132. Virginia Tech; 2009. https://pubs.ext.vt.edu/442/442-132/442/132.html.

Gerrish J. How to Set up Good Stock Tanks. On Pasture; May 16, 2016. http://onpasture.com/2016/05/16/how-to-set-up-good-stock-tanks/.

Glauser A, Burger D, Van Dorland HA, Gygax L, Bachmann I, Howald M, Bruckmaier RM. No increased stress response in horses on small and electrically fenced paddocks. Appl. Anim. Behav. Sci. 2015;167:27−34.

Gourlay J. Choosing fencing. In: Stable Management; November 20, 2011. https://stablemanagement.com/articles/choosing-fencing.

Greene, E.A., 2007. Tools to help horse owners deal with muddy high-traffic areas. J. Ext. 45 (6), 71−74. https://www.joe.org/joe/2007/december/tt7.php. Also https://www.uvm.edu/~susagctr/resources/EquineGreenerPasturesColor.pdf.

HorseFencing.net. Horse Fencing: 11 Options and what to Consider when Buying. 2016. https://www.horsefencing.net/blogs/horse-fencing/17814567-horse-fencing-11-options-what-to-consider-when-buying.

Kammel DW, Pedersen JW. Preservative Treated Wood for Farm and Home. Agricultural Engineer's Digest. AED-30; 1989. https://www-mwps.sws.iastate.edu/sites/default/files/imported/free/aed_30.pdf.

Kencove.com. Volts Vs. Amps and Other Electric Fence Lingo. 2017. http://blog.kencove.com/volts-vs-amps-and-other-electric-fence-lingo/#more-2143.

Kencove.com. Volts Vs. Joules. 2017. http://www.kencove.com/fence/7_Volts+vs.+Joules_resource.php.

Marble J. Paddock Design and Stockmanship − Part 1. On Pasture; 2017. http://onpasture.com/2017/07/24/paddock-design-and-stockmanship-part-1/.

McDonnell SM, Freeman DA, Cymbaluk NF, Schott HC, Hinchcliff K, Kyle B. Behavior of stabled horses provided continuous or intermittent access to drinking water. Am. J. Vet. Res. 1999;60:1451−6.

McLeese JM, Patience JF, Wolynetz MS, Christison GI. Evaluation of the quality of the ground water supplies used on Saskatchewan swine farms. Can. J. Anim. Sci. 1991;71:191−203.

Mechanisms C. What Is Polyvinyl Chloride (PVC) and what Is it Used for? Everything You Need to Know about PVC Plastic. 2016. https://www.creativemechanisms.com/blog/everything-you-need-to-know-about-pvc-plastic.

Moors E, Crönert D, Gauly M. Paddocknutzung des Pferdes in Abhängigkeit von der Umzäunungstechnik. Züuchtungskde 2010;82:354−62.

Morgan SE. Water quality for cattle. Vet. Clin. N. Am. Food Anim. Pract. 2011;27(2):285−95.

NAHMS (National Animal Health Monitoring Service). NAHMS Equine Studies. 2017. Modified October 26, 2017, https://www.aphis.usda.gov/aphis/ourfocus/animalhealth/monitoring-and-surveillance/nahms/nahms_equine_studies.

NAHMS. Death Loss in U.S. Cattle and Calves Due to Predator and Nonpredator Causes, vol. 2015; 2017. https://www.aphis.usda.gov/animal_health/nahms/general/downloads/cattle_calves_deathloss_2015.pdf.

NRC (National Research Council). Nutrients and Toxic Substances in Water for Livestock and Poultry. Washington, D.C.: National Academy Press; 1974. p. 93.

NRC (National Research Council). Nutrient requirements of beef cattle. Seventh Revised Edition. Washington, D.C.: National Academy Press; 1996. p. 81.

NRC (National Research Council). Water and water quality. In: Nutrient Requirements of Horses 6th Revised Edition. Washington, D.C.: The National Academies Press; 2007. p. 132.

NRCS Natural Resources Conservation Service. Chapter 5 Management of Grazing Lands. in: National Range and Pasture Handbook. United States Department of Agriculture; 1997. https://www.nrcs.usda.gov/Internet/FSE_DOCUMENTS/16/nrcs143_023884.pdf.

NRCS Natural Resources Conservation Service. Watering Systems for Serious Grazers. United States Department of Agriculture; 2006. p. 1−40. https://prod.nrcs.usda.gov/Internet/FSE_DOCUMENTS/stelprdb1144213.pdf.

Olkowski AA. Livestock Water Quality, a Field Guide for Cattle, Horses, Poultry and Swine. 2009. http://www5.agr.gc.ca/resources/prod/doc/terr/pdf/lwq_guide_e.pdf.

OMAFRA (Ontario Ministry of Agriculture Food, Rural Affairs). Water management on pastures. In: Publication 19 Pasture Production; 2000. p. 45. http://www.omafra.gov.on.ca/english/crops/pub19/Publication19.pdf.

OMAFRA. Alternative Livestock Watering Systems. 2004. Revised 2015, http://www.omafra.gov.on.ca/english/engineer/facts/04-027.htm.

Page D. Understanding wire gauges used in welded and woven wire mesh and fence. In: The Fence Post. Louis E. Page Inc; 2009. http://www.louispage.com/blog/bid/7075/understanding-wire-gauges-used-in-welded-woven-wire-mesh-fence.

Pavlis R. DIY natural backyard pond. Mother Earth News October/November 2015. Issue 272, http://www.motherearthnews.com/diy/garden-and-yard/backyard-pond-zm0z15onzmar.

Petcha.com. Goldfish in Cattle Troughs. 2017. https://www.petcha.com/goldfish-in-cattle-troughs/.

Robertson Gt, Mowrey RA. Fences for Horses. North Carolina Cooperative Extension Service; 2005. https://equinehusbandry.ces.ncsu.edu/equine-factsheets/.

Saskatchewan. Fencing Costs. Government of Saskatchewan; 2015. https://www.saskatchewan.ca/business/agriculture-natural-resources-and-industry/agribusiness-farmers-and-ranchers/livestock/cattle-poultry-and-other-livestock/cattle/fencing-costs.

Sevenster A. PVC's physical properties. In: PVC; 2016. http://www.pvc.org/en/p/pvcs-physical-properties.

Smith Thomas H. Understanding Equine Insurance Policies. The; 2017. Horse.com. http://www.thehorse.com/articles/38048/understanding-equine-insurance-policies?utm_source=Newsletter&utm_medium=farm-barn&utm_campaign=08-13-2017.

Statistics Canada. Inventory of Sheep and Lambs. 2016. http://www5.statcan.gc.ca/cansim/a26?lang=eng&id=0030094&p2=33.

The Engineering Toolbox. Hazen-Williams Equation − Calculate Head Loss in Water Pipes. 2017. http://www.engineeringtoolbox.com/hazen-williams-water-d_797.html.

Thoen CO. Tuberculosis in Horses. Merck Veterinary Manual. 2016. https://www.merckvetmanual.com/generalized-conditions/tuberculosis-and-other-mycobacterial-infections/tuberculosis-in-horses.

Thunes C. Keeping Water Troughs Thawed with or Without a Heater. December 17, 2017. TheHorse.com. https://thehorse.com/137146/keeping-water-troughs-thawed-with-or-without-a-heater/.

Tregaskis S. Fish for garden ponds. Org. Gard. 2017;58(5). August/September 2011, https://www.rodalesorganiclife.com/garden/fish-garden-ponds.

Turner JW, Morrison ML. Influence of predation by mountain lions on numbers and survivorship of a feral horse population. Southwest. Nat. 2001;46(2):183−90.

Vitale MJ. Fence Law: The Law of Fences. 2009. http://www.duhaime.org/LegalResources/RealEstateTenancy/LawArticle-670/Fence-Law-The-Law-of-Fences.aspx.

Wheeler EF, Koenig B, Harmon J, Murphy P, Freeman D. Horse Facilities Handbook. MidWest Plan Service: Iowa State University; 2005. MWPS-60.

Williams S, Horner J, Orton E, Green M, McMullen S, Mobasheri A, Freeman SL. Water intake, faecal output and intestinal motility in horses moved from pasture to a stabled management regime with controlled exercise. Equine Vet. J. 2015;47(1):96−100.

WindRiver Fence. Horse Fencing 101. 2017. InfoHorse.Com. http://www.infohorse.com/horse_fencing_101.asp.

Worley JW, Heusner G. Fences for Horses. University of Georgia Extension; 2015. Bulletin 1192, extension.uga.edu/publications/detail.html?number=B1192.

Wright CL. Management of water quality for beef cattle. Vet. Clin. N. Am. Food Anim. Pract. 2011;23.1:91−103.

Chapter 16

Pasture-Related Diseases and Disorders

Bridgett McIntosh[1], Tania Cubitt[2] and Sherrene Kevan[3]

[1]*Virginia Tech, Department of Animal and Poultry Sciences, Middleburg, VA, United States;* [2]*Performance Horse Nutrition, Jeffersonton, VA, United States;* [3]*Enviroquest Ltd., Cambridge, ON, Canada*

INTRODUCTION

While pastures provide the most natural and healthy environment for horses, diseases and disorders related to the animals, plants, or management may occur. The quality and quantity of forage in pastures can contribute to the incidence and severity of some equine diseases. Poor management of pastures can contribute to metabolic disorders in horses such as laminitis, Cushing's disease, pituitary pars intermedia dysfunction, and insulin resistance (IR). Poor pasture management can also increase populations of certain weeds that when ingested are toxic to horses. Toxins that adversely affect horses can be found in some woody plants, forbs (herbaceous, broad-leaved, nonwoody, nonleguminous plants), legumes, and some grasses, leading to sickness and even death. Most of the problems associated with horses on pasture can be avoided by knowing the risk factors involved. Pasture-related diseases and disorders affect a variety of physiologic systems of a horse, as shown in Table 16.1.

COLIC

Colic is not a disease, but rather a generic term that refers to a combination of signs that indicate abdominal pain in the horse, which may be due to an accumulation of gas, fluid, or impacted solids. Colic can involve irritation of the intestinal mucosa, blood supply alterations, or torsion of the intestines. Pasture-related risk factors for colic include decreased access to quality pasture, abrupt changes in access to pasture, and seasonal changes in pasture nutrients. Many of the conditions that cause colic can become life threatening in a relatively short period of time. Increased concentrate intake and a change in hay source are known to make colic more likely. With decreased levels of water intake, the risk of colic increases, especially impaction colic.

A major problem for horse owners is being able to identify the signs of colic, which can vary greatly among individuals and may also depend on the severity of the pain. The following signs of colic are common:

- turning the head toward the flank
- pawing
- kicking or biting at the abdomen
- stretching out as if to urinate
- repeatedly lying down and getting up, or attempting to do so
- lack of appetite
- lack of bowel movements/few or no droppings
- absence of, or reduced, digestive sounds.

When a horse is in more severe pain the aforementioned symptoms will be apparent, along with sweating, rapid respiration, elevated pulse rate, and depression. A colicky horse may be dehydrated, and this should be suspected if its gums are tacky rather than moist to the touch and if the skin, pinched into a tent, takes more than a second to flatten back to normal appearance.

A horse with colic should be kept as calm and comfortable as possible while waiting for the veterinarian. Allowing the animal to lie down is acceptable if it appears to be resting and is not at risk of injury. Sometimes the old adage of walking

TABLE 16.1 Physiologic Systems of the Horse Affected by Pasture-Related Diseases

Primary System Affected	Disease or Disorder
Digestive	Colic, pasture-associated laminitis, selenium toxicity
Endocrine	Pasture-associated laminitis
Respiratory	Pasture-associated obstructive pulmonary disease, selenium toxicity
Nervous	Equine grass sickness, various toxicities
Musculoskeletal	Seasonal pasture myopathy, pasture-associated laminitis, selenium deficiency
Circulatory	Nitrate poisoning
Integumentary	Laminitis, photosensitization from plant toxins

them can inflict more harm than good. There is also a risk of injury to the handler when walking a horse with colic. Conversely, if the horse is rolling or behaving violently, attempting to walk the horse slowly may be beneficial. It is important to not administer drugs unless specifically directed to do so by your veterinarian.

Managers and owners can assess a horse's general status by listening to its heart rate (normal 28 to 45 bpm) and its gut sounds, assessing capillary refill time, and its skin pinch return time. If a horse's heart rate is persistently high (i.e., above 75 bpm), colic is considered severe. Complete absence of gut sounds also indicates a more severe type of colic. Mucous membrane color and capillary refill times also indicate severity. Horses with purple mucous membranes may have compromised circulation. Feed should be withheld from the horse until further instruction from the veterinarian. Follow your veterinarian's advice exactly while awaiting his or her arrival.

When the veterinarian arrives, he/she will establish the severity of the colic and identify its cause. Examination and treatment may consist of the following procedures:

- obtaining a history of the horse
- observation of behavior, sweating, abdominal distension, rapid breathing, flared nostrils
- monitoring vital signs, including temperature, pulse, respiration, color of the mucous membranes and capillary refill time
- rectal palpation for evidence of intestinal blockage, distension, or other abnormalities
- fecal samples can be examined for their firmness, and some feces may be mixed with water in a plastic bag to observe whether sand settles out.
- analgesics or sedatives to relieve pain and distress
- laxatives to help reestablish normal intestinal function
- passage of a stomach tube to determine presence of excess gas, fluids, and ingesta
- abdominal tap to evaluate protein level and cell type in the peritoneal fluid
- continued observation to determine response to treatment
- possible surgery referral
- administration of fluids and pain-controlling medication.

Only by quickly and accurately recognizing colic (and seeking qualified veterinary help) can the chance for recovery be maximized (Tinker et al., 1997; Hillyer et al., 2002; Scantlebury et al., 2015).

Measures to prevent colic include the following:

- Establish a set daily routine, including feeding and exercise schedules.
- Feed a high-quality diet comprised primarily of roughage.
- Avoid feeding excessive grain and energy-dense supplements.
- At least half the horse's DMI should be supplied through hay or forage, which should comprise 1.25% of body weight daily. Divide daily concentrate rations into at least two feedings.
- Follow a regular parasite control program.
- Provide exercise and/or turnout on a daily basis.
- Provide fresh, clean water at all times.
- Avoid putting feed on the ground, especially in sandy soils.

- Make dietary and other management changes as gradually as possible.
- Reduce stress.
- Maintain accurate records of management, feeding practices, and health

Pasture management is important in preventing colic and includes avoiding overgrazing and intakes of forage with high sugar contents. Maximizing intake of pasture and other forages while minimizing grain intake helps to avoid colic. Good pasture management to reduce colic involves slow changes to diets and turnout times. Good quality hay and lots of water should be available when grazing is no longer an option. Reducing the incidence of horses getting sand colic is mainly by preventing horses from eating on sandy surfaces. Feeding grain in buckets, having enough water, not allowing horses access to pastures that are overgrazed or sandy, and feeding high-fiber hay are recommended (Rose and Hodgson, 2000).

PASTURE-ASSOCIATED LAMINITIS

Pasture-associated laminitis (PAL) is associated with elevated intakes of nonstructural carbohydrates (NSC), which can cause acute digestive disturbances via rapid fermentation and chronic equine metabolic disorders, including IR (Menzies-Gow et al. 2016; Frank et al., 2010). Ingestion or physical contact with parts of black walnut trees or their wood can also lead to laminitis. A notable feature of PAL is that within.a population of horses or ponies, some individuals are susceptible to recurrent episodes of laminitis, while others grazing the same pasture remain healthy, suggesting that additional factors contribute to laminitis risk. A tendency to have elevated blood insulin and glucose concentrations has been associated with PAL cases in several studies (Karikoski et al., 2011; Pleasant et al., 2013; Menzies-Gow et al., 2016). Laminitis is also an inflammatory disease that culminates in the hoof, where eventual failure of the laminar tissue results in rotation and sinking of the coffin bone. Clinical signs of laminitis include increased digital pulse, lameness, shifting weight to the hind limbs while stretching forelimbs forward (Fig. 16.1), reluctance to walk or pick up the feet, plus heat and inflammation of the lower leg and hoof.

Obesity, hyperinsulinemia, and IR are involved in a predisposition for laminitis that may interact with NSC intake during certain times of the year (Pleasant et al., 2013). Fig. 16.2 shows a horse with significant obesity, including a slight dip where the neck joins the withers, the crest with a sausage-like fat deposit, and poor definition of the shoulders. The annual incidence of laminitis in the United States is reported to be 2%, but rises to 5% in the spring and summer, and nearly half of all reported cases of laminitis in the United States occur in animals at pasture (Kane et al., 2000; USDA, 2000). Elevated NSC concentrations in temperate pastures are typically observed in the spring and autumn when environmental conditions favor rapid growth of cool season forages. Concentrations of NSC are linked to sunlight, where they are reported to be lower in the overnight and early morning hours, and increase throughout the day, with the most elevated concentrations occurring in the late afternoon and early evening. Mean insulin concentrations in grazing horses displayed a circadian pattern that correlated to NSC levels (Fig. 16.3) (McIntosh, 2006). Pasture NSC content and alterations in carbohydrate metabolism and digestion increase the risk of laminitis via exacerbation of IR. Strategies for management

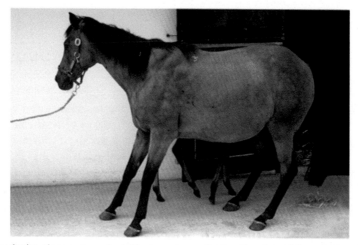

FIGURE 16.1 A horse showing obesity, plus a common laminitis stance, in an effort to reduce the weight on the front feet. These are typical features of pasture-associated laminitis.

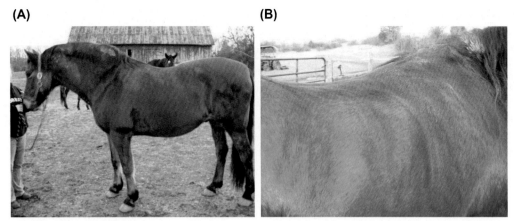

FIGURE 16.2 A horse exhibiting obesity, a fat-filled neck that crests, and regional fat deposits, linked to pasture-associated laminitis; (A) a side view of the whole horse; (B) a close-up of the topline.

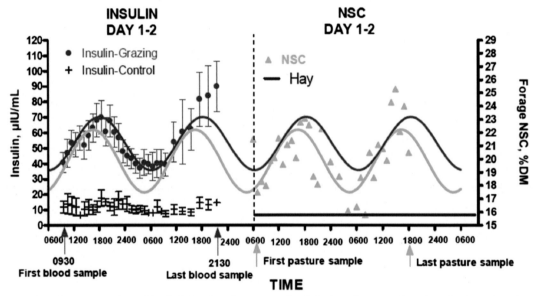

FIGURE 16.3 Fluctuating concentrations of insulin in horses and nonstructural carbohydrates in pasture throughout the day. In a study conducted in Virginia in a cool season grass pasture, NSC was greatest in April with lowest concentrations at 6:00 a.m. and most elevated at 6:00 p.m. Mean insulin concentrations in grazing horses displayed a circadian pattern that correlated to NSC levels in the pasture ($r^2 = .601$, $P = .008$).

practices are needed to decrease intakes of pasture NSC by horses at risk of developing metabolic disorders. Susceptible horses can be identified through oral glucose tests (Meier et al., 2018).

Restricted access to pasture, especially when NSC concentrations are elevated, is the leading recommendation for reducing the risk of PAL (Longland and Byrd, 2006). The use of grazing muzzles is also effective at reducing overall pasture intake, and thus NSC (Longland et al., 2016). Risk for elevated intakes of NSC can also be modulated through forage species selection and management. Warm season grasses are often lower in NSC than cool season grasses, but time of day and stage of maturity also play a role in NSC accumulation in grasses. While pastures that are managed for optimum production are ideal for most horses, they can be a problem for at-risk horses. Determining if a horse is truly at risk for digestive and metabolic disorders should validate concerns for PAL.

Carefully managing pasture turnout and forage and grain intake in horses and ponies that are at risk for developing laminitis or are currently affected should help to reduce the incidence of PAL. We also understand that horses suffering from IR and/or Cushing disease, as well as horses and ponies with the "easy keeper" phenotype that are often overweight or obese, and may be persistently hyperinsulinemic, should also be managed carefully with regard to their carbohydrate intake. The following points summarize current advice regarding strategies for avoiding high NSC intakes by horses and ponies at risk for pasture laminitis:

- Animals predisposed to laminitis should be denied access to grass pastures, particularly during the spring.
- At other times of the year, limit the amount of turnout time each day (e.g., 1–3 h) and turn horses out late at night (after 10:00 p.m.) or early in the morning, removing them from pasture by midmorning at the latest (before 10:00 a.m., because NSC levels are likely to be at their lowest late at night through early morning).
- Alternatively, limit the size of the available pasture by use of temporary fencing to create small paddocks or use a grazing muzzle.
- Do not turn horses out onto pasture that has been exposed to low temperatures in conjunction with bright sunlight, such as occurs in the autumn after a flush of growth or on bright, cool winter days, because cold temperatures reduce grass growth, resulting in the accumulation of NSC.

PASTURE-ASSOCIATED OBSTRUCTIVE PULMONARY DISEASE

Equine asthma is similar to human asthma and is commonly referred to as equine recurrent airway obstruction (RAO). Recently, RAO, because it is similar to asthma, has been referred to as severe equine asthma (Costa et al., 2016). Severe equine asthma or RAO diseases are based on similarities of bronchoconstriction or narrowing of airways, where the horse secretes excessive mucus and its airways are obstructed. Airway obstruction makes it difficult for the horse to breathe and also causes coughing and wheezing. Factors that cause RAOs are inefficiency of lung clearance, genetic predisposition, and environmental exposure to excessive levels of "respirable dust particles" such as aeroallergens (i.e., pollen and mold spores), which often can be found in moldy hay and pastures. Pasture-associated obstructive pulmonary disease (PAOPD), along with its causes, symptoms, and methods of prevention, can be divided into two forms of RAO:

1. hay-associated RAO, also known as chronic obstructive pulmonary disease (COPD), chronic airway disease, broken wind, heaves, chronic obstructive lung disease, or chronic bronchitis
2. summer pasture-associated obstructive pulmonary disease (SPAOPD).

CHRONIC OBSTRUCTIVE PULMONARY DISEASE

COPD is a severe form of equine asthma that occurs after a horse has been exposed to high levels of dust particles or aeroallergens when fed moldy hay, whether the horse lives indoors or on pasture. Clinical signs of COPD are the same as described previously: airway obstruction, mucus oversecretion, heavy breathing, coughing, and wheezing. Management of this disease requires that the horse has good quality hay (e.g., alfalfa hay vs. grass hay) and that it is checked daily for mold. The most common mold spores in hay are *Thermoactinomyces vulgaris*, *Microspolyspora faeni*, and *Aspergilus fumigatus*. Pasture management to reduce dust and exposure to allergens is critical, especially during times of the year when temperature and humidity are high. A COPD horse may have to be removed from pasture until late fall and winter.

SUMMER PASTURE-ASSOCIATED OBSTRUCTIVE PULMONARY DISEASE

SPAOPD, also referred to as heaves, summer heaves, summer pasture-associated heaves, pasture-associated pulmonary disease, summer pasture allergy, or summer pasture-associated RAO, is a severe form of equine asthma (Costa et al., 2016). This form of asthma mostly affects horses exposed to pasture environments during the warmer months of the year for more than 12 h per day. SPAOPD is most commonly diagnosed in the southeastern United States, especially in Florida, Georgia, Louisiana, and Mississippi, although it has been described in other parts of the world, including England, Scotland, and Canada. In the southeastern region of the United States, affected horses are generally older than 6 years of age. Similar obstructive pulmonary disease is seen in the following three videos.

https://www.youtube.com/watch?v=KGBXt5VVltU
https://www.youtube.com/watch?v=oKpxhgzGNKc
https://www.youtube.com/watch?v=Bist8Tcbk1Q

Allergies develop when a horse inhales certain particles, including dust mites, storage mites, molds, and pollens, and the immune system reacts extra strongly to them. Some allergies are present from birth, but others develop over time. This is why the disease most often occurs in horses 6 years and older. When airborne allergens get down into the airways, they irritate the cells and cause mucous secretion, which will trigger a snort or cough. In addition, if the horse is allergic to one

FIGURE 16.4 These horses are in an environment that appears to have plenty of leafy, vegetative, green, cool-season grass and sunny weather, conditions that are unlikely to trigger SPAOPD.

or more of these particle types, inhaling them will also cause inflammation. Large numbers of white blood cells move into the affected area, and some of these cells secrete chemicals that cause swelling. Others produce antibodies to the allergen(s), causing even more inflammation. Due to the mucus and inflammation, less air can get through. Smooth muscles constrict the walls of the lower airways to prevent the allergens from passing further down into the lungs, which reduces the amount of total air space in the airways and lungs. Wheezing and coughing occur, which then worsen the irritation and inflammation in the lungs. It is a vicious cycle in which the body's own defenses ultimately cause the most harm (Costa et al., 2006).

Clinical onset and exacerbation of SPAOPD include ambient temperatures above 86°F (30°C) and daily minimum dewpoint temperature greater than 63°F (17°C), thus heat and humidity. Mold spores and grass pollens are the primary triggering agents. The mold spores associated with SPAOPD are Alternaria, Basidiospore, Cercospora, Cladosporium, Curvularia, and Nigrospora. SPAOPD is a seasonal airway disease, and it can be managed by controlling a horse's environment prior to the onset of clinical signs. Restricting access to pastures and keeping affected horses in a low-dust environment in the summer months is ideal (Fig. 16.4). Clipping pastures short may also reduce the risk of exposure to allergens and triggering agents (Seahorn et al., 1996). Pastures with certain grasses cause more problems than others. For example, perennial ryegrass is less likely to be associated with clinical exacerbation of the disease, in part because it grows better in colder climates. The owners of affected horses are encouraged to investigate the best grasses for their pastures and their affected horses.

Elimination of all known triggering agents is imperative, as well as minimizing the horses' exposure to dust. Several feeding options include offering a complete pelleted diet to avoid feeding dry hay. If hay is offered, it should be soaked instead of simply wetting it (Mack et al., 2017). Another preventive measure is to avoid feeding moldy hay to horses.

EQUINE GRASS SICKNESS

Equine grass sickness (EGS), also known as equine dysautonomia, is thought to be associated with *Clostridium botulinum type C* organisms, which are ubiquitous in soil and commonly encountered by horses. EGS has been found in Britain and northern Europe since 1907, but only two cases have been reported in North America (Milne and McGorum, 2006). Horses with lower levels of antibodies to the organisms and certain soil types appear to be risk factors for the disease. EGS is also thought to be triggered by a change in nutrition, which is followed by a vast increase in the number of *C. botulinum* type C organisms, possibly causing massive production of a neural toxin in the gut. The result is that the horse's immune system is unable to cope (Hunter et al., 1999).

Soil containing high levels of nitrogen along with soil that has been disturbed (e.g., mechanical removal of manure) has also been indicated. Another risk factor to consider is temperature. Survey results from areas in Britain indicated that temperatures between 45 and 52°F (7−11°C) were a significant factor 2 weeks prior to multiple outbreaks of the disease.

The disease affects the central and peripheral nervous system in grazing horses, resulting in paralysis of the digestive tract, sometimes starting with the esophagus and ending with intestinal stasis. While in acute cases, neuronal degeneration is severe and the disease is fatal, some reported chronic cases are less severe and exposed horses survive.

In acute grass sickness, the symptoms are severe, appear suddenly, and are fatal. Severe gut paralysis leads to signs of severe colic, including rolling, pawing at the ground, and looking at the flanks, difficulty in swallowing, and drooling of saliva. The stomach may become enlarged and contain putrid fluid, which flows out of the nostrils. Impaction colic occurs, and if any manure is passed, it is typically coated with mucus. Muscle tremors and patchy sweating may occur.

In subacute grass sickness, the symptoms are similar to those of the acute disease but are less severe. Accumulation of fluid in the stomach may not occur, but the horse is likely to show difficulty swallowing, mild to moderate colic, sweating, muscle tremors, and rapid weight loss. Small amounts of food may still be consumed, but affected horses may die or require euthanasia within a week.

Symptoms of chronic grass sickness usually develop more slowly, and only some cases show mild, intermittent colic, reduced appetite, and varying degrees of difficulty in swallowing. One of the major symptoms is rapid and severe weight loss, which may lead to emaciation.

Diagnosis of EGS horses is difficult, as many symptoms may not be apparent, making it difficult for a veterinarian to determine causation. A definitive diagnosis for this disease is usually determined after the horse has died by conducting a microscopic examination of the nerve ganglia, or by examination of tissue that is sampled from the small intestine (Milne and McGorum, 2006). Treatment of grass sickness for acute and subacute levels is euthanasia. For chronic cases, where the pain is not severe and the horse can eat small quantities of food, treatment can be done by feeding the horse smaller quantities of grass, chopped vegetables, and high-energy foods that are soaked in molasses. Treatment also involves constant management by the owner where the horse is in daily human contact and by grooming. Adding appetite stimulants to food is also recommended.

Specific measures to prevent EGS are not well defined. However, managing horses by stabling them in early spring and summer, especially during times when there is drier weather with temperatures between 7 and 11°C for 10 consecutive days, can be beneficial. On farms where the disease is known to occur, stabling new horses and removing others from those affected fields is suggested. Neurodegenerative changes have been found in equine victims of EGS, similar to findings in humans that have died from Alzheimer disease (McGorum et al., 2016).

At this time, EGS is poorly understood, and more research is needed to further investigate the cause and possible treatments (Pirie et al., 2014).

SEASONAL PASTURE MYOPATHY

Seasonal Pasture Myopathy (SPM) is a highly fatal muscle disease caused by the toxin hypoglycin A in the seeds of the box elder tree (*Acer negundo*) (Fig. 16.5) and likely some other *Acer* species in North America and Europe (McKenzie et al., 2016; Westermann et al., 2016). In the United Kingdom and Ireland, box elder is also known as ashleaf maple. Its range extends from Canada to Guatemala (in the mountains), from New York in the east to Florida, west to southern Texas, and northwest across the great plains into Canada. In Canada, box elder is commonly called Manitoba maple. Further west in Colorado and California, box elder occurs on slopes, in valleys, and associated waterways. Although native to North America, it is considered an invasive species in some areas of that continent. It can quickly colonize both cultivated and uncultivated areas, and the range is therefore expanding both in North America and elsewhere. In Europe, where box elder was introduced in 1688 in parks, it is able to spread quickly in places and is considered an invasive species in parts of Central Europe (Germany and the Czech Republic, middle Danube, Vistula River valley in Poland), where it can form mass growth in lowlands, disturbed areas, and riparian biomes on calcareous soils. It has also become naturalized in eastern China and can be found in some of the cooler areas of the Australian continent, where it is listed as a pest invasive species (van Gelderen and van Gelderen, 1999). Box elder trees are not common in Europe, but the sycamore maple (*Acer pseudoplatanus*) is common there and is a potent source of the toxin hypoglycin A, which is known to be produced in the seeds of box elder and to cause atypical myopathy.

FIGURE 16.5 Seeds from the box elder tree (*Acer negundo*) contain the toxin hypoglycin A that causes seasonal pasture myopathy (Valberg et al., 2013). *Credit: W. John Hayden, University of Richmond.*

When seeds containing the toxin hypoglycin A of *A. negundo* are ingested by horses, symptoms of SPM occur, such as muscle stiffness, difficulty walking and standing, dark-colored urine, and rapid breathing, with death occurring within 72 h. SPM is fatal in 90% of horses. What the toxin does physiologically is prevent the metabolism of fat, and this contributes to the destruction of respiratory and postural muscle cells. Most cases are reported in the autumn but rarely in the spring and summer. Not all horses that are exposed to the seeds develop the disease. Horses that develop SPM often have limited access to forage and supplemental feed and are housed on overgrazed pastures with box elder trees, fallen leaves, and branches. Proper pasture and feeding management, in addition to controlling access to the box elder, reduces the risk for SPM (Finno et al., 2006; Valberg et al., 2013). Keeping horses off overgrazed pastures and providing hay where pasture forage is limited can reduce the ingestion of seeds and other toxic plants. If removal of the tree is not practical, then removing seed-bearing lower branches and branches hanging over pastures is a reasonable alternative.

NITRATE POISONING IN HORSES

While nitrate poisoning is considered to be very rare in horses, certain growing conditions increase nitrate accumulation and subsequent risks. Nitrate poisoning is more commonly seen in ruminants than horses because rumen microflora readily convert nitrate to toxic nitrite, whereas nitrate reduction to nitrite only occurs in the cecum of horses but not to the same extent as in ruminants. Compared to ruminants, a much larger dosage of nitrate is required to cause clinical signs in horses. However, horses are very sensitive to nitrite. Ingestion of nitrite can occur when nitrates in forages or water have been converted to nitrite by environmental microbes prior to ingestion.

Cereal grasses such as oats, rye, sorghum, and millet can concentrate dangerous levels of nitrate. Several different weeds found in pastures can also be nitrate accumulators, including pigweed, lamb's quarters, thistles, jimson weed, fireweed, smartweed, dock, and Johnson grass. Plants take up nitrogen from the soil, store it as nitrates, and later convert it to protein. If the plant becomes stressed due to drought, too much fertilizer, fertilizer applied at the wrong time, herbicide application, or prolonged periods of low sunlight, nitrates continue to accumulate to elevated concentrations, and the plant does not convert it to protein (Davidson et al., 1941). Runoff from manure piles and fertilizer can increase nitrate concentration in groundwater that is subsequently taken up by plants. Decreased light intensity on cloudy days and shading restrict the rate of turning nitrate into nitrite, which is further reduced to ammonium in chloroplasts. Ammonium is then used to create amino acids and other nitrogenous compounds. In horses and other mammals, elevated nitrate concentrations interfere with the ability to carry oxygen in the blood by binding to hemoglobin and preventing oxygen binding. As a result, animals may present with clinical signs including labored or rapid breathing, tachycardia, blue-brown discoloration of the mucosal membranes, tremors, ataxia, convulsions before death, and abortion (Oruc et al., 2010). Slow intravenous injection of methylene blue is a common treatment for nitrate poisoning of horses and other livestock. Prevention of nitrate poisoning includes restricting nitrogen fertilizer applications to small amounts through the growing season rather than one large application and keeping animals off pastures that have been growing rapidly but are then frozen or exposed to several very cloudy days. Managers should be aware of nitrate sources and ensure that large amounts of nitrate are not taken up by forage plants.

PASTURE-ASSOCIATED LIVER DISEASE IN THE HORSE

The liver is an organ that helps to rid the body of foreign substances, like toxins, and assists in the regulation of energy metabolism (Elfenbein and House, 2011). Pasture-associated liver disease (PALD), also known as chronic megalocytic hepatopathy, is a disease that is caused by the ingestion of toxic plants that contain pyrrolizidine alkaloid (PA), such as *Amsinckia* spp. (fiddlenecks, yellow burr weed), *Crotalaria* spp. (rattlepod, rattleweed, brown hemp), *Cynoglossum officinale* (houndstongue, houndstooth, gypsy flower), *Echium plantagineum* (purple viper's-bugloss, Paterson's curse), and *Heliotropium europaeum* (heliotrope, European turn-sole) over time. Horses tend not to ingest these plants unless there is little to graze on in pastures during times of drought or cooler weather in the fall, resulting in frost conditions. The PA in plants is still toxic when dried and mixed in with baled hay, so ingestion of the alkaloid can occur from foraging on pastures or when stabled. Toxicity from consumption of these plants occurs over the horse's lifetime and is found in any breed and age, and to cause liver toxicity, the horse usually has consumed 2%−5% of its body weight.

Clinical signs of PALD are variable and may not develop until weeks or months after the horse has ingested the plants. Symptoms include anorexia, intermittent colic, icterus (jaundice of the eyes and skin), and lethargy. Photosensitization, caused by phylloerythrin, a byproduct of the bacterial breakdown of chlorophyll from eating plants, appears as crusty skin lesions on the horse, especially in areas of the skin that are white, looking like sunburn (Barrington, 2018). Other clinical signs of a diseased liver are bilateral laryngeal paralysis (difficulty in breathing that can be heard as a "roaring" noise), diarrhea, deep-red or congested mucous membranes, excessive drinking, and swelling along the abdomen (House and Elfenbein, 2011). The physical changes in horses may not be apparent until 80% or more loss of liver function occurs, leading to veterinary intervention when the disease is most severe. To administer treatment, blood samples must be taken and evaluated in the lab for an increase in liver-specific enzymes such as γ-glutamyl transferase and alkaline phosphatase. Serum bile acids are made in the liver and stored in the gall bladder until they are needed to break down food during digestion. Increases in bile acids of 50 umol/L or greater have been found in horses whose chance of survival is low and when the clinical signs of PALD are apparent.

At present, there is no specific treatment for liver disease. Horses that have it will benefit from good nutrition and supportive care. However, the best way to prevent this disease is to remove all of the plants that have PA in the pastures, and if applicable, in hay.

Two other plants, alsike clover (*Trifolium hybridum*) and red clover (*Trifolium pratense*), both legumes, can produce signs of PALD if ingested from foraging on pasture or from baled hay (House and Elfenbein, 2011), but the toxin that contributes to the liver disease is not known. Clinical signs of clover toxicity are similar to PA plants, and diagnosis is assisted by conducting blood tests that show increases in liver enzymes, serum bile acids, and bilirubin levels. Treatment is also similar to PA plant toxicity in that an affected horse is supported by good nutrition (low-protein, high-energy diet) and management of the pasture by removal of these legumes.

PASTURE-ASSOCIATED STRINGHALT

Pasture-associated stringhalt (PAS) or Australian stringhalt is a neurologic condition affecting horses worldwide (Huntington et al., 1989). The first case was reported more than 120 years ago of affected horses living in southeastern Australia. Horses with stringhalt have grazed on the pasture plant catsear (*Hypochaeris radicata*) or flatweed with the specific toxin unknown. Fungal contamination may also contribute to the disease but has yet to be identified.

Symptoms of PAS are extreme hyperflexion of the hind limb, making contact with the belly when the horse steps forward. Because of this involuntary movement of hind limbs, the horse develops an abnormal gait. The risk factors for PAS are older horses, draft horse breeds, horses taller than 17 hands, long and dry summer periods, and horses with a history of grazing on pastures poorly maintained and pastures containing flatweed.

Treatment is by removal of flatweed and by putting affected horses on high-quality grass.

OTHER PLANTS THAT ARE TOXIC TO HORSES

There are many plants that have been reported to be toxic to horses to varying degrees. In some cases the clinical problem may not be directly caused by toxin(s) in a plant. The severity of clinical signs and symptoms is affected by the concentration of toxin in the plant, the amount the horse has eaten, how long the horse has been eating the plant, what else the horse has been eating, and whether the plant has been treated with herbicide. Horses with plenty of familiar, nutritious food are less likely to eat toxic plants than horses that are running out of good, familiar things to eat. Some toxins can be absorbed through contact with hooves or skin rather than through ingestion. Four separate tables (16.2 through 16.5) provide common names in North America for toxic trees, forbs, legumes, and grasses, respectively. These are not the only plants that can cause poisoning of horses. Readers are advised to find websites or other resources for their geographic area that list common causes of plant toxicity and to learn to recognize these plants.

Trees

The cause of SPM was unknown as recently as 2011, so it was not considered a typical toxicity before the discovery of the relationship among the box elder and sycamore maple tree seeds, the toxin hypoglycin A, and the disease. There are several other trees that are toxic to horses (Table 16.2).

During the spring of 2001 and 2002, in the Ohio River Valley of the United States, about 25% of pregnant mares aborted. Several causes, including cyanide poisoning from cherry trees were considered. Research eventually revealed that an outbreak of eastern tent caterpillars on cherry trees enabled mares to eat these caterpillars, either because the caterpillars were consumed along with black cherry leaves from trees bordering the pastures or the caterpillars dropped off the cherry trees onto the pastures below. Eastern tent caterpillars have sharp bristles along their backs. Mares that ingested these caterpillars had tiny lesions in their gastrointestinal tracts. Bacteria from the mares' mouths and from lower in their digestive tracts entered their blood through these lesions. Bacteria entered and infected the uterus of these mares and caused placental separation, leading to abortion. Similar types of abortions have been reported from other states and from Australia. Removing cherry and apple trees from the edges of horse pastures and preventing horses from ingesting tent caterpillars appears to prevent this abortion problem.

TABLE 16.2 Trees and Other Woody Plants Reported to be Toxic to Horses

Common Names	Toxic Parts	Symptoms
Avocado (*Persea americanum*)	Leaves and fruits	Muscle tremors, convulsions, death
Black cherry (*Prunus serotina*)	Bark, leaves, seeds, shoots	Breathing rapidly and with effort, pupils dilated, muscles not moving or trembling, death
Black locust, false acacia (*Robinia pseudoacacia*)	Bark, seeds, leaves, twigs	Appetite reduced, breathing labored, depressed, diarrhea with blood, esophagus irritated, heart rate elevated, limbs cold and paralyzed, muscles weak
Black walnut (*Juglans nigra*)	Bark, nuts, roots, wood shavings	Laminitis, lower leg edema, colic
Box elder (*Acer negundo*)	Fruits	Muscle stiffness, difficulty walking and standing, dark-colored urine, rapid breathing, and eventual death
Boxwood (shrub) (*Buxus* spp.)	Bark, leaves	Colic, convulsions, diarrhea (sometimes bloody), gastroenteritis, respiratory distress, seizures, death
Burning bush (*Euonymus alatus*)	Bark, fruit, leaves, seeds	Abdominal pain, diarrhea, heart rhythm abnormal, weakness

TABLE 16.2 Trees and Other Woody Plants Reported to be Toxic to Horses—cont'd

Common Names	Toxic Parts	Symptoms
Buckeye (*Aesculus spp.*) (California, Ohio, Red, Yellow)	Leaves, sprouts, seeds	Colic, hyperglycemia, glucosurea, proteinurea, muscle twitching, weakness
China berry (*Melia azedarach*)	Leaves and especially the fruits	Colic, diarrhea, seizures, death
Choke cherry (*Prunus virginiana*)	Bark, leaves, flowers, roots, seeds, twigs	Breathing labored, collapse, distress, muscles not moving, seizures, weakness, death
Golden chain tree (*Laburnum anagyroides*)	Leaves and fruits	Muscle tremors, convulsions, death
Horse chestnut (*Aesculus hippocastanum*)	Leaves, nuts, sprouts	Diarrhea, depression, gastrointestinal irritation and pain, muscle twitching, hopping gate, weakness, pupils dilated, seizures, coma
Kentucky coffee (*Gymnocladus dioicus*)	Leaves, pods, seeds, sprouts	Colic, gastrointestinal irritation, diarrhea, heart rate decreased, muscle paralysis or convulsions, respiration rate decreased, salivation, death
Mountain laurel (shrub) (*Kalmia latifolia*)	Flowers, leaves, stems	Abdominal pain, breathing is difficult, collapse, coma, depression, diarrhea, heart rate elevated or slowed, respiratory failure, salivation, weakness, death
Oak (*Quercus* spp.)	Buds, branches, green acorn hulls, leaves	Anorexia, constipation, colic, diarrhea, dehydration, edema, urination frequent and bloody
Oleander (*Nerium oleander*)	Flowers, fruit, leaves, roots, sap, stems, seeds	Breathing labored, colic, diarrhea, muscles not moving or have tremors, inability to stand, depressed or agitated, death
Persimmon (*Diospyros virginiana*)	Fruits	Gastric impaction, ulceration, and perforation
Red maple (*Acer rubrum*)	Wilted leaves	Red blood cell breakdown, colic, fever, laminitis, hemoglobin in urine
Rhododendron (azalea) (*Rhododendron* spp.)	All plant parts, especially leaves	Anorexia, colic, coma, coordination loss, diarrhea, depression, drooling, heart failure, paralysis, death
Wild cherry (*Prunus avium*)	Bark, leaves, seeds	Agitation, breathing labored, coordination loss, muscles trembling, nostrils flared, uncontrolled urination and defecation
Yew (*Taxus baccata*)	Needles, seeds	Breathing difficulty, coordination loss, convulsions, diarrhea, heart rate reduced, muscle trembling, nervousness, sudden death

Forbs

Forbs are herbaceous, nonwoody, vascular plants that do not include the grasses or sedges, and for the purpose of this discussion, that do not include legumes. The largest number of plants that are toxic to horses falls within the forbs category. Most are considered weeds because they are not part of a cultivated pasture or hayfield (Table 16.3).

Legumes

Legumes are vascular, nonwoody, nongrass plants that produce seeds in pods. The family name is Fabaceae and was formerly called Leguminosae. It includes peas, beans, clovers, peanuts, some trees, and several forage species. Most legumes found on horse farms are cultivated species and varieties, and most of these are nontoxic. Some legumes are included in this section because of toxic compounds created when the legumes are infected with fungi (Table 16.4).

TABLE 16.3 Toxic Forbs (Herbaceous, Nonwoody, Nongrass Plants, Excluding Legumes Here)

Common Names	Symptoms
Angel's trumpet (*Brugmansia*)	Colic, abnormal behavior, death
Bracken fern (*Pteridium aquilinum*)	Nervousness, circling, staggering, muscle spasms or tremors, blindness, dilated pupils, convulsions, death
Brassicas (mustard, kale, rape, turnips, etc.)	Pain, salivation, diarrhea, irritation in mouth, colic, head shaking
Buttercup, smallflower and buttercup, tall	Blisters on lips, swollen face, excess salivation, mild colic, diarrhea, twitching, paralysis, convulsions, death
Castor bean	Dermatitis, colic, diarrhea, muscle weakness, tremors, incoordination, swollen abdomen, anorexia, depression, sweating, elevated body temperature, death
Cockleburr	Appetite decreased, depression, colic, muscle spasms or ataxia, irregular breathing, weakness, convulsions, death
Curly (curled) dock	Appetite decreased, depression, recumbent, bluish mucous membranes, lacking coordination, slobbering, tremors, weakness, erratic behavior, abortion, coma, death
Field bindweed (morning glory)	Weight loss, colic, diarrhea, abdominal pain, dry mouth, weak and rapid pulse, dilated pupils, convulsions, death
Giant hogweed, giant cow parsnip (*Heracleum maximum*)	Phototoxic dermatitis (photosensitization), rashes and blisters that may become infected, eye damage
Hemlock, poison (spotted) (*Conium maculatum*)	Nervous tremors, muscle weakness, incoordination, salivation, rapid pulse and respirations, birth defects, frequent urination, colic, diarrhea, pupils dilated, nervousness, coma, death
Hemlock, water (*Cicuta maculata*)	Found dead, seizures, anxiety, muscle twitching, teeth grinding, frothy saliva, tongue lacerations
Hoary alyssum	Edema in lower legs, body temperature >103°F (39.5°C), mild diarrhea, laminitis, founder, depression, death
Horsetail (*Equisetum*)	Scruffy coat, weight loss, diarrhea, incoordination, loss of muscle control, staggering, nervous, unable to rise, seizure, death
Jimson weed, thornapple (*Datura*)	Increased thirst, pupils dilated, agitation, heart rate increased, lowered body temperature, frequent urination, respiratory distress, incoordination, diarrhea, trembling, convulsions, coma, death
Leafy spurge	Salivation, diarrhea, colic, weakness, blistering and hair loss around hooves, irritation to skin and eyes from sap, temporary blindness, cold extremities, painful urination, inflamed urethra, death
Lupine	Nervousness, excess salivation, depression, diarrhea, gait changes, reluctant to move, muscle spasms or twitching, convulsions, agitation, loss of muscle control, difficulty breathing, wandering aimlessly, coma, death
Milkweed	Depression, excess salivation, reluctance to stand, irregular heartbeat, colic, dilated pupils, muscular weakness, tremors, incoordination, labored breathing, death
Monkshood	Diarrhea, colic, tremors, muscle weakness, staggering, recumbency, paralysis, cardiac arrhythmia, heart and respiratory rates elevated, convulsions, death
Nightshade, black (*Solanum nigrum*)	Stomach pain, pupils dilated, diarrhea, weakness, incoordination, hemorrhagic gastroenteritis, respiratory difficulty, drooling, unconsciousness
Nightshade, climbing (also woody, bitter, European bittersweet, poisonous, blue, and deadly nightshade) (*Solanum dulcamara*)	Gastric upset, abdominal pain, drooling, loss of appetite, confusion, drowsiness, lowered heart rate, diarrhea, dilated pupils, muscle spasms, muscle twitching, paralysis, coma, death, birth defects
Nightshade, deadly (*Atropa belladonna*)	Dry mouth, anorexia, dilated pupils, irregular and rapid heart rate, sensitivity to light, nervousness, muscle tremors and convulsions, constipation or diarrhea, disorientation, incoordination, recumbence, death

TABLE 16.3 Toxic Forbs (Herbaceous, Nonwoody, Nongrass Plants, Excluding Legumes Here)—cont'd

Common Names	Symptoms
Parsnip, wild (poison) (*Pastinaca sativa*)	Skin develops: redness, swelling, serum oozing, hard crusts and pain on touching; furanocoumarins cause photosensitivity (appearing like severe sunburn in white-skinned areas) following ingestion and exposure to sunlight in horses and people; dermal contact causes similar skin blistering in people
Rattlebox, showy (*Crotalaria albida*)	Appetite loss, colic, circling, depression, diarrhea, photosensitization, hyperexcitability, incoordination, jaundice, liver disease, nervousness, pushing against objects, wandering, and respiratory disease
Rattlebox, smooth (*Crotalaria pallida*)	Aimless wandering, depression, food refusal, loss of condition, head pressing, incoordination, jaundice, photosensitization with red, swollen skin, yawning, death
Snakeroot, white (Joe Pye weed, deerweed, deerwort) (*Eupatorium rugosum, Ageratina altissima*)	Blood in urine, depression, difficulty swallowing, jaundice, wide stance with head close to the ground, incoordination, muscle trembling, stiff gait, sweating, teeth grinding, unable to stand, rapid and irregular heart rate, swollen lower neck, death
Sneezeweed, yellowdicks, Spanish daisy, bitterweed (*Helenium autumnale*)	Difficulty breathing, diarrhea, salivation, dehydration, restlessness, blind staggering, rapid pulse, weakness, loss of muscle control, very sensitive to touch, spasms, convulsions, sudden death
St. John's wort (*Hypericum perforatum*)	Diarrhea, head rubbing, hind limb weakness, panting, photosensitization (unpigmented skin becomes red, swelling, cracking, weeping fluid), sensitivity to light, listless, dilated pupils, lameness, itching
Tansy ragwort, stagger wort, stinking willy, St. James wort (*Senecio jacobaea*)	Appetite loss, colic, depression, diarrhea, weight loss, jaundice, liver damage, photosensitization of unpigmented skin, eyes red and watery in sunlight, depression, compulsive and aimless walking, head pressing, apparent blindness, convulsions, weakness, death
Tobacco	Drooling, loss of muscle control, tremors, collapse, seizures, mucous membranes pale, conjunctivitis, coma, sudden death

TABLE 16.4 Legumes and Associated Conditions Producing Toxicity in Horses

Common Names	Conditions	Symptoms
Alfalfa (*Medicago sativa*), white clover (*Tifolium repens*) and other legumes infected with a fungus	*Rhizoctonia leguminicola* fungus infection produces slaframine toxin; high humidity, drought and continuous grazing favor the fungus; infected plants have brown or black spots on leaves	Excess salivation, sometimes tearing, skin lesions, difficult breathing, increased urination, and feed refusal
Alfalfa, infested with blister beetles (family Meloidae, with four subfamilies)	Many species of blister beetles move from other plants into alfalfa field; 5 to 10 beetles can be fatal due to toxin cantharadin; poisoning is more common with hay than pasture; more common in the southern United States than northern states	Appetite loss, blisters on lips, tongue and through digestive tract, lethargy, colic, diarrhea, increased or irregular heart rate, damage to kidneys and heart, death
Alsike clover (*Trifolium hybridum*)	Part of pasture mixes for ruminants, grows well in cool temperatures and moist soil	Severe liver damage, photosensitization (blister, peeling skin), weight loss, jaundice, depression, neurological problems, death
Crown vetch (*Securigera varia, Coronilla varia*)	Disturbed soils, dry conditions, control of soil erosion; beta-nitropropionic acid from leaves and stems causes methemoglobinemia	Decreased appetite, incoordination, staggering gait, irregular heart rate, rapid respiration rate, difficulty breathing, paralysis, death

Continued

TABLE 16.4 Legumes and Associated Conditions Producing Toxicity in Horses—cont'd

Common Names	Conditions	Symptoms
Leucaena (*Leucaena leucocephala*)	A tropical tree planted for green manure, soil conservation, firewood, charcoal production and livestock feed, which produces the goitrogenic amino acid mimosine	Reduced appetite, excess salivation, enlarged thyroid gland, hair loss (mainly in mane and tail), weight loss and lethargy, loss of coordination, decreased concentrations of thyroid hormones and swelling of the thyroid gland
Loco weed, milk vetch (*Astragalus* and *Oxytropis* spp.)	Many species across the United States; some species, when infected with the *Undifilum* spp. fungus produce an alkaloid called "Swainsonine" in all plant parts; highly palatable to horses	Depression, lethargy, weight loss, erratic behavior, overreaction to stimuli, including rearing and flipping backwards, loss of control of movements, head bobbing, high-stepping gait, violent behavior, permanent neurological damage, abortions, birth defects
Lupines (*Lupinus* spp.)	Several varieties in northern and southern hemispheres; leaves, seeds, and seedpods contain highest quantities of alkaloids	Nervousness, depression, diarrhea, difficulty breathing, muscle twitching, wandering, reluctance to move, loss of muscle control, salivation, agitation, convulsions, coma, respiratory arrest, death
Red clover (*Trifolium pratense*)	*Rhizoctonia leguminicola* fungus infection produces slaframine toxin; high humidity, drought, and continuous grazing favor the fungus; infected plants have brown or black spots on leaves	Excess salivation, sometimes tearing, skin lesions, difficult breathing, increased urination and feed refusal; just one of several causes of slobbering
Sweet clover (*Melitolus alba* and *officinalis*)	When damaged or cut, coumarins in sweet clover are converted to dicoumarol, an anticoagulant that interferes with vitamin K; this is much more common in moldy sweet clover hay or silage (especially round bales) than in pasture; sweet clover can be safely used as horse hay in dry regions, such as prairies, where it can be dried quickly; in humid regions (Great Lakes and coastal), it should not used for hay or pasture; may also produce slaframine and cause slobbers if infected by *Rhizoctonia*	Prolonged blood clotting times, hemorrhages, sudden death, stiffness, lameness, hematomas, nosebleed

Grasses

Grasses are the monocotyledon plant types within the family Poaceae (formerly called Gramineae) that horses and other ungulate herbivores eat in the greatest quantities. The action of eating grasses and adjacent plants is grazing. In some situations, grasses are harmful to horses. Table 16.5 lists grasses that have documented links to poisoning of horses.

OTHER CONDITIONS THAT ARE TOXIC TO HORSES

Selenium Deficiency

Selenium deficiency occurs when horses ingest selenium-deficient forages without any supplementation. Forages can contain relatively low concentrations of selenium due to a low concentration in the soil or due to a high soil selenium concentration that is mostly unavailable. White muscle disease, also known as nutritional muscular dystrophy, is a noninflammatory degenerative disease affecting skeletal and cardiac muscle in foals from birth to 11 months (Löfstedt, 1997). The cause of white muscle disease is related to selenium deficiency, but vitamin E deficiency has been also implicated in the disease, and both are used as treatment. Not all selenium-deficient foals show clinical signs, however. White muscle disease can manifest as two clinical syndromes: acute fulminant disease, resulting in cardiovascular collapse and death within hours or days, and subacute disease, which can cause severe muscular weakness and dysphagia (inability

TABLE 16.5 Grasses and Associated Conditions Producing Toxicity in Horses

Common Names	Conditions	Symptoms
Any grass with high nitrate accumulation	High soil nitrate from manure, fertilizer, or septic system; accumulation in plant during drought or after frost	Tucked abdomen, frequent urination, diarrhea, abnormal breathing, rapid heart rate, brown mucus membranes
Perennial rye (*Lolium perenne*)	Fungal endophyte infection causes toxic alkaloid production and ryegrass staggers	Loss of coordination, staggering, dragging back feet, collapse, tremors, head nodding, jerky movements, easily spooked by sounds
Reed canary (*Phalaris* spp.)	Wild varieties produce alkaloids, especially during drought, nitrogen fertilization, hot, cloudy days, and rapid leaf growth	Incoordination, staggering, muscle spasms, convulsions, death
Rye (*Secale cereale*), triticale (*x Titicosecale*), and wheat (*Triticum* spp.) grains, other grass seeds	*Clavliceps purpurea* fungus produces alkaloids following cool, damp weather	Convulsions, delirium, gangrene, low appetite, tremors, weakness, lameness, vasoconstriction, abortion, prolonged gestation, agalactia, uterine contractions
Sorghum, Sudan grass and Johnson grass (*Sorghum* spp.)	Mostly in southwest United States and Australia; nitriles, nitrates and cyanogenic glycosides release cyanide; drought, rapid growth after drought, and frost cause accumulation of nitrates; usually requires days or weeks of exposure	Posterior incoordination from nerve damage, cystitis, urinary incontinence, hair loss on hind legs, fetal deformities, abortion, photosensitization blisters
Tall fescue (*Lolium arundinaceum*)	Endophytic fungus in early varieties of tall fescue produces alkaloids	Prolonged gestation, abortion, difficult delivery, retained placenta, agalactia, abnormal foal maturation

to swallow). Complications include aspiration pneumonia, failure of passive transfer, and failure to thrive. Treatment includes exercise restriction and supplementation of vitamin E and selenium.

Selenium is an essential part of selenoproteins (three glutathione peroxidase enzymes, selenoprotein P, and deiodinase). Vitamin E and glutathione peroxidase work as antioxidants and protect lipid-containing organelles and cellular membranes from peroxidative damage caused by reactive oxygen metabolites or free radicals. Selenium also participates in the immune response of animals, and a selenium deficiency can lead to chronic infections. Vaccination response is reported to be improved with supplementation of vitamin E and Se (Baalsrud and Overness, 1986). When mares are kept on selenium-deficient forages with a concentration of less than 0.1 mg selenium/kg DM, their foals may develop white muscle disease. It is, however, rarely reported because the primary disease may be masked by clinical signs of failure of passive transfer, septicemia, aspiration pneumonia, and failure to thrive.

Equine motor neuron disease (EMND) is a neurodegenerative disorder of the somatic lower motor neuron system in adult horses. Clinical signs include weight loss (muscle loss), muscle fasciculations, and recumbency (Divers et al, 1994). Experimentally induced deficiency developed after 21 months of vitamin E—deficient diet (Divers et al., 2006). Four out of eight deficient horses developed clinical signs of EMND, and all eight had decreased concentrations of vitamin E. There appears to be individual susceptibility to oxidative stress, since not all horses maintained on a low—vitamin E diet with no access to pasture develop the disease. It is thought that dietary deficiency of vitamin E will cause increased generation of oxygen-derived free radicals that cause oxidative stress of tissues affected (Mohammed et al., 2012). A supply of fresh pasture is an important source of vitamin E (α-tocopherol) in the prevention of EMND (Finno et al., 2016).

The distribution of low-selenium and adequate-selenium plants in North America was shown on a map by Lewis (2005). Most of the southern United States and the central plains have adequate selenium concentration (described as 80% of all forages and grain contain >0.1 ppm of selenium). This area of adequate selenium continues north through the central plains into the three Canadian prairie provinces. A large area of low selenium concentration in forages and grains includes all of northern and eastern Canada, plus the province of British Columbia, with an extension southward through the western half of Washington and Oregon and into the northern half of California. In the east, the low-selenium zone surrounds the Great Lakes and extends into most of the area of Michigan, Wisconsin, Illinois, Indiana, and Ohio. All states from Pennsylvania going northeast to Maine are also in the low-selenium zone. A narrower projection from this zone includes the area of western North Carolina and eastern Tennessee. The state of Florida and a narrow belt extending north along the Atlantic coast through Georgia, the Carolinas, and Virginia joins the low-selenium zone in Maryland. Small pockets of intermediate and variable selenium concentration are scattered through the central adequate zone. In the United States, four states have not reported any selenium deficiency problems: Delaware, Rhode Island, West Virginia, and Wyoming (Lewis, 2005). Adding limestone to acidic soils and adding selenium and phosphorus to fertilizers can improve selenium content in forages.

SELENIUM POISONING

In areas of North America where alkaline soils predominate and rainfall is minimal (California, Oregon, Montana, Wyoming, Colorado, New Mexico, Idaho, South Dakota, and Utah), pasture grasses and other plants can accumulate high levels of selenium that will cause poisoning in horses grazing such pasture or rangeland (Van Vleet and Ferrans, 1992; Mikklesen et al., 1989). Chronic selenium poisoning in horses manifests itself as lameness due to defective hoof wall formation. Circular ridges and cracks develop in all hooves, leading to severe lameness. Frequently affected horses also have loss of the long hairs of the mane and tail. Signs of ingestion of toxic doses include severe dyspnea (difficulty breathing), diarrhea, incoordination, prostration, and death within a few hours of ingestion. Serum selenium levels of 2—2.5 mg/dL are indicative of acute poisoning. It is rare for herbivores to consume primary selenium accumulator plants (*Astragalus, Stanleya, Haplopappus, Machaeranthera* spp.) unless deprived of other feed sources (Kingsbury, 1964). Blind staggers, involving apparent blindness, head pressing, perspiration, abdominal pain, colic, diarrhea, and lethargy (Rosenfeld and Beath, 1964) can result from the consumption of selenium accumulator plants with less than 200 ppm. In areas where high-selenium soils are prevalent, horses should be kept off these pastures, especially in dry years.

GETTING HELP

The diversity of symptoms of plant poisoning of horses makes it very difficult to remember which abnormality in a horse may indicate a particular plant species as the cause. Thus, excellent knowledge of the normal health conditions of horses is an asset. Table 16.6 lists possible signs of plant poisonings in horses by body functions and systems to aid horse managers in determining whether poisoning has likely occurred.

Collecting samples of the suspected toxic plants and taking photographs of them can provide valuable evidence for veterinarians in making a diagnosis. Once these signs or symptoms have been observed, recorded, and reported to a veterinary clinic, another aid in the diagnosis and selection of appropriate treatment is available. The American Society for the Prevention of Cruelty to Animals (ASPCA) maintains an Animal Poison Control (APC) hotline at (888) 426-4435. The responder at the ASPCA APC number can take your information about the horse and its condition and use that to provide information that will assist the attending veterinarian in making a diagnosis and beginning treatment quickly. The fee for the service is US $65.00 and can be charged to a credit card.

TABLE 16.6 Possible Signs of Plant Poisoning in Horses

Functions/Systems	Possible Observations
Neurological	Nervous, paralysis, pupils dilated, depression, tremors, convulsions, collapse, deranged, aimless wandering
Behavioral	Appetite poor, depressed, tucked abdomen, stiff gait, spasms, slobbering, posture abnormal, pawing the ground, looking at flanks, rolling
Heart, Circulation	Rapid, slow or irregular beat, heart stops, brown blood
Respiration	Rapid, slow, difficult
Manure	Diarrhea, bloody, clay colour
Urine	Brown, dark, more frequent
Skin	Blue, red or yellow color where hair is sparse, blisters, hair loss
Mucous Membranes	Chocolate, yellow, pale
Skeleton	Lameness, big head
Limbs/Extremities	Cold extremities, laminitis stance, wide stance, edema, lameness
Reproduction/Glands	Prolonged gestation, abortion, stillbirths, agalactia, swollen thyroid

REVIEW QUESTIONS

1. Which of a horse's physiologic systems is primarily affected by colic?
2. What are the pasture-related risk factors for colic?
3. What are the clinical signs of laminitis?
4. What nutrient is linked to pasture-associated laminitis?
5. What is the name of the disease that is similar to asthma in humans?
6. What are the risk factors for SPAOPD?
7. What pasture-related disease is thought to be caused by *Clostridium botulinum* type C organisms?
8. Where are *Clostridium botulinum* type C organisms found?
9. What are the clinical signs of seasonal pasture myopathy?
10. What types of trees produce the seeds associated with seasonal pasture myopathy?
11. What causes nitrates to accumulate in plants?
12. What are the clinical signs of nitrate toxicity in horses?

REFERENCES

Baalsrud KJ, Overness G. Influence of vitamin E and selenium supplement on antibody production in horses. Equine Vet. J. 1986;18:472.

Barrington GM. Overview of Photosensitization. Merck Veterinary Manual; 2018. https://www.merckvetmanual.com/integumentary-system/photosensitization/overview-of-photosensitization.

Costa LR, Johnson JR, Baur ME, Beadle RE. Temporal clinical exacerbation of summer pasture-associated recurrent airway obstruction and relationship with climate and aeroallergens in horses. Am. J. Vet. Res. 2006;67(9):1635—42.

Costa LR, Johnson JR, Swiderski CH. Managing Summer Pasture-Associated Obstructive Pulmonary Disease, an Asthma-like Disease of Horses. 2016. https://aaep.org/horsehealth/managing-summer-pasture-associated-obstructive-pulmonary-disease-asthma-disease-horses.

Davidson WB, Doughty JL, Bolton JL. Nitrate poisoning of livestock. Can. J. Comp. Med. Vet. Sci. 1941;5(11):303—13.

Divers TJ, Mohammed HO, Cummings JF, Valentine BA, De Lahunta A, Jackson CA, Summers BA. Equine motor neuron disease: findings in 28 horses and proposal for pathophysiologic mechanism for the disease. Equine Vet. J. 1994;26:409—15.

Divers TJ, Mohammed HO, Hintz HF, de Lahunta A. Equine motor neuron disease: a review of clinical and experimental studies. Clin. Tech. Equine Pract. 2006;5(1):24—9.

Elfinbein JR, House AM. Review of pasture-associated liver disease. AAEP Proc. 2011;57:206—2009.

Finno CJ, Valberg SJ, Wünschmann A, Murphy MJ. Seasonal pasture myopathy in horses in the midwestern United States: 14 cases (1998—2005). J. Am. Vet. Med. Assoc. 2006;229(7):1134—41.

Finno CJ, Miller AD, Siso S, Divers T, Bianino G, Barro MV, Valberg SJ. Concurrent equine degenerative myeloencephalopathy and equine motor neuron disease in three young horses. J. Vet. Intern. Med. 2016;30:1344—50.

Frank N, Geor R, Bailey S, Durham A, Johnson P. Equine metabolic syndrome. J. Vet. Intern. Med. 2010;24:467—75.

Hillyer MH, Taylor FG, Proudman CJ, Edwards GB, Smith JE, French NP. Case control study to identify risk factors for simple colonic obstruction and distension colic in horses. Equine Vet. J. 2002;34:455—63.

House AM, Elfenbein JR. Pasture-associated Liver Disease in Horses. 2011. http://www.thepenzancehorse.com/2011/ARTICLES/Plants%20Liver%20Toxicities.pdf.

Hunter IC, Miller JK, Poxton IR. The association of Clostridium botulinum type C with equine grass sickness: a toxicoinfection? Equine Vet. J. 1999;31:492—9.

Huntington PJ, Jeffcotts LB, Friend SC, Luff AR, Finkelstein DI, Flynn RJ. Australian Stringhalt — epidemiological, clinical and neurological investigations. Equine Vet. J. 1989;21(4):266—73. http://www.ncbi.nlm.nih.gov/pubmed/2767028.

Kane AJ, Traub-Dargatz J, Losinger W, Garber L. The occurrence and causes of lameness and laminitis in the U.S. horse population. Proceedings of the Annual Convention of the American Association of Equine Practitioners. AAEP Proc. 2000;46:277—80.

Karikoski NP, Horn I, McGowan TW, McGowan CM. The prevalence of endocrinopathic laminitis among horses presented for laminitis at a first-opinion/referral equine hospital. Dom. Anim. Endocrinol. 2011;41:111—7.

Kingsbury JM. Poisonous Plants of the United States and Canada. Englewood Cliffs, NJ: Prentice Hall; 1964.

Lewis LD. Feeding and Care of the Horse. second ed. Ames Iowa: Blackwell Publishing; 2005. p. 29—30.

Löfstedt J. White muscle disease of foals. Vet Clin North Am equine prac 1997;13:169.

Longland AC, Byrd BM. Pasture nonstructural carbohydrates and equine laminitis. J. Nutr. 2006;136:2099S—102S.

Longland AC, Barfoot C, Harris PA. Effects of grazing muzzles on intakes of dry matter and water-soluble carbohydrates by ponies grazing spring, summer, and autumn swards, as well as autumn swards of different heights. J. Equine Vet. Sci. 2016;40:26—33.

Mack SM, Mansour TA, Eddy AL, Mochal CA, Claude AK, Robbins ML, Wenzel CJ, Cooley JA, Johnson ME, Costa LR, Bowser JE, Swiderski CE. The equine pasture asthma-restricted lung transcriptome: a tool to decipher the pathophysiology of airway hyper-responsiveness in horses. J. Equine Vet. Sci. 2017;52:75.

McGorum B, Scholes S, Milne E, Eaton S, Wishart T, Poxton I, Moss S, Wernery U, Davey T, Harris J, Pirie S. Equine grass sickness, but not botulism, causes autonomic and enteric neurodegeneration and increases SNARE protein expression within neuronal perikarya. Equine. Vet. J. 2016;48(6):786−91.

McIntosh BJ. Circadian and Seasonal Variation in Pasture Nonstructural Carbohydrates and the Physiological Response of Grazing Horses. Virginia Tech. Ph.D. Thesis; 2006.

McKenzie RK, Hill FI, Habyarimana JA, Boemer F, Votion DM. Detection of hypoglycin A in the seeds of sycamore (Acer pseudoplatanus) and box elder (A. negundo) in New Zealand; the toxin associated with cases of equine atypical myopathy. N.Z. Vet. J. 2016;64(3):1−17.

Meier AD, de Laat MA, Reiche DB, Pollitt CC, Walsh DM, McGree JM, Sillence MN. The oral glucose test predicts laminitis risk in ponies fed a diet high in nonstructural carbohydrates. Dom. Anim. Endocrinol. 2018;63:1−9.

Menzies-Gow NJ, Harris PA, Elliott J. Prospective cohort study evaluating risk factors for the development of pasture-associated laminitis in the United Kingdom. Equine Vet. J. 2016;49(3):300−6.

Milne E, McGorum B. Grass Sickness in Horses. The Equine Grass Sickness Fund. Scotland: The Moredun Foundation; 2006. http://www.grasssickness.org.uk/advice/grass-sickness-in-horses.

Mikklesen RL, Page AL, Bingham FT. Factors affecting selenium accumulation by agricultural crops, Soil Science Society of America and American Society of Agronomy. Special publication 23. 1989. p. 65.

Mohammed HO, Divers TJ, Kwak J, Omar AH, White ME, de Lahunta A. Association of oxidative stress with motor neuron disease in horses. Am. J. Vet. Res. 2012;73(12):1957−62.

Oruc HH, Akkoc A, Uzunoglu I, Kennerman E. Nitrate poisoning in horses associated with ingestion of forage and alfalfa. J. Equine Vet. Sci. 2010;30(3):159−62.

Pirie RS, Jago RC, Hudson NP. Equine grass sickness. Equine Vet. J. 2014;46:545−53.

Pleasant RS, Suagee JK, Thatcher CD, Elvinger F, Geor RJ. Adiposity, plasma insulin, leptin, lipids, and oxidative stress in mature light breed horses. J. Vet. Intern. Med. 2013;27:576−82.

Rose RJ, Hodgson DR. Manual of Equine Practice. Elsevier Health Sciences; 2000.

Rosenfeld I, Beath OA. Selenium: Geobotany, Biochemistry, Toxicity, and Nutrition. New York: Academic Press; 1964.

Scantlebury CE, Archer DC, Proudman CJ, Pinchbeck GL. Risk factors for recurrent colic in UK general equine practice. Equine Vet. J. 2015;47:202−6.

Seahorn TL, Groves MG, Harrington KS, Beadle RE. Chronic obstructive pulmonary disease in horses in Louisiana. J. Am. Vet. Med. Assoc. 1996;208(2):248−51.

Tinker MK, White NA, Lessard P, Thatcher CD, Pelzer KD, Davis B, Carmel DK. Prospective study of equine colic risk factors. Equine Vet. J. 1997;29:454−8.

USDA. Lameness and Laminitis in U.S. Horses. #N318.0400. Fort Collins, CO: USDA, APHIS, VA, CEAH, National Animal Health Monitoring System. C. Fort Collins, CO: USDA: APHIS:VA, National Animal Health Monitoring System; 2000. #N318.0400.

Valberg SJ, Sponseller BT, Hegeman AD, Earing J, Bender JB, Martinson KL, Patterson SE, Sweetman L. Seasonal pasture myopathy/atypical myopathy in North America associated with ingestion of hypoglycin A within seeds of the box elder tree. Equine Vet. J. 2013;45:419−26.

van Gelderen CJ, van Gelderen DM. Maples for Gardens: A Color Encyclopedia. Portland, Oregon: Timber Press; 1999.

Van Vleet JF, Ferrans VJ. Etiologic factors and pathologic alterations in Selenium-vitamin E deficiency and excess in animals and humans. Biol. Trace Elem. Res. 1992;33:1.

Westermann CM, van Leeuwen R, van Raamsdonk LWD, Mol HGJ. Hypoglycin A concentrations in maple tree species in The Netherlands and the occurrence of atypical myopathy in horses. J. Vet. Intern. Med. 2016;30:880−4.

Chapter 17

Coexisting With Wildlife

Paul Sharpe[1] and Daniel J. Undersander[2]

[1]University of Guelph, Guelph, ON, Canada (retired); [2]College of Agriculture and Life Sciences, University of Wisconsin, Madison, WI, United States

INTRODUCTION

Wildlife habitat is any land that can be used as a shelter, breeding ground, or a food source for wildlife (Jeswiet and Hermsen, 2015). Many farms, whether used to raise horses, other animals, or crops, contain some land that can be considered wildlife habitat. Of all the agricultural land in Canada, 22.7% is considered "natural land for pasture," and 7.6% is considered "woodlands and wetlands," and together, this 30.3% of agricultural land is considered to be wildlife habitat, according to the 2011 census (Jeswiet and Hermsen, 2015). Tame (improved) pasture and hayland make up unspecified portions of "cropland" in that census, and it is reasonable to include these areas as wildlife habitat. Thus the amount of agricultural land that could be considered wildlife habitat is about one-third of the total. Grasslands provide habitat for many species of wildlife, including rodents, carnivores, birds, and large herbivores. Technically, amphibians, reptiles, insects, bugs, fungi, algae, and plants could be considered "wildlife," but this chapter will deal mainly with larger animals. The management of grasslands can have large effects on the quality of wildlife habitat and its suitability for various species. The distribution of agricultural land and thus wildlife habitat is uneven across the country (Jeswiet and Hermsen, 2015), depending on soil type, topography, water resources, and climate.

Humans have demonstrated contrasting tendencies in their relationship to wildlife, depending upon the stage of their societal development. Relatively primitive human societies with low populations tended to live in harmony with nature and did not endanger populations of wild organisms. As populations grew, more tools were used, and people became more efficient at harvesting wild plants and animals. As North America and Australia were developed, increasing numbers of wild organisms were either overharvested or suffered collateral damage as people changed the environments that they were developing. North America, Australia, New Zealand, and parts of Africa have many examples of wild populations that have shrunk and many that have become extinct. Sometimes the endangerment and extinction of a species is due to competing or predatory species that have been introduced. In other cases, extinction or near extinction has been due to overhunting or overfishing.

In well-developed societies, people's demands for resources become somewhat satiated, and they have time and opportunity to look at the environment and realize what has been lost. That is when ideas of conservation, preservation, and reestablishing wildlife habitat get going. Governments enact regulations on seasons, bag or catch limits, and licenses to capture and kill wild creatures. Parks and preserves are set aside. In contrast, much earlier in societal development, governments have imposed bounties to encourage the killing of animal species that society felt were dangerous, scary, or harmful to the economy. Attempts to regulate the numbers of individuals of many species have become the focus of whole government departments and volunteer organizations.

THE RELATIONSHIP BETWEEN THE ABUNDANCE OF SPECIES OF WILD PLANTS AND ANIMALS AND THE HEALTH OF AN ECOSYSTEM

Lists and facts about endangered species are available for the United States (U.S. Fish & Wildlife Service, 2017a,b) and Canada (Government of Canada, 2017). The more species contained in an environment or habitat, the richer and healthier it is deemed to be. Every species of organism is thought to have intrinsic value, and many species are easily recognized for their value as food, clothing, medicines, or trade goods. Contributions of individual species to food webs and ecosystems lead to their consideration as contributors to ecosystem services.

According to the Food and Agriculture Organization (FAO) of the United Nations, ecosystem services are "the multitude of benefits that nature provides to society" and biodiversity is "the diversity among living organisms, which is essential to ecosystems function and services delivery" (FAO, 2018). Examples of ecosystem services on farms include crop pollination, breakdown of organic matter to release nutrients, contaminant degradation, and agricultural pest control (Jeswiet and Hermsen, 2015). Some bats can eat 600 mosquitos per hour; hawks control gophers and mice; and lady beetles eat soybean aphids.

Landowners can be proactive or reactive regarding wildlife on their property. Choices can be made to encourage some species of wildlife through enhancing habitat and to discourage other species through habitat changes, trapping, hunting, and exclusion. Due to interactions among species that result from an initial encouragement or discouragement of a single species, the equilibrium result may be either an increase or a decrease in biodiversity.

Habitat

The environmental components that wild creatures use and take advantage of to thrive are collectively called its "habitat." It is from habitat that an animal, plant or microorganism obtains its needs for food, water, cover (a place to seek shelter and to hide), and a place to reproduce. The ideal habitat for a certain species of grass involves soil type, soil moisture, weather, latitude, and altitude. The ideal habitat for a species of bird includes the food and shelter (partly provided by grasses, trees, and other plants, as well as available water and geographic features) plus available space not already used by competing species. Land managers can alter habitats to discourage undesirable species and encourage desirable species. Analyzing a habitat involves surveying horizontally along the ground and vertically through the vegetation and topographic features to determine what is present, before deciding what changes to make. Within the vertical dimension is a basement zone next to the soil that contains lichens and mosses. Progressing upward is a floor containing low-growing groundcovers, then an understory of vines, shrubs, and small trees, and uppermost is a zone of tree tops. Some species will live only in one zone, and some species thrive in two or more zones.

When assessing the overall habitat in a horizontal space, consider the condition of the plants and their locations, how much area is shaded, whether trees are deciduous or coniferous, and whether they provide food for animals. Diversity of plants will encourage a diversity of animals, so to encourage biodiversity, plant ground covers, flowering perennials, shrubs, and trees.

To encourage certain species of birds, landowners can learn the components of their favored habitat and food. They can provide ideal nesting sites, which might be tall grass prairie, coniferous trees, or dead trunks of deciduous trees. Some mammals, amphibians, fish, plants, and fungi also benefit from dead trees. Increasing or decreasing the amount of water has a dramatic effect on the diversity of species that thrive in a habitat (Natural Resources Conservation Service, NRCS). The mature sizes of trees, shrubs, and forages will be altered by the amounts of wind, water, and nutrients in their location.

Lists of plant species that enhance habitat for birds, bees, and butterflies are available from NRCS. Bats can be attracted by building and erecting bat houses on poles about 15 feet in the air in a sunny location. Rocks piled in sunny and shady spots are good habitat for amphibians and reptiles. Shade-tolerant groundcovers, stumps, and thick layers of leaves under trees provide cool shelter for amphibians and reptiles (NRCS).

Water in ponds, streams, and livestock waterers will be used by wildlife and help them thrive where there is no other water source, as long as the source is clean and frequently refreshed. Within a water source, logs, rocks, and some vegetation provide shelter, resting places, and shade for a variety of aquatic and terrestrial animals (NRCS).

Cover provided for birds, amphibians, and reptiles will also attract rabbits and mice. Squirrels, raccoons, and opossums are attracted to trees. Bird seed attracts more than just small birds. Squirrels, raccoons, chipmunks, bears, and mice will try to access bird seed. Garden vegetables and flowering plants are often targeted as food by rabbits, woodchucks, and deer, which also nibble at trees, shrubs, hay, and grain (NRCS).

The National Wildlife Federation (NWF) sponsors certification programs to help people develop and apply a certified wildlife habitat plan for a small acreage (NWF, 2018).

Farm practices that enhance wildlife habitat include the following:

1. rotational grazing
2. planting windbreaks (to reduce soil erosion, provide shelter, food and breeding sites, provide travel corridors)
3. planting winter cover crops
4. maintaining buffer zones around water bodies (to reduce soil erosion, provide travel corridors)

5. tillage practices that retain most of the crop residue on the surface (Jeswiet and Hermsen, 2015).

With agricultural land providing much wildlife habitat and with wildlife providing ecosystem services, the relationship between agriculture and wildlife is codependent (Jeswiet and Hermsen, 2015).

Biodiversity and Ecosystem Services

In the Endangered Species Act of 1973, the US Congress explained that endangered and threatened species of wildlife and plants are "of esthetic, ecological, educational, historical, recreational and scientific value to the Nation and its people." The intent of the Act was to "conserve the ecosystems upon which endangered and threatened species depend" (U.S. Fish & Wildlife Service 2017b). Some people only consider the value of a species if it has some direct effect on them. For example, few people feel affected by the presence or absence of certain mosses, reptiles, insects, bats, clams, mussels, bumble bees, or warblers in their communities. Some species gain a significant public profile, such as whooping cranes, Atlantic salmon, right whales, and monarch butterflies. Public pressure and donations ensure that significant efforts and dollars are contributed to attempts to preserve these high-profile species.

Management of forage and grazing lands involves provision of ecosystem services including reduced soil erosion and improvements in water quality, wildlife habitat, and air quality. Even if land managers receive no direct economic return, society recognizes that these ecosystem services are important for the public good (Sanderson et al., 2012). Recognition of ecosystem services such as soil conservation, water quality protection, and pleasing aesthetics were recognized and published by the United States Department of Agriculture (USDA) in 1948 (USDA. 1948). Ecosystem services include the benefits human populations derive directly or indirectly from ecosystem functions (the habitat, biologic or system properties, or processes of ecosystems) (Costanza et al., 1997). Thus the efforts of individual landowners to increase biodiversity and implement conservation measures provide ecosystem services to the public. Different plant species absorb different nutrients at different rates through the growing season and provide different collections of nutrients to herbivores in different parts of the growing season, smoothing out nutrient cycling (Voth, 2018a).

Ecosystem goods and services have four categories:

1. provisioning services, including products such as food, fiber, and fuel
2. supporting services, such as primary production and nutrient cycling that enable all other ecosystem services
3. regulating services such as climate and flood regulation
4. cultural services, including nontangibles such as aesthetic, spiritual, educational, or recreational experiences (Sanderson et al., 2012)

Rewards are given to farmers by the USDA NRCS Conservation Stewardship Program for multiple ecosystem services such as soil conservation, water quality protection, and carbon sequestration. A USDA Grassland Reserve Program indicates that ecosystem services from grasslands include: "domestic animal productivity, biological productivity, plant and animal richness and diversity and abundance, fish and wildlife habitat (e.g., for pollinators and native insects), water quality and quantity benefits, aesthetics, open space and recreation" (Federal Register, 2009; Sanderson et al., 2012). In addition, fishing and hunting on pasture and grazing lands provide revenue through sales of licenses, sporting equipment, and access rights, while contributing to healthy wildlife populations. Permanent vegetation cover on forage and grazing lands reduces soil erosion, protects water quality, supports symbiotic relationships such as those involving rhizobia and mycorrhizae with plant roots, and provides an aesthetically pleasing landscape. Biodiversity provided by grasslands is also a desired ecosystem service, not only involving plants but insects and smaller organisms. Enhanced bird diversity is an ecosystem service that facilitates bird watching and thus provides economic dividends in tourism, access rights to lands, environmental dividends in promotion of healthy wildlife populations, and pleasure involved in observing wildlife. The ecosystem good, "clean water" satisfies needs of households, agriculture, and industries and provides income from sale of water and recreation. Clean water enhances aquatic habitat, drinking water for terrestrial wildlife, and rejuvenation of riparian areas. Viewing clean water and playing in it are aesthetically pleasing (Sanderson et al., 2012).

One of the best examples of an organism that promotes biodiversity and supplies ecosystem services is the beaver. Beavers cause accumulations of water above their dams. Some of the water infiltrates the soil, while some is absorbed by plants, algae, microscopic bacteria, archaea, and fungi. Beaver ponds become habitat for small invertebrates (including insects), fish, mollusks, amphibians, reptiles, birds, and other mammals. Unfortunately, for beavers, in agricultural areas, farm managers are not impressed by the services provided by beavers, compared to the inconveniences they cause through flooding. High water tables restrict farm vehicle access to land and delay planting and harvesting of crops.

In 2008 the USDA developed the Conservation Effects Assessment Project (CEAP) to quantify scientifically the environmental outcomes of conservation practices used by private landowners that are supported by the USDA and other conservation programs (Duriancik et al., 2008). The purpose of CEAP is to "help implement existing and design new conservation programs to more effectively and efficiently meet the goals of the U.S. Congress and the Administration" (James and Cox, 2008). Assessments were conducted on effects of conservation programs on wildlife in croplands, wetlands, and grazing lands. CEAP assessment reports from 2004 to 2016 can be found at https://www.nrcs.usda.gov/wps/portal/nrcs/detailfull/national/technical/nra/ceap/na/?cid=nrcs143_014151.

Wildlife Population Imbalance

Some of the greatest challenges in managing sustainable agriculture in the United States are created by the multiple resource demands on the arid lands extending from western Texas through New Mexico and Arizona to Nevada and California, then north through Oregon to eastern Washington and central British Columbia. Competition occurs among people with conflicting priorities and between people and animals for water and land. The animals simply want habitat, a home. As a nation develops, different groups of people want water and land initially for agriculture, then for cities and industry, and later for conservation. The low annual precipitation and greater rates of evaporation than precipitation in these western states mean that these desert and semidesert lands can support only small numbers of large herbivores per acre. Where irrigation and well drilling occur, very productive agricultural operations thrive. Irrigated alfalfa fields are an example of very desirable forage to herds of pronghorn antelope or deer (Putnam and DelCurto, 2007). From the air, it is remarkable to see bright green circles or squares of irrigated cropland within the dull brown arid brushlands.

After arid lands are developed for agriculture, wild animals often take advantage of increases in available water and nutritious plant growth, and their populations grow. During droughts the artificially high populations of animals sometimes cannot be sustained, and they move if they can to find more forage and water. This sort of migration of wild animals is common on a seasonal basis in Africa. In North America, populations of large migratory herbivores have included bison, pronghorn antelope, and caribou.

There are many examples of unintended conflicts between wild animals and humans who are trying to develop land so that they can make a living. Emus are flightless birds similar to ostriches that stand up to 6 feet (1.8 m) tall and are native to arid parts of Australia. Prior to the 1920s, after their breeding season in the interior of Western Australia, emus normally migrated west toward the Pacific coast. Following World War I (1914–18), many Australian military veterans were given arid land in Western Australia, near the Pacific coast, for the purpose of farming. Native bushy vegetation needed to be cleared, and watering points for livestock (both dams and wells with stock tanks) were developed on new farms. During the Great Depression years of the early 1930s, there was government pressure to increase wheat production. At the same time, emus were thriving on the growing areas of cleared land with crops, pastures, and increased water availability. Beginning in 1932, long fences were erected in the central part of Western Australia to keep emus, dingoes, and rabbits out of the agricultural areas. An estimated 20,000 emus on the move made breaks in the new fences, allowing the birds to eat and trample wheat crops. Rabbits and dingoes also passed through the broken fences. Following complaints by farmers, a military team with a machine gun was recruited to control the emu population in early November 1932. After 6 days of operation, the machine gun team had fired 2500 rounds of ammunition and killed between 50 and 500 emus. The emus were apparently very wary, evasive, and difficult to kill. After a second campaign that year the military officer in charge claimed 986 emus killed and 2500 wounded from the use of 9860 rounds of ammunition. Public pressure stopped the 1932 "emu war," but farmers in the region requested military assistance to deal with destructive emus again in 1934, 1943, and 1948 but were refused (Wikipedia, 2017a).

Bounties were paid for emus killed in Western Australia until the 1960s. Emus are now protected in Australia. The estimate of the Australian wild emu population in 1992 was 630,000–725,000, and their population is considered stable with a conservation status of "least concern" (IUCN Red List, 2017). In 1932, governments and farmers thought that they were doing something good for Australia, for veterans and their citizens in promoting farming and wheat production in a formerly wilderness area. Wells and water troughs helped wild emus thrive. Fences built across a migration route to keep emus in the wild areas failed. The extreme measure of shooting large numbers of emus had very low success initially. Then when it became more successful, public opinion stopped it.

If there is a lesson from the emu wars perhaps it is to include wildlife biologists in plans to greatly change land use and to continue monitoring wildlife populations as the land use change continues.

BENEFITS OF A DIVERSE ECOSYSTEM

By their existence, all organisms contribute to biodiversity, which is considered to have economic value (TEEB, 2010). This value includes regulation of water and air quality, decomposition, nutrient cycling, pollination, and flood control. It may sometimes be difficult to determine the value of animals that are causing a problem on your farm. The degree of biodiversity in an ecosystem is considered an indicator of the health of that ecosystem. All or most of the organisms in an ecosystem make a contribution to a food web, and when an organism is taken out of that ecosystem, other organisms that depended upon it will need to adapt to that absence, and that may be costly to them in terms of time and energy expended.

In certain locations and times, many people consider wild creatures as sources of enjoyment and recreation, sometimes as a source of food, and having value in the natural world presently and for future generations. An important component of tourism is the willingness of people to pay for opportunities to view, study, photograph, listen to, smell, hunt, and collect souvenirs of wild creatures. Some are willing to do this as part of farm vacations, and some will pay to do it on horseback.

Grazing management decisions that reduce soil erosion, maintain or enhance soil organic matter, increase biodiversity, and provide wildlife for people to enjoy all contribute to the value of a property. Managers can make management decisions that provide new income streams in the short term and in the long term, allow them to pass on the land in as good or better condition than when it was first acquired (Krausman et al., 2009). The ecosystem services discussed before can contribute indirectly to the economic situation of a private property or a public conservation area. The items described next relate more to direct economic benefits to private properties, mainly from attracting fee-paying clients to a property.

Accommodating Hunters or Other Consumers of a Diverse Ecosystem

Wild organisms are resources that people use to their benefit and sometimes abuse. In most developed countries, there are laws and regulations on how wild animals may be used and prohibitions on some uses. To add income to a farm enterprise, many managers attract clients who want to hunt on their land. In some communities, hunting privileges are free, and in others the expectation is that hunters pay for the ability to hunt at a certain time and place. Like any part of a business, a hunting enterprise can be managed poorly or well with appropriate investments in wildlife habitat (Voth, 2017a).

South Dakota has a "Walk-In Area Program" that provides public hunting access to privately owned lands with wildlife habitat. Landowners are paid and given immunity from nonnegligent liability, and hunting is free to the public. Payments vary from about $1 to $5/acre/year. In a "Controlled Hunting Access Program," landowners are paid $2 to $10 per hunter per day with acreages over 1000 eligible for $250/year up to $12,000 per year. In addition to big game, South Dakota has a large pheasant population that attracts many visiting hunters.

Other regulated hunting programs on private land include private leases, self-guided hunts, semiguided hunts, and fully guided hunts, some offering room and board to hunters.

Landowners considering a hunting enterprise need to consider the strengths and weaknesses of their resources and assess what wildlife species are present plus how a hunting enterprise would fit with the existing enterprises and labor. Any equine enterprise that allows hunting would need to be very careful about separating hunters and riders in time and in space. Zones where horses live need to be safely separated from zones where hunting is allowed and where stray projectiles might land. Most regulated hunting occurs during specific seasons, some as short as 1 week and some extending over a number of months. Thus combining hunting with equestrian pursuits becomes more manageable as the size of the property increases and separation of enterprises is easier.

Written lease agreements between landowners and hunters using their land protect both parties. Hunters can benefit from access to animals and habitat, while landowners can benefit from reduced damage to crops by deer and game birds.

The University of Maryland Extension division provides sample hunting leases (Kayes, 2017) and Texas AgriLife Extension has a Hunting Lease Checklist (Lashmet, 2017). More sample hunting lease forms are available through the Michigan Farm Bureau, the Mississippi State University Extension, and the Kansas State University AgManager.info. A "permission to hunt" form with no lease is available from the Maryland Department of Natural Resources (2017). Similar documents can be drafted by a landowner or a prospective hunter. To ensure that all the information conforms to laws of the appropriate state, a new lease agreement should be examined by an attorney. A landowner should ensure the following components are included in a hunting lease: specifications of the property, rental rate, uses of the property such as use of vehicles, closing of gates, cleaning of animals, disposal of animal parts, building of blinds or stands, requirements of hunters for liability insurance, abiding by state and federal laws, safety practices such as where shooting should and should not occur, types of weapons prohibited, retaining the right of the landowner to hunt, dispute resolution clause, and assigning payment of attorney's fees in case of a dispute (Goeringer, 2017).

Where landowners recognize markets for similar enterprises that capitalize on a diverse wildlife population and the desire of people to experience it, a separate business plan can reveal whether providing a service to clients can fit with the rest of their enterprise and not be a financial burden. The service can simply be guiding clients around a property for birdwatching, photography, fishing, fox hunting, or trail riding. More complex services can include providing vehicles, equipment, horses, accommodation, and meals.

Fox hunting on horseback with hounds was declared illegal in Scotland in 2002 and England in 2005. Fox hunting with horses and hounds is still legal in Northern Ireland, Canada, and the United States and still practiced by some clubs in England. North American hunts sometimes prefer coyotes or bobcats because they are more numerous and more of a local nuisance than foxes. Managing a private property to promote a population of carnivores to hunt and ensuring a safe environment for fox hunting could be an expensive proposition, and it may restrict some other activities, but it has earning potential (Doggart, 2011; Masters of Foxhounds, 2017).

EFFECTS OF ADDING OR SUBTRACTING WILD SPECIES ON AN ECOSYSTEM

Each species of organism has a potential influence on at least one other species in its ecosystem. As the number of species grows, the number of possible interactions among species grows. In complex ecosystems, the sum of interspecies relationships can be considered a web. Thus a rise or fall in the number of individuals of one species can affect a few other species directly, and those affected species can affect others.

Subtracting and Adding Animal Species

Wolves were common throughout North America north of about 20 degrees north latitude before the 1600s. As settlers moved west, their fencing, if any, was ineffective at keeping large carnivores from their livestock. Wolves found cattle, sheep, and the occasional horse to be easier prey than moose, elk, and deer, so farmers and ranchers killed wolves where and when they could. Eventually, wolves were eliminated from much of the northwestern United States, so there was no longer much predation pressure on these wild ungulates.

In 1995, it was recognized that large areas of Yellowstone National Park were overgrazed and overbrowsed by large numbers of elk. Wildlife managers realized the park had too many elk, partly because the park had no wolves. The park population of about 15,000 elk killed many of the cottonwood, aspen, and willows there, particularly along rivers and streams. The lack of these trees in riparian areas contributed to lower songbird and beaver populations and higher water temperatures, which were unfavorable to native trout. Wolves were imported from Canada to reestablish a large predator that would control the elk population. As wolves reduced elk numbers, the survivors became more wary and spent less time eating tree shoots near lakes and rivers.

Trees regrew in the riparian areas, and the shade helped improve bird, beaver, and fish habitat. Populations of many species of birds, mammals, fish, and reptiles increased again. With more beavers building dams and creating ponds, numbers of waterfowl rose.

The wolf population grew rapidly, and because they range widely, they spread to the plains within and beyond Yellowstone Park. There the wolves found cattle, sheep, and coyotes to kill. With a decline in the coyote population, small mammals, foxes, ravens, eagles, and pronghorn antelope became more numerous. Some biologists believe that wolves would not waste time searching for pronghorn fawns. In areas with high coyote populations, only about 10% of pronghorn fawns survived, but where wolves reduced the number of coyotes, pronghorn fawn survival rose to 34%, and the pronghorn population doubled from 1995 to 2017 (Voth, 2017b).

The spread of wolves beyond the Yellowstone Park boundaries has hit ranchers hard, with more significant kills of beef cattle than coyotes could manage. This is not the only example of consequences to populations of many animal and plant species when people have decided to subtract or add just one.

North America has experienced significant changes in habitats due to initial small introductions of grasses, forbs, birds (starlings, English house sparrows), earthworms, cats, and pathogenic fungi and viruses. Supposedly, civilized North Americans of European descent purposely shot large numbers of birds and mammals in the 19th and early 20th centuries, partly to supply meat to growing city populations. Unforeseen consequences were the extinction of the passenger pigeon and near extinction of bison and whooping cranes. Cougars and elk have been declared extinct from almost all of eastern North America, probably due to human activities.

Australian natural habitats have been transformed due to introduction of dingoes about 4000 years ago, then planned importation of foxes (1855), rabbits (1859), the Myxoma virus to control rabbits (1950s), Calicivirus to control rabbits (1995), and a new strain of Calicivirus (2017). Some other introductions of species to Australia were possibly not intentional. Starlings and English house sparrows, along with mice, rats, and many weed seeds probably hitchhiked aboard ships. The rapid growth of the rabbit population stimulated the growth of the fox population.

The following numbers of Australian native species have been declared extinct since 1788: 23 birds, 4 amphibians, 1 reptile, 27 mammals and marsupials, and 6 invertebrates (Wikipedia, 2017b). The extinctions of four of these bird species are thought to be due to overhunting and fires set by man. Five of the bird species became extinct on Lord Howe Island due to black rats that came ashore from a grounded ship in 1918. Many small birds and marsupial mammals on the extinct and endangered lists are there largely because of introduced foxes and feral cats. Many domestic cats have gone feral in Australia, where the climate does not bother them, food is plentiful, and they breed freely. It is estimated that an average feral cat in Australia kills about 1000 animals and birds per year. The estimated population of feral cats is about six million in years with normal rainfall. The federal government of Australia announced a plan in 2017 to kill two million feral cats to help save about 124 species of native wildlife (Power, 2017). Feral populations of pigs, goats, camels, water buffalos, horses, and domestic cats continue to have damaging effects on Australian wildlife habitat. In 2014, in Australia, 10 mammals, 4 birds, 13 fishes, 7 reptiles, 15 amphibians, and 37 arthropod species were endangered (Australian Geographic, 2014). A 2013 report indicated that free-ranging domestic cats killed 1.3−4 billion birds and 6.3−22.3 billion mammals per year in the United States (Loss et al., 2013a), whereas single-tower wind turbines killed only about 140,000−328,000 birds per year (Loss et al., 2013b).

A native grey-backed cane beetle, a scarab beetle, and a French's cane beetle ate and reduced yields of sugar cane in northeastern Australia for many years, so the cane toad (about 15 cm long, weighing close to 1 kg) was imported in 1935 to eat these beetles (Commonwealth of Australia, 2010; Australian Museum). There were no natural predators or diseases of cane toads in Australia, so they spread rapidly and widely, far beyond the sugar cane growing region, with an estimated population of about 1.5 billion in 2017 (Mercer, 2017). To make matters worse, apparently cane toads do not eat significant numbers of these cane beetles. Currently, they are spreading at 31−37 miles (50−60 km) per year. They eat native terrestrial and aquatic insects and snails. Cane toads exude poisonous venom through their skin when provoked, and many animals that attempt to eat them die rapidly. A few Australian animals that can eat cane toads and sometimes survive include the snapping turtle, the freshwater snake, the saltwater crocodile, the water rat, and some birds such as ibises and the Torresian crow (Queensland Museum, 2018). Goannas, freshwater crocodiles, several snake species, dingoes, quolls, and domestic dogs have died from eating cane toads (Australian Museum, 2016). In Western Australia, within 5 years of the arrival of cane toads, the decline of three species of goanna lizards called water monitors caused increased survival of crimson finches (Doody et al., 2015; Slezak, 2015).

The natural habitat of cane toads is humid tropics, but they seem to have adapted to Australia's inland deserts also. Cane toads have been "leap-frogging" from one livestock dam (water hole) to the next. Some native predators and ground-nesting birds have declined in numbers where cane toads have moved in. In such large numbers, cane toads compete with several native amphibians, reptiles, and birds for habitat.

Any time a biologic control of a nuisance species is considered, there needs to be detailed examination of possible consequences before a control agent is introduced, or more disasters to native populations of organisms similar to those in Australia may occur.

Subtracting and Adding Plant Species

Farming is arguably the basis of civilization. Because of farming, wandering tribes of people could settle in one place, near where food was grown. Not having to search for food every day allowed people the time to become proficient and profitable at other things. A consequence of farming is that native plants are removed from an area of soil and plant species that provide high yields of nutrients for people and livestock are planted in their place. Much production of grain, vegetable, oilseed, and fruit crops has displaced vegetation that was less palatable to many large herbivores. Occasionally, these crops need protection from nuisance animals, or there will be significant damage and reduction in yield. Forage crops may be palatable to a variety of large wild herbivores, particularly deer, elk, and bison. Native grasses have been replaced by tame grasses such as timothy, orchard grass, Bermuda grass, tall fescue, and brome grass over much of North America. In the arid western United States, large range areas produce mostly native grasses that are well-suited to cattle, horses, and wild animals.

Habitat Fragmentation

As civilization and development expand, wildlife habitats shrink, and areas of ideal habitat for a particular species that were once continuous are broken up or fragmented. Small areas of habitat are in some cases too small to support the species that once flourished there. For species that require large ranges and feeding patches (bison, caribou, cougars), habitat fragmentation is particularly damaging. Fragmentation results in longer perimeters or edges, so penetration by competing species which cross habitat boundaries is more frequent. For example, brown-headed cowbirds penetrate shrunken habitats of black-capped vireos (Robinson et al., 1993). This is damaging not just because of competition for food but because brown-headed cowbirds lay their eggs in the nests of smaller birds such as vireos, and the cowbird chicks are larger and more aggressive than vireo chicks. Susceptible animals within fragmented habitats may be unwilling to cross from one fragment through nonhabitat to the next habitat, so inbreeding and its resultant loss of vigor occurs (Young et al., 1996). Agriculture, fences, highways, railroads, pipelines, power lines, border walls, and urbanization are all causes of fragmentation. The effects of habitat fragmentation can be partly undone by creating corridors and patches of habitat in a stepping stone pattern that help to connect fragments. A remarkable example of corridors is the installation of wildlife crossings under and over highways in Banff National Park, Alberta. Purchasing or donating lands to be set aside for habitat preservation and creating zones around them where development is prohibited can help fragmented lands to recover (Jannota, 2014).

Horse Parasite Control and Unintended Effects

An unintended effect of the use of a pest eliminator was discovered in the 1980s when ivermectin was studied for its ability to control internal parasites in horses. It was discovered that the manure piles on a pasture from horses treated with ivermectin did not break down as quickly as manure piles from horses treated with oxibendazole or no wormer (Herd, 1995). In overgrazed pastures, this phenomenon contributes to large latrine areas where manure accumulates, its breakdown is slowed, and the forage among the manure pats is not grazed.

Ivermectin residues in feces are toxic to fly larvae, which aerate dung pats. Aeration normally attracts earthworms, which help to disperse manure (Herd, 1995). Dung from ivermectin-treated cattle inhibits development of flies and dung beetles and apparently repels dung beetles from their normal dung-dispersal efforts (Lumaret et al., 1993).

Dung dispersal, nutrient recycling, soil aeration, humus content, water percolation, and nematode control in turn have beneficial effects on plant growth.

Ivermectin persists with a half-life of 111−260 days at 68°F (20°C) (Nessel et al., 1983). Toxicity of ivermectin to aquatic *Daphnia magna* was extremely high (Garric et al., 2007), with adverse effects on growth and reproduction at very low concentrations. Acute and long-term (>229 days) toxic effects were discovered in aquatic sediment-dwelling organisms at or below predicted environmental concentrations of ivermectin (Sanderson et al., 2007).

Herd (1993, 1995) recommended that a careful balance between overuse and underuse would protect the environment, reduce parasite burdens, and encourage development of immunity. Negative collateral effects of ivermectin on ecologically important organisms could be minimized by a combination of the following:

1. using only ecologically safe anthelmintics outside the breeding season of dung fauna Woodward (2005)
2. keeping ivermectin-treated horses off pasture for 3 days posttreatment
3. use of quantitative fecal egg count kits and a selective chemotherapy program to target only those parasites that are evident (Herd et al., 1993)

TECHNIQUES OF ENCOURAGING MORE WILDLIFE TO VISIT OR INHABIT YOUR FARM

Diversity of Forage Species

Benefits of a diverse ecosystem are discussed in Chapter 5 Pasture Ecology and Chapter 10 Mixed Species Grazing. Establishing several compatible perennial forages provides opportunities for a relatively large number of nonplant species to coexist with them. A perennial forage sward protects soil from erosion, allows water infiltration, reduces runoff, retains nutrients, builds organic matter (sequestering carbon) and nitrogen content in the soil, and provides shelter for wildlife (Barker et al., 2012).

Forage Management and Habitat

While the primary reason for producing forage on horse farms is providing suitable quantities of forage with appropriate nutritional content, trade-offs could be made if improving wildlife habitat becomes profitable. It is difficult to make recommendations for specific forage mixtures that would encourage growth of specific wildlife populations in a variety of ecosystems. However, it is known that each species of large herbivore has its own preference order for various forages and browse plants (shrubs and trees). As seasons change, wild herbivores tend to move to locations where they can easily obtain the most palatable and energy-dense feed materials. Consulting with local expert wildlife managers and pasture or range managers could help horse farm managers to make decisions that encourage wildlife of interest to them. Harper et al. (2007) described warm season grasses that encouraged mixed wildlife populations. While dense stands of grass provide plenty of forage for large herbivores, these stands are not as desirable as more open stands for ground-nesting birds (Vickery et al., 2001; Barker et al., 2012).

Biodiverse vegetation with plenty of flowering plants supports many insects and birds. Changing management of forage stands such as altering harvest dates can influence the availability of flowers or the length of time that birds can safely remain on their nests (Barker et al., 2012).

Jim Gerrish increased biodiversity in pastures that were predominantly endophyte-infected tall fescue by overseeding legumes, aggressive grazing in spring, and longer summer recovery periods (Voth, 2017d).

Grasslands are natural habitats for many songbirds, and each species has its specific preferences for habitat. A mosaic or diversity of grassland habitats provides a spectrum from tall and dense to short and sparse vegetation, which will accommodate the habitat preferences of many bird species (Krausman et al., 2009). For example, chestnut-collared longspurs prefer short, sparse vegetation, and grasshopper sparrows prefer tall, dense vegetation such as ungrazed productive pasture. These preferences can be revealed in numbers of birds and in clutch sizes (number of eggs per nest).

Removal of a significant amount of vegetation from a productive grassland will greatly alter the number and diversity of insects that find it attractive. As a result, insectivorous birds, amphibians, reptiles, and mammals will move to places where prospects of finding insects are better. Carnivores will also follow. If the forage is allowed a long rest period, it grows tall and again becomes good habitat for certain insects. Continuous grazing or overgrazing can keep forage in a short, sparse condition that will not support the same diversity of insects, insectivores, and carnivores as taller, denser forage.

In more than 500 years of settlers from Europe managing forages in the Americas, many native plants have been unable to adapt to the new management or to compete against the vigorous annual weeds that quickly fill voids or against the many introduced forage species. Thus, the composition of invertebrate and vertebrate animal species has also shifted. Before settlement in the Americas, riparian zones along waterways would be visited occasionally by deer, elk, or moose, with minimal impact from hoofs, manure, and urine. When a herd of bison came along for a drink, the impact would be dramatic but brief. Plants would be trampled, contributing to soil organic matter, but not killed. Manure and urine would seep into the water for a time but then be depleted. The native plants recovered due to a long regrowth time, probably not being grazed or trampled again for months. Rotational and deferred grazing systems mimic these natural systems in terms of short grazing periods and long regrowth times.

Today, some people in rural and urban areas purchase hay, grain, and commercial wildlife mixes to feed to wild animals, and this is widely recommended for small birds. However, such feed does not become part of the habitat. A longer-lasting measure is planting perennial or annual plants, such as bushes and trees that produce nuts or berries, in addition to attractive forages. Planting grains and oilseeds such as corn, soybeans, and sunflowers is somewhat expensive, depending on the area and the amount of machinery used, but it can be very effective in luring animals to a particular location. If some income can be earned by having nature photographers or hunters pay to come to these locations, it can be economically feasible. Planting annual crops just for livestock grazing is probably not sustainable in the long term, but if the yield is high and it is part of crop rotation with other benefits, it may be worthwhile. If hunting is to take place near the lure crop, then hunting regulations regarding lure crops should be consulted and followed. In many jurisdictions the use of lure crops or bait for hunting is strictly forbidden.

Haying Practices

The practice of management-intensive grazing or rotational grazing, as promoted in this book, encourages minimizing the production and use of hay (or silage) and maximizing the use of grazing to feed horses. The length of the grazing season is largely limited by climate and weather but may be modified by elevation, soil characteristics, choice of forage species and varieties, and other management practices. Wherever conditions do not allow year-round grazing of forage that meets nutrient requirements, some processed forage will need to be fed, so the usual practice is to make hay (see Chapter 11). Fields of hay are attractive to a large number of organisms including bugs, insects, amphibians, reptiles, small mammals, and many species of birds. Small perching birds such as bobolinks, red-winged blackbirds, meadowlarks, and loggerhead shrikes may nest in the same fields as pheasants, wild turkeys, or mallard and pintail ducks.

Harvesting hay of high nutrient concentration involves cutting it when the plants are developing from a vegetative to a reproductive stage, before flowers turn into mature seed heads. If higher yields of lower quality, more mature hay are desired, harvest is delayed. Wildlife conservation organizations in many locations encourage delaying the first harvest of hay until the bird nesting season is over (Undersander et al., 2000; USDA NRCS, 2010). This would allow chicks to hatch and learn to fly before the harvest. The disadvantage to the farm and its horses is that hay is not made until its nutritional quality is significantly reduced.

Jimmy Doyle of South Dakota State University Extension suggested cutting hay from the center of a field toward the edges or from one side of the field to the other, preferably toward suitable cover. This provides more opportunities for small animals and birds to move through the cover of existing tall hay plants and only have a short distance to cross from the edge of the hayfield to adjacent cover than the traditional method of circling from the outside to the center. If there is no other cover near the hayfield, leaving an unharvested strip can maintain some cover and provide some standing or stockpiled forage for winter grazing (Doyle, 2017). Tall fescue and Russian wild rye grasses maintain their quality through stockpiling better than most grasses. The unharvested strip can also trap snow to retain moisture and reduce drifting of snow across a field onto a road.

Where wildlife has been observed, slowing machinery can provide more opportunities for broody hens and young chicks to escape before a haying implement reaches them. Some wildlife organizations including Ducks Unlimited and Pheasants Forever can lend flushing bars to farmers for installation on tractors ahead of hay mowers. Flushing bars are often used during a nesting season so that hay can be cut when it has high quality. Chains dangling from the flushing bar scare hen ducks and pheasants off their nests, which are often destroyed. However, after flushing to save their lives, the hens may nest again. Deer fawns left in hayfields may also be flushed from their hiding places by these bars. Fawns up to 4 or 5 weeks old will bed down in grass near their mother for most of the day and only rise to nurse four to six times per day. It may often seem as if a fawn has been abandoned, but the fawn conserves energy by hiding and resting. By choosing open areas with tall grass, such as hayfields away from bushes, fawns are hidden from view at eye level from common predators.

Land owners make decisions about encouraging versus discouraging wild species. If part of the farm income can be derived from wildlife-oriented activities, then there may be a reasonable economic trade-off between preserving the wildlife and accepting lower quality hay. Hay for most horses does not need to be of the highest concentration of protein and nonstructural carbohydrates (NSC), so producing some relatively mature hay of moderate protein and NSC concentrations is probably not an economic loss (Voth, 2017e). The rate of decline in nutrient concentration of hay crops varies with latitude, as does the date of fledging of various bird species, and there is some scope for making hay of reasonable quality (e.g., for horses) after allowing birds to safely raise their young (Brown and Nocera, 2017).

Grazing Management Techniques

The quality of wildlife habitat has been degraded in many environments due to inappropriate grazing management for the natural conditions. Continuous grazing, overstocking, and overgrazing appear to have caused considerable harm, especially in arid and semiarid rangelands. Overgrazing reduces productivity, impedes plant survival, alters plant species composition, and reduces plant species diversity (Krausman et al., 2009). In a series of wildlife conservation studies reviewed by Krausman et al. (2009), rotating rested pastures maintained productive cover and forage for elk, and prairie grouse nesting success increased. Furthermore, a rest-rotation grazing system provided an open habitat desired by young greater sandhill cranes, and rotational grazed areas provided habitat for a prey base of small mammals for hawks and owls.

Krausman et al. (2009) proposed the following:

1. Instead of continuous grazing, grazing areas should be appropriately stocked and managed to provide blocks of undisturbed cover at times that allow for plant reproduction and energy storage plus wildlife reproduction and survival.
2. Managers establish grazing management systems that help to achieve long-term native habitat conservation.
3. Lengthy rest of forages should be used for sustainability and recovery of grazed ecosystems.

Riparian zones within pastures and ranges are particularly vulnerable to overgrazing by livestock. It is generally recommended that grazing animals be excluded from riparian zones by permanent fences and only be allowed into them for short periods of grazing during dry soil conditions. Where farm management requires that animals have occasional access to a stream crossing, a well-constructed crossing with a firm rocky base will help to minimize riparian damage and water contamination (Undersander and Pillsbury, 1999).

Increased plantings of row crops and windbreaks have decreased habitat for grassland birds such as meadowlarks, bobolinks and upland sandpipers. Rotational grazing with many paddocks means that long resting periods between grazing periods are possible and this provides opportunities for grassland birds to be undisturbed through much of their nesting time. Alternate grazing of paddocks, larger paddocks, longer resting periods and leaving at least four in (10 cm) of residual forage help to increase undisturbed time for grassland birds in preferred cover.

Providing Drinking Water, Feed, Cover, and Breeding Habitat

A ranching family east of Burns, Oregon, decided to practice holistic management, which meant big changes in philosophy and land management. Holistic management guides managers in three components: environmental, human, and financial, and it encourages lifelong learning. Annual precipitation at Burns is just 10.92 inches (277 mm). The average winter minimum temperature is 15.6°F (−9.1°C), and average summer maximum temperature is 82°F (28°C). The ranch only has about 70 frost free days. Practicing holistic management meant changing the focus from raising beef cattle to being grass farmers, harnessing the sun's energy, and harvesting it with cattle, which could be sold for beef. In that environment, the most successful native forage is bunchgrass. To provide drinking water for 3000 to 4000 cows in all grazing areas, the owners dug wells, developed additional water holes, and installed pipelines and large water troughs in key areas. Frequent movement of animals in their intensive rotational grazing system means that at any given time, 98% of the forage is being rested. A paddock grazed in spring in 1 year is grazed in a different season the next year. A surprising amount of forage is purposely trampled, not to waste it but to increase soil organic matter. In 10 years, they have been able to double their cattle carrying capacity. Over a 15-year period, due to more appropriate habitat, the sage grouse population increased by 30%. From 1999 to 2009, redband trout numbers spawning in the streams rose by 80%, migrant neotropical bird numbers were up 19%, greater sandhill crane numbers were up 251%, and the mule deer population had also grown (Voth, 2017c). Similarly, converting marginal croplands in arid landscapes to grasslands with stock water tanks and wildlife-friendly ramps to allow wildlife to escape has been so successful in Montana that in 2015, the US Fish and Wildlife Service concluded that the greater sage grouse no longer needed protection under the Endangered Species Act (Voth, 2017c).

Subtracting and Adding Prolific Omnivores

Wild turkeys were common in much of the United States when settlers first arrived. Overhunting and logging of their forest habitat through the 1700s to about 1950 reduced their number to about 30,000 (Miller, 2013). By 2001, the population of wild turkeys in the United States had risen to about 6.7 million (Baggaley, 2017). Between 2001 and 2014, wild turkeys could be found in all 48 contiguous states, Hawaii, the southern parts of four Canadian provinces, and along a mountain range into central Mexico (Eaton, 2014). Some writers called this the greatest wildlife conservation success story ever (Miller, 2013; Eaton, 2014). To accomplish this recovery, prominent conservationists petitioned governments to set bag limits, seasons, and other hunting regulations. Protected areas of prime turkey habitat (especially oak forests) were established. A National Wild Turkey Federation was formed, and it worked with state and federal agencies to protect turkeys and their habitat. Regrowth of logged forests provided good nesting habitat. Early attempts to raise wild turkeys on farms and release young turkeys to the wild were failures. Predators, hunters, disease, bad weather, and probably release into inappropriate habitat were the problems (Miller, 2013). Beginning in the 1950s, programs of trapping wild turkeys from successful flocks and releasing them in reforested habitat were very successful, even in parts of the United States and Canada where wild turkeys had not lived since the last ice age. With relatively low predator populations in many areas, a typical egg clutch size of 10−18, and a diverse, omnivorous diet, turkey populations grew rapidly, especially in the 1990s.

Over 200,000 turkeys were reintroduced by 2011, and populations appear to have peaked (Eaton, 2014). Since the rise of the US wild turkey population, turkeys have moved into suburban and sometimes urban areas, where they are reported to have eaten all the feed at bird feeders, chased children, challenged adult pedestrians, picked at their reflections in car doors, and blocked traffic (Miller, 2013). They are also responsible for collisions with vehicles that smashed windshields and injured drivers. In some regions, turkeys are blamed for a decline in ruffed grouse numbers.

Since about 2001, wild turkey populations have begun to shrink, and some game commission managers claim that they did not see this trend coming (Baggaley, 2017). Competition from squirrels and deer for acorns and beech nuts, decline of beech trees due to beech bark disease, and predation by bobcats, coyotes, fishers, and raccoons contribute to the downturn in turkey numbers. One possible reason for the rise in predator numbers is that their numbers were suppressed by a rabies epidemic in the 1990s, and the disease incidence is now low (Baggaley, 2017). The Audubon Society (2014) predicts that climate change is already contributing to shrinkage of wild turkey habitat and that the species will lose 87% of its current winter range by 2080. More frequent storms in springtime cause more turkey hens and poults to be soaked by rain, making the hens smell more, and the miserable, wet poults more vocal, so they are easier for predators to locate (Baggaley, 2017). More information is available on 314 bird species threatened by climate change at http://climate.audubon.org.

Some state game seasons have been shortened and bag limits reduced due to the turkey population decline. A survey of the wishes of New York turkey hunters indicated that primarily they want to see and hear more turkeys. Secondarily, they want a chance to be in the woods, and the third thing they want is to actually kill a bird (Baggaley, 2017).

The increase of numbers of wild animals to levels that consume large amounts of food and use large amounts of habitat, in competition with other animals, is often considered a nuisance by some people. This can occur in agricultural, suburban, and urban areas, as mentioned for turkeys. Crows, deer, geese, coyotes, and raccoons have all been vilified for their effect on North American farms and communities. In his book, *Nature Wars: How Wildlife Comebacks Turned Backyards into Battlegrounds*, Jim Sterba (2012) wrote that the high degree of proximity of wildlife to human populations has caused, "an animal-lover's dream-come-true … into a sprawl-dweller's nightmare."

Tools and Resources to Help You Coexist With Wildlife

Farmers and ranchers need open space for grazing. Wildlife biologists, managers, and conservationists want to maintain or enhance wildlife values on farms and ranches. Open space provides habitat for much wildlife, so there is a natural partnership among farmers, ranchers, wildlife managers, and conservationists in advancing wildlife and habitat management (Krausman et al., 2009). Wildlife populations go through natural ups and downs. Numbers of horses or cattle on rural properties also fluctuate, for a variety of reasons, so the amount of surplus grazing area that a manager would not mind giving up to wildlife could fluctuate. Trying to coordinate these fluctuations along with year-to-year weather differences and changing economic situations are challenges for any farm manager. One of the benefits of practicing holistic management is that it helps you look at the big picture and use financial tools to keep farm businesses economically sustainable.

The Natural Resource Conservation Service (NRCS) of the USDA provides America's farmers and ranchers with financial and technical assistance to voluntarily put conservation on the ground, helping the environment and agricultural operations. Services include technical expertise, conservation planning, and incentives to landowners to put wetlands, crop lands, grasslands, and forests under long-term easements. Agricultural and wet lands can be conserved, and nonagricultural uses of land can be limited. Landowners can promote recovery of endangered or threatened species, improve plant and animal biodiversity, and enhance carbon sequestration. Conservation efforts benefit soil, water, air, plants, and animals for productive lands and healthy ecosystems. NRCS experts from many disciplines work together to ensure effective conservation practices.

NRCS Field Office Technical Guides (FOTG) are primary scientific references containing technical information about conservation of resources. Guides are specific to each state. A general section of each FOTG provides maps, descriptions of land resource areas, watershed information, and links to other NRCS information sources. Links are provided to researchers, agencies, and universities that partner with the NRCS and landowners. Information sources include costs, laws, regulations, resources, and details about protected animals and plants. Detailed descriptions of soils and adjacent water, plant, and animal resources are included.

The guides can be found at https://www.nrcs.usda.gov/wps/portal/nrcs/main/national/technical/fotg/.

The NRCS Conservation Stewardship Programs help farm managers build on their existing conservation efforts (Voth, 2017f). Once managers have established conservation goals, the stewardship programs can assist in accomplishing those goals. Some of the benefits participants have realized include these:

1. better productivity of forages and other crops
2. decreased inputs required
3. improved wildlife populations
4. more resilience to weather extremes

Local Conservation Districts in the United States and watershed Conservation Authorities in Ontario are also good sources of information on protection of the natural environment on working farms.

EXPLAINING SITUATIONS IN WHICH WILD ANIMALS CAN BE DETRIMENTAL TO THE GOALS OF YOUR FARM, INCLUDING PREDATION AND CROP DAMAGE

Wild animals are frequently considered "nuisance animals," especially when a landowner perceives that they have caused a problem. The same landowner might be happy to see the same animals in a zoo, game park, or national park. Table 17.1 lists some North American animals that are often considered a nuisance plus perceived problems and benefits associated with them.

TABLE 17.1 North American Wild Animals and Perceived Problems for Horse Farm Managers

Animal Species	Perceived Problems	Potential Benefits
Arthropods (flies, mosquitos, ticks)	Annoyance, discomfort, and disease for horses and people	Provide food for birds, amphibians, reptiles, fish, and small mammals
Badger	Dig holes and tunnels, causing injuries to horses	Eat insects and small rodents
Bat	Rabies vector	Eat mosquitos, pollinate flowers
Beaver	Flood fields and roads, reduce land area, remove valued trees, frighten horses	Maintain water table, encourage vegetation, provide aquatic habitat, increase biodiversity
Bear	Frighten horses and people, damage property	Help to turn over organic material and recycle nutrients in ecosystems
Birds, small (e.g., starlings and sparrows)	Deposit feces that look messy and may carry pathogens (toxic *E. Coli* and *Salmonella* bacteria, *Histoplasma* spores), fleas, mites, lice and ticks	Pollinate flowers, eat insects, spread seeds, pleasant to view and hear
Cat, domestic and feral	Kill birds that may be desired	Kill and eat small rodents
Coyote, wolf	Kill and eat young horses and other livestock, spread mange mites	Keep herbivore populations in check
Crows, ravens, jays	Large crow flocks may be noisy and messy, spread West Nile virus, *Salmonella*	Clean up the environment, sometimes pleasant to view and hear
Deer, elk, moose, pronghorn	Eat forages designated for horses and livestock, break fences, collide with vehicles, spread Lyme disease and chronic wasting disease	Pleasant to view, may provide income through accommodation of paying hunters
Dog, domestic and feral	May attack young horses or other livestock	Help control small rodent populations
Donkey, feral	Eat forages designated for horses and livestock	Pleasant to view
Fisher, mink, weasel	Kill cats, wild birds, and poultry	Rare opportunities to view
Fox	Kill poultry, spread mange, rabies	Pleasant to view
Goose	Large flocks may eat legume and grass forages	Pleasant to view and hear, may provide hunting income
Hare, rabbit	Strip bark from planted sapling trees	Pleasant to view, fit into forest and prairie ecosystems, provide hunting opportunities
Horse, feral	Compete with domestic horses for food and water, cause injuries	Tourists may pay to view
Mouse and rat	Damage buildings and commercial feeds, spread diseases including hantavirus and plague	Contribute to control of plant populations, provide food for carnivores
Muskrat	Scare and injure horses	Grazing vegetation in aquatic and riparian zones, nutrient cycling
Owl, hawk, eagle, osprey, vulture	Kill or injure young livestock and desirable birds and small mammals	Keep populations of small birds and mammals in check
Pigeon	Spread manure, West Nile Virus, toxic *E. Coli* and *Salmonella* bacteria, *Histoplasma* spores	Clean up spilled grain
Porcupine	Damage wooden implements and structures by chewing, deposit quills, frighten horses, injure dogs	Thins vegetation by eating leaves, buds, fruit, and bark
Prairie dog, ground squirrel (gopher)	Dig holes and tunnels that may cause injuries to horses	Provide burrows for other species, cycle nutrients, eat insects
Raccoon	Damage buildings, spread roundworms, leptospirosis, parvovirus and rabies, eat wild bird eggs	Eat smaller pests, including wasp larvae

Continued

TABLE 17.1 North American Wild Animals and Perceived Problems for Horse Farm Managers—cont'd

Animal Species	Perceived Problems	Potential Benefits
Skunk	Unpleasant smell, spread rabies and tularemia, eat wild bird eggs	Eat insects and mice
Snake	Some are poisonous, some scare animals and people	Help control small rodent populations
Snapping turtle	Rarely scare or bite people or animals	Clean up decaying material in aquatic and riparian environment, reduce numbers of insects, fish, birds, amphibians
Tree squirrel and chipmunk	Minor damage to buildings and wiring, can spread mange, take feed from bird feeders	Spread seeds, berries and nuts from trees, eat some insects
Wild turkey	Take feed from bird feeders, compete with smaller ground-nesting birds, collide with vehicles, frighten horses	Pleasant to view and to hunt, may provide hunting income
Woodchuck (groundhog)	Eat forage legumes, grasses and garden plants, dig holes and burrows that injure horses	Increase soil porosity and drainage, mix soil layers, nutrient turnover

Predation, Injuries, and Scaring

Since most adult horses are larger than most predators and most properties that have horses are managing more than one horse, there are not many predator attacks on horses, compared to those on sheep, goats, or cattle. Attacks on humans by coyotes are extremely rare, but there are recorded cases of coyotes killing horses in Ontario. The number of horses killed by predators was less than 8 per 100,000 in 2015. Cougars may be an occasional predator on feral horses in the western states and provinces. See Chapter 15 for more details on predator kills. Bears rarely attack horses, and well-maintained electric fences are good bear deterrents, being especially effective in protecting bee hives. The most vulnerable horses will be foals and miniature horses, in small numbers or alone, possibly sick or injured, with poor fencing and a significant predator population nearby. A single coyote of 15–45 lbs (7–20 kg) is unlikely to harm a single 1000-lb (454 kg) horse. There are plenty of techniques used to minimize the probability of predation on horses. Several are described later in this chapter. Injuries to horses from animals that are not trying to kill and eat a horse are also rare, but horses have been injured directly by porcupines, muskrats, and other surprisingly small animals. An animal that frightens a horse can also cause the horse to injure itself.

The scariest wild animals tend to be large carnivores (wolf, mountain lion) and omnivorous bears. Other animals that may sometimes scare horses include any medium-size animal that is initially hidden from a horse's view. Then both animals frighten each other by sudden movements. Snakes that suddenly appear or rattle and birds such as grouse, pheasants, and turkeys that are noisy when flushed can also frighten horses. A nuisance wildlife control professional in Ontario had a client complain of a muskrat that grabbed the neck of her horse when it bent to drink at a pond. When the horse reared up, the muskrat was unable or unwilling to let go, causing further distress to the horse. Presumably, beavers and snapping turtles could also surprise horses and cause fear (McKenzie and Alkerton, 2008).

Another indirect way that an animal causes an injury to a horse is by digging holes and tunnels in which a moving horse catches a leg.

Crop Damage

Forage crops can be damaged by a variety of bacteria, fungi, insects, and other arthropods. This chapter focuses on larger organisms that can damage crops, particularly mammalian herbivores and birds. The large mammalian herbivores that are most likely to eat pasture and hay plants are deer, elk, pronghorns, moose and bison. The large variability in sizes of these animals means that if fences are used to exclude any of these species or to allow them to pass through, it is important to know which species predominates and customize fence height and style to them.

While a great diversity of birds can be found in pastures and hay fields, their numbers are usually small, and the length of their stay is usually short. However, large flocks of Canada geese will eat and trample legume and grass hay fields in late summer and early autumn, causing a significant loss of yield (Fig. 17.1).

FIGURE 17.1 Flocks of Canada geese graze and trample legume/grass hayfields on this farm next to an equestrian center in Ontario, Canada. *Photo by Paul Sharpe.*

Digging animals such as prairie dogs, ground squirrels, woodchucks, European rabbits, and badgers can damage an area of pasture or hay field. The results can include loss of hayland and pasture, damage to machinery, and injury to horses.

Alfalfa is a very attractive crop to wildlife. An estimated 27% of California's 627 wild species use alfalfa for cover, breeding, or feeding (Putnam et al., 2001; Putnam and DelCurto, 2007).

Every geographic region can have a unique set of nuisance animals, many of them feral. The continent that seems to have more examples of nuisance feral animals than any other is Australia. Certainly, Australian farmers have tried to control native kangaroos for years in the belief that the kangaroos are eating the pastures that are intended for sheep and cattle. Elsewhere in this chapter is a story of problems with native emus. Depending on the location, Australian farms and wildlife habitats could also be negatively influenced by 73 species of introduced vertebrates, including wild dogs, dingoes, foxes, European rabbits, and the following feral animals: cats, pigs, goats, horses, rose-ringed parakeets, cane toads, water (swamp) buffalo, camels, and deer of six species. Negative effects of introduced species can include crop damage, preying on livestock, soil erosion, stream turbidity, spread of weeds, competition with native species, destruction of habitat, and spreading diseases (Invasive Plants and Animals Committee, 2016). This continent, which is so well-known for droughts and bush fires, has now had another native nuisance revealed in a journal article entitled, "Intentional fire-spreading by 'firehawk' raptors in Northern Australia" (Bonta et al., 2017; Warnica, 2018). The authors documented indigenous and nonindigenous accounts of three species of raptors picking up burning sticks and dropping them into unburned grass, which then burned. The birds then waited for rodents and reptiles fleeing the flames to capture and eat them. Later the birds picked through remains of burned small animals.

Arthropod Pests

Several arthropods cause annoyance, discomfort, and/or disease in horses and people who work with them.

Horses sometimes demonstrate annoyance to large numbers of mosquitos, and some horses develop hypersensitivity with reddening of skin, swelling, and itching. Mosquitos transmit viruses that cause Eastern Equine Encephalitis (EEE), Western Equine Encephalitis (WEE), and Venezuelan Equine Encephalitis (VEE). Horses infected with EEE and VEE often die, but survival rates of horses infected with WEE can be 70%−80%. People, pigs, llamas, bats, reptiles, amphibians, and rodents can also be infected. Birds are virus reservoirs but often show no signs of the disease. A few bird species (pheasants, emus, whooping cranes, and partridges) can develop illness and die from these viruses. Vaccines are available for horses against the EEE, WEE, and VEE viruses. Humans can protect themselves by decreasing mosquito exposure through covering exposed skin, using mosquito repellent containing DEET, and avoiding outdoor activities when mosquitos are most active (dawn and dusk) (Wright, 2017).

West Nile virus (WNV) is also carried by mosquitos and can cause disease in horses and humans. Up to a third of horses showing signs of WNV disease will die (Wright, 2006). Crows, jays, and humans are at high risk for West Nile disease. Two vaccines against WNV are available for horses. The first WNV vaccine for use in humans entered safety trials in 2015, and no commercial product was available as of 2017. Many urban communities have mosquito monitoring and control programs, but in rural areas, this is up to landowners. On a farm, tasks in a mosquito reduction plan include the following:

1. Identify where water accumulates and empty, cover, reduce, or fill them in if possible.
2. Use mosquito larvicide on drains or catch basins that hold water by design.
3. Old tires used as weights should have holes drilled in them to drain them.
4. Put aerators (wind driven) in ponds to reduce mosquito breeding.
5. Artificial ponds and stock tanks can have small fish stocked in them to eat larvae.

Various species of flies cause annoyance to horses, and the bot fly is well-known to most horse owners for laying its eggs on the hair of horses (PetMD. 2018). The horse licks the eggs and some hatched larvae, which stay in the mouth for days to weeks. The larvae pass to the stomach, where they hatch from remaining eggs and attach to the lining, staying for 8 to 10 months. Finally, the larvae pass through the digestive tract and are ejected in manure. Larvae burrow into the ground and mature into pupae, then adults. Large numbers of bots can contribute to colic, stomach ulcers, blockage, damage to the gut, loss of nutrients, and poor appetite. Secondary problems in the esophagus and stomach may follow a large infestation. Adult flies can cause annoyance and fright to horses. People working with horses may sometimes have bot larvae enter their skin, causing inflamed tracks that cause irritation and itching (McLendon and Kaufman, 2007).

Other arthropods that are a significant pest for horse owners are lice and ticks. Lice tend to spread from horse to horse, so will not be considered wildlife here. However, ticks hook onto horses from vegetation as horses pass by. Within a few hours, ticks attach to the skin and suck blood for a few days, then drop to the ground where they molt and lay eggs. Very small tick larvae are capable of hooking onto another host very close to where they hatched. Thus, tick concentrations can build where there is a high population of livestock or wildlife, such as at feeding and watering points and resting places. Stock horses in Central America tend to get the highest tick burdens. Life cycles of different tick species vary. Horses tend not to become annoyed by ticks until they are infested with very many. Blood loss from hundreds of ticks can lead to anemia. A mild toxin from tick saliva can affect a horse's immune system, leading to toxicosis and paralysis. Almost all tick species can transmit a disease, depending on the geographic region and the tick species. Not all individuals in a tick population are infected. Tick-borne diseases may be caused by protozoa or bacteria of the spirochaete or rickettsial types (Junquera, 2017). The tick-borne disease most worrisome for people in North America is Lyme disease, caused by a bacterium carried by the black-legged tick. Other diseases that may be transmitted by ticks to mammals include piroplasmosis, ehrlichiosis, and equine infections anemia (Lenz, 2017). Wild birds may spread ticks over surprising distances.

Where ticks are known to exist, horses should be inspected within 12 h of being in tick-infested vegetation (Junquera, 2017). This may mean daily inspections of horses on pastures.

Inspections of horses for ticks may reveal the following:

1. engorged larvae 0.04−0.08 in long (1−2 mm), too small to be seen by the naked eye
2. engorged nymphs 0.08−0.2 in long (2−5 mm), hard to see
3. unfed adults 0.08−0.4 in long (2−10 mm)
4. engorged adults 0.4−0.8 in long (5−20 mm)

If you are not experienced in dealing with ticks, you should learn which types of ticks to expect in your geographic area, which diseases they likely carry, and how to remove them. There are a number of specific tools available for tick removal, including tweezers, scoops, hooks, and loops. The tick should be grasped firmly where the head enters the skin and pulled straight out. Twisting or squeezing the tick body causes the tick to regurgitate blood back into your horse, increasing the chance of infection or disease transmission. You can kill the ticks and learn something about them by dropping them into a bottle of alcohol or a dry bottle with an airtight lid and taking them to your veterinarian. Wearing disposable gloves during tick removal can help to prevent developing an allergy to the ticks.

Insect repellents containing DEET and designed for humans have very little effect on ticks, and this lasts just a few hours if at all. Some parasiticides that kill ticks can have a slight repellent effect. Biologic control methods for tick control are not very effective yet. Pathogenic fungi have been tried with modest results. Homemade recipes based on plant extracts have not been demonstrated to be highly effective as tick repellents. Horses that are usually kept in barns or bare soil yards and only occasionally go into longer vegetation are a target market for tickicide sprays, pour-ons, or shampoos marketed to repel or kill ticks. It is advisable to check with a veterinarian before using a tickicide or tick repellent product because many of these products have not been tested or approved for use on horses. One active ingredient called "amitraz" is used for tick control on cattle, goats, sheep, and dogs but is toxic to horses (Junquera, 2017).

Tick-specific repellents include coumaphos spray or power, permethrin applied as a wipe, spray or spot-on, and zeta-cypermethrin dusting powders. Check labels to be sure they are effective against ticks. These should be applied to a horse's mane, tail head, chest, and underbelly before riding or turning horses out to pasture (Lenz, 2017). Often, it is easier to feel ticks than see them.

Running your fingers over a horse's skin can reveal a tick, and a local skin reaction to a tick often appears as a firm nodule. Grooming every day helps in tick detection and removal. Applying baby oil, petroleum, or a lit match to a tick are not effective removal methods. The number of ticks on your property may be reduced by removing brush and discouraging wildlife, such as deer, that carry ticks. Free-range chickens and guinea fowl find and eat ticks (Lenz, 2017).

DETERMINING WHICH DETRIMENTAL SPECIES MAY BE PRESENT OR RESPONSIBLE FOR DAMAGE

Predation and Signs of Predators

Direct evidence of the recent presence of animals includes tracks, scat, hair on fences, photographs, written descriptions, and parts or all of an animal. Diagrams and photographs of animal tracks and scats are available in books and websites (Cabrera, 2016; Evans, 2016; McWelch, 2014). Hair found on fences or bushes can be analyzed for DNA. Analysis of animals that were killed by predators can indicate probabilities of the predator species (Alberta Agriculture, 2010; ICWDM, 2015). For example, when a small herbivore is found dead, bites in several parts of the body, injuries to several animals, and no evidence of eating the carcass are probable indications of dogs or young, inexperienced coyotes. If an animal was killed by a single bite to the throat, the killer was likely an experienced adult coyote or wolf. Coyotes, foxes, mountain lions, and bobcats usually feed behind the ribs and consume some of the organs. Bears are more likely to eat meat and an udder from lactating females (Sheep101, 2014). Mountain lions usually eat the front quarters and neck (ICWDM, 2015).

To reduce the time required to obtain photos of animals, especially in the dark, a game or trail camera can be used. Binoculars, spotting scopes, and telephoto lenses on cameras are useful aids to identification when you have time to be observing. The result of trapping can prove the presence of a certain species of animal but does not indicate which animal has done the damage.

Evidence of nonpredatory wildlife causing damage to a farm can often be found where animal feed is stored. Pigeons, rats, raccoons, wild pigs, bears, and other animals may damage buildings and containers to get at grain and other concentrate feeds. Deer sometimes break and scatter hay bales. Silage bags can be damaged by mice at ground level, allowing air in and silage to spoil. Coyotes can cause further damage to silage bags in their search for mice.

Location of damage done by wildlife can provide a clue as to the species of the culprit, since each species has preferred habitats. Animals that prefer forest are not likely to go into the middle of large open pastures to cause damage, but animals that like to hunt the edges of forest and follow low, bushy cover are more likely to go into an open field. Wide open spaces also attract mule deer, elk, and bison. In forested or bushy areas, moose, whitetail deer, wild pigs, wolves, and mountain lions find plenty of cover for hiding.

Information can be gathered from neighbors, local trappers, hunters, agricultural extension workers, and wildlife management officials about which predatory animals have been frequently observed nearby and how many episodes of livestock kills there have been recently. Learning more about the life of possible nuisance animals can improve your probability of accurately identifying which species has been a nuisance and help you make decisions on how to manage the situation better. A group of neighbors discussing their problems with wildlife may lead to a group or regional management program that shares information and resources and meets goals over a large geographic area. This can help all of the neighbors become proactive rather than reactive. With a proactive approach to wildlife, few domestic animals are likely to be injured or killed, wildlife management costs are lower, the species you prefer to support will be more numerous, and your public image may be enhanced. If you just wait before doing anything about wildlife habitat and management, your first need to do something will likely be a case of an animal being injured or killed, expenses may be higher, and your public image will be less positive than if you were known as a supporter of wildlife.

TECHNIQUES TO DISCOURAGE CERTAIN ANIMAL SPECIES FROM VISITING YOUR FARM

In the section on encouraging wildlife to visit a property, the question of planting lure crops was discussed. If there is already a population of animals that is a nuisance on your property, perhaps because they damage a valuable crop of hay, a strip of an even more attractive crop in another part of the property or on a neighboring property may prevent or reduce the most costly damage.

Scaring Nuisance Animals

Scare devices such as bright flashing lights, sirens, alarms, propane-powered bangers that sound like shotguns, plastic flags on fences, owl decoys, and scarecrows are all used to keep some types of nuisance animals away from pastures, crops, gardens, or livestock, for a while. Animals are able to observe these items from a safe distance and learn whether they are actually harmful, so their long-term effectiveness may be limited.

Laws in many North American jurisdictions allow protection of property, so persons believing that wildlife is damaging or about to damage their property can legally, on their own land, harass the wildlife to deter it from damaging property or capture or kill the wildlife. Local, state (provincial), and federal laws should be checked before you act. Each of these three actions may have a permanent effect on one animal, but none of them will likely have a long-term effect on other members of the same species. Animals move from place to place to find food, shelter (cover), and a place to breed and raise young. If one member of a species has found at least one of these resources on a farm, then it is likely that others will follow. A management plan that identifies the resources present and includes a combination of techniques, implemented with appropriate timing, can be partially to largely successful in preventing visits by nuisance animals.

Integrated Methods

Agronomists and farmers have found success in dealing with weed, fungal, and arthropod pests in crops by combining a number of control techniques. A system of combining control techniques is called "integrated pest management" (IPM) (UC Davis, 2018). For example, in weed control, IPM programs have six major components:

1. pest identification
2. monitoring pest numbers and damage
3. guidelines for timing of actions
4. prevention
5. using a combination of biologic, cultural, physical, and chemical tools
6. assessing the effect of actions (UC Davis, 2018)

If IPM is done well, efforts are directed only at the problem organisms, at the ideal rate (dose), time, and location, with minimal cost and time required. Assessment after action helps the manager learn from mistakes and successes, with increases in knowledge and skill. Similar programs can be developed to manage animal pests.

Fences

Chapter 15 provides details on fencing and watering systems for horse farms. Old-style fences composed of hedgerows and stone walls or zig-zag styles of cedar rail fences can become wildlife habitats themselves, with little effect as barriers, especially when woody vegetation grows within and around them. In Australia, individual fences stretching thousands of miles have been built to restrict movements of single species of nuisance animal (Invasive Plants and Animals Committee, 2016). Wire fences have been used since the 1880s in Australia to keep small wild dogs called dingoes out of agricultural regions. One of these is 3488 miles (5614 km) long. In June 2017, a new fence was announced for the 994-mile (1600 km) perimeter of a former cattle station in central Australia to protect endangered native marsupials from feral cats. The Australian government is attempting to reduce its feral cat population by two million from its estimated current 6 to 11 million. It is estimated that a feral cat can kill five to seven animals every night (Long, 2017).

In the Intermountain and Pacific Northwest regions of the United States, wolves and grizzly bears are significant pests on ranches. Common fences used to keep wolves and grizzlies out include the following:

Six-strand high-tensile electrified wire with either alternating hot and ground wires or all hot wires. Voltage of 7000–10,000 is recommended.

A three-wire electric fence is frequently used to keep black bears from beehives. The energizer battery is recharged by solar panels. This fence carries 8000–9000 V. It has also been used to keep grizzly bears from small fields of corn on a Montana dairy farm. On that farm in 2015, 27 bears were counted in the corn field, and in 2016, after installation of this fence, there were only three or four bears. Bright red plastic flagging tape about 2 inches (5 cm) wide and 18 inches (46 cm) long attached at intervals of about 30 inches (76 cm) along light polymer rope is referred to as "fladry." If the rope is electrified (polywire) the product is called "turbo fladry." It is recommended by an organization called "People and Carnivores" and ranchers in Montana and nearby states to improve the effectiveness of existing fences against wolves and bears (Voth, 2018b).

A farm in Montana kept a herd of 400 wild elk away from a large stack of hay bales with a two-wire electric fence running about 4 feet (122 cm) away from the stack. The top wire was located 24 inches (61 cm) above frozen ground and was electrified. The bottom wire was located 18 inches (46 cm) above ground and was grounded. These wires were about knee-high on adult elk. The combination of hot and grounded wires ensured that elk touching the fence were shocked, even though frozen and snow-covered ground does not carry much current (Feight, 2018).

Based on the data in Table 17.1, the risk of predation of horses is quite low in Ontario and eastern American states with similar geography and levels of agriculture. If any fence protection is planned for horses in eastern North America, it should be aimed at coyotes and probably domestic dogs, which sometimes chase horses.

In parts of western North America where horses are on tame, fenced pasture, there is a slight risk of coyote, wolf, and cougar attacks.

In a highland forest area of southeastern Brazil where pumas (*Puma concolor*, same as mountain lion or cougar) are the predominant large predator, horses were the most common prey, comprising 51% of the domestic animals attacked over 3 years (Palmeira et al., 2015), and pumas killed 8.6% of the horses in the region over those years. Numbers of cattle and sheep living in the area were much higher, but the percentages of their losses to pumas were just 1.9% and 0.9%, respectively. Predation was expected to be higher as the distance from an urban center increased and as the elevation increased (Palmeira et al., 2015).

Wolves were the most significant killers of horses in extensively managed areas of Mongolia and China. In Mongolia, near Hustai National Park, domestic horses were preferred prey over cattle, goats, and sheep. As the distance from the park increased and the number of tents in the area decreased, there was an increase in numbers of domestic horses killed, all by wolves. Reintroduced Przewalski horse foals were also killed by wolves, and their kill sites were more numerous closer to the forest, at higher altitudes, with lower shrub cover, higher forest cover, and higher red deer density. The authors suggested that increased conservation of deer would reduce predation on the Przewalski horses (Van Duyne et al., 2009). In northwest Yunnan, China, over 2 years, wolves were responsible for 79% of livestock predation, and black bears killed 15% of the animals. The percentages of donkeys and horses in the livestock population were 7% and 5%, respectively, but the percentages killed by predators were 14% and 12%, respectively, indicating either a preference by wolves to attack equids or relative ease of access to donkeys and horses. During the wet season, when livestock were left to graze unattended in alpine meadows, there was a peak in predation (Li et al., 2013). The nomadic lifestyle of the farmers was considered to be a significant contributing factor in the predator losses.

Trapping and Hunting

Some farm managers will feel a strong emotional reaction to damage by nuisance animals and feel very justified in trying to kill the nuisance animal. The first tool that some will try is a gun to shoot the animal. Eventually, many managers realize the following:

1. This is not as easy as I thought it would be.
2. The animal lives in this habitat all the time and knows it better than I do.
3. I have a limited amount of time to devote to guard duty and trying to find this nuisance animal and its family and friends.
4. I am not as good a shot as I thought I was.
5. There must be a better way to remove the nuisance animal(s).

Similar principles can by followed to trap or hunt any of the animals in Table 17.1. In addition to identifying the species of nuisance animal from tracks, scat, hair on fences, photos, or live observation, it is worth learning the biology of the species, so you can determine when and where the animals will be most susceptible to your efforts. Important aspects of the biology of a target animal include its normal geographic range, its normal and abnormal behaviors, common feeds it chooses at various times of year, and what type of habitat it uses for cover. You might learn some of this while monitoring their numbers and determining the extent of damage they cause. Few farm managers have enough spare time and hunting skills to effectively kill off animals that arrive on their own schedules. A high success rate at trapping requires training, patience, and specialized equipment. If trapping is to be a part of your IPM program, hire a reputable trapper, preferably one certified by the National Wildlife Control Operators Association (NWCOA.com) or the National Pest Management Association. Many people feel that for small mammals, they can simply buy a box trap (marketed as a "live trap" or "humane trap") from a farm supply store, bait it, and leave it where they saw a nuisance animal, and their problems will be solved. Unfortunately, problems can grow once an animal is captured by the trap. It is only a "live" trap if you check it frequently enough to prevent the animal dying due to distress from a combination of restraint, dehydration, lack of food, overheating, or hypothermia. Before buying the trap, persons with a nuisance animal problem should ask themselves what they can competently, ethically, and legally do once they have trapped an animal. In the Canadian province of Ontario, once you have trapped an animal, provincial law does not allow you to move the animal more than 5/8 of a mile (1 km) before releasing it. Every state and province and many local governments may have their own laws regarding capture, release, and killing of nuisance animals. In Florida, "live-captured bobcats must be released and other wildlife must be released or killed within 24 hours of capture or trap inspection. If transported, bobcats must be released within the county of capture and within a minimum of 40 contiguous acres if the animal is a native species. The releaser must have written permission from the owner of the release site and the transportation of the wildlife must not violate any rabies alert or area quarantine issued by a county health department or county animal service" (FFWCC, 2018).

Some wildlife experts feel that the biggest disadvantage to trapping and relocating nuisance animals is that wherever you put them, there is probably another individual or family of the same species already using that home range, with the consequence that one animal may be driven away and die over competition for resources. Relocated animals sometimes carry diseases that spread to other animals, and many will wander for long distances in search of an unoccupied habitat or their old habitat. What looks like a good place to a human to release an animal may not have the basic needs for the animal to survive (FFWCC, 2018; WDFW, 2018).

A single trapped and relocated animal may have babies that are left behind, and there is a chance that they may starve to death. If this happens in a house or barn, the smell will be most unpleasant and sometimes hard to find. Remember that in the northern hemisphere, the birthing season for many wild animals is April through June. If you are planning to erect fences that are very wildlife-resistant or to barricade openings in buildings, delay this until young animals and birds have left, so you do not accidentally separate mothers and offspring or close some into a building (Ottawa-Carleton Wildlife Centre, 2010). If part of the problem with nuisance wildlife is that they have set up a home in one of your farm buildings or your home, there are several things you can do to discourage their presence:

1. Use stronger, animal-resistant containers for garbage and compost.
2. Install motion-sensitive floodlights or sprinklers to frighten the animals.
3. Do not leave pet food outside, and keep pet doors locked at night.
4. Keep barbecues clean and closed.
5. Do not feed wildlife, especially near the buildings in question.
6. Cover window wells and fish ponds with tough wire mesh.
7. Control grubs in your lawn.
8. Inspect buildings, especially attics, roofs, chimneys, and vents for points of entry.
9. Cap animal entry points with tough sheet metal or wire mesh.
10. To prevent climbing of posts or tree trunks, cover them with smooth metal sheets up to 6.5 feet (2 m) high.
11. If animals have dug under buildings, install heavy L-shaped sheet metal or tough wire screen against the outer wall and into a trench about 15 inches (38 cm) deep around the building (Ontario SPCA, 2018).

Skilled trappers have a variety of traps to suit specific situations. Certified wildlife control professionals will do what they can to reduce the suffering of any animal they trap. This means either using a trap that kills quickly as soon as the animal contacts it or a trap that causes as little pain as possible while the animal is trapped. Sometimes, it will be appropriate to use a box trap similar in design to the "live" or "humane" traps that are widely sold to consumers. Professional versions have the box made of solid plastic, so entrapped animals cannot clearly see the trapper, and if a skunk is the trapped animal, the trapper is protected from its spray. These will be checked every morning before the air temperature becomes hot. A relatively modern trap that looks like a combination of a leg-hold trap and a snare is well-suited to catch animals as small as foxes or bobcats and as large as bears. The Belisle foot snare trap and similar traps have a mechanism that is triggered when an animal steps on it, and then the spring mechanism falls away, leaving a snare around the leg just above the foot. This foot snare holds but does not cut off the circulation to a foot and avoids the severe injury and pain of older style leg-hold traps (ITTM, 2018). It also avoids the strangling by wire snares set on animal trails. The trapped animal is not injured, and the trapper can then either relocate or quickly and humanely kill the trapped animal. Where skilled trappers do want a trapped animal to be instantly killed, they will use a body-crushing trap.

In some cases, word of mouth can help a landowner find hunters willing to hunt nuisance animals. It is worthwhile for the landowner to consult local hunting regulations to determine whether certain seasons, times of day, types of weapons, or luring animals is illegal. While bait is not allowed for hunting most animals, it is allowed for hunting bears in some places.

Recorded calls are allowed for hunting snow geese in some parts of Canada where population control efforts are in effect but not for any other migratory birds. Powered predator calls may be allowed in some areas.

Skunks can be repelled by bright lights, cotton rags soaked in ammonia, a loud radio, and moth balls. To prevent unpleasant encounters with coyotes on your property, the Government of Ontario recommends the following:

1. Limit attractants.
   a. Keep pet food indoors.
   b. Use garbage containers with locking lids in enclosed structures.
   c. Put garbage out the morning of scheduled pickup.
   d. Use enclosed composters rather than exposed piles.
   e. Pick ripe fruit and seed from trees and remove fallen fruit from the ground.
   f. Protect vegetable gardens with heavy-duty fences.
2. Discourage coyotes from entering your property.
   a. Clear away bushes and dense weeds where coyotes find cover and small animals.
   b. Use motion sensor lights.
   c. Close off spaces under porches, decks, and sheds.
   d. A fence should be 6.5 feet (2 m) high and extend at least 8 inches (20 cm) underground.
   e. Electric fencing can also help.
3. To prevent predation of livestock:
   a. Bring animals into barns or sheds at night.
   b. Use guard animals such as donkeys, llamas, or dogs and help them develop bonds with the livestock.

Guardian Animals

The animal on which the most effort has been spent to protect it outdoors is the sheep. Across North America the animal that causes the most predation on sheep and larger animals is the coyote. Much can be learned from the efforts to protect sheep from coyotes. For example, fences are not predator proof, but fences that are well built and well maintained with a particular predator in mind can be highly predator resistant. As mentioned earlier, an integrated pest (predator) management program will be more successful at reducing predation than depending on only one technique. Horse owners are fortunate in that horses are better equipped to defend themselves and their offspring against predators than sheep or goats are. Horses are quite vigilant and can run faster than sheep. Horses are generally taller than sheep, so they can see farther across a tall grass pasture to detect predators. Most adult horses are much larger and stronger than coyotes or wolves and can use their large hoofs to deliver damaging and fatal blows to predators. So you may be wondering why protection of horses against predators is even included in this chapter. It is simply to help horse managers understand how to protect their valuable animals against a rare, tragic occurrence.

Guardian animals that have varying degrees of success in protecting livestock against predators include dogs, llamas, donkeys, and adult cattle. Some farms use a combination of donkeys and llamas. The types of guardian dogs used are specific breeds that were developed centuries ago in Europe and Asia to protect sheep against wolves. These are large dogs that are not used for herding, and thus they are discouraged from chasing and barking at livestock. They will normally bark at predators that they see. Popular breeds include Great Pyrenees, Anatolian Shepherd, Akbash, Maremma, and Komondor. Guardian dogs are raised from a very young age with the animals that they will eventually guard, so a strong protective bond forms. They act independently from people, and to be effective, they must be self-confident, alert, and ready to bark aggressively at approaching predators. The most frequent problems with guardian dogs are biting and aggression to people and livestock (Green and Woodruff, 1988). A single guardian dog is sometimes overwhelmed by superior numbers of coyotes or wolves (Braithwait, 1996).

Guardian donkeys and mules appear to naturally hate dogs and apparently love to intimidate them. Features that improve the probability of a donkey or mule working well as a guardian include training to associate with livestock from a young age, using females rather than males, and selecting from medium- to large-size stock. A single donkey can protect up to 200 smaller animals in a field up to 80 acres in size. An advantage of donkeys over dogs is that they eat the same pasture as the animals to be protected, whereas guardian dogs need to have dog food provided daily (Braithwait, 1996; OMAFRA Staff, 2016).

Llamas are naturally aggressive against dogs and coyotes, attempting to chase, kick, paw, and kill them. Males, geldings, and female llamas can all be effective, and they are not as likely as dogs or donkeys to be aggressive to livestock or people. Many llamas have proven effective as guardians even though they were not introduced to sheep until they were

2 years old. One gelded male llama can protect up to 300 sheep on 300 acres. One llama per group seems to be better than two per group, perhaps because llamas will bond and spend time with one of their own in preference to the species they are to guard. Llamas are used to guard cattle as well as sheep (Braithwait, 1996).

Cattle are known to help protect sheep from predators. Initially, confining sheep, goats, and cattle in a small pen for a few days causes a more effective bonding than first combining them in a larger pasture. In one USDA study the best biologic method of protecting livestock from predators was bonding sheep and goats to cattle and using a guard animal. Other benefits of mixed species grazing are covered in Chapter 10.

Eliminating Attractive Plants, Water, Feed, or Cover

If the addition of more nutritious and more palatable plants, drinking water, and cover will attract animals to a location, it is logical that eliminating some or all of these will decrease the attractiveness of the property. However, if property managers require all of these resources for the horses and other animals under their care, then eliminating them is not practical. One of the NRCS projects aimed at increasing and improving sage grouse habitat along with native grasses for cattle grazing included the removal of shrubs that are not palatable to cattle or grouse. These shrubs may have been a component of preferred habitat for skunks or foxes, which prey on sage grouse nests, because the sage grouse population rose dramatically after the shrubs were removed. In the case of the emus of Western Australia thriving where dams were built and wells were bored and connected to open stock tanks, it is possible that restricting access to those water sources may have a negative effect on the emu population. One measure that seems to be slowing the spread of cane toads into the northern part of Western Australia is building fine mesh fences around the dams that hold water for livestock or replacing open water storage with enclosed polymer tanks. Cane toads need water sources close to each other to move into new territory. Preventing their access to dams in the dry season caused excluded toads to die within 2 days. In the following wet season the fences also kept toads from the water and reduced toad populations for a year after exclusion (Letnic et al., 2015). These researchers referred to the dams as "invasion hubs." Creative thinking can be used in preventing access of other nuisance animals to critical components of habitat, to stop their mass movement into farms. For decades, a required measure to reduce rabbit populations on farms in a number of states was to use heavy machinery to rip up rabbit warrens. This was considered a more effective population control than the introduction of diseases to rabbits for a number of years. Fumigation and baiting with poisons are also used for rabbit control.

It is already well-known that numbers of mice, rats, squirrels, sparrows, and pigeons in barns can be kept to a minimum if they do not find livestock feed that they can eat. Storing concentrate feeds in rodent-proof bins with latched covers, cleaning up spilled feed after every feeding, and only bringing enough hay into a barn for a few days of feeding prevent rodents from taking up residence.

Dead Animal Disposal

If there is a local service that picks up dead livestock for a reasonable fee, call it when you have had an animal killed by predators. Send along any parts removed from the animal, so odors attractive to predators are minimized. If dead stock pickup is not available or too expensive, check local regulations on alternatives. These usually include digging a hole deep enough to hold the animal and cover it with at least 1 foot (30 cm), preferably 2 feet (60 cm), of soil or composting it. Some government regulations may allow composting within a manure pile and others prohibit this. If you have neighbors that compost dead calves, lambs, chickens, or pigs, ask whether they will accept your rare offering into their compost bin.

A Model to Watch

After suffering feral animal plagues since the mid-1800s, Australia enacted the 2007 Australian Pest Animal Strategy. Through the strategy, teams of people ascertained what was needed if they were to have a chance of overcoming the country's multitude of pest animal problems. Partnerships were formed and plans were made (Invasive Plants and Animals Committee, 2016). Over the next 10 years, progress was made in the prevention of new pest animal problems and in coordinating efforts in eradication and containment. Achievements of the strategy were not publicized as numbers of pest animals were found and exterminated. Instead, they included agreed challenges, needs for strategic action and collaboration, first steps toward more integrated approaches, developing frameworks and principles, revision of guidelines, prioritization, coordination, development of action plans, and maximizing national consistency. Australia's Proposed Pest Animal Strategy for 2017−27 laid out eight principles of effective pest animal management that are worth reviewing and adapting to other continents.

1. Prevention and early intervention to avoid the establishment of new pest animal species is generally more cost-effective than ongoing management of established populations.
2. Pest animal management is a shared responsibility between landholders, community, industry, and government.
3. Management of mobile pest animals requires a coordinated approach across a range of scales and land tenures.
4. Management of established pest animals should focus on the protection of priority assets but also usually requires a "buffer" management area around the asset to account for pest animal mobility.
5. Pest animal management should be based on actual rather than perceived impacts and should be supported by monitoring to measure whether impact reduction targets are being achieved.
6. Best practice pest animal management balances efficacy, target specificity, safety, humaneness, community perceptions, efficiency, logistics, and emergency needs.
7. Best practice pest animal management integrates a range of control techniques (including commercial use where appropriate), considers interactions between species (such as rabbits and foxes), and accounts for seasonal conditions (for example, to take advantage of pest animal congregations during drought) and animal welfare.
8. The cost of pest animal management should be borne by those who create the risk and those who benefit from its management. Governments may coinvest where there is a net public benefit from any such intervention.

These principles should be incorporated into pest animal prevention, eradication, and management strategies, plans, and actions across all management levels (Invasive Plants and Animals Committee, 2016).

REVIEW QUESTIONS

1. Explain the relationship between the abundance of species of wild plants and animals and the health of an ecosystem. Include discussion of habitat, biodiversity benefits, and ecosystem services.
2. Explain possible benefits of a diverse ecosystem on a horse farm.
3. Describe possible effects of adding or subtracting wild species on the ecosystem of a horse farm.
4. Describe techniques of encouraging more wildlife to visit or inhabit a horse farm.
5. Explain two situations in which wild animals can be detrimental to the goals of a farm. One situation should involve predation and the other should involve crop damage. Describe your existing prevention strategy and a possible consequence if you do not do something about the situation. Use the same situations when you answer the next two questions (6 and 7).
6. Describe how to determine which detrimental species may be present or responsible for damage to a farm.
7. Describe techniques to remove and discourage coyotes from a farm.

REFERENCES

Australian Geographic. Australian Endangered Species List. 2014. http://www.australiangeographic.com.au/topics/science-environment/2014/06/australian-endangered-species-list.

Alberta Agriculture. Rancher's Guide to Predator Attacks on Livestock. Government of Alberta; 2010. http://aep.alberta.ca/fish-wildlife/wildlife-damage-control-programs/documents/RanchersGuideToPredatorAttacks-May2010.pdf.

Australian Museum. Animal Species: Cane Toad. 2016. https://australianmuseum.net.au/cane-toad.

Audubon Society. Climate Threatened Wild Turkey. Audubon; 2014. The Climate Report, http://climate.audubon.org/birds/wiltur/wild-turkey.

Baggaley K. Wild turkeys are in trouble and this time hunting and clear-cutting aren't to blame. Popular Sci. November 22, 2017 https://www.popsci.com/wild-turkey-decline#page-3.

Barker DJ, MacAdam JW, Butler TJ, Sulc RM. Chapter 2 forage and biomass planting. In: Nelson CJ, editor. Conservation Outcomes from Pastureland and Hayland Practices: Assessment, Recommendations and Knowledge Gaps. USDA. NRCS. ARS. Allen Press, Lawrence, Kansas; 2012.

Bonta M, Gosford R, Eussen D, Ferguson N, Loveless E, Witwer M. Intentional fire-spreading by "firehawk" raptors in Northern Australia. J. Ethnobiol. 2017;37(4):700−18.

Braithwait J. Using Guard Animals to Protect Livestock. Missouri Department of Conservation; 1996. http://www.predatorfriendly.org/how-to/how-to-pdf-docs/Using%20Guard%20Animals%20to%20Protect%20Livestock.pdf.

Brown LJ, Nocera JJ. Conservation of breeding grassland birds requires local management strategies when hay maturation and nutritional quality differ among regions. Agric. Ecosyst. Environ. 2017;237:242−9. https://www.sciencedirect.com/science/article/pii/S0167880916305412.

Cabrera KA. Beartracker's Animal Tracks Den. 2016. www.bear-tracker.com.

Commonwealth of Australia. The Cane Toad (Bufo marinus). Department of the Environment, Water, Heritage and the Arts; 2010. http://www.environment.gov.au/biodiversity/invasive-species/publications/factsheet-cane-toad-bufo-marinus.

Costanza R, D'Arge R, De Groot R, Farber S, Grasso M, Hannon B, Limburg K, Naeem S, O'Neill RV, Paruelo J, Raskin RG, Sutton P, Van Den Belt M. The value of the world's ecosystem services and natural capital. Nature 1997;387:253−60.

Doggart S. Tally-ho Pardner! Foxhunting in America. The Telegraph; 2011. http://www.telegraph.co.uk/expat/expatlife/8362003/Tally-ho-pardner-Foxhunting-in-America.html.

Doody JS, Soanes R, Castellano CM, Rhind D, Green B, McHenry CR, Clulow S. Invasive toads shift predator-prey densities in animal communities by removing top predators. Ecology 2015;96(9):2544–54.

Doyle J. Haying with Wildlife in Mind. SDSU Extension; 2017. http://igrow.org/livestock/land-water-wildlife/haying-with-wildlife-in-mind/.

Duriancik L, Bucks D, Dobrowolski JP, Drewes T, Eckles SD, Jolley L, Kellogg RL, Lund D, Makuch JR, O'Neill MP, Rewa CA, Walbridge MR, Parry R, Weltz MA. The first five years of the conservation effects assessment project. J. Soil Water Conserv. 2008;63:185A–97A.

Eaton SW. Wild Turkey Meleagris gallopavo. Birds of North America; 2014. https://birdsna.org/species-Account/bna/species/wiltur/introduction.

Evans J. Nature Tracking. 2016. www.naturetracking.com.

FAO (Food and Agriculture Organization of the United Nations). Ecosystem Services and Biodiversity. 2018. http://www.fao.org/ecosystem-services-biodiversity/en/.

Feight L. Winter Electric Fence Solutions. On Pasture; January 8, 2018. https://onpasture.com/2018/01/08/winter-electric-fence-solutions/.

FFWCC (Florida Fish and Wildlife Conservation Commission). Relocating Wildlife; 2018. http://myfwc.com/wildlifehabitats/profiles/relocating-wildlife/.

Federal Register. Grassland reserve program; final rule. Fed. Reg. 2009;74:3855–79.

Garric J, Vollat B, Duis K, Péry A, Junker T, Ramil M, Fink G, Ternes TA. Effects of the parasiticide ivermectin on the cladoceran *Daphnia magna* and the green alga *Pseudokirchneriella subcapitata*. Chemosphere 2007;69:903–10.

Goeringer P. Hunting Lease Examples. On Pasture; September 25, 2017. http://onpasture.com/2017/09/25/hunting-lease-examples/.

Government of Canada. Species at Risk Public Registry. 2017. https://www.registrelep-sararegistry.gc.ca/default.asp?lang=En&n=24F7211B-1.

Green JS, Woodruff RA. Breed comparisons and characteristics of use of livestock guarding dogs. J. Range Manag. 1988;41(3):249–51.

Harper CA, Bates GE, Hansbrough MP, Gudlin MJ, Gruchy JP, Keyser PD. Native Warm-Season Grassis Identification, Establishment and Management for Wildlife and Forage Production in the Mid-South. PB 1752. Knoxville: University of Tennessee Extension; 2007. https://extension.tennessee.edu/publications/Documents/PB1752.pdf.

Herd RP, Stinner BR, Purrington FF. Dung dispersal and grazing area following treatment of horses with a single dose of ivermectin. Vet. Parasit. 1993;48:229–40.

Herd RP. Control strategies for ruminant and equine parasites to counter resistance, encystment and ecotoxicity in the USA. Vet. Parasit. 1993;48:327–36.

Herd R. Endectocidal drugs: ecological risks and counter-measures. Int. J. Parasitol. 1995;25(8):875–85.

ICWDM. Internet Center For Wildlife Damage Management. Livestock and animal predation identification; 2015. http://icwdm.com/Inspection/Livestock.aspx.

Invasive Plants and Animals Committee. Australian Pest Animal Strategy 2017 to 2027. Canberra: Australian Government Department of Agriculture and Water Resources; 2016. http://www.agriculture.gov.au/SiteCollectionDocuments/pests-diseases-weeds/consultation/apas-final.pdf.

ITTM. International Trapping Technical Magazine. 2018. http://www.ittm.info/?p=1051.

IUCN (International Union for Conservation of Nature). The IUCN Red List of Threatened Species. Dromaius novaehollandiae; 2017. http://www.iucnredlist.org/details/22678117/0.

James P, Cox CA. Building blocks to effectively assess the environmental benefits of conservation practices. J. Soil Water Conserv. 2008;63:178A–80A.

Jannota C. MSU study Proves that Wildlife Crossing Structures Promote 'gene Flow' in Banff Bears. MSU News Service; 2014. http://www.montana.edu/news/12440/msu-study/proves-that-wildlife-crossing-structures-promote-gene-flow-in-bears.

Jeswiet S, Hermsen L. Agriculture and wildlife: a two-way relationship. In: EnviroStats. Catalogue No. 16-002-X. Statistics Canada; 2015. http://www.statcan.gc.ca/pub/16-002-x/2015002/article/14133-eng.htm.

Junquera. Horse Ticks: Biology, Prevention and Control. 2017. Parasitipedia.net. http://parasitipedia.net/index.php?option=com_content&view=article&id=3134&Itemid=3570.

Kayes J. Landowner Liability and Recreational Access. University of Maryland Extension; 2017. https://extension.umd.edu/learn/landowner-liability-and-recreational-access.

Krausman PR, Naugle DE, Frisina MR, Northrup R, Bleich VC, Block WM, Wallace MC, Wright JD. Livestock grazing, wildlife habitat and rangeland values. Rangelands: October 2009;31(5):15–9.

Lashmet TD. Hunting Lease Checklist. 2013. http://denton.agrilife.org/2013/08/Texas-hunting-lease-checklist.pdf.

Lenz TR. Tick Control in Horses. American Association of Equine Practitioners; 2017. https://aaep.org/horsehealth/tick-control-horses.

Letnic M, Webb JK, Jessop TS, Dempster T. Restricting access to invasion hubs enables sustained control of an invasive vertebrate. J. Appl. Ecol. 2015;52:341–7.

Li X, Buzzard P, Chen Y, Jiang X. Patterns of livestock predation by carnivores: human-wildlife conflict in northwest Yunnan, China. Environ. Manag. 2013;52:1334–40.

Long C. Australia builds outback electrified fence to protect native animals. Reuters. Environ. June 13, 2017. https://www.reuters.com/article/us-australiawildlife-fence-idUSKBN1940X4.

Loss SR, Will T, Marra PP. The impact of free-ranging domestic cats on wildlife of the United States. Nat. Commun. 2013a;4:1396.

Loss SR, Will T, Marra PP. Estimates of bird collision mortality at wind facilities in the contiguous United States. Biol. Conserv. 2013b;168:201–9.

Lumaret JP, Galante E, Lumbreras C, Mena J, Bertrand M, Bernal JL, Cooper JF, Kadiri N, Crowe D. Field effects of ivermectin residues on dung beetles. J. Appl. Ecol. 1993;30:428–36.

Maryland Department of Natural Resources. Permission to Hunt Form. 2017. https://www.onpasture.com/wp-content/uploads/2017/09/MD-DNR-COMPASS-Permission-to-Hunt_Trap-Form.pdf.

Masters of Foxhounds. Association & Foundation. 2017. www.mfha.com/hunts-alphalist.html.

McKenzie J, Alkerton D. Notes on Equestrian and Wildlife Issues, Course Notes in Pasture Management. 2008. http://www.nwcinc.ca.

McLendon M, Kaufman PE. Horse Bot Fly. Featured Creatures; 2007. http://entnemdept.ufl.edu/creatures/livestock/horse_bot_fly.htm.

McWelch T. Animal Tracking: How to Identify 10 Common North American Species. 2014. https://www.outdoorlife.com/blogs/survivalist/2014/01/animal-tracks-gallery.

Mercer P. The Rapid Spread of Australia's Cane Toad Pests. Sydney: BBC News; March 22 , 2017. http://www.bbc.com/news/world-australia-39348313.

Miller M. Wild Turkey Restoration: The Greatest Conservation Success Story?. Cool Green Science; November 26, 2013. https://blog.nature.org/science/2013/11/26/wild-turkey-restoration-the-greatest-conservation-success-story/.

Nessel RJ, Jacob TA, Robertson RT. The human and Environmental safety aspects of ivermectin. In: Proceedings MSD Agvet Symposium on Recent Developments in the Control of Animal Parasites. In association with 22nd World Veterinary Congress, Perth, Australia; 1983. p. 98−108.

NRCS. Wildlife Habitat. USDA Natural Resources Conservation Service; 2018. https://www.nrcs.usda.gov/wps/portal/nrcs/detail/national/home/?cid=nrcs143_023553.

NWF. National Wildlife Federation. Home. Habitats; 2018. https://www.nwf.org/Home/Our-Work/Habitats.

Ontario SPCA. Wildlife Proofing Your Property. Ontario SPCA and Humane Society; 2018. http://ontariospca.ca/63-fact-sheets/fact-sheets/1187.html?_ga=2.160690941.1571680800.1515616277-1676375472.1515616277.

OMAFRA Staff. Guidelines for Using Donkeys as Guard Animals with Sheep. Ontario Ministry of Agriculture, Food and Rural Affairs; 2016. www.omafra.gov.on.ca/english/livestock/sheep/facts/donkey2.htm.

Ottawa-Carleton Wildlife Centre. Human/wildlife Conflicts. 2010. http://www.wildlifeinfo.ca/conflicts.html.

Palmeira FBL, Trinca CT, Haddad CM. Livestock predation by puma (*Puma concolor*) in the highlands of a southeastern Brazilian Atlantic forest. Environ. Manag. 2015;56:903−15.

Pet MD. Botfly Infection in Horses. 2018. https://www.petmd.com/horse/conditions/skin/c_hr_bots_parasites.

Power J. War on Feral Cats: Australia Aims to Cull 2 Million. Sydney Morning Herald; February 19 , 2017. http://www.smh.com.au/national/war-on-feral-cats-australia-aims-to-cull-2-million-20170214-gucp4o.html.

Putnam DH, DelCurto T. Ch 22, forage systems for arid areas. In: Forages. The Science of Grassland Agriculture. sixth ed., vol. II. Blackwell Publishing; 2007. pp323−339.

Putnam DH, Russelle M, Orloff S, Kuhn J, Fitzhugh L, Godfrey L, Kiess A, Long R. Alfalfa, Wildlife and the Environment, the Importance and Benefits of Alfalfa in the 21st Century. Navato, CA: California Alfalfa & Forage Asssociation; 2001.

Queensland Museum. Cane Toad. 2018. http://www.qm.qld.gov.au/Find+out+about/Animals+of+Queensland/Frogs/Cane+Toad#.WpxF5ejwbIU.

Robinson SK, Grzybowski JA, Rothstein SI, Brittingham C, Petit LJ, Thompson FR. Management implications of cowbird parasitism on neotropical migrant songbirds. In: Finch DM, Stangel PW, editors. Status and Management of Neotropical Migratory Birds: September 21-25, 1992, Estes Park, Colorado. Gen. Tech. Rep. RM-229. Fort Collins, Colo., Rocky Mountain Forest and Range Experiment Station. U.S. Dept. of Agriculture, Forest Service; 1993. p. 93−102. http://www.treesearch.fs.fed.us/pubs/22891.

Sanderson H, Laird B, Pope L, Brain R, Wilson C, Johnson D, Bryning G, Peregrine AS, Boxall A, Solomon K. Assessment of the environmental fate and effects of ivermectin in aquatic mesocosms. Aquat. Toxicol. 2007;85:229−40.

Sanderson MA, Jolley LW, Dobrowolski JP. Pastureland and Hayland in the USA: land resources, conservation practices and ecosystem services. In: Nelson CJ, editor. Conservation Outcomes from Pastureland and Hayland Practices: Assessment, Recommendations and Knowledge Gaps. Lawrence Kansas: Allen Press; 2012. p. 25−40.

Sheep101. Predators. The Wily Coyote; 2014. www.sheep101.info/predators.html.

Slezak M. Cane Toad Has Surprise Effect on Australian Ecosystem. Daily News; March 19 , 2015. https://www.newscientist.com/article/dn27199-cane-toad-has-surprise-effect-on-australian-ecosystem/.

Sterba J. Nature Wars: How Wildlife Comebacks Turned Backyards into Battlegrounds. New York: Crown Publishers; 2012.

TEEB (The Economics of Ecosystems and Biodiversity). The economics of ecosystems and biodiversity. In: Kumar P, editor. Ecological and Economic Foundations. London and Washington: Earthscan; 2010. http://www.teebweb.org/our-publications/teeb-study-reports/ecological-and-economic-foundations/#.Ujr1xH9mOG8.

U.S. Fish & Wildlife Service. Endangered Species. Endangered Species Database. 2017. https://www.fws.gov/endangered/species/us-species.html.

U.S. Fish & Wildlife Service. Endangered Species. Endangered Species Database. Why Save Species. 2017. https://www.fws.gov/endangered/species/why-save-species.html.

UC Davis (University of California), Davis. What Is IPM?. Statewide Integrated Pest Management Program; 2018. http://www2.ipm.ucanr.edu/WhatIsIPM/.

Undersander D, Pillsbury B. Grazing Streamside Pastures. University of Wisconsin Extension. A3699; 1999. https://learningstore.uwex.edu/Grazing-Streamside-Pastures-P102.aspx.

Undersander D, Temple S, Bartlet J, Sample D, Paine L. Grassland Birds: Fostering Habitats Using Rotational Grazing. University of Wisconsin Extension A3715; 2000. https://learningstore.uwex.edu/Grassland-Birds-Fostering-Habitats-Using-Rotational-Grazing-P103.aspx.

United States Department of Agriculture (USDA). Grass. The Yearbook of Agriculture. Washington, DC: U.S. Gov. Print. Office; 1948.

USDA NRCS. Management Considerations for Grassland Birds in Northeastern Haylands and Pasturelands. Washington, D.C.: Wildlife Insight No. 88; 2010. https://www.bobolinkproject.com/docs/NRCS_Grassland_leaflet.pdf.

Van Duyne C, Ras E, De Vos AEW, De Boer WF, Henkens RJHG, Usukhjargal D. Wolf predation among reintroduced Przewalski horses in Hustai national park, Mongolia. J. Wildl. Manag. 2009;73(6):836−43.

Vickery JA, Tallowin JR, Feber RE, Asteraki EJ, Atkinson PW, Fuller RJ, Brown VK. The management of lowland neutral grasslands in Britain: effects of agricultural practices on birds and their food resources. J. Appl. Ecol. 2001;38:647–64.

Voth K. Can Wildlife Improve Your Bottom Line?. On Pasture; 2017a. https://onpasture.com/2017/09/18/can-wildlife-improve-your-bottom-line/.

Voth K. More Wolves = More Trees and More Pronghorn Antelope. On Pasture; 2017b. https://onpasture.com/2017/11/13/more-wolves-more-trees-and-more-pronghorn-antelope/.

Voth K. Assistance for Adding More Grass, Reducing Hay Feeding. On Pasture; 2017c. http://onpasture.com/2017/09/25/assistance-for-adding-more-grass-reducing-hay-feeding/.

Voth K. Biodiversity Through Grazing Management. On Pasture; 2017d. http://onpasture.com/2017/biodiversity-through-grazing-management/.

Voth K. Haying with Wildlife in Mind. On Pasture; 2017e. https://onpasture.com/?x=0&y=0&s=haying+with+wildlife.

Voth K. Have You Checked Out the Conservation Stewardship Program?. On Pasture; 2017f. https://www.nrcs.usda.gov/wps/portal/nrcs/main/national/programs/financial/csp/.

Voth K. Why Should You Grow Multiple Types of Forage Grasses for Grazing Animals?. On Pasture; 2018a. https://onpasture.com/2018/01/08/why-should-you-grow-multiple-types-of-forage-grasses-for-grazing-animals/.

Voth K. Fladry Fencing Can Protect Cattle from Wolves. On Pasture; 2018b. https://onpasture.com/2018/02/05/fladry-fencing-can-protect-cattle-from-wolves/.

Warnica R. Birds of Prey...And Flames. National Post. News, World; January 9, 2018. http://nationalpost.com/news/world/australian-birds-have-weaponized-fire.

WDFW (Washington Department of Fish & Wildlife). Trapping Wildlife. Nuisance Wildlife. Living With Wildlife; 2018. https://wdfw.wa.gov/living/nuisance/trapping.html.

Wikipedia. Emu War. 2017. https://en.wikipedia.org/wiki/Emu_War.

Wikipedia. List of Extinct Animals of Australia. 2017. https://en.wikipedia.org/wiki/List_of_extinct_animals_of_Australia.

Woodward KN. Veterinary pharmacovigilance. Part 3. Adverse effects of veterinary medicinal products in animals and on the environment. J. Vet. Pharmacol. Therap. 2005;28:171–84.

Wright B. West Nile Virus – Protect Yourself and Your Horse: Practical Advice for Horse Owners. Ontario Ministry of Agriculture, Food and Rural Affairs; 2006. http://www.omafra.gov.on.ca/english/livestock/horses/facts/brochure.htm.

Wright B. Controlling Mosquitoes on Horse Farms and Rural Properties. Ontario Ministry of Agriculture, Food and Rural Affairs; 2017. http://www.omafra.gov.on.ca/english/livestock/horses/facts/info_mosq.htm.

Young A, Boyle T, Brown T. The Population Genetic Consequences of Habitat Fragmentation for Plants. 1996. http://www.sciencedirect.com/science/article/pii/0169534796100458.

FURTHER READING

Adams P. RHDV–K5: New Strain of Pest Rabbit-killing Calicivirus Disease Given Green Light for 2017 Release. Australian Broadcasting Corporation; 2016. http://www.abc.net.au/news/2016-04-29/new-strain-of-rabbit-killing-calicivirus-disease-approved/7367142.

CSIRO. Myxomatosis to Control Rabbits. CSIROpedia; 2015. https://csiropedia.csiro.au/myxomatosis-to-control-rabbits/.

Inaturalist. North Animal Tracking Database. 2013. https://www.inaturalist.org/projects/north-american-animal-tracking-database.

Moon R. Tick Control in Horse Pastures. Stable Management. University of Minnesota; 2016. https://stablemanagement.com/articles/tick-control-horse-pastures-53219.

South Dakota Game Fish, Parks (SDGFP). 2017. http://gfp.sd.gov/wildlife/private-land/docs/HuntingAccessProgramOverview.pdf. http://gfp.sd.gov/wildlife/private-land/walk-in.aspx.

USDA NRCS. Articles on Wildlife. United States Department of Agriculture. Natural Resources Conservation Service; 2018a. file:///C:/Data/Pasture%20Book/REference%20articles/Wildlife/Home%20_%20NRCS.html.

USDA NRCS. Conservation Stewardship Program. 2018. https://www.nrcs.usda.gov/wps/portal/nrcs/main/national/programs/financial/csp/.

Chapter 18

University of Kentucky Horse Pasture Evaluation Program

Krista L. Lea, S. Ray Smith and Thomas (Tom) Keene

Department of Plant and Soil Sciences, College of Agriculture, Food and Environment, University of Kentucky, Science Center North, Lexington, KY, United States

INTRODUCTION

Horse farms in Kentucky vary tremendously in size, type, breed, and resources. The most well-known are large-scale thoroughbred breeding operations, surrounded by four-plank board fencing and rolling green pastures. These farms generally have low stocking rates, maintaining foaling mares, growing horses, and stallions on 3–5 acres per horse. Surrounding these farms are smaller operations that may include small breeding, sport horse, or private boarding operations. Finally, the area is dotted with private family farms with backyard horses. Small and private farms often maintain a much higher stocking rate and have limited resources, but they form the backbone for the industry in Kentucky. Thoroughbreds are the most prominent breed in Kentucky; however, the 2012 Kentucky Equine Survey found a significant population of many other breeds including the American Quarter Horse, Tennessee Walking Horse, American Saddlebred, and others (KY Equine Survey, 2012; Table 18.1).

The University of Kentucky Horse Pasture Evaluation Program began in 2005 in response to greater demand for pasture management information in the region. The program works with horse farms of all sizes to improve pasture management and utilization while reducing the need for stored feeds and the negative impacts of horses on the environment. Environmental impacts of horse grazing can include soil erosion and nutrient runoff due to close grazing, water contamination from direct access, and the spread of noxious weeds by allowing seed production.

HISTORY OF THE UNIVERSITY OF KENTUCKY HORSE PASTURE EVALUATION PROGRAM

The University of Kentucky has a long history of successful extension programing. Unlike other livestock and commodity groups, however, the horse industry did not historically have a strong relationship with the university. That began to change in the 2000s due, in part, to the crisis created by Mare Reproductive Loss Syndrome (MRLS).

TABLE 18.1 Equine Breeds in Kentucky

Thoroughbred	54,000
Quarter horse	42,000
Tennessee walking horse	36,000
American Saddlebred	14,000
Donkeys and mules	14,000
Mountain horse breeds	12,500
Standardbred horses	9,500
Total horses in Kentucky	242,400

Mare Reproductive Loss Syndrome

In early 2001, breeding farms in Kentucky began reporting large numbers of morbid foals. By the end of the foaling season in June of 2001, it was reported that about 25%–30% of the foal crop was lost (Thalheimer and Lawrence, 2001), resulting in ~$330 million of loss in revenue (Webb et al., 2004). The cause at the time was unknown, but termed MRLS. The equine industry reached out to the University of Kentucky for help in this crisis and research began soon after. The 2002 foaling season again brought significant reproductive losses, although not to the extent as the previous year. One area of research at the University of Kentucky focused on tall fescue toxicosis, known to occasionally cause late term abortions in horses consuming infected tall fescue. Eventually, researchers concluded that MRLS was actually a product of the accidental ingestion of the eastern tent caterpillar (*Malacosoma americanum*) by broodmares (Webb et al., 2002). Today, farms take a number of steps to prevent MRLS by removing host trees, treating caterpillar infestations, or moving mares away from caterpillar nests. The MRLS crisis demonstrated to the industry and the university that both sides were open to working with the other, and that the relationship would benefit all involved (Fig. 18.1).

University of Kentucky Equine Initiative

An overarching program was needed to encompass all equine activities on the University of Kentucky campus. The University of Kentucky Equine Initiative was announced in 2004 to serve as an umbrella for ongoing research at the Gluck Equine Research Center and the Department of Animal and Food Sciences, as well as classes and competitive student riding teams and clubs. It would cover all things equine, including research throughout the College of Agriculture, a new Equine Science and Management Degree Program, and extension programing throughout the state. In 2014, it was renamed the University of Kentucky Agriculture Equine Programs.

In 2005, as part of the Equine Initiative, The University of Kentucky Horse Pasture Evaluation Program was launched. Dr. Ray Smith, Forage Extension Specialist, and Tom Keene, Forage Agronomist, designed the farm service program to assist farms in pasture establishment and management as well as evaluation of the tall fescue toxicosis risk in the Bluegrass region. The program began small, evaluating just a few pastures on 10 farms in and around Lexington. Today the program has grown, working on thousands of acres each year on farms of all sizes across the state.

University of Kentucky Horse Pasture Evaluation Program Today

The Horse Pasture Evaluation program has enjoyed 13 years of success working on horse farms across the state. As of 2017, 211 evaluations have been completed on 142 unique farms across 21 counties in Kentucky. A number of farms have evaluations scheduled every year to 2 years to monitor changes in pasture composition. Farms range from small, private acreages to large, commercial breeding operations to nonprofit rescue operations and everything in between. Educational programing is also a major focus; extension events include the winter annual *Pastures Please!!* event and the summer

FIGURE 18.1 MRLS and the eastern tent caterpillar. The eastern tent caterpillar is blamed for significant losses in the 2001 and 2002 foal crops of central Kentucky after mares inadvertently ingested caterpillars moving across the pasture from their nest, seen here.

annual *Equine Farm and Facilities Expo*. Educational articles are also submitted to the Bluegrass Equine Digest. Finally, the program maintains an email listserv with participating farms to disseminate timely information.

Tall Fescue Toxicosis

Tall Fescue [*Schedonorus arundinacea* (Schreb.) Dumort.] is a cool season grass naturally occurring throughout the southeastern United States, estimated to cover over 35 million acres (Ball et al., 2007). It is known to have good forage quality, as well as grazing and drought tolerance, pest and disease resistance, and a longer growing season compared to other common cool season grasses (Arachevaleta et al., 1989; Gwinn and Gavin, 1992). This is due primarily to widespread infection of plants by the endophyte *Neotyphotium coenophialum* (Morgan—Jones and Gams). The tall fescue plant and the endophyte form a symbiotic relationship, the endophyte living inside the plant and producing ergot alkaloids, while the plant hosts and provides nourishment to the endophyte (Bacon and Siegel, 1988; Fig. 18.2).

Endophyte-infected tall fescue was planted widely throughout the southeastern United States, before studies suggested that grazing infected tall fescue negatively affects animals including cattle and horses. The first studies measured cattle responses on high- and low-infected pastures and documented decreased average daily gain (Hopkins and Alison, 2006), rough hair coats (McClanahan et al., 2008), increased body temperatures (Aldrich et al., 1993), and leg tenderness, and gangrene loss of hooves, ears, or tails (Ball et al., 2007). Later research identified ergovaline as the primary ergot alkaloid produced, which acts as an active vasoconstrictor in both horses and cattle (Lyons et al., 1986; Belesky et al., 1998; Klotz et al., 2007). While other ergot alkaloids likely play a role in fescue toxicity, ergovaline (and its epimer ergovalinine) is measured to quantify toxicosis potential in seed, hay, and pastures.

Vasoconstriction in nonpregnant horses often causes little to no physiologic effects (Aiken et al., 1993; McCann et al., 1991; Redmond et al., 1991). However, ergovaline is a prolactin inhibitor and is known to cause significant issues in pregnant mares. Classic tall fescue toxicity symptoms in late-term mares include prolonged gestation (Monroe it al., 1988), difficulty foaling, (Putnam et al., 1991) and decreased milk production (Kosanke et al., 1987). Foal and mare mortality are possible due to these complications. Early-term mares may also be sensitive to high levels of ergovaline, causing poor reproductive performance and early-term fetal losses (Brendemuehl at al., 1994).

Establishing threshold values for ergovaline consumption in late- and early-term mares is difficult for a variety of factors: (1) Ergovaline concentration within tall fescue plants varies by year, season, and stage of growth (Rogers et al., 2011; Greene et al., 2013). (2) Ergovaline is an unstable compound in tissues and changes in concentrations during sample transport and storage (Lea, 2014). (3) Laboratory analysis and reporting methods vary widely with no oversite at this time. (4) Individual animal responses to ergovaline likely vary. Several states have adopted conservative threshold values based on reported research and anecdotal evidence. The University of Kentucky Horse Pasture Evaluation Program considers 200 ppb ergovaline + ergovalinine in the total diet a risk to late-term mares and 500 ppb in total diet as a risk to early-term mares.

FIGURE 18.2 Tall fescue in a horse pasture. Tall fescue is a cool season, perennial grass common throughout Kentucky and the southeastern United States. Most tall fescue is infected with an endophyte toxic to late-term pregnant mares.

PASTURE SAMPLING

On-farm evaluations are performed by trained technicians and undergraduate students from late March through November on horse farms in any part of the state. Pastures are evaluated for species composition by visual estimation of a 2-sq. ft (0.6 × 0.6 m) quadrat placed randomly in 10−20 locations (depending on pasture size) in the pasture (Fig. 18.3).

Percentages, rounded to 5%, are recorded for tall fescue, Kentucky bluegrass (*Poa pratensis* L.), orchardgrass (*Dactylis glomerata* L.), and white clover (*Trifolium repens* L.), as well as weeds and bare soil. Perennial ryegrass (*Lolium perenne* L.) and timothy (*Phleum pretense* L.) are also common pasture forages for horses, but they occur infrequently in Kentucky due to hot summer conditions. A collective percentage of weeds is recorded, and a list of all weeds identified in descending order of prevalence is noted. Bare soil percentages include warm season annual grasses that will die off in the fall, leaving bare soil in the winter months, and include crabgrass (*Digitaria sanguinalis* L.), foxtail [*Setaria pumila* (Poir.), Roem. and Schult.] and goosegrass [*Eleusine indica* (L.) Gaertn]. Species composition numbers from each sample location are averaged across the pasture to generate pasture composition numbers. Photographs of sample grids are taken with GPS-equipped cameras to mark the location of samples. Maps are subsequently generated showing these locations.

Farms with equine breeding stock may also choose to sample for tall fescue to determine endophyte infection rates (%) and ergovaline concentration (ppb). Twenty tillers of tall fescue are collected from across the pasture and submitted for testing to University of Kentucky Regulatory Services. Endophyte presence is determined using Agrinostics Immunoblot test kits and reported as a percentage of total plants infected. Grab samples, mimicking the grazing behavior of horses, are also collected randomly throughout the pasture and submitted to the University of Kentucky Veterinary Diagnostic Laboratory. This lab reports ergovaline plus ergovalinine on a dry matter basis, and the results are determined by uHPLC (ultrahigh-performance liquid chromatography) with fluorescence detection.

FIGURE 18.3 Sampling locations for a small paddock. Paddocks less than 5 acres are sampled in 10 locations randomly throughout the paddock. Each location is documented using a GPS-equipped camera and numbered to correspond with the datasheet.

DATA REPORTING AND RECOMMENDATIONS

Participating farms receive comprehensive reports containing all data collected on their farm as well as maps, recommendations, and a wealth of publications. This report is presented by a team member who walks the farm manager through in detail, going over findings, answering questions and discussing any follow up needed.

Soil Maps

Soil maps and pasture yield maps are produced for each farm using the USDA-NRCS Web Soil Survey online program. Soil maps may be used to understand differences in topography and how these differences affect land use. Pasture yield maps give recommendations for the carrying capacity of a farm based on soil type and slope and predict pasture yield. Web Soil Survey provides "soil ratings" that recommend the Animal Unit Months (AUM) of a given soil type (see Appendix 5). Experience suggests that this is a difficult unit to work with; therefore these numbers are converted to acres per horse per year, suggesting how many acres are needed to carry a full-grown horse for an entire year.

Pasture Maps, Data, and Photographs

Data collected on pastures, the photographs taken, and the map of sample locations generated are included for each pasture sampled in the report. Pasture data sheets contain detailed information regarding species and weeds present in each sample grid. Grid locations allow farms to identify trends within certain areas of the pasture. For example, if tall fescue is a concern in the pasture, a quick glance might suggest that it is prominent only in grids 2 through 9. Looking at the map helps the farm manager determine what part of the pasture the tall fescue is concentrated in and allows for a targeted control option in that area only (Fig. 18.4).

Recommendations

For each pasture sampled, detailed recommendations on seeding, weed control, soil fertility, and tall fescue analysis are generated using species composition numbers and tall fescue analysis. Seeding recommendations are based on bare soil numbers and are "an option" when 10%–25% bare soil is observed or "highly recommended" for pastures containing more than 25% bare soil. Weed control options are recommended based on weeds identified and may include mowing or specific herbicidal control. Weed control is "an option" with 10%–25% weeds and "highly recommended" at >25% weeds. Tall fescue risks are based on the percentage of infected tall fescue present. Greater than 25% infected tall fescue or >200 ppb in the total diet are considered a significant risk to late-term mares. Recommendations are often adjusted to better meet the needs of individual farms or unique situations. All recommendations are summarized on a chart to provide a quick glance of recommendations for all pastures (see Appendix 6.)

Farm Data Summary Sheet

Shown in Fig. 18.5, The Farm Data Summary Sheet is the most important piece of the report. It contains all data collected from all pastures evaluated. Ergovaline and endophyte test results are also reported here to allow for easy comparison from one pasture to the next. Several additional calculations are made and reported here to help farm managers better understand pastures. "Available Forage" is the sum of percentages of tall fescue, bluegrass, orchardgrass, and white clover and represents the useful or productive portion of the pasture. This number is often surprising to managers who may not realize how much of their land is not productive due to weeds and bare soil. "Tall Fescue in Available Forage" provides a percentage of tall fescue in available forage and therefore in the diet. "Ergovaline in Available Forage" gives an ergovaline concentration (ppb) in the total diet of animals housed in each pasture. This accounts for the dilution of other grasses found in the pasture and provides a realistic comparison of toxicity from pasture to pasture. Although tall fescue is a coarser textured grass than orchardgrass or bluegrass, recent research has shown the ingestion rate of these three species is similar (Morrison, 2008).

Publications

The notebook contains a number of extension publications containing valuable information on seeding, fertility, weed control, grazing management, and other concepts. The publication list is updated regularly to contain the most up-to-date and relevant information possible. A complete list of publications is available on the University of Kentucky Horse Pasture Evaluation Program website under "Additional Resources."

Date Sampled 6/3/2016
Evaluator Savannah
Farm Central Kentucky Horse Farm
Field # Pasture 1
Approx. Acre. 7.5
Sample ID 16054

Ground cover estimates of percent species composition

Grid #	Tall Fescue	KY Blue-grass	Orchard-grass	White Clover	Weeds	Bare Soil	Observations
1	10	45	0	30	5	10	SG, LQ
2	5	45	15	10	20	5	BP, NP, C, SW, RW, CW, BW
3	0	75	15	5	5	0	BW, NP, SW
4	0	75	5	10	10	0	BW, SG
5	10	60	0	25	5	0	BW, CW
6	0	65	0	10	20	5	BW, SM, SW, NW
7	55	35	0	0	5	5	BW
8	5	60	25	5	5	0	SW, D
9	40	50	0	5	5	0	SW
10	10	55	15	5	15	0	BW, MW
11	35	40	10	0	10	5	SG, D, T
12	10	45	15	15	10	5	NW, T
13	35	40	10	5	10	0	BW
14	80	20	0	0	0	0	-
15	0	55	0	5	40	0	BW, NW, D, JG, SW
Average	**20**	**51**	**7**	**9**	**11**	**2**	

Other Forages		Weeds					
AL	Alfalfa	BA	Barnyard grass	LQ	Lamb's Quarters	SG	Sedge
BM	Black Medic	BE	Bergamot	MA	Mallow	SP	Shepherd's Purse
BR	Bromegrass	BW	Bindweed	MW	Milk Weed	SM	Smartweed
C	Chicory	BP	Broadleaf Plantain	MG	Morning Glory	SW	Speedwell
HC	Hop Clover	BD	Burdock	NP	Narrowleaf Plantain	SU	Spurge
LZ	Lespedeza	BC	Buttercup	NI	Nightshade	SB	Star of Bethleham
RC	Red Clover	CA	Carpet Weed	NW	Nimblewill	ST	Strawberry
RG	Ryegrass	CW	Chickweed	NS	Nutsedge	TH	Thistle
T	Timothy	CQ	Cinquefoil	NT	Nutweed	UK	Unknown
VE	Vetch	CU	Cudweed	PC	Pennycress	VE	Velvet Leaf
		CD	Curly Dock	PW	Pepper Weed	V	Veronica
Summer Annuals		D	Dandelion	PG	Pigweed	VC	Virginia Copperleaf
CG	Crabgrass	FB	Fleabane	PA	Pineappleweed	WC	Wild Carrot
FX	Foxtail	GC	Groundcherry	PH	Poison Hemlock	WD	Wild Daisy
GG	Goosegrass	HG	Hairy Galinsoga	PI	Poison Ivy	WG	Wild Geranium
		HB	Henbit	PL	Prickly Lettuce	WM	Wild Mustard
		HN	Horse Nettle	PS	Prickly sida	WO	Wild Onion
		HW	Horseweed	PD	Purple Deadnettle	WV	Wild Violet
		IW	Ironweed	PU	Purslane	WS	Woodsorrel
		JG	Johnsongrass	QA	Queen Anne's Lace	WW	Woody Weed
		KW	Knotweed	QG	Quackgrass	Y	Yarrow
		LT	Lady's Thumb	RW	Ragweed	YR	Yellow Rocket

FIGURE 18.4 Sample field datasheet. Species composition and a list of observed weeds are recorded for each sample location.

University of
Kentucky
College of Agriculture,
Food and Environment

Central Kentucky Horse Farm
Horse Pasture Evaluation Results 2016

Evaluation Month	Ground cover estimates of percent species composition						Tall Fescue Analysis		Calculations		
Pasture	Tall Fescue	KY Blue-grass	Orchard-grass	White Clover	Weeds	Bare Soil	Ergovaline (ppb)	Endophyte (%)	Available Forage	Tall Fescue in Available Forage	Ergovaline Conc. in Available Forage
Pasture 1	20	51	7	9	11	2	557	77	87	23	128
Pasture 2	15	44	11	5	16	9	230	74	75	20	46
Pasture 9	40	40	4	1	6	9	109	50	85	47	51
Pasture 10	45	32	5	1	4	13	655	71	83	54	355
Pasture 11	31	45	5	3	13	3	<100	15	84	37	-
Pasture 12	52	26	3	3	4	12	1120	82	84	62	693
Pasture 13	23	37	8	7	22	3	687	64	75	31	211
Pasture 15	38	45	2	4	8	3	440	73	89	43	188
Pasture 16	26	40	12	5	7	10	522	84	83	31	164
Pasture 17	9	55	27	2	4	3	450	71	93	10	44
Pasture 18	9	49	19	5	11	7	293	44	82	11	32
Pasture 19	38	32	12	5	6	7	581	70	87	44	254
Pasture 20	25	19	13	17	18	8	351	80	74	34	119
Pasture 21	38	18	2	5	15	22	169	95	63	60	102
Pasture 22	22	33	6	7	14	18	345	58	68	32	112
Pasture 23	28	22	2	5	22	21	417	67	57	49	205
Average	**29**	**37**	**9**	**5**	**11**	**9**	**462**	**67**	**79**	**37**	**180**

Summary of All Farms Evaluated 2005-2015
Representing 176 Evaluations and ~ 35,000 acres in 21 Kentucky Counties

	Ground cover estimates of percent species composition						Tall Fescue Analysis	
	Tall Fescue	KY Blue-grass	Orchard-grass	White Clover	Weeds	Bare Soil	Ergovaline (ppb)	Endophyte (%)
Average	17	26	11	10	22	13	476	74
Minimum	1	1	0	0	2	0	-	-
Maximum	55	56	28	44	65	53	-	-
Standard Deviation	11	12	6	9	12	11	-	-

FIGURE 18.5 Sample farm datasheet. Species composition and tall fescue analysis numbers are summarized on a farm datasheet.

CASE STUDIES

The first 13 years of the program have generated a tremendous data set. However, analyzing this data as a whole is challenging as pastures and farms show tremendous variability. Positive changes on individual farms better illustrate the success of the program; therefore, this section contains case studies of three farms that participated in the program over the last 13 years. Care is always taken not to disclose the identity of participating farms for privacy reasons. Therefore, they will be identified by size, type, and number.

Farm #1: Large-Scale Commercial Breeding Farm

The University of Kentucky Horse Pasture Evaluation Program team first met with the managers of Farm #1 in 2012 after several unexplained late-term foal losses coupled with a mild winter on the over 400-acre farm. While the initial investigation did not suggest tall fescue toxicity to have caused the losses, it was discovered that the farm contained large populations of both tall fescue and nimblewill (*Muhlenbergia schreberi*), an aggressive warm season grass not grazed by horses or other livestock. Management requested a full evaluation of the farm that spring to evaluate the amount of tall fescue and nimblewill on the farm.

Populations of both undesired grasses in question varied widely on the farm, but generally were a concern in the majority of the pastures. The decision was made to begin small and renovate a few pastures, including Pasture 3 that commonly houses late-term broodmares. Spraying of targeted herbicide Imazapic (10−12 oz/acre) removed the majority of the tall fescue in the pasture; however, nimblewill quickly took over many of the open spaces before seeding could be

FIGURE 18.6 (A) Nimblewill. Began taking over the field when tall fescue was selectively removed (October 2012). (B) Complete spray down. The pasture was completely killed and reseeded with a Kentucky bluegrass/orchardgrass mix in the fall (December 2012). (C) New seeding. By spring, orchardgrass made up more than 80% of the pasture (March 2013). (D) Healthy pasture. The pasture was managed carefully and in 2016 was found to contain nearly 90% orchardgrass and Kentucky bluegrass (May 2016).

done. It was decided that the pasture should be killed completely to combat the nimblewill. Two rounds of glyphosate were sprayed before reseeding a Kentucky bluegrass and orchardgrass mix in late September. Evaluation of the pasture the following year revealed the pasture to contain predominately orchardgrass. Farm managers commented that foals were growing so fast on this pasture that concentrate feeds were reduced. With good management, Pasture 3 was still highly productive with few weeds or tall fescue in 2016 (Fig. 18.6A–D).

Farm #1 experienced similar success throughout the farm and completely renovated additional pastures throughout the operation. They have begun to harvest some recently renovated pastures for hay before spring grazing, so they often produce enough hay to feed their broodmares all winter long. The reduced need for hay and concentrates along with elimination of tall fescue has encouraged the farm to continue investing time, money, and effort into improved pasture management.

Farm #2: Medium-Scale Commercial Breeding Farm

Farm #2 participated with the University of Kentucky Horse Pasture Evaluation Program in its inception year of 2005 and has repeated evaluations annually. This 300-acre farm breeds and foals thoroughbred mares, preps and sells yearlings, breaks and trains 2 year olds, and stands stallions. Their major concern is endophyte-infected tall fescue in their pregnant mare pastures.

Based on early evaluations, the farm removed tall fescue from many mare pastures and replaced it with endophyte-free tall fescue varieties. Over the years, Farm #2 has continued monitoring of pastures to ensure that endophyte infection rates remain low. As infection rates increase, removal and replanting were completed or changes in management were made to remove mares from highly infected pastures. Pasture A was found to be 71% infected in 2008. Applications of a selective herbicide (Imazapic) removed tall fescue, and Pasture A was replanted in endophyte-free tall fescue that same year. With

FIGURE 18.7   No fescue. This farm identifies safe pastures for broodmares with "no fescue" signs by the gates of pastures.

careful management, endophyte-free tall fescue was maintained, and infection rate in 2016 was found to be just 14%. Pasture B was found to be 37% infected in 2008. Careful management and reseeding with endophyte-free seed helped maintain this pasture with a low infection rate, only 20% in 2016. Pastures can be deemed a low risk when less than 20% of available forage is infected tall fescue. Pastures with low levels of infected tall fescue are marked at the gate with "No Fescue" signs, indicating to everyone from grooms to managers that this pasture has been deemed low risk to pregnant mares (Fig. 18.7).

Farm #3: Small Private Farm

In 2015, a small private farm with just three paddocks contacted the University of Kentucky Horse Pasture Evaluation Program. The local county agent had suggested that tall fescue was present in sufficient amounts to pose a significant risk to the farm's one broodmare, expecting her first foal in 2016. The evaluation found that the paddock that had been designated to house the broodmare the last 60 days before foaling contained significant tall fescue (26%) but low infection (10%). Subsequent rechecks of this and an adjacent paddock confirmed the low infection rate of both pastures, thereby confirming that risk to the broodmare was low.

During the evaluation, a large percentage of bare soil was identified in another paddock on the farm. Because tall fescue was no longer a worry, the manager turned her attention to addressing this issue. Diligent pasture rotation between the better pastures and overseeding followed by rest and fertilization of this pasture resulted in better forage productivity for the farm as a whole. While subsequent evaluations were not performed, the manager reported that the mare foaled a healthy filly, and all human residents of the farm were enjoying more grazing and feeding less hay (Fig. 18.8).

CHALLENGES FACING UNIVERSITY OF KENTUCKY HORSE PASTURE EVALUATION PROGRAM

Any public program will face challenges as it grows, and the University of Kentucky Horse Pasture Evaluation Program is no different. Several challenges and obstacles have been identified that could impact the longevity and viability of the program moving forward.

Soil Sampling

Agronomists agree that regular soil sampling is a key aspect to successful pasture management. At its inception, the University of Kentucky Horse Pasture Evaluation Program chose not to include soil sampling as part of its service package to horse farms but instead encourage farms to work with their local county extension agents to obtain soil samples. For county extension agents, soil sampling is a primary tool used to establish relationships with farms, and the staff of this program did not want to jeopardize that relationship. Our team has found that most farms have subsequently built strong relationships with their county extension agent.

FIGURE 18.8 Small farm success. Evaluation of tall fescue risk and changes in management allowed this small farm to deliver a healthy filly, "Ruby," in 2016. *Credit: Jennifer Huhn.*

Competition With Commercial Businesses

Comparison of services provided by the University of Kentucky Horse Pasture Evaluation Program and farm consulting or farm supply stores might suggest that the program directly competes with industry. While some overlap is unavoidable, we find that each sector provides unique services to farms, negating competition between groups and in many cases complementing each other to support complete pasture management. Consulting firms offer soil sampling and drafting skills to provide farms with detailed maps. Farm supply stores offer custom fertilizing and seeding work to farms. Our program focuses primarily on vegetative cover and tall fescue analysis when needed. We find that many farms work with two or more of these groups to obtain comprehensive pasture data.

Extension Versus Fee for Service

The Smith-Lever Act of 1914 established federal funding for the Cooperative Extension Service. This service is charged with dissemination of new ideas from research at the Land Grant Universities to the public, a free service supported by federal funds. The University of Kentucky Horse Pasture Evaluation Program team regularly visits farms across the state and offers recommendations free of charge. Fees charged for detailed sampling and laboratory work allow us to offer more detailed information. Fees cover about one-third of total costs, with grants covering the remaining costs. Additionally, a small fee ensures that valuable time is spent working only with farms that are genuinely interested in the information and are invested in making positive changes in their management (Fig. 18.9).

Securing Funding

As mentioned in the previous section, the University of Kentucky Horse Pasture Evaluation Program is funded two-thirds by grants. Federal, state, and corporate grants are more competitive than ever and therefore more difficult to secure. Funding for a program like this, especially the cost of labor and transportation, will continue to become more and more difficult and does threaten the long-term viability of the program.

FIGURE 18.9 Horses enjoying crabgrass. While crabgrass is a high-quality forage for horses, its annual lifecycle and limited growing season in Kentucky make it less desirable. The UK Horse Pasture Evaluation Program works with horse farms to reduce the abundance of crabgrass and other warm season annual grasses and replace them with cool season, perennial forages. *Credit: Amanda Cole.*

Labor and Quality Control

When the program began, only a few small pastures on 10 farms were evaluated over the course of 3 months, requiring minimal labor. Today, the team covers roughly 3000 acres in the same time frame. Covering more ground with a relatively small team requires streamlining procedures and attention to detail to ensure accuracy and timely turnaround. The program has developed a series of checklists and templates to reduce office time and allow more time on farms. Data and recommendations are reviewed before being delivered to the farm, ensuring accuracy and the ability to tailor each document to the specific needs of the farm.

OTHER IMPACTS OF UNIVERSITY-BASED FARM SERVICES

The goal of the University of Kentucky Horse Pasture Evaluation Program is to improve pasture management on horse farms, therefore improving the economic viability of the farm, reducing soil erosion, and protecting the equine industry as a whole. However, we have found a number of additional benefits of the program.

Training Students

Most students hired to assist with the program have little or no background in agronomy or pasture management. All leave the program with new knowledge and appreciation for the equine industry. A few have found research and extension to be exciting and decided to pursue a career in agronomy through graduate education or a position within the industry (Fig. 18.10).

Relationships With Farms

As mentioned in the introduction of this chapter, the equine industry and the University of Kentucky Cooperative Extension Service historically had a limited relationship. Programs like the University of Kentucky Horse Pasture Evaluation Program help to bridge the gap between these industries and the research institutions that serve them. Our team maintains a long list of farm contacts and regularly shares new information with these farms via email blasts, social media, and one-on-one communication.

Developing Resources

Maintaining close relations with farms has given us valuable insight to the needs and concerns of the farms. Using this knowledge, we have worked to develop many resources for other farms including extension publications, a new website, an email listserv, and popular press articles.

FIGURE 18.10 Training students. Undergraduate students, both from the University of Kentucky and surrounding schools, provide much of the labor for this program. In the process, students are exposed to research and leave with a better understanding of agriculture in general. Several have switched their majors to plant science or changed their career path as a result of this program. *Pictured: 2015 team, left to right. Christine Voll, Sydney Beidleman, AnnMarie Kadnar, and Megan Baker.*

Future Grants

The program has generated a wealth of data, as well as a long list of cooperators, and has been used to illustrate the need for additional research and outreach when applying for grants. In 2016, the University of Kentucky Forage Extension Program secured a federal grant from USDA-NRCS to fund educational programs and cost share pasture improvements on horse farms in central Kentucky.

In opposition of Fee for Service Programs: Programs like this are in direct competition with local, for-profit companies and have an unfair advantage in being subsidized by grants. All Cooperative Extension programming should be free to state residents and supported by grants and other lines of funding.

REVIEW QUESTIONS

1. Describe environmental impacts of horse grazing.
2. Discuss what you know about the event, MRLS, which first built a line of communication between the Kentucky horse industry and the University of Kentucky.
3. Discuss the negative impacts of infected tall fescue on cattle and horses and how these species respond differently.
4. Based on this program, describe the ideal species composition of a pasture for broodmares, yearlings, and idle horses.
5. Imagine being a manager of Farm #1. Identify 4−5 bullet points highlighting the successes in pasture management under your leadership.
6. Managers of Farm #2 felt compelled to post signs to illustrate the amount of fescue in a pasture. How do you suppose farm employees would use this information? What if it changes over time?
7. Discuss how, over time, the infection rate of the broodmare pasture at Farm #3 could increase to unsafe levels.
8. The Smith-Lever Act of 1914 established federal funding for the Cooperative Extension Service. Some argue that this federal regulation makes Fee for Service Extension programing conflicting. Write a one-paragraph statement in support or opposition for this type of program.

REFERENCES

Aiken GE, Bransby DI, McCall CA. Growth of yearling horses compared to steers on high and low-endophyte infected tall fescue. J. Equine Vet. Sci. 1993;13:26−8.

Aldrich CG, Paterson JA, Tate JL, Kerley MS. The effects of endophyte-infected tall fescue consumption on diet utilization and thermal regulation in cattle. J. Anim. Sci. 1993;71:164−70.

Arachevaleta M, Bacon CW, Hoveland CS, Radcliffe DE. Effect of the tall fescue endophyte on plant response to environmental stress. Agron. J. 1989;81:83–90.

Bacon CW, Siegel MR. Endophyte parasitism of tall fescue. J. Prod. Agric. 1988;1:45–55.

Ball DM, Hoveland CS, Lacefield GD. Southern Forages. fourth ed. Norcross, GA: International Plant Nutrition Institute; 2007.

Belesky DP, Stuedemann JA, Plattner RD, Wilkinson SR. Ergopeptine alkaloids in grazed tall fescue. Agron. J. 1988;80:209–12.

Brendemuehl JP, Boosinger TR, Pugh DG, Shelby RA. Influence of endophyte-infected tall fescue on cyclicity, pregnancy rate and early embryonic loss in the mare. Theriogenology 1994;42:489–500.

Greene E, Smith SR, Cotten KL, Davis D. Comparison of ergovaline concentrations in BarOptima Plus E34 tall fescue and control varieties. In: Proceedings of the American Forage and Grassland Council Annual Conference; 2013.

Gwinn KD, Gavin AM. Relationship between endophyte infestation level of tall fescue seed lots and Rhizoctonia zeae seedling disease. Am. Phytopathol. Soc. 1992;76:911–4.

Hopkins AA, Alison MW. Stand persistence and animal performance for tall fescue endophyte combinations in the south central USA. Agron. J. 2006;98:1221–6.

Kentucky Equine Survey. University of Kentucky; 2012. https://equine.ca.uky.edu/kyequinesurvey.

Klotz JL, Bush L, Smith DL, Shafer WD, Smith LL, Arrington BC, Strickland JR. Ergovaline-induced vasoconstriction in an isolated bovine lateral saphenous vein bioassey. J. Anim. Sci. 2007;85:2330–6.

Kosanke JL, Loch WE, Worthy K, Ellersieck MR. Effect of toxic tall fescue on plasma prolactin and progesterone in pregnant pony mares. In: Proceedings of the Tenth Equine Nutrition and Physiology Symposium; 1987. 663–668 Lea, K.L., Smith, L., Gaskill, C., Coleman, R., and Smith, S.R. 2014. Ergovaline stability in tall fescue based on sample handling and storage methods. Front Chem. 2014 Sept. 8; 2:76.

Lea KL, Smith L, Gaskill C, Coleman R, Smith SR. Ergovaline stability in tall fescue based on sample handling and storage methods. Front Chem. 2014;2:76.

Lyons PC, Plattner RD, Bacon CW. Occurrence of peptide and clavine ergot alkaloids in tall fescue grass. Science 1986;232:4749.

McCann JS, Heusner GL, Amos HE, Thompson DL. Growth rate, diet digestibility and serum prolactin of yearling horses fed non-infected and infected tall fescue hay. In: Proceedings of the 12th Equine Nutrition and Physiology Society Symposium; 1991.

McClanahan LK, Aiken GE, Dougherty CT. Influence of rough hair coats and steroid implants on the performance and physiology of steers grazing endophyte-infected tall fescue in the summer. Prof. Anim. Sci. 2008;24:269–76.

Monroe JL, Cross DL, Hudson LW, Hendricks DM, Kennedy SW, Bridges WC. Effect of selenium and endophyte-contaminated fescue on performance and reproduction in mares. Equine Vet. Sci. 1988;8:148–53.

Morrison JI. Using Microhistological Techniques to Predict Botanical Composition of Horse Diets on Cool-season Grass Pasture. University of Kentucky Master's Theses; 2008. p. 504.

Putnam MR, Bransby DI, Schumacher J, Boosinger TR, Bush L, Shelby RA, Vaughan JT, Ball D, Brendemuehl JP. Effects of the fungal endophyte Acremonium coenophialum in fescue on pregnant mares and foal viability. Am. J. Vet. Res. 1991;52:2071–4.

Redmond LM, Cross DL, Jenkins TC, Kennedy SW. The effect of Acremonium coenophialum on intake and digestibility of tall fescue hay in horses. J. Equine Vet. Sci. 1991;11:215–9.

Rogers WM, Roberts CA, Andrae JG, Davis DK, Rottinghaus GE, Hill NS, Kallenbach RL, Spiers DE. Seasonal fluctuation of ergovaline and total ergot alkaloid concentrations in tall fescue regrowth. Crop Sci. 2011;51:1291–6.

Thalheimer RR, Lawrence G. The Economic Loss to the Kentucky Equine Breeding Industry from Mare Reproductive Loss Syndrome (MRLS) of 2001. 2001. Available from: http://cbpa.louisville.edu/eip/Newsletters/research/MRLS.asp.

Webb BA, Barney WE, Dahlman DL, Collins C, Williams NM, McDowell KJ. Induction of mare reproductive loss syndrome by directed exposure of susceptible mares to Eastern tent caterpillar larvae and frass. In: Workshop on Mare Reproductive Loss Syndrome. Lexington, KY: University of Kentucky College of Agriculture; 2002.

Webb BA, Barney WE, Dahlman DL, DeBorde SN, Weer C, Williams NM, Donahue JM, McDowell KJ. Eastern tent caterpillars (Malacosoma americanum) cause mare reproductive loss syndrome. J. Insect Physiol. 2004;50(2-3):185–93. https://www.sciencedirect.com/science/article/pii/S002219100300249X.

Appendix 1

Units of Measurement and Conversion Factors

LENGTH

Imperial (United States [US] and United Kingdom [UK]) Units of Length and Metric Equivalent

1 in. = 2.54 cm, about the width of a thumb
12 in. (1 ft) = 30.548 cm, length of a size 10 shoe
3 ft (1 yard) = 0.9144 m, from your nose to your outstretched finger
1 furlong (1/8 mile) = 1610 m, length of 10 sq. acres
1 mile (5280 ft) = 1.6093 km
1 mile = 1760 yards

Metric Units of Length

1 m = 100 cm, one ten-millionth of the length of the Earth's meridian along a quadrant through Paris, that is, the distance from the Equator to the North Pole
1 m = 3.281 ft = 1.0936 yards
1 km = 1000 m = 0.621371 miles
1 cm = 0.3937 in.

AREA

Imperial (UK) Units of Area

144 sq. in. = 1 sq. ft = 0.093 sq. m
1 sq. ft = 0.09 sq. m
9 sq. ft = 1 sq. yard = 0.8361 sq. m
4848 sq. yards = 1 acre = 4047 sq. m
43,560 sq. ft = 1 acre (208.71 ft × 208.71 ft)
1 acre = 4840 sq. yards = 0.4047 hectares
1 sq. mile = 640 acres = 2.59 sq. km

Metric Units of Area

1 sq. m = 1 m × 1 m = 10.7639 sq. ft	
1 hectare = 10,000 sq. m = 2.4711 acres	
100 hectares = 1 sq. km	

MASS OR WEIGHT

Imperial (UK) Units

1 ounce = 28 g, weight of a tablespoon of sugar
16 ounces (1 pound) = 454 g
1 pound = 0.4536 kg, 450 barley grains
1 ton (1 long ton) = 2240 pounds

US Units of Mass or Weight

1 ton (1 short ton) = 2000 pounds = 0.9 metric tonne

Metric Units of Mass or Weight

1 g = weight of 1 cubic cm of pure water
1 mg = weight of 1/1000 of a gram
1 kg = weight of 1000 g (1000 cubic cm) = 1 L of water = 2.2046 pounds
1000 kg = 1 metric tonne = 2204.62 pounds

VOLUME

US Units of Volume

1 fluid ounce = volume of water with a mass of about 1 ounce
1 cup = 8 fluid ounces
1 pint (US) = 2 cups = 16 fluid ounces = 473 mL
1 quart = 2 pints = 32 fluid ounces = 946 mL
1 gallon (US) = 4 quarts = 128 fluid ounces = 3.785 L
1 cubic in. = a cube 1 in. on each side (cu in or in^3)
1 peck = 2 gallons (8 dry quarts) = 8.81 L (US)
1 bushel = 4 pecks = 537.6 cubic in.

Imperial (UK) Units of Volume

1 pint (UK) = 20 fluid ounces = 568 mL, an English beer
2 pints = 1 quart = 1.1 L, a German beer
4 quarts = 1 gallon
1 cubic ft = 1728 cubic in. = 0.028 cubic m
27 cubic ft = 1 cubic yard = 0.76 cubic m
1 peck = 2 gallons (8 dry quarts) = 9.09 L (UK)
1 bushel = 4 pecks = 554.8 cubic in.

Appendix 2

Measuring Forage Dry Matter Yield Using Clipped Forage Samples

Select or make a sampling quadrat, a square, or rectangular frame of convenient size to throw onto a population of plants for sampling. An inside measurement of 1 ft × 1 ft provides 1 sq. ft of sampling space. These can be made of stiff wire, wood with metal-reinforced corners, or plastic water pipe. Measure the inside dimensions, and multiply length × width to determine the area. Record this number and its units (either square inches or square feet).

Walk across the pasture to be sampled in a zigzag pattern. Gently and randomly throw the quadrat. To just learn the procedure, two or three throws would be enough. Accuracy of the estimated forage yield of the whole pasture increases with an increasing number of throws and samplings. The first time that a field is sampled, about 10−30 throws and samplings would be reasonable.

At each place where the quadrat lands, gently pull up on the forage plants to straighten them. Use a ruler, a pasture stick, or similar device to measure the plant height in at least four and preferably 10 places within the quadrat. Record these height measurements and average them. Use grass clippers or electric hedge clippers to cut the forage plants as near to the ground as is practical (about 1/4−1/2 inch). Collect all the clipped forage from each quadrat in a separate paper or plastic bag. Fine-mesh laundry bags have good air flow and do not take up water. Ensure that the bags can handle the heat of your drying system without burning. Label the bags with the pasture and paddock name and/or number and its type of forage.

What you do next depends on the size and type of oven you have for drying samples. Commercial and research laboratories use large convection ovens to dry forage samples at temperatures around 100−176°F for 8−72 h. Microwave ovens, Koster testers, and food dehydrators are used. Table A2.1 indicates features of alternative devices for drying forages.

TABLE A2.1 Features of Various Forage Drying Devices

Devices	Accuracy	Relative Cost	Time Required	Number of Samples	Disadvantage
Convection oven	High	High	Overnight to 3 days	Depends on size	High cost
Microwave oven	Medium high	Low	5−30 min	1	Needs constant monitoring
Heater/fan (Koster)	Low	Low	25−50 min	1	Highest dry matter loss
Food dehydrator	Medium high	Low	2−10 h @ 165°F	<10	
Electronic probe	Low	Moderate	A few seconds	1	Lowest accuracy

No matter which drying oven you use, an accurate, reliable balance such as a postal or dietetic scale is recommended. It should have a capacity of at least 100 g and precision to 1 g or less. Weigh one of the sample bags. If it is porous, dry it for a short time before weighing. Weigh the whole wet, freshly cut forage sample. After drying, weigh the dry sample in its container. Record all these weights. Alternatively, use an electronic balance with enough capacity to weigh your biggest samples and the container used for weighing and drying samples. Set an empty sample container (paper bag, pie plate, etc.) on the balance and press the "tare" button. The reading should go to zero. If all of your containers have the same mass, you do not need to subtract the mass of the bag or plate from your sample, and your calculations will be a little simpler. If your balance loses the tare setting, just put another empty container on it and press tare again.

If your sample is larger than the capacity of your drying device, then you need to use a subsample for dry matter analysis and multiply the dry matter percent by the wet weight of the sample. If your entire sample fits easily into your oven (such as a convection oven) or other dryer, the dry weight of the sample minus the weight of the container is the amount of dry matter from your sample area. If you plan to subsample a large air-dried sample and oven-dry the subsample, remember to compensate for this in your calculations and that determining your actual dry matter yield will be delayed.

Heater/fan (Koster) moisture testers formerly came equipped with a spring scale, which is possibly not accurate enough or durable enough for long-term use. Choose an electronic scale (balance). Turn the scale on. Place 100 g of wet forage on the specimen container and spread it out. Place the specimen container on the evaporation unit, plug in the unit, and turn it on. Dry the sample for 30 min, then weigh it again. Continue drying for decreasing amounts of time and weighing the sample until there is no change in measurement. The final weight represents the dry matter of the sample.

For example: beginning (wet) weight = 100g.

End (dry) weight = 85 g. This is the amount of dry matter. The dry matter content is dry weight/wet weight = 0.85 in this case.

If you have a large *convection oven* (oven with a fan), poke about 20 holes in the paper bags with a pen or pencil. No sample chopping is required. Place the bags in the oven overnight at about 100−120°F.

A *microwave oven* should have at least 500-W capacity for drying forage samples. Containers for the forage should be microwave-safe. Glass pie plates work, but they become hot, and because of their weight, your balance must have sufficient capacity for them. Be cautious with paper plates or bags. If the container is porous, preheat it for about 10−20 s to drive out residual water, then weigh it, and record the empty oven-dry container weight (**A**). Place a chopped forage sample of about 100 g in an uncovered container. Record gross weight of wet sample + container (**B**). To prevent the sample from charring and protect the electronics, place a cup with about 6 oz of water in a back corner of the oven. Use full power to heat the sample for 3 min then weigh it. Stir the sample, dry for another 1−3 min, and reweigh. Repeat until the weight loss from each drying is less than 1 g. If charring occurs, use the previous weight. Record final gross weight of the oven-dry sample + container (**C**). Calculate the sample dry matter concentration as (C-A)/(B-A) × 100%.

Example:

Empty oven-dry container (**A**) = 20 g.
Wet sample + container (**B**) = 90 g.
OD sample + container (**C**) = 40 g.

DM concentration = (40 − 20)/(90 − 20) = 0.286 × 100% = 28.6%.

If you use a food dehydrator, weigh the dry sample trays, and then weigh the trays with wet forage samples on them. Dry (dehydrate) the samples according to instructions (8−12 h). Weigh the dry samples on their sample trays and calculate the dry matter concentration as for the microwave oven above.

The forage dry matter (FDM) weight from 1 sq. ft of sampled area is multiplied by 43,560 sq. ft/acre to give the weight of FDM per acre. If a dry sample from 1 sq. ft weighed 30 g, then it would represent.

30 g/sq. ft × 43,560 sq. ft/acre = 1,306,800 g/acre. This also equals 1306.8 kg/acre, and this is converted to lbs/acre by multiplying.

1306.8 kg/acre × 2.2046 lbs/kg = 2881 lbs of FDM/acre.

To determine the FDM yield in each inch of forage height per acre, divide the yield/acre by the number of inches of average forage height. For example, if the average forage height was 10 in.:

(2881 lbs FDM/acre)/10 in = 288.1 lbs FDM/acre/in.

Appendix 3

Graphic Representation of Changes in Sward Density and Forage Yield With Increasing Forage Height

Figure A3.1 illustrates a decline in total sward density for average and thick forage stands as plants grow taller. Differences among thin, average, and thick stands are described in Chapter 7, prior to Table 7.2. In average and thick swards, this is noticeable as a decrease of 32.5 and 73.3 lbs/ac/in for each inch of increasing plant height (as measured by falling plate meter). In thin forage stands, there appeared to be a slight tendency for sward density to increase with plant height. The graph reveals relatively large differences among thin, average, and thick stands at the lowest forage heights, emphasizing that subjective ranking of stands is consistent with differences in their measured densities. As forage plants grow taller, the decline in density is greatest for thick stands, intermediate for average stands, and possibly insignificant for thin stands. Well-established thick and average stands of grass provided plenty of forage to graze in each vertical inch.

Figure A3.2 illustrates how forage mass or yield increased with each inch of increase in forage height (shown here as falling plate height but also relevant to ruler height and rising plate height). In the thick and average forage stands, the increase of forage mass is curvilinear. Thick stands appeared to reach a maximum forage mass of slightly under 4000 lbs/ac at a falling plate height around 6 in. (which corresponded to 10 in. of ruler height). Beyond 6 in. of falling plate height or 10 in. of ruler height, there was no apparent increase in forage mass. As plants mature, hormonal changes inhibit leaf

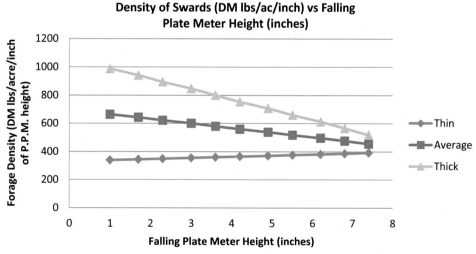

FIGURE A3.1 Changes in forage density (DM lbs/ac/in.) of thin, average, and thick forage stands with increasing forage height (as measured by a falling plate meter). *Rayburn, E., Lozier, J., 2003. Estimating Pasture Forage Mass from Pasture Height. Factsheet. West Virginia University Extension Service; Rayburn, E., 2016. Personal Communication.*

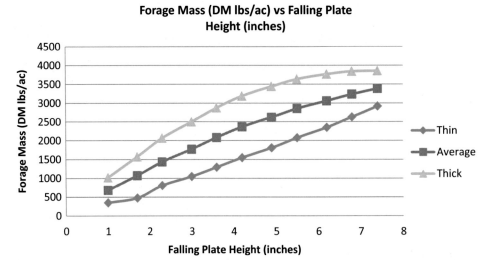

FIGURE A3.2 Changes in forage mass (DM lbs/ac) of thin, average, and thick forage stands with increasing forage height (as measured by a falling plate meter). *Rayburn, E., Lozier, J., 2003. Estimating Pasture Forage Mass From Pasture Height. Factsheet. West Virginia University Extension Service; Rayburn, E., 2016. Personal Communication.*

growth while stimulating flower and seed development. The trend for thin stands did not show any levelling off, perhaps because there was less competition for space among young and recently grazed plants. When forage yield is an important consideration, thick and average forage stands could be grazed at any falling plate height between 5 and 7.5 in. (8 and 12 in. ruler height) without a great increase in yield. However, it should be remembered that as forage plants mature, the concentration of proteins and nonstructural carbohydrates is declining, while the concentration of structural carbohydrates (cellulose, hemicellulose, and lignin) is increasing.

REFERENCES

Rayburn E, Lozier J. Estimating Pasture Forage Mass from Pasture Height. Factsheet. West Virginia University Extension Service; 2003.
Rayburn E. Personal Communication. 2016.

Appendix 4

Independent Evaluation of Falling Plate Meter and Rising Plate Meter

The falling plate meter (FPM) was developed in West Virginia, and initial data from its use there indicated that the dry matter (DM) yield in lbs of DM $= 432 \times$ average plant height (in) (Rayburn, 1997). The regression coefficient (R^2) was 0.78, indicating that plant height accounted for 78% of the variability in forage yield, and the standard error was 322 lb/acre. A device of the same design was made and used in Wisconsin at seven different locations where rotational grazing was practiced. Forage heights were measured with the FPM, and 70 samples were clipped and weighed to calibrate it. Pasture composition included some pure stands of cool season grasses and some mixed stands of cool season grasses with alfalfa, white clover, or red clover. A regression equation for the Wisconsin data indicated that DM yield (lbs/ac) $= 390 \times$ height (in) with $R^2 = 0.96$, which was considered in agreement with the West Virginia data (Cosgrove and Undersander, 2001). Both of these trials assumed that there would be no significant residual forage where the height was 0 inches.

The rising plate meter (RPM) has been evaluated several times in New Zealand, where it is mostly used on dairy farms, where budgeting feed supply is an important business practice. Most dairy farm pastures in New Zealand are a mixture of ryegrass and white clover, and most of the forage yield research there is based on this forage mixture. Potential users should have realistic expectations of this tool. It measures height, which allows estimation of pasture mass. The reliability of the estimate depends on the appropriate technique, calibration on the same forage type, use in appropriate environmental conditions, and keeping the device well-maintained. Observation of the grazing animals and postgrazing height of the residual forage should provide useful feedback on whether the estimates are reasonable.

The regression equation used by the RPM multiplies the (average compressed pasture height x the "multiplier") + the adder to give yield in kg of DM/ha. The adder is the point on a graph where the regression line crosses the vertical axis (which shows yield) when the RPM has a reading of 0 height. The multiplier is the slope of the regression line and accounts for the increase in DM yield for each increase in height (represented by 0.5-cm clicks). The standard regression equation for perennial ryegrass/white clover pastures in New Zealand is (height in clicks $\times$ 140) + 500 = kg DM/ha. The multiplier of 140 is a long-term average from results in all seasons. When grass is growing fastest, a multiplier of 115 works best, and in very dry conditions when growth is slow, 185 works best. A height of 0 clicks represents a residual forage yield of 500 kg DM/ha. For every 0.5-cm click of compressed forage height, the yield increases by 140 kg DM/ha. Anyone using either an RPM or an FPM should consult other users in their area and perform calibrations on the pasture species mixes they want to assess.

Any tool can be mismanaged. The problems with the RPM include grass and mud building up on the shaft and plate, corrosion on the shaft, a bent shaft, gear wheels not aligning on the shaft, and flat batteries. Before and during each use, the RPM needs to be checked for these problems and appropriate problems fixed before continuing to collect data. Operators can cause false readings if they apply too much pressure on the RPM, so that the shaft is pushed below the soil surface or pushed downward too rapidly so that the plate falls farther and compresses the forage more than it would just under its own weight. Frosted plants may be compressed and may prevent the shaft from reaching the soil surface. Strong wind can compress tall forage. Wet forage causes water and mud to accumulate on the plate, increasing its weight. Heavy rain and snow will compress the pasture height. So, adverse weather conditions are not good times to use the RPM. In pastures with significant weed infestations, especially tall, strong weeds, RPM readings can be artificially tall, so such weeds should be avoided. "Pug marks," which are depressions in soft soil from animal hoofs can cause false readings, so the RPM should not be used in such areas (DairyNZ, 2008).

REFERENCES

Cosgrove D, Undersander D. Evaluation of a Simple Method for Measuring Pasture Yield. University of Wisconsin-Extension; 2001. https://fyi.uwex.edu/forages/evaluation-of-a-simple-method-for-measuring-pasture-yield/.

DairyNZ. Using the Rising Plate Meter (RPM). Farmfact, pp. 1—15. https://www.dairynz.co.nz/media/253639/1-15_Using_the_Rising_Plate_Meter.pdf.

Rayburn EB. An Acrylic Pasture Weight Plate for Estimating Forage Yield. West Virginia University Extension Service; 1997. www.caf.wvu.edu/~forage/pastplate.htm.

FURTHER READING

Dyk, P. Monitoring feed dry matter: keep it simple and repeatable. Hoard's Dairyman (online). http://www.hoards.com/E_crops/cf21.

Gay SW, Grisso R, Smith R. Determining Forage Moisture Concentration. Publication 4424—106. Virginia Cooperative Extension; 2009. https://pubs.ext.vt.edu/442/442-106/442-106.html.

Griggs TC. Determining Forage Dry Matter Concentration With a Microwave Oven. Utah State University Cooperative Extension Factsheet; 2005. Ag/Forage & Pasture/2005-01, http://extension.usu.edu/files/publications/publication/AG_Forage_&_Pasture_2005-01.pdf.

Nennich T, Chase L. Dry Matter Determination. articles.extension.org/pages/11315/dry-matter-determination. 2007.

Pino FH, Heinrichs AJ. Comparison of on-farm forage-dry-matter methods to forced-air oven for determining forage dry matter. Prof. Anim. Sci. 2014;30(1):33—6.

Sanderson MA, Holden L, Rayburn EB, Soder KJ, Bryan WB. Assessing forage mass and forage budgeting. Chapter 2. In: Rayburn EB, editor. Forage Utilization for Pasture-Based Livestock Production. Ithaca, New York: Natural Resource, Agriculture and Engineering Service (NRAES) Cooperative Extension; 2006.

Appendix 5

Soil Maps from University of Kentucky Pasture Evaluation Program

Yields of Non-Irrigated Crops (Map Unit): Pasture (AUM)—Fayette County Area, Part of Fayette County, ...

Map Scale: 1:15,600 if printed on A portrait (8.5" x 11") sheet.

Meters
0   200   400   800   1200

Feet
0   500   1000   2000   3000

Map projection: Web Mercator   Corner coordinates: WGS84   Edge tics: UTM Zone 16N WGS84

Yields of Non-Irrigated Crops (Map Unit): Pasture (AUM)—Fayette County Area, Part of Fayette County, Kentucky

MAP LEGEND

Area of Interest (AOI)
☐   Area of Interest (AOI)

Soils

Soil Rating Polygons

▨   <= 6.70

☐   > 6.70 and <= 8.10

☐   > 8.10 and <= 8.60

☐   > 8.60 and <= 9.00

▨   > 9.00 and <= 9.50

☐   Not rated or not available

Soil Rating Lines

〜   <= 6.70

〜   > 6.70 and <= 8.10

〜   > 8.10 and <= 8.60

〜   > 8.60 and <= 9.00

〜   > 9.00 and <= 9.50

〜   Not rated or not available

Soil Rating Points

■   <= 6.70

☐   > 6.70 and <= 8.10

☐   > 8.10 and <= 8.60

☐   > 8.60 and <= 9.00

■   > 9.00 and <= 9.50

☐   Not rated or not available

Water Features

〜   Streams and Canals

Transportation

┼┼┼   Rails

〜   Interstate Highways

〜   US Routes

≈   Major Roads

〜   Local Roads

Background

▨   Aerial Photography

MAP INFORMATION

The soil surveys that comprise your AOI were mapped at 1:15,800.

Please rely on the bar scale on each map sheet for map measurements.

Source of Map: Natural Resources Conservation Service
Web Soil Survey URL: http://websoilsurvey.sc.egov.usda.gov
Coordinate System: Web Mercator (EPSG:3857)

Maps from the Web Soil Survey are based on the Web Mercator projection, which preserves direction and shape but distorts distance and area. A projection that preserves area, such as the Albers equal-area conic projection, should be used if more accurate calculations of distance or area are required.

This product is generated from the USDA-NRCS certified data as of the version date(s) listed below.

Soil Survey Area: Fayette County Area, Part of Fayette County, Kentucky
Survey Area Data: Version 12, Sep 15, 2015

Soil map units are labeled (as space allows) for map scales 1:50,000 or larger.

Date(s) aerial images were photographed: Data not available.

The orthophoto or other base map on which the soil lines were compiled and digitized probably differs from the background imagery displayed on these maps. As a result, some minor shifting of map unit boundaries may be evident.

Yields of Nonirrigated Crops (Map Unit): Pasture (AUM)— Summary by Map Unit—Fayette County Area, Part of Fayette County, Kentucky (KY643)

Map Unit Symbol	Map Unit Name	Rating	Acres in AOI	Percent of AOI
Hu	Huntington silt loam	9.50	42.3	9.3%
La	Lanton silty clay loam (dunning)	8.60	1.4	0.3%
MnB	McAfee silt loam, 2% to 6% slopes	8.10	67.3	14.8%
MnC	McAfee silt loam, 6% to 12% slopes	8.10	13.7	3.0%
MpB2	McAfee silty clay loam, 2% to 6% slopes, eroded	8.10	8.3	1.8%
MpC2	McAfee silty clay loam, 6% to 12% slopes, eroded	8.10	8.2	1.8%
MuB	Mercer silt loam, 2% to 6% slopes (nicholson)	8.60	13.5	3.0%
MuC	Mercer silt loam, 6% to 12% slopes (nicholson)	7.60	0.9	0.2%
Ne	Newark silt loam, 0% to 2% slopes, occasionally flooded	9.00	6.9	1.5%
ScC2	Salvisa silty clay loam, 6% to 12% slopes, eroded	6.70	5.7	1.3%
uBlmB	Bluegrass-Maury silt loams, 2% to 6% slopes	9.50	215.0	47.1%
uLbiB	Lowell-Bluegrass silt loams, 2% to 6% slopes	9.00	36.7	8.0%
uLfC	Lowell-Faywood silt loams, 6% to 12% slopes	8.60	12.2	2.7%
uMlmC	Maury-Bluegrass silt loams, 6% to 12% slopes	9.00	24.1	5.3%
Totals for Area of Interest			**456.2**	**100.0%**

DESCRIPTION

These are the estimated average yields per acre that can be expected of selected nonirrigated crops under a high level of management. In any given year, yields may be higher or lower than those indicated because of variations in rainfall and other climatic factors.

In the database, some states maintain crop yield data by individual map unit component and others maintain the data at the map unit level. Attributes are included in this application for both, although only one or the other is likely to contain data for any given geographic area. This attribute uses data maintained at the map unit level.

The yields are actually recorded as three separate values in the database. A low value and a high value indicate the range for the soil component. A "representative" value indicates the expected value for the component. For these yields, only the representative value is used.

The yields are based mainly on the experience and records of farmers, conservationists, and extension agents. Available yield data from nearby areas and results of field trials and demonstrations also are considered.

The management needed to obtain the indicated yields of the various crops depends on the kind of soil and the crop. Management can include drainage, erosion control, and protection from flooding; the proper planting and seeding rates; suitable high-yielding crop varieties; appropriate and timely tillage; control of weeds, plant diseases, and harmful insects; favorable soil reaction and optimum levels of nitrogen, phosphorus, potassium, and trace elements for each crop; effective use of crop residue, barnyard manure, and green manure crops; and harvesting that ensures the smallest possible loss.

The estimated yields reflect the productive capacity of each soil for the selected crop. Yields are likely to increase as new production technology is developed. The productivity of a given soil compared with that of other soils, however, is not likely to change.

RATING OPTIONS

Crop: pasture.
Yield units: AUM.
Aggregation method: no aggregation necessary.
Tie-break rule: higher.

WEB SOIL SURVEY SOILS MAP KEY

Yields of Nonirrigated Crops (Component): Pasture (AUM)

Central Kentucky horse farm: for soil types greater than 2% of total.

Soil Survey Map Symbol	Soil Type	AUM Horses	Acres per Horse per Year	Total Acres of Soil Type	Percent of Soil Type
Hu	Huntington silt loam	7.9	1.5	42.3	9.3
MnB	McAfee silt loam, 2% to 6% slopes	6.8	1.8	67.3	14.8
MnC	McAfee silt loam, 6% to 12% slopes	6.8	1.8	13.7	3.0
MuB	Mercer silt loam, 2% to 6% slopes (nicholson)	7.2	1.7	13.5	3.0
uBlmB	Bluegrass-Maury silt loams, 2% to 6% slopes	7.9	1.5	215.0	47.1
uLbiB	Lowell-Bluegrass silt loams, 2% to 6% slopes	7.5	1.6	36.7	8.0
uLfc	Lowell-Faywood silt loams, 6% to 12% slopes	7.2	1.7	12.2	2.7
uMlmC	Maury-Bluegrass silt loams, 6% to 12% slopes	7.5	1.6	24.1	5.3
Total				456.2	100.0

Appendix 6

2016 Field Recommendations: Central Kentucky Horse Farm

PASTURE 1

Weed Control

<u>**Spraying is an option**</u> in Pasture 1, since it contained a moderate amount of weeds, **11%.** As a general guideline, spraying is an option for fields with a weed population of **11%-25%**, depending on the operation.

Most prominent weeds: **bindweed and sedge**.

All weeds found:

Bindweed
Broadleaf plantain
Chickweed
Chicory
Dandelion
Johnsongrass
Lamb's quarters
Narrowleaf plantain
Nimblewill
Ragweed
Sedge
Speedwell
Timothy

The most effective herbicide for **bindweed and sedge** is **Dicamba+2,4-D,** according to the label recommendations. See the herbicide charts (Tables 1 and 2) in the previous recommendations section for more details and costs of application.

There were several patches of nimblewill found in Pasture 1. Since there are not available herbicides for this weed, the only control method (without complete spray down and replanting) is good pasture management to maximize competition of desirable grasses.

Establishment

<u>**Seeding is not necessary**</u>, since seeding recommendations are based on percent bare soil in fields and Pasture 1 contained **2%** bare soil*. Pasture 1 contains an average of **51%** Kentucky bluegrass; therefore there is a good chance that the Kentucky bluegrass will fill in the bare soil and provide future cover. The percentage of bare soil will vary over time based on environmental conditions, grazing intensity, and grass competition**.

*Note: Warm season annual grasses such as crabgrass, foxtail, and goose grass are counted as bare soil for recommendation purposes. These grasses will die off each fall, leaving bare soil unless seeded.

**Note: Seeding especially encouraged around gates and other bare spots.

Fertility

Soil test pastures every 3 years, submit the samples to your local county agent, and apply recommended levels of lime, phosphorus, potassium, and micronutrients. We recommend *fertilizing with nitrogen in the fall*. Two applications would be ideal; however, one application will pay big dividends the following year.

Apply the first application of actual nitrogen (30–40 pounds per acre) around the first of September. This will promote tillering and vegetative growth of the cool season grasses as they begin to prepare for dormancy. Growth of summer annual grasses is waning, and this N application gives the desired cool season grass species an opportunity to compete for sunlight and nutrients. It is also good for any newly seeded grass that may have been sown in late August.

Another application of actual nitrogen (30–40 pounds per acre) around the middle of October to early November prepares the plants for winter by stimulating tillering and root growth without excessive top growth. The plants will be healthier going into the winter, and they will stay green longer into the winter and green up earlier in the spring. A healthy and strong plant has a much better opportunity to compete with the summer annual weeds such as ragweed, yellow foxtail, etc.

If only one nitrogen application can be made, it should done around mid- to late October with a rate of 40-60 pounds of actual nitrogen per acre. Remember that urea is only 46% actual nitrogen and ammonium nitrate is only 34% actual nitrogen.

Tall Fescue

Current Risk

Pasture 1 **poses some risk to late-term pregnant mares**. Ergovaline in available forage is 128 ppb, below the UK threshold of 200 ppb. Ergovaline concentrations will change based on season and management.

Long-Term Risk

Pasture 1 contains **20% tall fescue and possesses a small risk** to late-term mares, but there are *no guarantees*. Remember that fescue toxicity is based on how much fescue your horses are consuming, not on how much is present in the pasture. One risk time would be during severe stress periods (for example, in a hot, dry summer) when the tall fescue may be growing and the Kentucky bluegrass is dormant.

Appendix 7

Answers to Review Questions

CHAPTER 1

1. Polysaccharides in plants include starches, fructans, cellulose, and hemicellulose. Cellulose and hemicellulose are structural. Starches and fructans are nonstructural.
2. An indigestible polymer that binds other cell wall components is lignin.
3. Plasmodesmata are microscopic channels through plant cell walls that allow transport of large molecules between cells, and their benefit is that they allow transfer of nutrients for growth and communication among cells, for example, to cause a response to an attack.
4. Similarities between xylem and phloem cells are that they are both hollow tubes, and they both carry water and dissolved substances. Differences are that xylem tissue is dead, while phloem tissue is alive; the xylem carries water with minerals from the soil, while the phloem carries sugar from leaves to other plant parts; movement of sap in the xylem is upward, while movement of phloem sap is generally downward.
5. Buds exist at the uppermost tip of a central stem (apical bud) and at the joint of a petiole (leaf stem) with the main stem (axillary bud). The apical bud produces the hormone auxin, which circulates in the plant and prevents axillary buds from developing. The hormones cytokinin and strigolactone are also involved. When auxin levels are reduced, axillary buds become active and produce new stems, branches, or tillers.
6. Cuticle is a layer of wax made from fatty acids, covering the epidermis of aerial plant organs to protect tissues and prevent evaporation of water. Mesophyll is a plant tissue just under the upper epidermis of leaves that carries on photosynthesis. It consists of two cell types, the palisade layer and the spongy parenchyma. Stomata are openings on leaves, mostly on the underside. They allow carbon dioxide in and both oxygen and water vapor out.
7. On a sunny day, sunlight stimulates the combination of carbon dioxide from the air and water from the soil to be combined through photosynthesis into sugar. The sugar becomes dissolved in phloem sap and circulates in the plant. The amount of sugar produced is greater than the amount used up in plant growth and development through the day, so sugar accumulates during the sunny hours. During the dark hours, sugar will continue to be used up in plant metabolism, without being replaced, so the concentration of sugar in the plant declines.
8. Differences between C3 and C4 plants are these: In C3 plants, carbon from carbon dioxide is fixed into three carbon compounds, whereas carbon dioxide is fixed into four carbon compounds in C4 plants. The C3 pathway is used in cool season plants, and the C4 system is used in warm season plants. C3 grasses store energy as fructans, whereas C4 plants store energy as starch. C3 plants are sweeter and more palatable than C4 plants.
9. Several mineral nutrients are absorbed into roots and then to the rest of the plant, and they have various roles. **Nitrogen** becomes a component of amino acids, proteins, RNA, and DNA. It promotes production of chlorophyll. **Potassium** stimulates opening and closing of stomata, activates enzymes, and stimulates energy transfers through ATP. **Phosphorus** is a component of nucleic acids, which regulates protein synthesis and, as such, is involved in photosynthesis, respiration, energy storage and transfer, cell division, cell enlargement, and development of new tissue. **Sulfur** is an important component of plant proteins and enzymes. **Boron** is involved in cell wall development, cell division, seed development, sugar transport, and hormone development. **Calcium** is responsible for holding cell walls together. **Magnesium** is important in photosynthesis because it forms the central atom of chlorophyll.

CHAPTER 2

1. Leaf blades, leaf sheaths, ligules, auricles, and rhizomes vary widely across the various grass species.
2. Use the "pull" test. If a seedling pulls out easily and roots all come out, it is likely an annual weed. If the seedling pulls out with difficulty or breaks off at the ground level, it is likely one of the desired perennials.
3. Bunch grasses: both ryegrasses, both fescues, orchardgrass, timothy. Sod formers: bluegrass, bromegrass, reed canarygrass, quackgrass.
4. At the three-leaf to four-leaf stage.
5. A stolon is an aboveground runner (stem capable of producing new plants), and a rhizome is an underground stem capable of producing new plants.
6. Alsike clover, hairy vetch.

CHAPTER 3

1.
   a) The simplest carbohydrates are sugars (also called monosaccharides), almost all containing either five or six carbon atoms, plus hydrogen and oxygen, whereas lipids are composed of the three carbon structure called glycerol with two or three fatty acids attached to it.
   Sugars are produced in plant cells by photosynthesis, using energy from the sun to join carbon dioxide and water into glucose, whereas glycerol and fatty acids do not directly require the sun's energy.
   Complex carbohydrates are polysaccharides; that is, they are made from long chains (and sometimes branched chains) of a few different sugar molecules, whereas fatty acids are quite diverse in their carbon chain length (2 to about 22 carbons) and in the number and placement of double bonds in the carbon chain.
   b) Carbohydrates are composed of just carbon, hydrogen, and oxygen, whereas proteins contain these three atoms plus nitrogen.
   Carbohydrates are major sources of energy storage in plants and animals, whereas proteins are mainly used for nonenergy functions.
   In plants, there are structural carbohydrates that provide support, shape, and strength and nonstructural carbohydrates that are mainly for storing energy. In plants, protein functions are mainly for metabolic purposes, such as acting as enzymes to speed reactions.
   c) Fatty acids consist of carbon chains 2 to 22 carbons long, with a methyl group at one end and a carboxyl (COOH) organic acid group at the other end, whereas amino acids have a central carbon with a carboxyl group, a hydrogen atom, an amino (nitrogen-containing) group, and an R group attached. Twenty different R groups provide variation among the 20 different amino acids.
   Fatty acids do not form very large polymers, but amino acids join in long polymer chains called proteins.
   d) Disaccharides consist of two monosaccharides, whereas polysaccharides consist of medium-length chains of monosaccharides (e.g., fructans are medium-length chains of fructose monosaccharides) and very long chains of glucose, fructose, and other sugars to form starches, cellulose, and hemicellulose. Several disaccharides have a sweet taste, but polysaccharides do not.
2. Alpha-1,4 bonds between sugars can be broken by enzymes produced in the digestive systems of herbivorous animals, but beta-1,4 bonds cannot. However, certain species of microbes that normally inhabit the digestive systems of herbivores do produce the enzymes to break beta-1,4 bonds, allowing them to digest cellulose.
3. Negatively charged portions of some amino acids form hydrogen bonds with positively charged portions of other amino acids in a protein. The sulfur atoms of two sulfur-containing amino acids in a protein chain will join together in a disulfide bond. Both of these bonds contribute to protein shape, which affects their function.
4. Four components of plant cell wall fiber are cellulose, hemicellulose, lignin, and pectin.
5. A bomb calorimeter is used to determine the gross energy concentration of an organic substance. A small ground, dried amount of an organic substance is placed in the combustion chamber of the bomb calorimeter with oxygen added. A spark between ignition wires causes the sample to explode. The resulting temperature of a water jacket surrounding the combustion chamber indicates how much gross energy was produced from the sample. The temperature rise is put into a formula to determine how many calories of gross energy were produced per gram of sample.

6. The gross energy concentration does not represent how much energy is available to an animal. So animals can be fed a test feed, and their manure is collected. The difference in gross energy concentration between the feed and the manure is called the digestible energy (DE), and it is less than the gross energy. DE represents the amount of energy that the animal extracted from the feed.

7. Neutral detergent fiber (NDF) is the result of digesting a feed sample in neutral detergent and filtering it. The NDF stays on the filter and can be weighed and compared to the sample weight. The percentage NDF represents cell wall components (cellulose, lignin, and most of the hemicellulose). The NDF percentage is inversely proportional to voluntary intake of feeds, so a low NDF value indicates that animals will eat it readily, and a high NDF value indicates that animals will not eat much of it. NDF can be used in equations to calculate dry matter intake, relative feed value, and relative forage quality.

8. A subsample of the NDF is digested in acid detergent, and the residue is collected on a filter and weighed. The acid detergent fiber (ADF) percentage includes lignin, cellulose, silica, and insoluble nitrogen but not hemicellulose. ADF is useful to put into a formula to determine the digestible dry matter content, relative feed value, and DE concentration of a feed.

9. The following increase in their absolute dry weight within a forage plant as it matures to the heading stage: lignin, protein, nonstructural carbohydrates (but they start to decline after the boot stage), cellulose, and dry matter.

10. The following decrease as relative values (percentages) while a forage plant matures: minerals percentage, percentage of leaves, and percent protein.

11. The feed that should be the most valuable per ton is b) CP 22%, ADF 25%, NDF 45% because it has the highest crude protein percentage and the lowest ADF and NDF percentages. The lowest ADF indicates the highest DE, and the lowest NDF indicates the highest voluntary intake.

CHAPTER 4

1. Obtain a soil survey report for your county and locate your pasture soil on the soil survey map. The report should provide a general description of the area including the geology of the parent material and underlying bedrock, native vegetation, climate, topography, and drainage. Review the classification of soils in the area and identify the specific classification name of your soil. The classification will describe in more detail the features of your soil such as soil texture and characteristics of a typical soil profile.

2. Abundant large trees and lush, diverse understory vegetation.

3. Loam
   Sandy loam
   Silty loam
   Clay

4a. Calculate the volume of soil in an acre-furrow-slice

Area of an acre is 43,560 ft^2

$$\text{Volume of soil to 0.5 ft depth} = 0.5 \text{ ft} \times 43,560 \text{ ft}^2$$
$$= 21,780 \text{ ft}^3$$

$$\text{Mass of soil in an acre} - \text{furrow} - \text{slice} = 81.2 \text{ lbs/ft}^3 \times 21,780 \text{ ft}^3$$
$$= 1,768,536 \text{ lbs}$$

4b. Calculate the volume of soil in a hectare-furrow-slice

Area of a hectare is 10,000 m^2

$$\text{Volume of soil to 0.15 m depth} = 0.15 \text{ m} \times 10,000 \text{ m}^2$$
$$= 1500 \text{ m}^3$$

$$\text{Mass of soil in a hectare} - \text{furrow} - \text{slice} = 1300 \text{ kg/m}^3 \times 1500 \text{ m}^3$$
$$= 1,950,000 \text{ kg}$$

5. The topsoil of the pasture would be rich in organic matter (plant residues, roots, and humus) and contain an abundance of macroaggregates. As a result, the surface soil would have a low bulk density and a high porosity, which allow ease of movement of water and air.
6. In the soil micropores.
7.

Soil water characteristic	Soil water potential (kPa)	Sandy loam	Loam	Clay loam
Saturation	0	47	50	53
Field capacity	−33	16	29	36
Permanent wilting point	−1500	6	11	22
Air-dry soil	−3100	4	7	14
Plant available water (%)		10	18	14
Plant available water* inch (cm)		1 (2.5)	2.2 (5.4)	1.7 (4.2)

8. Water that saturates soil macropores drains out faster, leaving the soil better aerated. Also, the drainage system lowers the water table of the soil profile down to the depth of installation.
9. A high soil CEC increases retention of nutrient cations, thereby minimizing the losses of nutrients by leaching. Soils with high CEC have a high reservoir of plant-available nutrients.
10. Applications of agricultural lime will reduce soil acidity and increase soil pH.
11. Rhizobia bacteria contained in nodules located on the roots of leguminous plants.
12. Mycorrhizal fungi colonize the roots of most plants and assist with the uptake of nutrients such as phosphorus.
13. Decomposition of plant residues and soil organic matter and converting the organic matter into inorganic forms (carbon dioxide, ammonium, phosphate) into forms that plants can use. Microbial growth during decay of soil organic matter also produces glues that promote formation of soil aggregates that help to stabilize the structure of the soil.
14. Pasture vegetation protects the surface soil against erosion, thereby minimizing losses of soil particles, nutrients, and amendments. Actively growing vegetation, especially during the early spring and fall time periods when excess precipitation events often occur, minimizes leaching losses of nutrients and reduces emissions of greenhouse gases.

CHAPTER 5

1. Sunlight, plants, soil, and animals.
2. Plants.
3. Photosynthesis, respiration.
4. Sunlight, water, nutrients.
5. Timing and intensity of defoliation/grazing.
6. The goal is the long-term health and productivity of the pasture to provide optimal quantity and quality forage that meets the nutritional needs of the grazing animal.
7. The animal's age, production state, and rate.

CHAPTER 6

1. Proper fertilization, species and variety selection, seeding date and rate, seeding method and control of weed competition.
2. Blue tags indicate certified seed, with guaranteed seed quality and purity. White tags are for common seed and list only the germination percentage, inert matter, and weed seed. Seed is placed in firm contact with the soil at a depth from which the seedling can emerge.
3. Seed is placed in firm contact with the soil at a depth from which the seedling can emerge.

4. Completed chart

Species	Season	Positive Attribute	Negative Attribute
KY bluegrass	Cool	Sod-forming, high palatability, vigorous growth	Dormant during summer drought conditions
Orchardgrass	Cool	Quality forage, high yield, good for hay production	Does not tolerate close continuous grazing
E− or novel tall fescue	Cool	Adapted to more frequent grazing than orchardgrass, tolerant to drought and water-logged conditions, novel varieties are long lived	Goes dormant under high temperatures, rough leaf texture so not as palatable as other cool season grasses
Perennial ryegrass	Cool	High seedling vigor, high quality	Short lived, high sugar content
Bermudagrass	Warm	Good grazing tolerance, heat and drought tolerance, ability to carry high stocking rates during the summer	Requires high levels of nitrogen and potassium, does not survive north of the transition zone
Timothy	Cool	Good hay production, tolerates poorly drained soils	Short stand life, poor survival under close grazing
White clover	Cool	Low growth habit, tolerant of frequent and close grazing	Lower yielding than red clover and alfalfa, goes dormant under high temperatures and dry conditions
Alfalfa and red clover	Cool	High quality, high yielding	Poor tolerance to frequent close grazing

5. Pastures should be sampled every 3 years to track changes in pH and fertility. Hay fields should be sampled yearly because a large amount of nutrients is removed in the hay crop.

6. When a visual assessment determines the pasture contains less than 50% desirable species.

CHAPTER 7

1.
  a. 6 horses x 1200 lbs = 7200 lbs; 7200 lbs x 0.022 = 158.4 lbs forage DM/day.
  b. 8 months x 30.44 days/month = 243.52 days; 158.4 lbs/d x 243.52 d = 38,573.57 lbs of forage DM as hay; 4 months x 30.44 = 121.76 d, 158.4 lbs/d x 121.76 d = 19, 286.78 lb of forage DM as pasture.
  c. 2 months x 30.44 = 60.88 days, 158.4 lbs/d x 60.88 d = 9643.39 lbs forage DM as hay; 10 months x 30.44 = 304.4 days; 304.4 d x 158.4 lbs/d = 48,216.96 lbs forage DM as pasture.
2. 5 months x 30.44 = 152.20 days; 158.4 lbs/d x 152.20 d = 24,108.48 lbs; 24,108.48 lbs ÷ 3000 lbs/ac = 8 acres.
3. Here are two options. Use a ruler to measure forage height. Why? Because you can look up the approximate forage yield for different heights, if you know the species of forage and its approximate density. Use a plate meter or a capacitance probe. Why? Because it will give you an estimate of forage yield after placing the end of the meter on the pasture 20−30 times.
4. Initially, use a sampling quadrat, clippers, a balance, and a drying oven to gather forage samples from 20−30 sites in a pasture with a locally common mixture of forage species. Calculate yield from your results. Also, measure forage height in the selected sampling sites prior to cutting them. Study your data and determine a relationship between forage height and yield. Why? You should gain experience with local forages and conditions, so you can have confidence in future estimates of yield, based on forage height and density.
5. Forage density is 371 lbs DM/ac/in. Forage mass is 1809 lbs DM/ac.
6. Young plants start very low in height and dry matter yield. The increase in height is approximately linear from the leafy stage, through boot and heading stages to the bloom stage. The increase in dry matter is curvilinear, similar to a lazy "S", with a levelling off at the bloom stage, prior to a decline.

7.

    a. From the leafy stage to the bloom stage, the weight or content of cellulose has a curvilinear increase, levelling off in the bloom stage. NSC increases as rapidly as cellulose up to the end of the boot stage, levels off, and declines through the heading and bloom stages. The increase in protein content is slow to the end of the heading stage; then it declines. Lignin content increases slowly in a straight line through all phases.

    b. The rate of increase of fiber and lignin percentage is steep and in a straight line, with no levelling off. Stem percentage closely follows fiber and lignin percentage. The percentage of protein, leaves, and minerals all decrease in straight lines through the growth stages, with protein percentage having the steepest slope and minerals the shallowest.

8. Increased tillering will most likely increase the yield of FDM/acre.

9. Dimensions of one side of each of the following are:

    a. one foot (12 inches)

    b. three feet

    c. four feet

    d. the square root of 43,560 = 208.71 feet

10. 2000 kg/ha ÷ 1.121 = 1784 lbs/ac.

CHAPTER 8

1. A typical range of percentage of hours spent grazing per day is 42%−62% of a day or as one-half to three-quarters of the daytime and half to three-quarters of the night. An inactive horse needing only a maintenance diet can consume it in 8−10 h/day, but some will choose to eat longer. Higher forage quality and forage yield will enable horses to take bigger bites and more frequent bites. Conversely, lower forage yield, including shorter plant heights, will cause bites to have smaller volume and mass, and the horse will attempt to overcome this by taking more frequent bites. Season and weather may also affect the times a horse chooses to graze.

2. Intake rate declines with a decline in forage quality. Immature, leafy forage has higher nutritional quality. The leafier the forage is, the faster the bite rate and the bigger the bites. Intake rate is limited by the time required to handle larger bites, so larger horses can consume more per minute than ponies with smaller mouths.

3. At the bite level, preference for various forage species is evident but complicated by other factors, including maturity, prior experience, and current levels of various chemicals. Generally, horses prefer grasses to forbs and browse. Younger plants, higher in sugars, are preferred to mature plants with lower sugar content. Horses generally prefer familiar forages to novel feeds, and some horses are more efficient grazers than others. At the patch level, negative postingestive feedback will increase the probability that horses will move to a different feeding site the next day, and rough terrain, requiring extra amounts of energy to walk away from, will decrease the probability that horses will move to another site before the forage at the first site is depleted. Sward height, distance to water, and the presence of manure may affect the site that a horse selects to graze. At the home range level, feral horses grazing rangeland may move to different home ranges due to effects of altitude or elevation. Cover for thermoregulation may also affect selection of home range, as shelter helps protect horses from extreme heat and cold. Home ranges closer to plentiful supplies of drinking water will be grazed more heavily than ranges far from good water. While it may be subconscious, herbivores practice cost-benefit analysis in choosing where to graze. They attempt to eat the most palatable, most digestible, and nutritious feed while spending the least time and the least energy doing it.

4. Using its sense of smell, a horse avoids grazing next to horse manure. This may help horses to avoid parasite larvae from manure. The length of aversion to grazing where manure was dropped may extend from one grazing season well into the next. In fenced pastures, particularly those with long grazing periods, latrine or rough areas form where horses do most of their defecation and much tall vegetation grows around manure piles. Horses in wild, unfenced environments apparently do not form latrines.

5. Concentrations of chemicals in blood and fetal fluids, due to intake by pregnant mares, could condition foals in later life to quickly accept feeds that contain those same chemicals. Similarly, chemicals detected by suckling foals in their mothers' milk will be more readily accepted in feeds in later life than if there has been no exposure to them. Watching what their mothers eat and mimicking their mothers' feed choices leads to first experiences with foods, which leave lasting impressions.

6. Horses choose forages more nutritious and less toxic than the average forage because of conditioned flavor aversions and preferences. Animals learn to prefer foods that previously provided positive digestive consequences, such as overcoming nutrient deficiencies, and they develop aversions to foods when postingestive feedback causes negative feelings such as nausea. Foraging decisions are likely to result in nutritional benefits with low probability of toxic or poor-quality plant material. External influences such as social interactions with peers and plant attributes plus internal factors (animal attributes that influence ingestion, digestion and metabolism that direct foraging decisions) contribute to learning. As animals mature, they sample various foods and develop flavor aversions and preferences. Flavors and presumably odors of plants are associated with postingestive feedback. The preference that herbivores show for mixed species diets may help them maximize intake of what they perceive to be lacking.

7. Grazing horses leave urine and feces on the pasture, which contributes nutrients useful for plant growth, and organic matter, which improves soil structure. However, these nutrients and other substances can also be considered pollutants if they should be transported off the field and into a waterway. Horses also damage pasture plants by trampling, eventually leading to soil erosion. This can be controlled to an extent by managing grazing. Overgrazing occurs when horses repeatedly graze the same grass plants until they can no longer recover. Controlled grazing can be beneficial to a grass plant by removing older leaves, allowing light to reach younger leaves, and stimulating plant metabolism in response to defoliation.

8. Presence of horse manure discourages grazing by horses at the level of bite and plant. Formation of latrines discourages grazing at the level of patch and feeding site. The plant itself—its type and species—may discourage grazing if it is not preferred by horses. They are most likely to consume forages that are familiar to them, so a novel forage may not be grazed. Horses often select forages based on maturity, so tall, stemmy swards with low nutritional quality will deter grazing. If the plant is in a location with environmental conditions like high altitude, rough terrain, and steep slope, horses may be discouraged from grazing it.

9. Maturity of the tall fescue plant being grazed will impact horses' preference. Immature, leafy tall fescue may be more palatable than mature plants that have gone to seed. Grass species also come in a number of varieties, so it could be that different studies compared different varieties of tall fescue. Additionally, preference ratings depend on which grasses are offered to horses at the same time. If tall fescue is offered with a selection of highly preferred grasses, it will have a low rating. If it is offered with a selection of less preferred grasses, then its palatability score will be higher. The horses could have higher nutrient needs while on pasture due to exercising more.

10. Horse-sick pasture occurs on relatively small pastures with high stocking rates and long grazing periods. These are overgrazed. The appearance of a horse-sick pasture includes roughs, lawns, and bare patches. The roughs are areas of long forage that may look palatable and nutritious, but because horses do most of their defecating in roughs, they will not graze there. The lawns are areas of very short forage that horses repeatedly graze. Lawns may have a few tall weeds and some manure in them. Bare patches develop from lawns once repeated grazing and hoof action have killed forage plants and weeds there. Horse-sick pastures can be avoided by making paddocks large enough for the number of horses and ensuring reasonable stocking rates. Follow good rotational grazing management, which includes short grazing periods of 3—5 days and long rest periods of about 14 days when forage is growing fast and 25—40 days or more when forage is growing slowly. Monitor pasture height to ensure that grazing starts at about 6—10 in (15—25 cm) and adequate residual forage height of 3—5 in (7.5—12.5 cm) for plant recovery is left after horses are removed from a paddock. When overgrazing is observed, move horses to another paddock, and if necessary, feed them hay to supplement or replace pasture. When horses are removed from a paddock, trim residual tall forage, and remove or harrow the manure to break it up and spread it uniformly over the paddock.

CHAPTER 9

1. $(9 * 750) + (16 * 1100) + (5 * 1500) = 31{,}850$ lbs animal weight
   31,850 lbs/1000 = 31.9 AU
   31.9 AU/35 acres = 0.9 AU/acre
   35 acres/31.9 AU = 1.1 acres/AU

2. Sometimes, it might sound odd to speak of animals in fractions (i.e., 0.6 AU/acre), so we use acres/AU. However, when farms have high stocking rates, it is more understandable to speak in terms of animals than acres, for example, 6 AU/acre instead of 0.17 acre/AU.

3. Overstocked farms can make use of stress lots to preserve pasture grasses when there is not enough fresh forage to feed animals. Setting up rotational grazing systems and managing them properly makes the land most efficient in terms of amount of feed provided. Understocked farms need to make use of mowing to control weeds and keep grasses in a vegetative state. Haymaking would be a good use of unneeded pasture land.

4. Trick question: neither is necessarily harmful if managed properly. However, an overstocked farm with suboptimal management is more likely to experience overgrazing and compacted soil, which reduces ground cover and water infiltration. This contributes to erosion and nutrient runoff.

5. Horses should not be allowed on pasture when the soil is wet from precipitation, when it is very dry and the plants are stressed, or when plants are dormant. In a rotational grazing system, horses should be confined to the stress lot when none of the grazing units have enough forage to be grazed.

6. Grass plants need leaf area to photosynthesize energy and metabolites for growth. When more than half of the plant's leaf area is removed, its ability to photosynthesize is drastically reduced and root growth stops. Growth is severely stunted, and it will take much longer for that plant to recover from a bout of grazing. Adequate leaf cover also prevents soil erosion and overheating of soil.

7. The rate of forage growth is variable throughout the grazing season. In addition, to seasonal differences, even the weather from one week to the next can affect the rate of forage growth.

8. Stockpiling can be used in the winter when forage in a pasture is rested starting in late summer and allowed to grow so that there is sufficient yield for horses to consume later in the winter. In the summer, irrigation of pastures of up to 1−2 in. per week helps to encourage grass growth when they would otherwise be dormant. Or, if in the transition zone of the United States, adding in warm season grasses to a cool season grass pasture can provide more forage in the hotter times of the year.

9. A minimum of 5 grazing units are needed at approximately 1.8 acres each, for a total of approximately 9 acres.
   - 28 days rest/7 days graze +1 = 5 grazing units
     1100 lbs × 2% × 8 horses × 7 days = 1.76 acres per unit
     200 lb/in/acre × 7 in × 0.50
   - Total acreage required: 5×1.76=8.8 acres

10. 3 days. The managers should remove one or two horses from the system because the grazing units will not get enough time to recover by the time the horses rotate through all 5five units. If that is not an option, then the horses will need to be confined to the stress lot and fed hay until a grazing unit has reached 5 in. The manager could also wait until the forage grows taller than 5 in. before grazing each unit. In the longer term, the pastures could be renovated using fertilizer and overseeding to improve the forage stand and increase the available forage estimate.
    - 5 horses × 1000 lbs × 2% BW = 100 lbs forage needed per day
    - 150 lbs/in/acre × 5 in × 4 acres × 0.5 (take half, leave half)/5 grazing units = 300 lbs of forage available in each grazing unit
    - 300 lbs forage/100 lbs needed daily = 3 days

11. All groups need 50% of their DM intake from pasture. For Groups 1 and 2, that comes out to 60 lb/day. For Groups 3 and 4, that comes out to 68 pounds per day. One grazing unit allows 600 lbs of available DM at the beginning of the grazing period. Groups 1 and 2 can stay on a unit for 10 days, and Groups 3 and 4 can stay on a unit for 8.8 days. Due to these long grazing periods, Linda might want to utilize temporary fencing to make her grazing units smaller when there is this much forage available.
    - Group 1 pasture consumption: 3600 lbs animal weight × 20%/2 = 36 lbs forage per day
    - Group 2: 2400 lbs × 20%/2 = 24 lbs forage per day
    - Groups 1 and 2 will need a total of 60 lbs pasture forage per day.
    - Group 3: 3000 lbs × 20%/2 = 30 lbs forage per day
    - Group 4: 2800 lbs × 20%/2 = 28 lbs forage per day
    - Groups 3 and 4 will need a total of 68 lbs pasture forage per day.
    - 200 lbs/in/acre × 6 in × 4 acres × 0.5 (take half, leave half)/4 grazing units = 600 lbs of forage available in each grazing unit
    - Groups 1 and 2: 600 lbs available forage/60 lbs per day = 10 days on each unit
    - Groups 3 and 4: 600 lbs available forage/68 lbs per day = 8.8 days on each unit

CHAPTER 10

1. Domestic livestock species: sheep, goats, cattle. Wild livestock species: deer, elk, pronghorn antelope.
2. Benefits of mixed species grazing include these:
   a. improved forage utilization efficiency
   b. enhanced energy efficiency of a soil-forage-herbivore complex in three stages: increased conversion of radiant solar energy through photosynthesis into forage plants; the increased solar energy in forages is consumed by grazing animals; then animals convert the energy from plants into animal products usable by humans

    c. improved ability to optimize quantities of forage resources and animal resources

    d. better control of weeds and brush

    e. better control of parasites with potential to reduce amounts of ecotoxic anthelmintics

    f. enhanced ability to efficiently manage pastures and rangelands that have many plant species and a diversity of terrain

    g. providing diversity of income and more uniform cash flow

    h. developing mutually beneficial interrelationships between animal species

    i. utilization of forage affected by feces of other animal species

    j. elevating some plant species from weed to forage status

    k. increasing grazing capacity per unit of area

3. Reducing the loss of pasture land near dung piles: free-roaming horses, cattle and others, with plenty of forage rarely eat near manure, especially near manure of their own species. If feed supply is short, they may eat near manure of another animal species, but not of their own species. During day or night, grass growing through cow dung will not be grazed by cattle but will be grazed by horses. If some of this is cut and placed on clean pasture, then cows will eat it. Thus the trigger to refuse forage near manure is caused by the smell rather than the sight of manure.

4. Mixed species grazing can help to reduce parasite infections as follows. Virtually all parasites are species-specific. With only mild exposure and without frequent reintroduction to their own typical parasites, horses and other herbivores will develop immunity to those parasites and have only a low level of parasitism. Do not keep horses in environments where they are being frequently reinfected. Either remove the manure from the pasture, plow up and reseed the pastures about every 3 years, or have enough pasture paddocks and an appropriate rotation of paddocks that horses are not placed into paddocks with high levels of infective parasite larvae in them. After horses have grazed a paddock, leave it ungrazed for an appropriate time for good regrowth, then graze it with livestock other than horses. The longer the interval between grazings, the more parasites will die. Alternately grazing and harvesting hay from paddocks will also lengthen the time between grazings by horses. To simply dilute the concentration of parasites on a pasture, graze other livestock in the same paddock with horses to dilute the horse manure and reduce the chance that horses grazing near horse manure become reinfected. Long, hot, dry weather and ultraviolet light from sunshine will help to dry out manure and kill parasite larvae in it.

5. Organisms that may be harmed by anthelmintic residues include arthropods (dung beetles), flies, aquatic Daphnia, and aquatic sediment-dwelling organisms.

6. Recommended parasite control practices include the following: Do not allow young horses to follow slightly older horses on a pasture. Alternate horses with one or more of sheep, cattle, and goats in rotational grazing. Test manure of all animals with fecal egg count kits to determine which animals need treatment with anthelmintics. Treat only the horses with significant parasite infections. Follow instructions from a veterinarian regarding the dose and timing of anthelmintic treatment. Keep anthelmintic-treated horses off pastures for 3 days after treatment so that live parasites and anthelmintic residues are not shed onto pastures.

7. Multispecies grazing aids in weed control because different animal species have different innate and learned preferences for eating certain plant species. Cattle, like horses, tend to eat mostly grasses but leave longer tufts of grasses after grazing than horses. Sheep naturally eat more forbs than horses and cattle. Goats and sheep eat many more weeds than horses and cattle. Goats eat more woody plants (browse) than horses, sheep, or cattle. Wild herbivores that eat considerable amounts of forbs and browse include deer and elk. Herbivores learn from their mothers, herd mates, and their own experiences what to eat and what to avoid. By taking advantage of learning through individual experience, herbivores can be trained to eat nontoxic weeds that are problematic in pastures.

8. MSG works to cause pastures to be utilized more efficiently because there are several influences on pasture utilization. Pastures are diverse, so a diversity of harvesting processes helps to use the pasture completely. Plants compete against other plants of the same and different species for resources. The more competitive the plant, the more offspring it will have. Grazing selectively in a low stocking rate environment stresses plant populations unevenly, further increasing diversity. The fates of pasture plants include being eaten, trampled, urinated on, manured on, or left untouched, so a grazing period can leave plants with a diverse ability to recover. Plants adjacent to feces will not be readily eaten by the animal species that defecated there, but if a second and a third animal species are present, they will eat that forage. Different animal species have different preference orders for plants they graze, and these preferences vary with time, location, and plant maturity. Sheep, cattle, and goats share a dietary overlap (the plants they like to eat) of 35%−53%, and wild herbivores are reported to share an overlap of 8%−55% with domestic livestock. Tall grass is most attractive to cattle and a little less so to horses. Sagebrush is most attractive to goats and deer. Short grass is most attractive to sheep. Pastures at high elevations are more attractive to sheep and goats than other animals, whereas low elevations and wetlands are preferred by cattle. The more diverse a forage stand is, the more efficiently it will be grazed by multiple species. Low grazing pressure (low stocking rate and stocking density) provides little competition, thus encouraging little diet overlap, but as grazing pressure increases, diet overlap increases due to increased competition among grazing animals and a weakening of their feed preferences. Thus with a diverse population of plants, a diverse population of animals is more likely to graze it efficiently and homogeneously.

9. Managing multispecies grazing integrates well with rotational grazing management.
   a. Examine plant populations and match new animal species to the plants they prefer. Establish goals for your land.
   b. Monitor yield and quality of forages. Budget feed for horses and for additional livestock and wild animals.
   c. Include wild herbivores in the feed budget if they frequently graze your pastures.
   d. Keep good records of plant species: percentage, condition, height, and/or yield.
   e. Record animal numbers (domestic and wild) and approximate weights.
   f. If adding small ruminants, add at least one electric wire to fences and ensure that all animals have easy access to good drinking water.
   g. If you do not want to own other species, rent pasture to someone who has them.
   h. Be careful that any feed supplements are compatible with all animal species.
   i. Manage economically (budgeting feed and money). Budget about 2.5% of grazing animal body weight to be supplied as FDM by the pasture. Ensure enough residual forage is left for rest and recovery of energy stores (i.e., take half, leave half). Look for opportunities to cover extra costs with extra income.
   j. Manage humanely. Ensure staff members have knowledge of how to manage all species. Try grazing mixed species together, but if there are conflicts, graze different species sequentially. Train animals to electric fences and handling facilities.
   k. Be open-minded to methods of managing different species.
10. Large livestock may injure or bully smaller ones, so extra management may be required to prevent this. More facilities, labor, and management may be required.
11. Horses evolved along with other ungulate herbivores, so they have some similar needs and behaviors. Interested parties can study and make informed judgements on which parts of the body of mixed species grazing research are applicable to horse management while planning and fund raising for future research that is more specific to horses.
12. The following are economic considerations of and potential economic consequences of mixed species grazing. You should know what your land can produce sustainably. Consider horses, forages, other animals, and animal products. There should be more efficient use of land and forage throughout the grazing season and thus lower annual feed costs. MSG may result in a more diverse product range and thus less risk of losing profits than with a single enterprise, and there will probably be income at more times during a year. Management will need adjusting to accommodate enhanced production, harvesting, and marketing. Faster growth of young horses and other livestock, plus a greater carrying capacity of grazing land should be benefits of MSG. Potential negative economic consequences include a need to account for increased costs of facilities, equipment, and labor for extra animal species. Marketing costs will likely be higher if you are starting to raise livestock to sell. If there is not enough diversity in the plant species that will grow, the potential for MSG may be limited. Toxic plants and predators may cause losses of valuable animals. Consider transforming some relatively nonproductive bush land into more productive pasture by grazing with goats and perhaps one other species in addition to horses. Consider renting grazing land to owners of other livestock species. Do not assume that markets for new products will appear without some effort or investment on your part.
13. Here are some recommendations on how to manage economic issues of adding additional grazing species. Produce optimum quantities of optimum quality forages and graze as much of it as is practical. Do not introduce small ruminants to areas with known predators without implementing proven predation prevention measures. Learn locations of potential markets, plus the costs of transport, feed, marketing, inspection, and grading. Look for opportunities to keep costs low by sharing or borrowing machinery or equipment, rather than buying things that you would use rarely.

CHAPTER 11

1. With advancing maturity, yields of forage crops increase while digestibility and intake decline.
2. Chemical characteristics are low ADF, low NDF, high crude protein, minerals, and freedom from antiquality compounds such as ergot alkaloids, cantharadin, slaframine, nitrates, and glycosides; visual/sensory characteristics are green color, freedom from mold, cleanliness, freedom from dust, leafiness, softness, and freedom from injurious plants and materials.

3. Antiquality/toxic factors and how to avoid them:

   a. Ergot alkaloids associated with the endophyte of tall fescue: Inspect pasture and hayfields for the presence of tall fescue, and test for presence/absence of fungal endophyte and/or the ergot alkaloids. Avoid as main source of forage for pregnant mares.

   b. Glycosides: Avoid using sorghum species for horses.

   c. Cantharadin: Manage hay sources to avoid contamination with blister beetles, using good weed control, insect scouting, and insecticides as needed.

   d. Mold and dust: Bale hay at appropriate moisture content, using hay preservatives, and store cured hay in dry barns without direct contact with the ground.

   e. Slaframine: Avoid consuming legumes in hay or pasture that have been infected with the blackpatch fungus (*Rhizoctonia leguminicola*). Red clover is the most common source of *Rhizoctonia*-induced slaframine toxicity.

   f. Nitrates: Avoid high-nitrate hays, which are most often made from warm season grasses subjected to growth stress following nitrogen fertilization.

   g. Physically injurious plants and materials: Avoid hay containing thistles, awns, or thorns that may injure the lips, tongue, or gums of horses. Prevent incorporation into hay of nonplant materials such as baling wire, twine, or net wrap. Inspect hay for foreign matter before feeding.

4. Two types of mowers: sickle bar and rotating disc. Sickle-type mowers are cheaper, require less horsepower per foot of cut, produce a cleaner cut (especially in thin crops), and tend to suck less dirt into the swath. Disadvantages include slow speed (especially in thick crops) and maintenance requirements. Rotary discs have greater capacity (especially in thick or lodged crops) and are simpler to maintain. Repair costs can be high, and they can suck dust into the swath under dry conditions.

5. Two types of mechanical hay conditioners: intermeshing rubber rollers and impellers. Goals of conditioning hay: crush or scrape stems of legumes and certain coarse grasses (e.g., pearl millet) to speed drying of stems relative to leaves and avoid leaf shatter in legumes.

6. Three types of hay rakes: wheel, rotary, and rotating bar. Wheel rake advantages include high-speed operation, low maintenance, and low horse power requirements. The ground-driven design of wheel rakes can introduce more dirt and foreign material than other types and can struggle with wet, high-moisture hay. Rotary rakes handle dry and wet material equally well and minimize foreign material and dirt in the windrow since the rotating tines do not touch the ground. They are more costly and complex than wheel or rotating bar rakes and require hydraulics and power takeoff (PTO) to turn the driveshaft. Rotating or roller bar rakes have simple, low-cost designs and are powered by the ground wheels, producing a clean crop. They have limited capacity and sometimes produce a ropey, twisted windrow.

7. Types of bale packages and their advantages and disadvantages:

   a. Small square bales: can be baled at higher moisture contents, small size is easier to handle, conveniently divides into feedable portions (flakes); require more labor to load/unload hay and have higher transportation costs because trucks cannot load to full legal weights.

   b. Medium square bales (three string, 100- to 150-lb bales): a more efficient package to produce and transport; bales flakes easily into feedable portions but are too heavy to easily handle.

   c. Round bales: widely available source of hay, can be baled at higher moisture contents relative to large rectangular bales. Must be handled mechanically, higher transportation costs, more difficult to stack, less efficient use of barn space. Feeding requires covered feeders to prevent excessive losses.

   d. Large rectangular bales: efficient to produce, load, and transport; too heavy to easily handle and too large to fit in traditional barn spaces (lofts); flakes are too large for individual feeding.

8. Characteristics of a good hay structure: keeps hay dry, prevents direct contact between hay and ground, easy to access by delivery trucks and any feeding vehicles/tractors, isolated from housing horses and equipment, minimizes sun exposure.

9. Considerations for choosing good hay for horses: cleanliness, nutrient value, type of horse being fed, price, availability, and package size.

10. Pros and cons of three methods of feeding horses individually:

   a. On the ground: places horse's head in the most natural position but can lead to greater waste.

   b. In hay bins or troughs: minimal loss but horse's nose will be more exposed to dust and mold; potential for the horse to put a leg inside the bin or trough, depending on the design.

   c. In hay racks or nets: keep hay contained (lowers waste), suspended high enough to prevent leg entanglement, but position allows hay particles and dust to fall into the horse's eyes and nose.

11. Best type of hay feeder for groups and why: covered hay feeders, to minimize waste.
12. Advantages and disadvantages of producing baleage:
    a. Advantages: low cost, less capital investment, higher quality feed, lower harvest and storage losses, bales are portable, small amounts can be ensiled, can utilize existing hay equipment.
    b. Disadvantages: (unchopped) forage crops are harder to ensile (less readily fermentable carbohydrates) than chopped forage, some balers cannot handle wilted (40%–50% dry matter) forage, bales can be very heavy, leading to larger tractor requirements, plastic wrap material can tear or puncture, leading to spoilage, and disposal of used plastic is necessary.
13. Conditions that favor the growth of *Clostridium botulinum* in baleage: baling forage at excessively high moisture content (above 70%), contamination of forage to be ensiled with dirt or fecal matter, improper fermentation due to punctured plastic, poor oxygen exclusion, too little or poor-quality plastic, low carbohydrates in forage, and loosely formed bales.

CHAPTER 12

1a. The similarity among months is that March and September have the same amount of solar radiation.

1b. The likely cause of the similar amount of solar radiation is that an equinox (a day of equal hours of light and dark) occurs in March and September.
1c. Monthly solar radiation is greater in a southern state. This difference is greatest in December and least in June.
2. As a young plant grows up to about four inches high, there is a rapid increase in the fraction of light intercepted.
3. An advantage of C3 forages is that C3 forages evolved to fit ecologic niches with relatively short growing seasons and freezing weather in winter. Cool season grasses have a greater temperature range for photosynthesis. Peak growth rate of cool season grasses and legumes occurs between 46 and 62°F (7.8–16.6°C); thus, C3 forages can thrive better than C4 forages in a temperate humid environment such as Vermont.
4. C4 plants have the Calvin cycle apparatus to use CO_2 and make glucose, plus they have a C4 pathway that takes in CO_2, uses ATP for energy, and delivers a high concentration of CO_2 to the Calvin Cycle. The enzyme bringing CO_2 into the C4 cycle has a higher affinity for CO_2 and no affinity for oxygen, so C4 plants are more efficient at fixing CO_2. C4 plants do not need to keep their stomata open as long as C3 plants to fix the same number of CO_2 molecules, so there is less time for water loss. Warm season grasses are more efficient in water use, thanks to deeper roots and the C4 cycle. Warm season grasses also produce more dry matter per unit of nitrogen fertilizer. C4 forages evolved to fit environments with longer growing seasons, mild winter weather, and summers with long periods of very hot weather. All of these features help C4 forages to thrive in the warm humid environment of South Carolina.
5. Warm season grasses produce more dry matter per unit of nitrogen fertilizer than cool season grasses. This apparent advantage is reflected in forage quality since the nitrogen taken up by C4 plants is diluted by the increased amount of dry matter, so protein concentration of warm season grasses is lower than cool season grasses (at the same stage of development), and concentrations of digestible energy and protein decline faster with maturity in warm than in cool season grasses.
6. GDD are useful to predict [NDF] of alfalfa or other forages and to determine a suitable stage of development of the forage before harvesting for the first cut of the season in temperate, nonarid environments.
7. Orchardgrass and tall fescue have medium winter survival and high horse preference.
8. Regional weather data is available from the National Weather Service, and average monthly data is available from the National Oceanic and Atmospheric Administration.
9. PET would be higher because it represents potential evapotranspiration rather than actual E. High temperatures and low soil moisture typical of warm arid regions promote evapotranspiration of more moisture than the amount that is available.
10. March to May is the period of greatest growth of cool season forages in the transition zone. The warm season forages have peak productivity from May through September.
11. Drought management and prevention measures can include the following:
    a. Plant drought-resistant plants, considering root depth and adaptation to local conditions.
    b. Plan for irrigation in the long term.
    c. Build soil carbon to increase water-holding capacity.
    d. Monitor feed inventory.
    e. Determine forage yield.
    f. Keep good records.

     g. Take photos of forage production.

     h. Plant annual forages if enough soil moisture.

     i. Graze stubble and crop aftermath.

     j. Send animals to areas with more forage.

     k. Sell least productive/least valuable animals.

     l. Buy feed.

     m. Substitute some alternative high-fiber feeds.

12. To improve winter hardiness or reduce winter damage to plants:

     a. Allow an accumulation of at least 500 GDD (calculated using a 41°F (5°C) base) to allow plant hardening

     b. Ensure that soils have relatively high levels of K but relatively low levels of N before dormancy

     c. Do not cut alfalfa in northern states during the 6-week autumn critical period.

     d. Select varieties and species that have been shown to tolerate local winter conditions.

13.

Measures in the Atmosphere	Expected Change by 2100
Temperature	↑
$[CO_2]$	↑
$[N_2O]$	↑
$[CH_4]$	↑
Precipitation	↕

14.

Pest Insect or Arthropod	Effect on Other Organisms
Midge	Bluetongue disease in sheep and antelope
Culex mosquito	West Nile virus disease
Mountain pine beetle	Destroys pine forests
Black-legged tick	Lyme disease in horses and people

15. Effects of increased temperature and $[CO_2]$ on a mixture of C3 and C4 plants in a plains region with no change in precipitation:

     a. increased photosynthesis around 58%, increased growth around 15%

     b. legume and C4 plant growth stimulated more than C3

     c. more NSC due to higher $[CO_2]$ and less [N]

     d. seasonal forage development changes

     e. optimal growth rate changes

     f. water availability changes

16. Combined effects of climate changes on horses might include the following:

     a. Horses are more frequently above their zone of thermal neutrality and THI, so they need more help in keeping cool and hydrated.

     b. Horses need more to drink.

     c. Nutritional quality of feed is slightly reduced.

     d. More health challenges from parasites, pathogens, and vector-borne diseases.

17. To minimize the contributions of horses to climate change and GHG emissions:

     a. Minimize fossil fuel use in machinery, the clearing of forests to make croplands, and pastures and tillage for cropping.

     b. Develop agronomic practices, higher quality diets, and feeding systems to reduce this methane production.

     c. In terms of grazing land management, abatement strategies suggested include introduction of improved plant species (including legumes) and preventing soil compaction, both of which are predicted to have small abatement potential.

     d. Look after one's own horses and land according to national, state, and local regulations and continue to educate oneself on ways to avoid and reduce GHG emissions.

18. Horse farm managers can look for opportunities to promote carbon sequestration into degraded soils and desertified ecosystems, follow recommended management practices on agricultural soils, ensure that grazing lands are covered by plants permanently, restore marginal lands to vegetative cover, and apply composted plant material and manure to pastures.

19. Professional resources available to help with matching management practices to your climate include these:

  a. state universities nearest to you that still provide reports of forage variety and management research trials,

  b. professionals who are Certified Forage and Grassland Apprentices (CFGA) and Certified Forage and Grassland Professionals (CFGP), with their certification coming from the American Forage and Grassland Council (AFGC),

  c. CFGPs who are also Technical Service Providers through the United States Department of Agriculture (USDA) Natural Resources Conservation Service (NRCS).

One can make use of these professional resources through participation in educational activities organized or promoted by AFGC and NRCS about forages and forage lands through conferences, conference proceedings, competitions, and grazing schools.

CHAPTER 13

1. Bermudagrass
2a. Tall fescue, reed canarygrass

2b. Kentucky bluegrass, smooth bromegrass, orchardgrass
3. Smooth bromegrass
4. White clover
5. Alfalfa, red clover
6. Tall fescue
7. Perennial ryegrass
8. Bluegrass white clover
9. Cool season forages grow best in cool moist environments.
10. Warm season forages grow best in warm to hot environments with adequate rainfall but tolerate droughts better than cool season forages.

CHAPTER 14

1.
  a. Eutrophication is occurring, and there is an algal bloom.
  b. Any fish in the pond may die as the oxygen levels in the water decrease. It may start to have a foul odor.
  c. Fence horses out of the stream, leave a filter strip between the pasture and stream on both sides, pick manure from the pasture (or at least near the stream and in areas where horses congregate), construct heavy use pads in areas where horses congregate, divert runoff in such a way that it outlets in a wide grassy space, limit horses' time on pasture or construct a stress lot.
  d. Fence off most of the stream and construct stream crossings in areas that horses and equipment need to cross.
  e. Horses accessing the stream to drink will erode the stream bank, and the water could be contaminated from an upstream landowner with livestock or crops.
2. Due to a loss of topsoil (the layer of soil where plants grow). Erosion also can decrease plant cover and alter plant species composition, both of which have an effect on infiltration and runoff.
3. Several methods to reduce parasites on pasture include removing manure from pastures, not rotating horses immediately after deworming, rotating pastures into hay or ruminant grazing for at least a year, and maintaining high-quality pastures. (Additional information: It is nearly impossible to eradicate small strongyle infection in horses that graze, so more effort should be focused on managing resistance. Most horses on a farm will have strong immunity to these parasites and require only one to two deworming treatments per year to maintain pasture refugia. The remaining high shedders can be treated more often as needed based on egg reappearance in fecal egg counts.)
4. Nutrient recycling: grazing allows for recycling of nutrients by depositing manure directly back onto pastures and providing nutrients for forage plants. Improves soil quality: manure provides carbon to the soil and provides food for soil microbes. This leads to increased organic matter and microbial activity. Reduces farm costs: if enough pasture land is available, keeping horses outside has a number of advantages, including less labor spent on stall cleaning.

5. Several guidelines include the following: Nutrients added are appropriate for grass so application rate must be limited to nutrient uptake rate. Pasture grasses will only take up so much nitrogen, phosphorus, and potassium in a grazing season, and applying too much manure will result in contaminated runoff and nutrient buildup in the soil. It is important to calibrate the equipment to ensure that the rate of application is correct. Location of spreading is important; for example, the manure must not be spread too close to any sensitive areas: surface water, wetlands, drinking water, sinkholes, or any other direct outlet to surface water.
6. Gutters are the first line of defense in rerouting roof runoff. From there, it can be collected in rain barrels and used for irrigation or arena watering, or routed via underground pipes to a suitable outlet such as a long grassed area where the water can infiltrate. French drain systems are another good way to reroute stormwater to a suitable outlet. A perforated pipe is laid underground and backfilled with gravel, so water can easily reach the pipe, plus the perforations allow water to infiltrate as it travels through the pipeline.
7. $30 \times 50 = 1500$ sq. ft, 0.6 gallons/sq. ft/in. of rain $\times$ 3 in. $= 1.8$, 1500 sq. ft $\times$ 1.8 gal/sq. ft $= 2700$ gallons.
8. Grazing animals in an area near a stream or river (riparian area) is a concern because the nutrients deposited in the manure can directly enter the waterway, increasing the chance for nonpoint source pollution.
9. Riparian buffers will allow sediment time to settle out of the water and infiltrate into the soil. They provide biologic and chemical filtration of nutrients and pesticides in runoff. Nutrients in runoff are utilized by the soil organisms and plants. Dissolved organic content and plant parts that fall into streams provide food for aquatic organisms. Shade from a forest riparian area keeps water from overheating and reduces oxygen use by fish and other organisms. Fallen trees and branches will create habitat for fish.
10. Zone 3 in a riparian forest can be intensively grazed for short periods of time as long as the soil is dry and firm.
11. See the Livestock Farmer Survey in Appendix 9. Answers will vary.

CHAPTER 15

1. Situations and locations where each type of fencing is appropriate on pasture:

Types	Situations and Locations
Permanent, nonelectric	Large properties, especially where sources of electrical power are far away Around the perimeter of the property, especially next to busy roads Along laneways where horses, vehicles, and machinery travel frequently Bordering major segments of a farm, dividing different functions Where physical security is of great importance Where eye appeal is very important
Semipermanent, electric	Within the boundaries of a farm, separating areas by geographic features, management functions, and other reasons that are subject to change Separating a large area of pasture into major paddocks for rotational grazing Allowing subdivision of major paddocks with temporary electric fences
Temporary, electric	Within boundaries of permanent and/or semipermanent fences Allowing pasture managers to allocate small amounts of pasture at a time to promote short grazing periods and long rest periods for desired forage plant recovery and persistence For flexibility, allowing quick responses to changes in forage conditions and weather

2. Visibility, strength, and durability.
3. Prevent injuries and being caught (safety); discourage chewing; keep dangerous animals out; add esthetic value (eye appeal); be reasonably priced.
4. Coyotes and cougars (mountain lions).
5. Moving horses, removing manure, transporting field equipment, snow storage and removal.
6. Problem (a): fence has no power. Check that all connections are consistent with the energizer manufacturer's instructions and are tight. Confirm that AC (main line) or DC (battery) power is reaching the energizer. Ensure that the number, distribution, and length of ground rods is consistent with the energizer manufacturer's instructions and that they are contacting moist soil. Look for power lines and buildings with AC power near the fence. Problem (b): voltage indicated by the fence tester is less than expected. Check the whole fence for vegetation and other conducting materials touching the fence and the ground or otherwise draining power. Remove grass and weeds by cutting or spraying herbicide. Remove trees or branches that can touch or fall on the fence when heavy with rain or snow or blown by wind. Problem (c): animals are getting through the fence. Check that voltage is satisfactory. If it is, add more wires to the

fence at heights that will block attempts to go under, through, or over the present fence. Check whether the current is ever turned off. Problem (d): animals in other parts of the farm change their behavior once the energizer is turned on. For example, some are reluctant to approach a feeder, waterer, or part of a building. Check for stray voltage in the problem area and insulate any leaks. Increase the grounding far from the problem area.

7.

Wire Types	Conductivity	Longevity	Visibility
Bare galvanized steel	High; 12.5 ga has higher than 14 ga	Very long, especially if Class 3 versus 1 or 2 galvanizing	Lowest, especially if 14 gauge or finer
Thin polymer string	Low, few fine wires	Moderate due to ultraviolet stabilizers in polymer and stainless-steel wires; not very strong compared to 12.5-gauge steel	Higher than bare galvanized steel
Thicker polymer rope	Moderate, more fine wires than string	Probably higher than polymer string	Higher than string
Polymer tape of varying widths	Low to moderate depending on the number of wires	Moderate, probably depending on physical stresses	High and wider tape has higher visibility
Steel coated with a conductive polymer	High; as much as uncoated galvanized	Very long, probably longer than uncoated steel	Higher than uncoated steel

8. Pasture gates should be as high as the fence attached to them, to prevent horses from jumping over. Their width depends on what is intended to go through them. For one person and one horse at a time, a gate should be at least 4 feet wide. To allow modern agricultural machinery through and to turn, widths of 16−25 feet (4.8−6 m) should be considered. If on a laneway, gates should be at least as long as the width of the laneway, so gates can be used to block the laneway. If horses are moved as a group through gates, locate gates in the middle of a fence, rather than at a corner. Gates should be in well-drained areas and if next to a roadway, at least 40−60 feet from a corner, to allow good visibility between the road and pasture for drivers.

9. It is safer to turn off the power to the electric fence and work on the wire with no danger of being shocked. Rubber gloves are available of different classes to handle different amounts of voltage, but they are expensive, require leather gloves over them, and will not protect the rest of your body from shocks.

10. Inspect the physical parts of the horse watering system, looking for signs of potential contamination. Consult an agricultural extension or public health agency about a water quality analysis, obtain appropriate containers, and collect water samples for quality analysis. Share the results with the extension or public health agency if the analysis results indicate any contamination above an acceptable level.

11. Any four of the following desirable features of a watering system for horses:
   a. Provide enough water for the needs of all horses in all weather and have reserve capacity in case of a change in needs.
   b. Water is good quality, which means no contaminants above recommended maximum concentrations for total dissolved solids, fluorine, selenium, sodium, blue-green algae (cyanobacteria), microsystins, and coliform bacteria or other harmful substances.
   c. Water is available in all paddocks, whether the arrangement is permanent or portable.
   d. The watering system is relatively inexpensive to purchase and maintain, using gravity, wind power, or solar power to move water, if possible.
   e. If the source is surface water, horses are fenced away from it and they drink away from it, so that manure and urine do not contaminate the water and the banks are not eroded.
   f. The capacity of the watering system is adequate, so horses do not need to wait for enough water to enter the troughs.
   g. Watering devices are well-suited to their location, protected from freezing if necessary, and easy to maintain.
   h. Location of waterers and construction of their bases prevents wet soil, mud accumulation, allows drainage, and prevents erosion.
   i. Watering devices stay clean by design or are easy to clean and regularly cleaned to prevent accumulation of cyanobacteria and other harmful organisms or substances accumulating in them.

12. Relative advantages and disadvantages of permanent versus temporary portable watering devices.

Types of Watering Devices	Advantages	Disadvantages
Permanent	Very little time required for operation or maintenance once installed Devices are reliable Pipes to them are durable and often well-protected underground	Initial cost is high Surrounding pasture can be trampled and muddy if base is not installed properly Cannot be moved, so even when grazing rotates through paddocks, horses are attracted to the waterer, and their traffic will destroy some forage around it
Portable	Do not need a device for every paddock, so the cost of hardware is low Some are so light and small that they can be carried by one person Others are conveniently moved by a tractor or four-wheeler High portability means that they can be placed in different parts of a paddock for each grazing, evening out the grazing pressure and preventing destruction of forage by horse traffic	Will require some time and effort to move them to paddocks where grazing will occur Usually are not built as tough as permanent waterers, so maintenance and replacement needs are greater

13. Look at places where forces on the fence will be greatest, such as at corners and gates or on hills and in valleys. Expect high horse traffic where they can watch the movement of other horses and people. See where trees or their branches can fall on fences. Look for wild animal trails near the fence, and inspect the fence near them for damage from animals attempting to go under, through, or over the fence. For electric fences, check all electrical connections between pieces of hardware. Look for vegetation and objects that contact live fence wires and carry current to the ground or to a building. Ensure that all electrical switches are in working order and in the correct position for paddocks containing animals. Ensure that voltage is adequate on fences, and use changes in voltage as clues to inspect possible reasons for the changes. Ensure that soil around ground rods is moist enough that current can return from soil to the rods and then to the energizer. Inspect the energizer and grounding system and check the fence voltage after seeing lightning or hearing thunder near the fence. Watch horses in relation to the fence to see if there are areas where they congregate and overgraze or areas that they avoid. Try moving water or salt/mineral sources to even out the traffic or shorten the grazing period. Check waterers to ensure that they and the water are clean and that water enters freely. If there are puddles under or near waterers, do what you can to reduce spillage and improve drainage from the area.

CHAPTER 16

1. The digestive system
2. Lack of access or abrupt changes in access to pasture and changes in pasture nutrient content
3. Lameness, reluctance to walk and shifting weight to hind limbs, increased digital pulse, heat and inflammation of the lower leg and hoof
4. Nonstructural carbohydrates
5. Summer pasture-associated obstructive pulmonary disease
6. Heat, humidity, mold spores, and grass pollen
7. Equine grass sickness
8. In the soil
9. Clinical signs include muscle stiffness, difficulty walking and standing, dark-colored urine, rapid breathing, and eventual death
10. Box elder tree (*Acer negundo*) and sycamore maple (*Acer pseudoplatanus*).
11. Stress, drought, prolonged cloudy weather, excessive fertilizer, improper timing of fertilization
12. Labored or rapid breathing, tachycardia, blue-brown discoloration of the mucosal membranes, tremors, ataxia, convulsions before death, and abortion

CHAPTER 17

1. As the abundance or diversity of species increases in an ecosystem, its health improves. Unrelated organisms tend to benefit from the presence of each other, providing benefits such as leftover food, hollow places to use as dens, or pathways through soil or vegetation. Each species has some modifying effect on the environment. As members of new species arrive in a location, if they thrive in the environment modelled by organisms before them, they will stay there. Thus, over time the population sizes of various species can change as they adapt to the presence of others. The benefits that are available to humans from biodiverse ecosystems are referred to as ecosystem services. These include "the multitude of benefits that nature provides to society" and include crop pollination, breakdown of organic matter to release nutrients, contaminant degradation, and agricultural pest control.

2. The degree of biodiversity in an ecosystem is considered an indicator of the health of that ecosystem. Many people consider wild creatures as sources of enjoyment and recreation, sometimes as a source of food and having value in the natural world presently and for future generations. An important component of tourism is the willingness of people to pay for opportunities to view, study, photograph, listen to, smell, hunt, and collect souvenirs of wild creatures. Some are willing to do this as part of farm vacations, and some will pay to do it on horseback.

   Grazing management decisions that reduce soil erosion, maintain or enhance soil organic matter, increase biodiversity, and provide wildlife for people to enjoy all contribute to the value of a property. Managers can make management decisions that provide new income streams in the short term and in the long term, allow them to pass on the land in as good or better condition than when it was first acquired (Krausman et al., 2009). The ecosystem services discussed earlier (crop pollination, breakdown of organic matter to release nutrients, contaminant degradation, and agricultural pest control) can contribute indirectly to the economic situation of a private property or a public conservation area.

   Direct economic benefits to private properties, mainly from attracting fee-paying clients to a property include accommodating hunters or other consumers of a diverse ecosystem. The service can simply be guiding clients around a property for birdwatching, photography, fishing, fox hunting, or trail riding. More complex services can include providing vehicles, equipment, horses, accommodation, and meals.

3. Direct effects of adding a species include more competition for related species or species that use the same resources. If the species is a predator, then prey species will decrease in number. If there are fewer prey animals, the vegetation they eat will be more abundant. If the added species is prey (herbivore), then some species of palatable vegetation will be eaten to a greater extent and may be less abundant.

   Indirect effects of adding a predator can be more abundant vegetation because the predator is reducing numbers of an herbivore species. This can add habitat including shade for other species. Indirect effects of adding an herbivore can be a reduction in some vegetation, possibly greater erosion, or infiltration by other species of vegetation.

4. To encourage more wildlife to visit or inhabit a farm, first increase the variety of plant species, which will provide a greater variety of feed sources and habitat, encouraging a greater diversity of animals to visit. Manage forages in a way that provides plenty of resting (growing and maturing) time for plants, so they can be effective habitat for breeding, nesting, brooding, raising young, sheltering, and feeding.

5. *Predation:* Small ruminant animals such as sheep and/or goats can be kept on pasture, following mares and foals, in a leader-follower rotation. One day you count the sheep and goats and discover that one is missing. It is found dead, apparently due to bites at the throat. Part of the animal has been eaten. The other members of the flock/herd (flerd) are more nervous than usual. The existing prevention includes a strong physical perimeter fence and electrified interior fences dividing paddocks. A consequence of inaction may be another attack within the next day and night.

   (Similar situations can include damage or death to a horse, death of a wild animal on your property, and several animals injured from being chased and attacked by dogs.)

   Crop Damage: Parts of a pasture have bare soil and holes dug in them. Before you do something about them a few days go by, and the grass surrounding them grows taller, so they are not so obvious, and then you notice another pasture has similar bare soil and holes. Decide whether you want to just scare and discourage the digging animals or have them killed. Consider whether you want to become a trapper and hunter or whether you should hire a certified pest animal control professional. Consider clearing any trees and bushes along your fence lines.

   (Similar situations can include a flock of Canada geese eating and trampling your alfalfa/grass hay field and pasture or large herbivores such as deer, elk, or moose damaging a fence to get to your forage fields.)

6. Describe how to determine which detrimental species may be present or responsible for damage.

   Predation: You want to prevent more attacks, so you take pictures, look for evidence of predator movement through the nearest fences, and call someone who is experienced at determining what species of predator is likely responsible. Look for hair on fences, damage to fences, and tracks crossing the fence, near the dead animal, and in moist or soft soil.

Write all evidence of the attack (damage to the animal, tracks, scat, hair, time of day or night) in a dated diary. Compare what you see in the way of damage to the dead animal, tracks, scat, and hair to reference sources. Call neighbors and a local wildlife agency to inform them of the attack and ask whether there have been similar attacks nearby and what predators have been seen recently.

    Crop Damage: To determine what the culprit is, look for scats and tracks in the soil, and spend some time with binoculars at a distance, watching to see what has caused the damage. Inspect fences for breaks but realize that a digging animal can dig under any pasture fence unless there is mesh that is partly buried in a trench during installation.

7. *Predation*: Review your fence maintenance schedule, and inspect all of the perimeter fence, repairing damage as you go. Determine whether there is a local program to have the predator attack inspected and whether you can make a claim for damage. If you can, call the inspector and ask plenty of questions. Decide whether you want to add additional electric wires or scare devices to your fences. Consider buying one or more guardian animals (donkey, mule, llama, or dog) from a reputable breeder. Be more vigilant for a few nights. Remove the dead animal, body parts, any injured animals, and any sick or weak animals from the pasture to remove as much attractive odor as possible. Dispose of the dead animal according to local regulations (dead stock removal service, composting, or burying). If attacks continue, call a certified pest animal control professional. Clear brush and trees away from fence lines. Consider shining lights and installing scare devices in areas where predators have crossed your fence.

    Crop Damage: Clear brush and trees away from fence lines. Fill in holes. Offer a responsible hunter or trapper the opportunity to hunt or trap the digging animals. Make an agreement on where and when driving, shooting, and trapping are allowed.

CHAPTER 18

1. Environmental impacts of horse grazing can include soil erosion and nutrient runoff due to close grazing, water contamination from direct access, and the spread of noxious weeds by allowing seed production.
2. In 2001, 25%−30% of the central Kentucky foal crop was lost, and quickly named Mare Reproductive Loss Syndrome. While a devastating loss for the industry, it encouraged the local horse industry and the University of Kentucky to work closely together to determine the cause and build lasting connections. The UK Equine Initiative and the UK Horse Pasture Evaluation Program both began as a response to this event. The eastern tent caterpillar was eventually blamed for MRLS.
3. Cattle grazing endophyte-infected tall fescue often experience increased core body temperature, decreased average daily gain, and rough hair coats. Most classes of horses are not affected, but pregnant mares can experience early-term pregnancy loss, prolonged gestation, foaling complications, and decreased milk production.
4. Horse pastures will ideally consist of a mixture of cool season grasses such as Kentucky bluegrass, orchardgrass, endophyte-free tall fescue (in the case of broodmares), and white clover. Less than 20% of the pasture would contain weeds or bare soil.
5.
   - Eradicated toxic tall fescue in broodmare pastures.
   - Removed nimblewill and other weeds from targeted pastures.
   - Reduced the need for grain feeding.
   - Produces on-farm hay, safe for pregnant mares.
   - Improved management to maintain renovated pastures for a number of years.
6. These signs provide a clear indication of what pastures are deemed safe for broodmares. Because the farm continues to monitor these pastures, signs can be taken down or added as needed when pasture composition changes.
7. Heavy grazing of endophyte-free pastures could result in the spread of endophyte-infected grasses, therefore increasing the risk to the broodmare. Careful pasture management and routine monitoring can reduce this.
8. In support of Fee for Service Programs: Federal and state budgets are being stretched more and more each year, limiting the funds available to Cooperative Extension. This evaluation program has been shown to be beneficial and valued by the local horse industry, but programs like this require labor, supplies, and travel. By asking farms to support a portion of the costs, farm managers are more likely to value the data and recommendations generated. The financial contribution allows programs like this to continue.

Appendix 8

Metric Equivalents for Hay Bale Sizes, as Described in Chapter 11

TABLE A8.1 Dimensions, Weights, Densities and Safe Baling Moistures for Various Bale Shapes (Metric Units)

Bale Shape	Height (cm)	Width (cm)	Length (cm)	Volume (m³)	Typical Weight (kg)	Density (kg m⁻³)	Safe Baling Moisture, %
Rectangular	37	46	98	0.16	27	167	18–20[a]
Rectangular	82	91	213	1.61	409	254	12–16[a]
Rectangular	122	122	244	3.62	817	225	12–16[a]
Round	122	–	122	1.42	227	159	15–18[a]
Round	122	–	152	1.78	386	217	15–18[a]
Round	152	–	122	2.22	454	204	15–18[a]
Round	152	–	152	2.78	590	212	15
Round	183	–	152	4.00	863	215	15

[a]The lower moisture range is preferred in areas of low humidity; the higher moisture % for other areas.
Adapted from Ball, D.M., Hoveland, C.S., Lacefield, G.D., 2016. Characteristics of hay bales. In: Sulewski, G. (Ed.), Forage Crop Pocket Guide, thirteenth ed. International Plant Nutrition Institute, Peachtree Corners, Georgia, p. 56.

Appendix 9

Environmental Risk Assessment Survey for Farms

Rutgers Cooperative Extension Fact Sheet FS1047

Livestock Farmer Survey: Is My Farm Environmentally Friendly?

Michael Westendorf, Extension Specialist in Livestock and Dairy

Livestock owners need to be aware of their effect on our environment and natural resources. Regardless of the kind of livestock you have, proper management of animal and land resources are important to limit potential impact on natural resources. Take this quiz to determine how eco-friendly your livestock operation is.

A. Grazing and Pasture Management

3 points for each "yes" answer

1.

Are your animals fenced at least 30 feet from ALL sensitive water features such as: well heads, creeks, streams, lakes, ponds, and wetlands?

The area between the animals and water is called a buffer strip. Steeper slopes need wider strips and all buffer or filter strips should have permanent vegetative cover. Livestock should also be kept off septic systems to prevent compaction and damage to the system.

○ Yes ○ No

2.

Is your buffer strip maintained in good vegetative cover like tall grass, not weeds?

Trees and shrubs along surface water are encouraged. If you have no water on or within 100 feet of property lines select "yes".

○ Yes ○ No

3.

Do you use fences, crossings, and limited access points to control animal access to sensitive waters? If you have no water on or within 100 feet of property lines select yes.

○ Yes ○ No

4.

Do you drag or harrow manure in your permanent pastures?

○ Yes ○ No

5.

Do you allow pasture grasses to regrow to at least 6" before regrazing?

○ Yes ○ No

B. Manure Storage

3 points for each "yes" answer

6.

Is your animal manure stored at least 100 feet from sensitive water features?

○ Yes ○ No

7.

Is the manure stored on a concrete pad or compact clay, or removed and disposed regularly (monthly)?

○ Yes ○ No

8.

Is your manure storage either covered, or does it have a grassed buffer around it?

○ Yes ○ No

9.

Do you spread or haul your manure away from your farm on a regular basis?

○ Yes ○ No

C. Nutrient Management

3 points for each "yes" answer

10.

Do you soil test every three years?

○ Yes ○ No

11.

Is manure application to soils managed to prevent phosphorus levels from getting too high (e.g., soil tests results are within Rutgers University guidelines)? If you don't spread manure, select "yes."

○ Yes ○ No

12.

Do you follow a formal written manure management plan?

○ Yes ○ No

13.

Is manure spread on relatively flat land (<8% slope), AND at least 100 feet from sensitive waters?

○ Yes ○ No

14.

Is manure spread only during the growing season and not on frozen soils?

○ Yes ○ No

D. Clean Water

2 points for each "yes" answer

15.

Is clean water from the roof or surrounding area directed away from the manure storage, animal lots, and bare ground?

○ Yes ○ No

16.

Is storm water from your property managed so it is not contaminated by manure and is it allowed to infiltrate into the soil? This helps recharge our ground water.

○ Yes ○ No

E. Erosion Control

2 points for each "yes" answer

17.

Are gullies on your property stabilized and soil erosion controlled? If no gullies select "yes". Someone could have no gullies but still have bad soil erosion.

○ Yes ○ No

18.

Is the amount of bare soil on the property minimized, possibly through pasture seeding and management?

○ Yes ○ No

19.

Is the runoff from bare and paved areas (e.g., arenas, driveways and parking lots) filtered through a vegetative buffer strip?

○ Yes ○ No

20.

Check all that are used:

☐ Buffers or borders around the fields

☐ Vegetative buffers to prevent runoff into open bodies of water

☐ Terraces to limit erosion

☐ Strip cropping or contouring of fields

☐ Use of winter cover crops to prevent erosion

F. Feed Management

2 points for each "yes" answer

21.

Do you manage milk house waste, silage waste, and excess or contaminated feed or hay disposals to prevent contact with stormwater and/or other water sensitive areas?

○ Yes ○ No

22.

Do you have feed bunks, mangers, hay feeders and other feeding areas that minimize feed contact with the ground.

○ Yes ○ No

23.

Do you reduce phosphorous level in the diet to minimize excretion of phosphorous?

○ Yes ○ No

24.

Do you monitor the feed intake of your animals to prevent over consumption and minimize waste?

○ Yes ○ No

25.

Do you balance diets to minimize overfeeding nutrients?

○ Yes ○ No

26.

What is the best description of how you feed your animals? (Check only two)

☐ Try to balance diets, with forages and concentrates

☐ Get advice from my feed store

☐ Get advice from Extension

☐ Use the services of a consulting nutritionist

☐ Veterinary consultant

[View Results]

Total Score: _____ Points Earned (57 possible)

Grade

0-35 Poor Serious attention should be paid to correcting "no" answers.

36-45 Fair I've seen worse, you can do better.

46-55 Good work - keep going.

56+ Very Good Outstanding Farm Management!

This survey is meant to give you a relative idea of your farm's environmental impact. It is not meant to be a score of your overall management. However, if you score very low on the test, you may want to implement a more thorough waste management plan on your farm

Reference and adapted from:

Gilkerson, B. 2006. Is my barn eco-friendly? University of Minnesota Extension Service, Hennepin County. extension.umn.edu/Agriculture/horse/care/is-my-barn-eco-friendly/.

Index

'Note: Page numbers followed by "f" indicate figures, "t" indicate tables and "b" indicate boxes.'

A

Acid detergent fiber (ADF), 55–57, 184, 185f, 212–213
Acid detergent solution, 55
Actual evapotranspiration (AE), 216
Aggregate destruction, 70–71
Air temperature
  cool season grasses, 210
  legumes, 210
  mean air temperature, 210–211, 211f
  peak growth rate, 211
  photorespiration, 210
  selecting forage species and varieties, 212
  transition zone, 212
  warm season grasses, 210
Alfalfa
  growing degree days (GDDs)
    acid detergent fiber (ADF), 212–213
    Neutral Detergent Fiber (NDF), 212–213
    predictive equations of alfalfa quality (PEAQ), 213
    relative feed quality (RFQ), 213
    relative feed value (RFV), 213
  Medicago sativa L. growth habit, 24–25, 25f
Alsike clover (*Trifolium hybridum* L.), 25, 26f, 26t
Alternating current (AC), 279b
American Forage and Grassland Council (AFGC), 223
Amino acids
  disulfide bridges, 47–48, 51f
  peptide bond, 47–48, 48f
  structure of, 47–48, 47f
Ampere, 279b
Amylase, 41
Amylopectin, 40–41
Amylose, 40–41
Animal unit (AU), 142
Annual ryegrasses, 13
Antiquality components, 87, 393
Apical bud, 5
Apical meristem, 9
Apparent digestibility, 52
Auricle, 12
Auxin, 5
Average annual precipitation, 214–215
Avermectins, 162
Aversions, 126–127
Awn, 12
Axil, 9
Axillary buds, 5, 9

B

Bagging equipment, 204
Bale
  conditioning, 192
  handling, 190–192
  packages for horses, 192–194
  storage, 192
Baleage, machinery requirements, 203–206
Bale-wrapping, 205
Band of feral horses, 37, 38f
Bermudagrass, 239–240
Biodegradation half-life, 162
Birdsfoot trefoil (*Lotus corniculatus* L.), 25–28, 27f, 28t
Bite/plant level
  familiar *vs.* novel forages, 129
  forage nutritional quality, 129
  individual variation, 129–130
  plant species preference, 129
Bomb calorimeter, 53, 59f
Boron, 49, 383
Breeding farms, 169
Buds, 5, 6f, 383
Bunch grasses, 12, 384

C

Calcium (Ca), 49, 383
Calvin cycle, 38, 210
Camp/home range level, 132
Cantharadin, 180–181
Capacitance probes, 109, 111f
Captive horses
  grazing patterns, timing of, 124
Carbohydrates
  digestion
    cecum and colon, 60–61
    stomach and small intestine, 60
  diversity in, 38–42
  fructans, 43
  nonstructural carbohydrates (NSCs), 1, 42
  structural carbohydrates, 2, 42
Carbon sequestration, 227
Carboxylase, 38
Carrying capacity, 143
Cation exchange capacity (CEC), 74
Cellobiose, 39–40, 40f
Cell solubles, 42
Cellulase, 41
Cellulolytic bacteria, 39–41
Cellulose, 2, 2f, 41, 44f
Cell walls, 42

Central Kentucky horse farm
  establishment, 381
  fertility, 382
  tall fescue, 382
  weed control, 381
Certified Forage and Grassland Apprentices (CFGA), 227–228
Certified Forage and Grassland Professionals (CFGP), 227–228
Chlorine (Cl), 50
Chloroplasts, 4, 38, 38f
Chromosomes, 4
Circuit, 279b
  breaker, 279b
Cis orientation, 43
Climate change
  on animals, 226–227
  breeding cultivars for, 225, 226t
  greenhouse gases (GHG), 223
    grazing and cycling of, 227
  on horses, 225–226
  measured elements of, 209
  methane (CH_4), 223, 224t
  nitrous oxide (N_2O), 223, 224t
  pests, 225
Cold temperature effects, 216
Collenchyma cells, 4
Companion cells, 4
Conductor, 279b
Contaminated water leaching/runoff, 247
Continuous grazing system
  benefit of, 145
  drawbacks, 145
  vs. rotational grazing, 153
  stress lot, 145, 145f
Control competition, 96
Cool season forages, 234
Copper (Cu), 49
Coulomb, 279b
C3 plants, 4, 83, 383
C4 plants, 4, 83, 383
Crown, 9
  bud, 5, 9
Crown vetch (*Coronilla varia* L.), 28, 29f, 29t
Crude fiber (CF), 52
Crude protein (CP), 184
Culm, 12
Current, 279b
Cuticle, 4, 9, 383
Cutin, 4
Cyathostomes, 160

D

Dietary crude protein (CP), 48
Dietary overlap, 164
Digestibility, of feeds, 52–53
Digestible dry matter (DDM), 56
Digestible energy (DE), 54
Digestion
  carbohydrate digestion
    cecum and colon, 60–61
    stomach and small intestine, 60
  fat digestion, 61
  nitrogen digestion, 61
Direct current (DC), 279b
Disaccharides, 1, 2f, 38
  diversity of, 39–40, 40f
Disulfide bridges, 47–48
Diverse ecosystem
  accommodating hunters, 333–334
  adding/subtracting effects, 334–336
  benefits, 333–334
  habitat fragmentation, 335–336
  horse parasite control, 336
  subtracting and adding plant species, 335
  unintended effects, 336
Domestic livestock species, 172, 390–392
Drinking water systems, 297–307
  quality, 298–299
  sources, 299
    pasture, 299–302
  watering devices, 303–307
Drought, 221f
  arid rangelands, 220
  conditions, 220
  3-phosphoglycerate (3PG), 220
  warm season grasses, 220
Drug-resistant parasites, 161
Dry matter basis (DM), 184

E

Ecologic niches, 234
Electrolysis, 279b
Electron, 279b
Elevation/topographic position, 214
Endophyte-infected tall fescue, 98, 99f
Energy costs, 158
Epidermis, 9
Ergot alkaloids, 180
Erosion, 245
Ester linkages, 43
"Estimating Pasture Forage Mass From Pasture Height", 111
Eutrophication/algae bloom process, 247

F

Falling plate meter (FPM), 108–109, 109f–110f, 375–376
Fat digestion, 61
Fat-soluble vitamins, 45
Fatty acids, 45
Feed choices, 125–126
Feed intake, 62

amount and rate of
  bite rates, 124–125
  dry matter intake (DMI), 124
  forage quality, 125
  fresh pasture to dried hay, 125
  factors, 123
Feed neophobia, 125, 129
Feeds
  digestibility of, 52–53
  nutrient analysis of, 50–52, 52f–53f
  selection preferences, 163
Fences
  calm horses, 263
  descriptions, 271–292
  designs, 271–292
  desired features, 261–262
  durability, 261
  electric fences
    connections and current, 284–285
    maintaining the current, 285–287
    maintenance, 285
    permanent, 283–284
    safety, 285
    semipermanent electric fences, 287–292
    temporary electric fences, 287–292
    training horses, 285
  electric fencing, 264t
  gates, 292–294
  locations, 263–264, 264f
  maintenance, 295–297
  materials, 265–271
    alkaline copper quat, 267
    concrete, 271
    copper azole, 267
    creosote, 267
    fiberglass, 271, 271f
    metals, 267–268, 268f
    plastic coatings, 271
    plastics, 269–270, 270f
    pressure-treated posts, 267
    stones and trees, 265, 265f
    wood, 265–271, 266f
  National Animal Health Monitoring System (NAHMS), 261–262
  perimeter fence, 263
  planning, 262–263, 271–292
  plastic polymer posts and rails, 282
  post and rail fences, 275–277
  purposes, 261–262
  rubber belt, 282–283
  strength, 261
  types, 271–292
  visibility, 261
  wire gauge sizes, 279b
  wood posts
    polymer-coated wire, 280–282
    polymer strand, 282
    steel strand wire, 277–279
    steel woven wire, 280
Feral goats, 165
Feral horses

diets of, 123, 123f
grazing patterns, timing of, 124
Fertility, 233
Flood, 221–222
Forage dry matter (FDM), 167, 371–372, 371t
Forage species
  characteristics, 236t
  ecologic niches, 234
  fertility, 233
  forage quality, 241–243
  growth distribution, 234, 240f
  growth habit, 234, 237f–238f
  management adaptation, 233–239, 236t
  pH status, 233
  species diversity, 238
  stand longevity, 237
Forage utilization efficiency, 158–160
Forage yield, 373–374, 374f
  acre and whole field, 116
  direct measurement of, 107
  dry matter intake, 106
  feed dry matter per day (FDM/d), 106
  feed horses need per day, 106
  vs. forage height
    canopy heights and structures, 113–116, 115f
    plant composition, of growing grass plants, 113, 114f
    relative composition, of protein and minerals, 113, 114f
    shoot and root structure, of bluegrass, 113, 115f
  forage species, 239
  grazing management, 239
  haying management, 239
  indirect measurement, 111–113
    capacitance probes, 109, 111f
    falling plate meters, 108–109, 109f–110f
    pasture rulers, 107–108, 108f, 108t
    pasture sticks, 107–108, 108f, 108t
    rising plate meters, 109, 110f
  nitrogen availability, 239
  pounds per acre and kilograms per hectare, 116
  precipitation, 239
  social media sites, 105
  soil fertility, 239
  soil type, 239, 106
  "take half and leave half", 106, 117
  voluntary dry matter intake (VDMI), 106
Foraging decisions, 388–389
  aversions, 126–127
  external factors, 126, 126f
  fermentation by-products, 127
  hierarchic continuum, 127–128, 128f
  minimal total discomfort, 127
  postingestive feedback, 126
  toxic feeds, 127
Foraging theory, 127
Freezing injury, 216
Fructan, 1
Fructans, 41, 43

Fructose, 1, 1f, 38
Fungal hyphae, 76
Fuse, 279b

G

Generator, 279b
Glomeromycetes fungi, 157
Glucose, 1, 1f, 38, 39f
    composition of, 39, 39f
Glume, 12
Glyceraldehyde-3 phosphate (G3P), 38, 38f
Glycerol, 43, 44f
Glycosides, 180
Glycosidic linkage, 39—40
α-1,4 Glycosidic linkages, 40—41
β-1,4 Glycosidic linkages, 40—41
Golgi apparatus, 4
Grasses
    cool season, 210
    establishment, 6, 8t
    growing degree days (GDDs), 213—214
    Kentucky bluegrass, 18, 18f, 19t
    legumes, 22, 23f
    meadow fescue, 14—15, 16t
    new tillers, 7
    orchardgrass, 15—16, 16f, 17t
    quackgrass, 20, 22f, 22t
    reed canarygrass, 19—20, 19f, 20t
    rushes, 13, 13f
    and rushes, 12, 13f
    ryegrasses, 13—14, 14f, 15t
    sedges, 13
    seedlings, 11—12
    smooth bromegrass, 20, 21f, 21t
    sod *vs.* bunch grasses, 12
    stems, 13, 14f
    structure and anatomy, 12, 12f
    tall fescue, 14—15, 15f, 16t
    timothy, 17, 17f, 18t
    warm season, 210
    whorl, 13
Grass/legume species, 239—240
Grass whorl, 13
Grazing animals, 88—90, 89t
Grazing behavior, 141—142
    effect of manure, 121, 122f
    on environment
        animal waste, 132
        overgrazing, 133
        positive effects, 133
        trampling, 132
    management
        limiting dry matter and NSC intake, 135
        manure and parasites, 134
        maximizing pasture intake and nutrition, 134—135
        multispecies grazing, 134
        rotational grazing, 133—134
    physiologic differences, 122
    tactile hairs and touch receptors, 121
Grazing plans, 153—154, 153b
Grazing season, 151
Grazing system
    continuous grazing

benefit of, 145
    drawbacks, 145
    stress lot, 145, 145f
graze and remove horses/length of grazing periods
    growth rate of grasses, 148—149, 148f
    lawn and rough pattern grazing, 150, 150f
    leaf volume removal on root growth, 149, 149f
    length of, 150
    "take half, leave half" rule, 149
grazing units
    number of, 147
    size of, 147—148
rotational grazing
    design, 146, 146f
    new pasture unit, 146, 147f
total acreage required, 148
Grazing units
    number of, 147
    size of
        additional factors, 148
        factors, 147
Greenhouse gases, 385
Ground, 279b
Ground rods, 284—286, 284f
Ground fault circuit interrupters (GFCI), 279b
Growing degree days (GDDs)
    alfalfa
        acid detergent fiber (ADF), 212—213
        Neutral Detergent Fiber (NDF), 212—213
        predictive equations of alfalfa quality (PEAQ), 213
        relative feed quality (RFQ), 213
        relative feed value (RFV), 213
    Celsius temperatures, 212
    diversity, varieties, 214
    grasses, 213—214
Growth restrictions, 217—218, 219f
Guardian animals, 158

H

Hairy vetch (*Vicia villosa Roth* L.), 28—29, 30f, 30t
Hay
    acid detergent fiber (ADF), 184, 185f
    bagging equipment, 204
    bale-wrapping, 205
    cellulose and hemicelluloses, 185
    chemical composition, 182—185
    commercial forage laboratories, 183
    crude protein (CP), 184
    definition, 177
    dry matter basis (DM), 184
    forage quality, 182—185
    forage requirements, 203
    horse numbers, North America, 178t
    horses, 196
        cleanliness, 196
        fed in groups, 199—200
        feeding, 198
        matching, 196
    individual bags, 204

individually wrapped bales, 205
in-line bale wrappers, 205
long tubes, 205
machinery requirements, baleage, 203—206
nutrient value, factors affecting, 177—185
    cantharadin, 180—181
    ergot alkaloids, 180
    glycosides, 180
    maturity, 178, 179f
    mold and dust, 181
    nitrates, 182
    plants and materials, 182
    plant species, 177—178, 178t
    slaframine, 182
    stages, 179t
    visual and physical characteristics, 179—182, 180f, 207
nutritionally balanced diet, 197—200
probes, 184f
quality, 177, 185—192
    bale. *See* Bale
    baling, 190
    harvesting, 186—187, 186f—187f
    preservatives, 190, 191f
    swath, 187—188, 188f
    windrow manipulation, 187—188
relative feed value (RFV), 185
silage baling
    advantages, 202—203
    disadvantages, 202—203
    forage crops for, 201—203
storage, 195—196
total digestible nutrients (TDN), 185
uses, 177
Head, 24
Hemicelluloses, 2, 41
Hexoses, 39
High temperature stress, 223
Horse farm, 166
Horse pastures
    constructing heavy use pads, 252
    contaminated water leaching/runoff, 247, 247f—248f
    designing storage, 252
    environmental concerns
        aesthetics, 249—250
        insects, 249—250
        salinity, 249
        weed seeds, 249
    erosion, 245—246
    grazing near streams, 257
    managing pastures, 252—257
    manure on pasture, 250
        guidelines, 251
        nitrogen availability, 251
        nutrient recycling, 250
        soil quality, 250
        spreading, 250—252, 250t
    other environmental concerns, 249—250
    parasite concerns, 248
    removing manure, 252
    riparian buffers, 257
    riparian forest buffers, 257—258
    vegetated buffers, 252—254

Hydrophilic, 44–45
Hydrophobic, 44–45

I

Individual bags, 204
Individually wrapped bales, 205
Induction, 279b
Inflorescence, 24
In-line bale wrappers, 205
Insulator, 279b
Intensive management strategies, 152
Intercalary meristem, 9
Internode, 9, 24
Inverter, 279b
Iron (Fe), 49
Irrigation systems, 151–152, 152f
Italian ryegrass, 13, 14f
Ivermectin, 162

J

Joule, 279b

K

Kentucky bluegrass (*Poa pratensis* L.), 18,
        18f, 19t
Kura clover (*Trifolium ambiguum* L.),
        30–31, 31f, 32t

L

Laminitis, 42
Law of diminishing returns, 81, 82f
Leaf, 9
    anatomy, 4
Leaf blade, 12
Leakage, 279b
Legumes, 8–9
    defined, 22
    structure and anatomy of, 22, 23f
Leguminous plants, 82
Levan, 43
Light interception, 83, 83f
Lignin, 2
    indigestible polymer, 383
Ligule, 12
Lipids, 43–46, 44f–45f, 47f
Load, 279b
Long tubes, 205
Lush pasture, 127, 128f

M

Macroaggregates, 70
Magnesium (Mg), 49, 383
Maltose, 39–40
Management adaptation, 236t
Managing, equine grazing, 133–135,
        141–155
Manganese (Mn), 49
Maturity, 178, 179f
Meadow fescue, 14–15, 16t
Mean Stage Count (MSC), 213
Measurement/conversion factors, units of

area, 369–370
    length, 369
    mass/weight, 370
    volume, 370
Meristem, 9
Mesophyll, 9, 383
Metabolizable Energy (ME), 54
Methyl group, 43
Minerals
    boron, 49
    calcium (Ca), 49
    chlorine (Cl), 50
    copper (Cu), 49
    iron (Fe), 49
    magnesium (Mg), 49
    manganese (Mn), 49
    molybdenum (Mo), 49
    nitrogen (N), 49
    phosphorous (P), 49
    potassium (K), 50
    selenium (Se), 50
    sodium (Na), 50
    soils, 66
        composition of, 66–68, 67f–68f
        organic matter, 66
    sulfur (S), 50
    zinc (Zn), 50
Minimal total discomfort, 127
Mitochondria, 4
Mixed or multispecies grazing (MSG),
        390–392
    advantages, 159t
    animal productivity, 157
    avermectins, 162
    benefits, 172
    benefits, 157–163
    biodegradation half-life, 162
    breeding farms, 169
    competition, 159
    consumption, 163–164
    cyathostomes, 160
    definition, 157
    dietary overlap, 164
    domestic livestock species, 172
    drug-resistant parasites, 161
    economic considerations, 167–168
    economic efficiency, 157
    energy costs, 158
    feed selection preferences, 163
    feral goats, 165
    forage dry matter (FDM), 167
    forage resources, 159
    forage utilization efficiency, 158–160
    glomeromycetes fungi, 157
    grazing animals, 159
    guardian animals, 158
    horse farm, 166
    ivermectin, 162
    limitations, 159t
    managing, 168–170
    monocultures of plants, 157
    mycorrhizas, 157
    negative economic consequences, 168
    nonchemical methods, 161

nonequine species, 167
    parasites control, 160–161
    pastured poultry production, 170
    pasture sweeping, 160
    pigs and poultry, 170
    positive economic responses, 167–168
    research, 160
    research, 172
    sequential grazing, 169
    sheep and cattle, 160
    stock rate (SR), 158
    tasks and actions, 170–171
    Trichostrongylus axei, 160
    vacuuming, 160
    weed control, 162–163
    workability influences, 166
Mob grazing, 152
Molybdenum (Mo), 49
Monosaccharides, 1, 38, 40f
Mowing, 152–153
Multispecies grazing, 134
Mycorrhizal fungi, 76, 385
Mycorrhizas, 157

N

Natural Resources Conservation Service
        (NRCS) standard, 223
Net Energy (NE), 54
Neutral detergent fiber (NDF), 55–57,
        212–213
Nitrates, 182
Nitrogen, 49, 76, 383
    availability, 239
    compounds, 2
    digestion, 61
Node, 9, 12, 24
Nonchemical methods, 161
Nonequine species, 167
Nonstructural carbohydrates (NSC), 42, 215
    disaccharides, 1, 2f
    monosaccharides, 1
    polysaccharides, 1, 2f
NSC. *See* Nonstructural carbohydrates (NSC)
Nutrients, 37
Nutritional value
    carbohydrates, 38–43
    chemical analysis
        acid detergent fiber (ADF), 55–57
        acid detergent solution, 55
        forage components, 57
        forage dry matter intake, 57, 58t
        neutral detergent fiber (NDF), 55–57
        neutral detergent solubles, 55
    concentrations of, 57–58, 59f
    digestion of, 60–61
    energy terminology, 53–54
    fatty acids, 45
    feeds
        digestibility of, 52–53
        intake, 62
        nutrient analysis of, 50–52
    lipids, 43–46
    minerals, 49–50
    proteins, 47–48

requirements of, 61–62
vitamin A, 46
vitamin D, 46
vitamin E, 46
vitamin K, 46
water-soluble vitamins, 48–49

O

Ohm, 279b
Ohmmeter, 279b
Ohm's Law, 279b
Open circuit, 279b
Orchardgrass
   characteristics of, 16, 17t
   grazing timing, 16
   heads and tall membranous ligule, 15, 16f
Overgrazing, 388–389
Overstocking, 143–144, 389–390

P

Palisade parenchyma, 4
Parasites control, 160–161
Parenchyma cells, 4
Pastured poultry production, 170
Pasture ecology
   cell division, 85
   cell enlargement, 85
   cell wall content changes, 86
   classical primary producer, 385
   competition between plants, 87–88
   energy reserves cycle with growth, 83–84,
      84f
   forage mass
      and dry matter intake, 90
      dry matter (DM) yield, 87, 87f
   forage quality and palatability, 87
   grazing animals, 88–90, 89t
   growing points, 85–86
   growth under rotational grazing, 84–85, 84f
   light interception, 83, 83f
   morphology, 88
   optimal environment vs. limiting factors,
      81–82, 82f
   parts of, 385
   plants, 82
      diversity, 88
   respiration, 82–83
   root growth, 85, 85f–86f
   selective grazing, 90–91, 90t
   tolerance to grazing, 88, 88t
   young plant cell, 85, 86f
Pasture establishment, 223
Pasture growth rate (PGR), 240f
Pasture plant establishment/management
   blue certified seed tag, 93, 94f, 386
   clipping/grazing, 98
   control competition, 96
   cool season and warm season forages, 95,
      95f
   endophyte-infected tall fescue, 98, 99f
   forage species for, 96
   frost seeding, 95
   grass species for

Bermudagrass, 96
endophyte-free/novel endophyte tall
   fescues, 96
Kentucky bluegrass, 96
orchardgrass, 96
perennial ryegrass, 96
timothy, 96
high-quality seed, 93
legume species for, 97
non-nitrogen soil amendments, 93
no-till seeding, 95
pasture renovation, 99, 100f
recommendations, 93
resting pastures, 100–102, 101f–102f
seeding rates and optimum seeding dates, 94,
   95t
soil fertility, 93
soil sampling and fertilizing, 97, 98f
weed control, 97–98
white seed tags, 93, 94f, 386
Pasture-related diseases
chronic obstructive pulmonary disease
   (COPD), 315
colic, 311–313
equine grass sickness (EGS), 316–317
equine motor neuron disease (EMND), 325
forbs, 321
grasses, 324, 325t
hyperinsulinemia, 313–314
legumes, 321
nitrate poisoning, 318–319
nonstructural carbohydrates (NSC), 313
obesity, 313–314
pasture-associated laminitis, 313–315
pasture-associated liver disease (PALD), 319
pasture-associated obstructive pulmonary
   disease (PAOPD), 315
pasture-associated stringhalt (PAS),
   319–320
pasture management, 313
physiologic systems, 311, 312t
seasonal pasture myopathy (SPM), 317–318
selenium deficiency, 324–326
selenium poisoning, 326
summer pasture-associated obstructive
   pulmonary disease (SPAOPD),
   315–316
toxic forbs, 322t–323t
toxics, 320t–321t
trees, 320
Pasture renovation, 99, 100f
Pasture rulers, 107–108, 108f, 108t
Pasture soils, 68, 69f
Pasture sticks, 107–108, 108f, 108t
   and meters
      average swards, 111–112
      calibrations for, 111, 112t
      pasture tiller density, 111
      ruler height, 111
      sward height measured, 111, 112t
      thick swards, 111–112
      thin swards, 111–112
      tiller density, 111
Pasture sweeping, 160

Patch/feeding site level
   altitude and elevation, 131
   bare areas, 130, 131f
   distance to water, 130
   forage nutritional quality, 131
   "horse-sick", 130, 131f
   lawns, 130, 131f
   roughs, 130, 131f
   sward height, 130
   terrain and slope, 131
Pectins, 2, 41
Peptides, 47–48
Perennial ryegrass, 13
Pericarp, 9
Peroxisomes, 4
Phospholipids, 43, 47f
Phosphorous (P), 49, 383
Photorespiration, 83
Photosynthesis, 4, 38
pH status, 233
Pith, 4
Plant available water, 73, 73t
Plant cells
   vs. animal cells, 3
   cytoplasm, 4
   middle lamella, 3
   photosynthesis, 4
   primary cell wall, 3
   secondary cell wall, 3
   structure, 3, 3f
Plant diversity, 88
Plant growth, 8–9
Plasmodesmata, 9, 383
Polarity, 279b
Polysaccharides, 1, 2f, 383
Polyunsaturated fatty acids (PUFAs), 43, 45f
Postingestive feedback, 126
Potassium (K), 50, 383
Potential evapotranspiration (PET), 216
Power, 279b
Precipitation, 214–215, 239
Predictive equations of alfalfa quality
   (PEAQ), 213
Primary cell wall, 42
Protective relay, 279b
Proteins, 47–48, 47f–48f, 51f
Pubescence, 24
Pull test, 11–12
Pulse, 279b
"Purchasing power calculator", 108–109

Q

Quackgrass, 20, 22f, 22t

R

Raceme, 24
Rachilla, 12
Red clover (Trifolium pratense L.), 32–33,
   32t, 33f
Reed canarygrass, 19–20, 19f, 20t
Regional climatic effects, 215–222
Regrowth, 9
Relative feed value (RFV), 56, 185

Relative forage quality (RFQ), 56
Resistance, 279b
Resistor, 279b
Respiration, 82—83
Rhizobia bacteria, 384
Rhizomes, 9, 12, 24, 234
Ribosomes, 4
Rising plate meter (RPM), 375
Rising plate meters, 109, 110f
Root growth, 85, 85f—86f
Roots
    of alfalfa plants, 5, 7f
    structure of, 5, 7f
    types of, 5
Rotational grazing, 133—134
    design, 146, 146f
    new pasture unit, 146, 147f
Rough endoplasmic reticulum, 4
Rubisco, 38
Ryegrasses
    annual (Italian) ryegrasses, 14, 15t
    endophytic fungus, 14
    heads of, 13, 14f
    perennial ryegrasses, 14, 15t
    tillers, 13

S

Saturated fatty acids (SFAs), 43
Sclerenchyma, 4
Seasonal changes, forage quality, 215
Secondary cell wall, 42
Selective grazing, 90—91, 90t
Selenium (Se), 50, 75
Sequential grazing, 169
Service, 279b
Sheath, 12
Short circuit, 279b
Sieve cells, 4
Silage baling
    advantages, 202—203
    disadvantages, 202—203
    forage crops for, 201—203
Slaframine, 182
Smooth bromegrass, 20, 21f, 21t
Smooth endoplasmic reticulum, 4
Sod grasses, 12
Sodium (Na), 50
Soil acidity, 75
Soil aeration, 74
Soil aggregation, 70—71, 70f
Soil alkalinity, 75
Soil chemistry, 74
Soil fertility, 75—76, 76t, 239
Soil humus, 68—70, 69f
Soil management, 66f
    acidity and alkalinity, 75, 75t
    aggregation, 70—71, 70f
    biology, 76—77
    chemistry, 74
    factors, 66, 67f
    fertility, 75—76, 76t
    formation, 66

mare and foal grazing pastures, 65, 65f
mineral soils
    composition of, 66—68, 67f—68f
    organic matter, 66
organic matter, 68—70
plant available water, 73, 73t
secrets, 77—78, 77f
soil aeration, 74
soil humus, 68—70, 69f
soil pore space, 71, 71t
soil water, 71—72, 72t
water drainage, 73
Soil organic carbon (SOC), 227
Soil pore space, 71, 71t
Soils
    amendments, 71, 93, 145, 152—153
    conditions, 96
    fertility, 75—76, 93, 97, 239
    fertilizing, 97
    health, 65, 78
    minerals, 56, 66—68
    moisture, 216, 220—221, 223, 284,
        286
    nutrients, 97
    organisms, 68—70, 74, 78, 81
    pH, 50, 75, 97
    poorly drained soils, 96, 234
    sampling, 97, 363
    soil-to-seed-contact, 95
    structure, 385—386
    testing, 76, 81, 93, 97
    types, 50, 239, 359
    water, 71—72, 72t
Solar radiation
    light intensity *vs.* plant canopy height, 210,
        210f
    light interception, 210
    sun's average energy, 209, 209t
Solute potential (osmotic potential), 9
Species diversity, 238
Spongy mesophyll cells, 4
Stand longevity, 237
Starch, 1, 2f, 38—39, 41f
Stem, 4, 5f, 9
Stipule, 24
Stocking density, 142
Stocking rate, 142, 142b—143b
Stockpiling, 151, 389—390
Stolon, 24
Stomates (stoma), 4, 9
Stress lot, 145, 145f
Strip grazing, 152
Structural carbohydrates, 42
    cellulose, 2, 2f
    hemicelluloses, 2
    pectins, 2
Structural strength, 42
Sucrose, 1, 2f, 38
Sulfur (S), 50
Summer pasture management, 151—152,
    152f
Sward density, 373, 373f

Swath, 187—188, 188f
Sweetness, 38

T

Table sugar, 1, 2f
Tall fescue, 14—15, 15f, 16t
Temperature humidity index (THI),
    225—226
Tendril, 24
Terminal bud, 5
Timothy (*Phleum pratense* L.), 17f
    characteristics of, 17, 18t
    corms, 17
    seedlings, 17
Total digestible nutrients (TDN), 53,
    185
Total nonstructural carbohydrates, 1
Tracheids, 4
Transition zone, 217, 218f
Trans orientation, 43
Transpiration, 9
Trichostrongylus axei, 160
Triglyceride lipids, 43, 44f

U

Umbel, 24
Understocking, 143—144, 144f, 389—390
United States Department of Agriculture
    (USDA) map, 222
University of Kentucky Horse Pasture
    Evaluation Program, 392—394
    case studies, 361—363
        large-scale commercial breeding farm,
            361—362
        medium-scale commercial breeding farm,
            362—363
        small private farm, 363
    challenges, 363—365
        commercial businesses, 364
        extension *vs.* fee, 364
        labor and quality control, 365
        securing funding, 364
        soil sampling, 363
    current scenario, 356—357
    data reporting, 359
        farm data summary sheet, 359, 361f
        pasture data, 359
        pasture maps, 359
        pasture photographs, 359
        publications, 359
        recommendations, 359
        soil maps, 359
    developing resources, 365
    equine breeds, 355, 355t
    future grants, 366
    history, 355—357
    mare reproductive loss syndrome (MRLS),
        356
    pasture sampling, 358
    relationships with farms, 365
    Tall Fescue, 357, 357f

training students, 365
University of Kentucky Equine Initiative,
     356

V

Vacuoles, 4
Vacuuming, 160
Variegation, 24
Vessel elements, 4
Vitamin A, 46
Vitamin D, 46
Vitamin E, 46
Vitamin K, 46
Volt, 279b
Voltage, 279b
Voltmeter, 279b

W

Water
     bodies, 246, 263, 299
     consumption, 298–299
     drainage, 73
     drinking, 297–307
     ground, 247, 257, 267, 305–306
     hardness, 298
     intake, 297–298
     lines, 302
     output, 300
     pipe, 220, 270, 299
     quality, 252–257
     requirements, 302
     sources, 299
     supply, 304–305
     system, 297–307

tank(s), 300–305, 302f
testing, 298
troughs, 299, 304
Water-soluble vitamins, 48–49
Watt, 279b
Weather, measured elements of, 209
Web soil survey soils map key, 379
Weed control, 162–163
White clover (*Trifolium repens* L.), 33–35,
     33t, 34f
Wild horses, 124
Wild plants and animals
     arthropod pests, 343–345
     biodiversity, 331–332
     cover, 350
     crop damage, 340–345
     dead animal disposal, 350
     detrimental species, 345
          predation, 345
          signs of predators, 345
     discourage animal species techniques,
          345–351
     diverse ecosystem, 333–334
     ecosystem, 329–332
     feed, 350
     fences, 346–347
     Food and Agriculture Organization (FAO),
          330
     guardian animals, 349–350
     habitat, 330–331
     hunting, 347–349
     injuries, 342
     integrated methods, 346
     landowners, 330
     model, 350–351

plants, 350
population imbalance, 332
predation, 340–345
scaring, 342
scaring nuisance animals, 346
techniques, 336–340
     adding prolific omnivores, 339–340
     breeding habitat, 339
     cover, 339
     drinking water, 339
     feed, 339
     forage management, 336–337
     forage species diversity, 336
     grazing management, 338–339
     habitat, 336–337
     haying practices, 337–338
     subtracting prolific omnivores, 339–340
     tools and resources, 340
trapping, 347–349
water, 350
Windrow manipulation, 187–188
Winter hardiness, 222
Winter pasture management, 151

X

Xylem tissue, 4, 9

Y

Yields of nonirrigated crops, 378t, 379

Z

Zinc (Zn), 50

Printed in the United States
By Bookmasters